Lower and Higher Plants for Biofuel Production

The global energy sector is rapidly evolving in response to climate change, environmental concerns, and the depletion of fossil fuels. Biofuels, particularly those derived from plant-based sources, offer a renewable and cleaner alternative that can integrate into existing energy systems with minimal disruption. *Lower and Higher Plants for Biofuel Production* explores the untapped potential of both microalgae and macroalgae (lower plants), as well as oilseed crops and high-biomass grasses (higher plants), as sustainable feedstocks.

It provides an interdisciplinary overview of plant-based biofuel technologies, covering biological mechanisms, processing methods, and system integration. Topics include genetic engineering, nanotechnology, bioreactor innovations, and integrated biorefineries. This book also addresses key environmental and socioeconomic considerations such as sustainable land use, water conservation, and food security.

This book offers a comprehensive resource for researchers and policymakers seeking practical, region-specific solutions to expand sustainable biofuel production globally.

Lower and Higher Plants for Biofuel Production

Edited by
Azamal Husen

CRC Press
Taylor & Francis Group
Boca Raton London New York

CRC Press is an imprint of the
Taylor & Francis Group, an **informa** business

Designed cover image: Shutterstock

First edition published 2026
by CRC Press
2385 NW Executive Center Drive, Suite 320, Boca Raton FL 33431

and by CRC Press
4 Park Square, Milton Park, Abingdon, Oxon, OX14 4RN

CRC Press is an imprint of Taylor & Francis Group, LLC

ISBN: 9781032966571 (hbk)
ISBN: 9781032987224 (pbk)
ISBN: 9781003600169 (ebk)

DOI: 10.1201/9781003600169

Typeset in Times
by codeMantra

Contents

About the Editor

Azamal Husen has served as Full Professor and Head of the Department of Biology at the University of Gondar, Ethiopia. He is currently affiliated as a Foreign Delegate with Wolaita Sodo University in Ethiopia and holds an Adjunct Professorship at the Department of Biotechnology, Graphic Era (Deemed to be University) in Dehradun, Uttarakhand, India. His earlier academic roles include Full Professor at Sankalchand Patel University in Visnagar, India, and Visiting Faculty at the Forest Research Institute and Doon College of Agriculture and Forest in Dehradun, India. With more than 25 years of experience in academia and research, he is highly regarded for his expertise in nanobiotechnology, plant physiology, environmental botany, medicinal plants, and plant propagation. His research has been supported by prominent national and international organizations, including the World Bank, the National Agricultural Technology Project, the Indian Council of Agricultural Research, the Indian Council of Forestry Research and Education, and the Japan Bank for International Cooperation. He is a prolific scholar, having published more than 350 academic works, including research articles, books, and book chapters. His editorial contributions are equally noteworthy. He serves on the editorial boards and peer-review panels of numerous prestigious journals published by Elsevier, Frontiers Media, Springer Nature, Taylor & Francis, RSC, Oxford University Press, the Royal Society, CSIRO, MDPI, PLOS, Sciendo, John Wiley & Sons, and UPM Journals. He is also a member of the advisory board of Cambridge Scholars Publishing in the United Kingdom. He is a fellow of the Plantae group under the American Society of Plant Biologists and holds a membership in several scientific organizations, including the International Society of Root Research, the Asian Council of Science Editors, and the International Natural Product Sciences Taskforce. Throughout his career, he has delivered invited lectures, keynote addresses, and technical presentations at numerous international conferences, symposia, and workshops. In the field of academic development, he has made significant contributions through the design and delivery of orientation programs for university faculty, the development of educational modules, and the formulation of curricula at the undergraduate, postgraduate, and doctoral levels. In recognition of his contributions to teaching, research, and community service, he was honored with a special award from the University of Gondar. He currently serves as the Editor-in-Chief of the *American Journal of Plant Physiology* and as the Series Editor for several major academic collections, including *Exploring Medicinal Plants* (Taylor & Francis, USA); *Plant Biology, Sustainability, and Climate Change* (Elsevier, USA); and *Smart Nanomaterials Technology* (Springer Nature, Singapore). He also leads the *Handbook Series on Nanotechnology and Its Applications* (Springer Nature, Singapore). His outstanding contributions to science have earned him international acclaim, including the prestigious distinction of being named among the "World's Top 2% Scientists" for five consecutive years (2020 onwards) by Stanford University, a recognition published in the Elsevier Data Repository. In addition to his regular academic responsibilities, he has served as a Guest Editor for several high-impact special issues, including "Advances in Plant Metabolomics: Insights, Applications, and Integration" (*Plant Physiology and Biochemistry*); "Exploring Nanotechnology in Medicinal Plants" (*Journal of Herbal Medicine*), and "Integrating Nanotechnology with Plant Compounds: Enhancing Abiotic Stress Resilience and Crop Improvement via Signaling Molecules" (*Frontiers in Plant Science*). His work continues to shape contemporary research at the intersection of plant science, biotechnology, and sustainable development. His academic leadership, research excellence, and international collaborations have positioned him as a prominent figure in global scientific circles.

Contributors

Sadia Abdullah
Department of Botany
Rawalpindi Women University
Rawalpindi, Pakistan

Sewunet Abebayehu
Department of Chemical Engineering
Debre Berhan University
Debre Berhan, Ethiopia
and
School of Chemical and Bio Engineering
Addis Ababa University
Addis Ababa, Ethiopia

Andualem Bahiru Abera
School of Chemical and Bio Engineering
Addis Ababa University
Addis Ababa, Ethiopia

Moonisah Aftab
Department of Botany
University of Kashmir
Srinagar, India

Nafees Ahmad
Department of Chemistry
Integral University
Lucknow, India

Naseem Ahmad
Department of Chemistry
Integral University
Lucknow, India

Mushtaq Ahmed
Department of Plant Sciences
Quaid-I-Azam University
Islamabad, Pakistan

Wasim Akhtar
Department of Botany
University of Azad Jammu and Kashmir
Muzaffarabad, Pakistan

Ambreen Ali
Department of Botany
University of Baltistan
Skardu, Pakistan

Ana Cristina Ramírez Anguiano
Chemistry Department
University of Guadalajara
Guadalajara, Mexico

Mohammad Israil Ansari
Department of Botany
University of Lucknow
Lucknow, India

Maria Adriana Delgado Armas
Chemistry Department
University of Guadalajara
Guadalajara, Mexico

Ayesha Ashfaq
Institute of Environmental Sciences and Engineering
National University of Sciences and Technology (NUST)
Islamabad, Pakistan

Mohd Azhar
Department of Chemistry
Kamlesh Kumari Laxmi Kant Mahavidyalaya
Kanpur, India

Megha Barot
Research and Development Cell
Parul University
Vadodara, India
and
Department of Environmental Science
Parul University
Vadodara, India

Tesfaye Kassaw Bedru
Department of Chemical Engineering
and
School of Chemical and Mechanical Engineering
Kombolcha Institute of Technology
Wollo University
Kombolcha, Ethiopia

Yi Chen
College of Chinese Materia Medica
Chongqing University of Chinese Medicines
Chongqing, P.R. China

Karen Christina
Division of Biotechnology
School of Agricultural Sciences
Coimbatore, India

Abayneh Getachew Demesa
LUT School of Engineering Sciences
LUT University
Lappeenranta, Finland

José de Jesús Luna Díaz
Chemistry Department
University of Guadalajara
Guadalajara, Mexico

Workisa Bacha Garuma
Department of Chemical Engineering
Haramaya University
Haramaya, Ethiopia

Desta Getachew Gizaw
Department of Chemical Engineering
Adama Science and Technology University
Adama, Ethiopia

Muhammad Baqir Hussain
Department of Soil and Environmental Sciences
MNS University of Agriculture
Multan, Pakistan

Sajjad Hyder
Department of Botany
Government College Women University
Sialkot, Pakistan

Nazish Iftikhar
Institute of Environmental Sciences and Engineering
National University of Sciences and Technology (NUST)
Islamabad, Pakistan

Anam Ilyas
Department of Botany
University of Education
Township, Lahore, Pakistan

Javed Iqbal
Department of Soil and Environmental Sciences
MNS University of Agriculture
Multan, Pakistan
and
Soil & Water Testing Laboratory for Research
Dera Ghazi Khan, Pakistan

Syed Uzma Jalil
Amity Institute of Biotechnology
Amity University Uttar Pradesh
Lucknow, India

Mani Jayakumar
Department of Biotechnology
Faculty of Engineering
and
Centre for Natural Products and Functional Foods
Karpagam Academy of Higher Education
Coimbatore, India

Zul Kamal
Department of Pharmacy
Shaheed Benazir Bhutto University
Sheringal, Pakistan
and
College of Chinese Materia Medica
Chongqing University of Chinese Medicines
Chongqing, P.R. China

Sang-Mo Kang
School of Applied Biosciences
Kyungpook National University
Daegu, South Korea

Noreen Kareem
Department of Botany
University of Education
Township, Lahore, Pakistan

Aimen Razzaq Khan
Department of Botany
Government College Women University
Sialkot, Pakistan

Isa Khan
Department of Pharmacy
Shaheed Benazir Bhutto University
Sheringal, Pakistan

Kashif Ali Khan
Department of Pharmacy
Shaheed Benazir Bhutto University
Sheringal, Pakistan

Mohd Arsh Khan
Department of Chemistry
Integral University
Lucknow, India

Muhammad Aaqil Khan
Department of Chemical and Life Sciences
Qurtuba University
Peshawar, Pakistan

Muhammad Saleem Khan
Department of Zoology
University of Okara
Okara, Pakistan

Suresh Kumar Krishnan
Division of Biotechnology, School of Agricultural Sciences
Karunya Institute of Technology and Sciences
Coimbatore, India

Neeraj Kumari
School of Biological and Environmental Sciences
Shoolini University of Biotechnology and Management Sciences
Solan, India

In-Jung Lee
School of Applied Biosciences
Kyungpook National University
Daegu, South Korea
and
LUT School of Engineering Sciences
LUT University
Kouvola, Finland

Ashutosh Kumar Mall
Division of Crop Improvement
ICAR-Indian Institute of Sugarcane Research
Lucknow, India

Saba Mazhar
Department of Botany
Government College Women University
Sialkot, Pakistan

Varucha Misra
Division of Crop Improvement
ICAR-Indian Institute of Sugarcane Research
Lucknow, India

Mamoona Munir
Department of Botany
Rawalpindi Woman University
Rawalpindi, Pakistan

Ghulam Murtaza
Department of Soil and Environmental Sciences
MNS University of Agriculture
Multan, Pakistan

Huu Hao Ngo
School of Civil and Environmental Engineering
University of Technology Sydney
Sydney, Australia

Selvakumar Periyasamy
Department of Chemical Engineering
Adama Science and Technology University
Adama, Ethiopia
and
Saveetha Dental College and Hospitals, SIMATS
Saveetha University
Chennai, India

Suraj Prakash
School of Biological and Environmental Sciences
Shoolini University of Biotechnology and Management Sciences
Solan, India

Ashok Pundir
Department of Civil Engineering
Graphic Era (Deemed to be) University
Uttarakhand, India

Radha
School of Biological and Environmental Sciences
Shoolini University of Biotechnology and Management Sciences
Solan, India

Qazi Inamur Rahman
Department of Chemistry
Integral University
Lucknow, India

Sandra Fabiola Velasco Ramírez
Chemistry Department
University of Guadalajara
Guadalajara, Mexico

Madiha Rashid
Department of Botany
University of Education Township
Lahore, Pakistan

Zinnabu Tassew Redda
Department of Chemical Engineering
Debre Berhan University
Debre Berhan, Ethiopia
and
School of Chemical and Bio Engineering
Addis Ababa University
Addis Ababa, Ethiopia

Nadia Riaz
Department of Botany
Lahore College for Women University
Lahore, Pakistan

Sahar Saleem
Department of Environmental Sciences
PMAS Arid Agriculture University
Rawalpindi, Pakistan

Shifa Shaffique
School of Applied Biosciences
Kyungpook National University
Daegu, South Korea

Maleeha Shafqat
Department of Biomass Supply Chain
Bulleh Shah Packaging Pvt. Ltd.
Lahore, Pakistan

Anis Ali Shah
Department of Botany
University of Education
Lahore, Pakistan

Leta Lemma Sharbe
Department of Chemical Engineering
Adama Science and Technology University
Adama, Ethiopia

Karan Singh
Department of Civil Engineering
Graphic Era (Deemed to be) University
Uttarakhand, India

Kavitha Subbaih
Division of Biotechnology
Karunya Institute of Technology and Sciences
Coimbatore, India

Muhammad Suliman
Department of Pharmacy
Shaheed Benazir Bhutto University
Sheringal, Pakistan

Yeit Haan Teow
Department of Chemical and Process Engineering
and
Research Centre for Sustainable Process Technology (CESPRO)
and
Cleaner Production – UKM Research Group
Universiti Kebangsaan Malaysia
Selangor Darul Ehsan, Malaysia

Leslie Andrea Colorado Torres
Chemistry Department
University Center of Exact Sciences and Engineering (CUCEI)
University of Guadalajara
Guadalajara, Jalisco, Mexico

Javier A. Trujillo-Rugamas
School of Applied Biosciences
Kyungpook National University
Daegu, South Korea
and
Plant Protection Laboratory
National Center for Agriculture and Forestry Technology
La Libertad, El Salvador

Bashir Ullah
Department of Pharmacy
Shaheed Benazir Bhutto University
Sheringal, Pakistan

Hanif Ullah
Department of Botany
University of Malakand
Chakdara, Pakistan

Ali Umar
Department of Zoology
University of Okara
Okara, Pakistan

Samuel Bernardo Perez Vega
LUT School of Engineering Sciences
LUT University
Kouvola, Finland

Karthik Velusamy
Department of Industrial Biotechnology
Government College of Technology
Coimbatore, India

Afifa Younas
Department of Botany
Baba Guru Nanak University
Nankana Sahib, Pakistan

Rabeea Zafar
Department of Environmental Science
Allama Iqbal Open University
Islamabad, Pakistan

Huiying Zhang
College of Life Sciences
Fujian Agriculture and Forestry University
Fuzhou, P.R. China

Min Zhang
Suzhou School of Environmental Engineering
Suzhou Polytechnic Institute of Agriculture
Suzhou, P.R. China

Preface

The global energy sector is undergoing a fundamental transformation in response to increasing environmental concerns, diminishing fossil fuel reserves, and the pressing need to mitigate climate change. As nations strive to reduce greenhouse gas emissions and transition toward more sustainable energy systems, biofuels have emerged as a promising alternative. Biofuels derived from plant-based sources are renewable and biodegradable and can be integrated into existing energy infrastructure with minimal disruption. Their potential to replace conventional fossil fuels, particularly in transportation and industrial applications, has made them a focal point of global energy research and policy development. This book, *Lower and Higher Plants for Biofuel Production*, explores the vast and largely underutilized potential of both lower and higher plant species as sustainable feedstocks for biofuel production. It provides a comprehensive examination of plant-derived biofuels, consolidating current scientific knowledge and technological advances across diverse disciplines. Despite the growing global emphasis on renewable energy, research related to plant-based biofuel sources remains fragmented and dispersed across academic and industrial domains. This book addresses that gap by bringing together contributions from leading researchers to present an integrated and up-to-date perspective on this critical field.

The appeal of biofuels lies in their many environmental and economic advantages. Compared to fossil fuels, biofuels produce significantly lower emissions of carbon dioxide, particulate matter, and harmful pollutants such as sulfur and aromatic hydrocarbons. Their cleaner combustion contributes to improved air quality and reduced public health risks, especially in urban environments. Moreover, many biofuels can be used directly in existing diesel engines and heating systems without requiring major modifications. This compatibility not only reduces the transition cost but also enhances their practicality as an immediate energy solution. At the scientific level, biofuels can be broadly categorized into three types: solid, liquid, and gaseous. Solid biofuels include materials such as firewood, crop residues, and wood pellets. Liquid biofuels encompass biodiesel, biogasoline, and ethanol, typically derived from plant oils or sugars. Gaseous biofuels, including biogas and syngas, are produced through anaerobic digestion or thermal processes. Each of these categories presents unique challenges and opportunities, ranging from feedstock availability and conversion efficiency to storage, transport, and emissions characteristics. This book examines these categories in detail, with a special focus on plant-based sources, offering insights into the biological mechanisms, processing technologies, and system integration strategies required to scale biofuel production sustainably.

The concept of using both lower and higher plants as biofuel feedstocks opens up a wide range of possibilities. Lower plants, such as microalgae and macroalgae, are particularly attractive due to their fast growth rates, high lipid content, and ability to thrive in non-arable environments including wastewater and saline water systems. In contrast, higher plants include oilseed crops such as soybean, sunflower, palm, and oilseed rape, as well as non-edible species like *Jatropha* and *Mahua*, and high-biomass grasses such as switchgrass, *Miscanthus*, and *Brachypodium*. These species benefit from well-established cultivation systems and a wealth of agronomic knowledge. Together, these plant types offer a diverse and flexible platform for biofuel production that can be adapted to a range of climates and land types.

Recent scientific advances have further strengthened the case for plant-based biofuels. Genetic engineering, plant cell wall modification, and synthetic biology are being used to enhance the oil content, stress tolerance, and growth efficiency of biofuel crops. Nanotechnology and advanced catalytic processes are improving the conversion of biomass into high-quality fuels. Innovations in bioreactor design, cultivation methods, and harvesting systems are making algal biofuels more economically viable. At the same time, integrated biorefinery concepts are being developed to co-produce biofuels alongside valuable byproducts such as bioplastics, fertilizers, and biochemicals, which increases the overall economic feasibility of these systems. This book is structured to guide readers through both foundational concepts and advanced topics. The initial chapters provide a global perspective on the current status and emerging trends in biofuel production, including policy developments, technological innovations, and market drivers. Following this, this book offers in-depth discussions on specific plant feedstocks, including algae, edible and non-edible oil crops, essential oils, lignin, lipids, and glycerol. Additional chapters focus on enabling technologies such as plant cell wall engineering, metabolic optimization, gene editing, and microbial interactions that support biofuel yield improvements. The role of nanotechnology in improving processing efficiency and storage stability is also addressed. This book presents detailed analyses of individual crop species that are widely recognized for their potential in biofuel production. These include *Jatropha*, palm, soybean, oilseed rape, sunflower, *Mahua*, and high-yield grasses such as switchgrass and *Miscanthus*. Each crop is examined in terms of its biological characteristics, agronomic requirements, processing potential, and long-term sustainability. These case studies offer valuable insights for researchers and practitioners seeking to identify region-specific solutions that align with local environmental and economic conditions. Equally important, this book explores the environmental and socioeconomic dimensions of biofuel production. It emphasizes that large-scale cultivation of energy crops must be carefully managed to avoid negative impacts such as deforestation, land degradation, food insecurity, and water resource depletion. Attention is given to sustainable land management practices, integrated agricultural systems, and life cycle assessments that support responsible implementation. The goal is to balance the benefits of biofuel production with the need to protect ecosystems and community livelihoods.

This book is intended for a broad and interdisciplinary audience. Researchers, graduate students, environmental scientists, agronomists, engineers, and policymakers will find this book to be a valuable resource. Its interdisciplinary approach encourages collaboration across fields such as plant biology, energy science, biotechnology, and environmental policy. By offering both theoretical frameworks and practical applications, this book aims to support informed decision-making, encourage innovation, and contribute to academic and industrial progress in renewable energy. The completion of this book would not have been possible without the collective efforts of dedicated contributors from diverse backgrounds. Each chapter reflects the unique expertise of its authors and presents knowledge based on extensive research and field experience. I am deeply grateful to all contributors for their valuable input, which has shaped the depth and breadth of this work.

Taken together, *Lower and Higher Plants for Biofuel Production* aims to serve as both a reference and a catalyst for future research and innovation. The insights presented here will, I hope, stimulate further exploration of sustainable plant-based bioenergy solutions. By advancing scientific understanding, promoting responsible practices, and supporting global efforts toward a clean energy future, this book aspires to make a meaningful contribution to the ongoing transition toward environmental sustainability and energy resilience.

Azamal Husen

1 Present Trend of Lower and Higher Plants for Biofuel Production

A Global Prospective

Varucha Misra, Ashutosh Kumar Mall, Megha Barot, and Azamal Husen

1.1 INTRODUCTION

Biofuels have emerged as a future substitute for fossil fuels (El-Araby, 2024). There has been steady growth in the world's consumption of biofuels with current projections estimating a rise from 257 to 500 billion liters per year by 2030. This will further be increased to 4,600 billion liters by 2045 (Dafnomilis et al., 2017; Global Bioenergy Statistics, 2019; IEA, 2019). Growing fleet sizes of vehicles, higher demand for transport services, and the transition to alternative energy sources are key drivers of growth (IEA, 2019). Several countries in Africa and Asia have increasingly focused on ethanol production for the transport sector, which is complemented by bioethanol blending requirements and policy measures. Ethanol output of 5.2 PJ in Asia and 220.1 PJ in Africa has been reported in 2021 (BP, 2022). Several countries have implemented bioethanol policies. For instance, National Policy on Biofuels has been implemented in India for targeting 20% blending of ethanol in gasoline by 2025–2026. The policy ensures the use of crops for the production of ethanol sustainably without sacrificing food security (Das and Rosmann, 2023). Likewise, South Africa has implemented a Biofuels Regulatory Framework with the aim of having a 4.5% market share of biofuels in the country's fuel market, of which 2% is reserved for first-generation biofuels. The policy specifically bans the utilization of food grade corn to ensure food security (Saravanan et al., 2020). Thailand, under its National Energy Plan, seeks to attain carbon neutrality in 2050 and net zero greenhouse gas emissions in 2065. It had been noticed that by 2022, ethanol supplied 12.6% of its gasoline fuel pool (Prasertsri, 2023). Its government intends to boost the mandate on ethanol blending to 20% with benefits such as increasing energy security and reduced particulate emissions (Syafiuddin et al., 2020). Both the Philippines and Vietnam in Southeast Asia had a great potential for biofuel production. Research indicates that Vietnam and the Philippines could supply 11% of their gasoline demand through agricultural residues and excess crops (USAID, 2009). Vietnam has aggressively marketed ethanol-blended fuel, with the E5 blend being sold throughout the country (Silva et al., 2024). Brazil witnessed historic ethanol and biodiesel production and is in the process of taking its biodiesel blending requirement from 11% to 15% (IRENA, 2016). Similarly, China and ASEAN member nations have rolled out E10 (10% ethanol) blending requirements to minimize fossil fuel reliance. On the other hand, nations such as Mexico and sub-Saharan Africa are yet to adopt biofuels, which was being held back by a lack of infrastructure, policy loopholes, and investment constraints (Sandesh and Ujwal, 2021). It has been estimated that global demand needs will enhance by 38 billion liters during 2023–2028 to a projected value of 200 billion liters in 2028. This can be fulfilled by ethanol and renewable diesel, with biodiesel and bio-jet fuel consumption expected to grow moderately.

DOI: 10.1201/9781003600169-1

Biofuels have been considered as a viable solution for curbing carbon emissions in the transport sector, with global consumption and production (BP, 2022). These renewable fuels improve energy security and also contribute to rural economies (Bhattarai et al., 2011; Vatsha and Alam, 2025). Biofuels are grouped into first, second, and third generations based on feedstock types (Gautam et al., 2024; Kumar et al., 2021; Malode et al., 2021). First-generation biofuels are obtained from feedstocks like sugarcane, maize, soybeans, and vegetable oils (Dharani et al., 2024). These feedstocks are edible, and the sugar and starch contents in these feedstocks are converted into ethanol or biodiesel by microbial/enzymatic fermentation. But nowadays, ethanol produced from these feedstocks has food security concerns and also the high production costs limit the long-term sustainability. Second-generation biofuels, on the other hand, are obtained from non-edible biomass such as lignocellulosic wastes, agricultural residues, and other byproducts obtained from forests (Liang, 2024). Biofuel produced from second generation has the potential to address the issue of food security of raw sources and expensive pretreatment used in the first-generation biofuels (Hirani et al., 2018). The second generation produced biofuels such as renewable diesel, and drop-in aviation fuels utilize cellulose and lignin as raw materials (Dharani et al., 2024). Third-generation biofuels are obtained from high-yield feedstocks such as microalgae and macroalgae (Deora et al., 2022; Dharani et al., 2024). The biofuels produced have high lipid content and carbon sequestration characteristics, but this biofuel commercial production has limitations like energy intensive harvesting, demands in nutrition, and increased cost of production (Dahman et al., 2019). The feedstock diversification in different types of biofuels is essential for its sustenance and scalability. Higher plants such as sugarcane, maize, and oilseed crop still lead to the supply of global biofuel feedstocks (El-Araby, 2024), due to established supply chain and conversion technology. For instance, the success of Brazil with sugarcane ethanol is the epitome of the effectiveness of biofuels derived from higher plants in tropical agroclimatic regions. Conversely, lower plants like microalgae can also be used for biofuel production with benefits like high growth rates, lipid yields, and lowland use demands (Wang et al., 2024). Algal biomass may be grown on non-arable land with wastewater, thus circumventing the food versus fuel conflict (Tahir et al., 2024). Commercial viability still poses a hurdle due to high processing costs, restricted strain improvement, and an absence of optimized large-scale bioreactors. In Africa, *Jatropha curcas*, a drought-tolerant shrub, is increasingly being investigated for use as a biodiesel plant, promising opportunities for marginal lands (Folaranmi, 2013; Ruatpuia et al., 2024). Nevertheless, inconsistency in yield, toxicity of the seeds, and water efficiency usage are challenges necessitating agronomical and biotechnological strategies. Taken together, this chapter focuses on the evolving role and emerging trends in biofuel production from both lower and higher plants, offering a comparative perspective on their feedstock potential.

1.2 EMERGING ROLE OF LOWER PLANTS IN SUSTAINABLE BIOFUEL PRODUCTION

Non-woody species like grasses, macroalgae, microalgae, and other herbaceous plants are being increasingly identified as a renewable feedstock for bioenergy because of higher growth rates, greater biomass productivity, and comparatively lower environmental footprints than conventional food crops (Table 1.1). Of special mention are lignocellulosic feedstock sources, i.e., perennial C4 grasses, e.g., *Miscanthus, Panicum virgatum*, and sorghum; agro-industrial residue, e.g., corn stover or wheat straw; and multiple algal taxa that have sufficient amounts of cellulose, hemicellulose, and lignin to be useful macromolecular resources for second-generation biofuels. Perennial grasses possess a number of agronomic benefits, e.g., high hectare biomass yield, low inputs, and multiple residues after harvesting.

In contrast, algal cultures, particularly microalgae, provide immense flexibility. They are able to be grown under a wide range of physicochemical conditions, with high photosynthetic efficiency and rapid biomass turnover while at the same time supporting CO_2 sequestration from the atmosphere. Rich in valuable lipids, algal biomass can also leave residual biomass fractions that can

TABLE 1.1
Potential of Lower Plants for Biofuels Production

	Biofuels Yield			
Plant Species	**Biogas**	**Biodiesel**	**Ethanol**	**Key References**
Anabaena variabilis	-	21.63%	-	Ismail et al. (2020)
Arthrospira platensis	481 mL/g VS	-	45.49%	Mussgnugm et al. (2010); Ismail et al. (2020)
Caulerpa prolifera	-	13.55%	-	Ismail et al. (2020)
Chlorella marina	-	-	23.24%	Ismail et al. (2020)
Chlorella kessleri	335 mL/g VS	-	-	Mussgnugm et al. (2010)
Chlorella minutissima	340 mL/g VS	-	-	Prajapati et al. (2013)
Chlorella pyrenoidosa	464 mL/g VS	-	28.07%	Prajapati et al. (2013); Yerizam et al. (2023)
Chlorella vulgaris	0.226 L methane/g COD	46%–57% dry weight of biomass[2]	-	Griffiths et al. (2012); Mendez et al. (2014); Stephenson et al. (2010)
Chlorococcum humicola	-	-	0.52 g ethanol/g substrate	Harun and Danquah (2011)
Chlorococcum infusionum	-	-	0.26 g ethanol/g substrate	Harun et al. (2010)
Chroococcus sp.	487 mL/g VS	-	-	Prajapati et al. (2013)
Dunaliella salina	505 mL/g VS	-	-	Mussgnugm et al. (2010)
Ecklonia kurome	-	-	21%	Sasaki et al. (2018)
Gracilaria	-	-	0.43 g/g sugars	Kumar et al. (2013)
Laminaria	615±7 mL/g VS	-	-	Abusweireh et al. (2023)
Porphyridium cruentum	-	-	0.60 g/g DCW	Kim et al. (2017)
Saccorhiza polyschides	227±4 mL/g VS	-	-	Pardilho et al. (2022)
Sargassum latifolium	-	-	29%	Soliman et al. (2018)
Scenedesmus abundans	0.2120 L biogas/g VS	-	0.10 g ethanol/g substrate	Guo et al. (2013); Passos et al. (2013)
Scenedesmus sp.	0.085 L methane/g COD	0.3 g/L	-	González-Fernández et al. (2012); Kandimalla et al. (2016)
Scenedesmus dimorphus	-	-	0.30 g/g DCW	Chng et al. (2018)
Scenedesmus obliquus	287 mL/g VS	28.44%	-	Ismail et al. (2020); Mussgnugm et al. (2010)
Synechococcus sp.	-	-	0.22 g ethanol/g substrate	Möllers et al. (2014)
Tetraselmis suecica	-	18%–23% dw	0.70 g ethanol/g substrate	Park et al. (2017); Reyimu and Ozcimen (2017)
Ulva linza	271 mL/g VS	-	12.01%	Bruhn et al. (2011); Ismail et al. (2020)
Ulva fasciata	-	88%	-	Panwar et al. (2025)
Ulva intestinalis	-	32.3 mg/g dw	8.1%	Osman et al. (2020)

be used for other options like animal feed or biofertilizers (Liu et al., 2017; Schenk et al., 2008). Third-generation bioethanol produced from microalgal biofuels is part of a continuum of sustainable, environmentally friendly energy sources to replace petroleum or fossil-based fuels and has environmental performance superiority over first- and second-generation biofuels in relation to their respective biosource (Zabed et al., 2017). Algae show biomass production due to their extensive cell growth with a high rate of proliferation, which completes its entire growth cycle in 1–10 days. Their production does not require arable land and can be done in open raceway ponds or in closed photobioreactors with different water sources like freshwater, seawater, or treated municipal effluents. This integrated system allows both nutrient recycling and water remediation at the same time (Malibari et al., 2018; Silva et al., 2017). Bioalcohols are biosynthesized by microbial fermentation of sugar-rich, starchy, or cellulosic substrates utilizing either *Saccharomyces cerevisiae* or other metabolically engineered consortia of microbes. Besides acting as renewable fuels, bioalcohols help toward improved combustion properties by raising octane ratings and reducing vapor pressure when added as fuel additives (Velazquez-Lucio et al., 2018). Standard bioethanol production takes advantage of carbohydrate-rich biomass and consists of three stages: pretreatment, enzymatic saccharification, and microbial fermentation, usually with yeast. Ethanol can also be produced via dark fermentation, in which photosynthesis is redirected in darkness to create hydrogen, organic acids, and alcohols (i.e., ethanol). Conversely, photo-fermentation exploits the natural metabolic apparatus of microalgae, i.e., cyanobacteria, and can be rewired by genetic engineering to boost bioethanol production. In the bioengineered pathway, light is used directly for the conversion of water and carbon dioxide without ethanol in one photobiological step, without requiring external carbon input or multistage downstream processing (Hernández et al., 2015; Magneschi et al., 2012). Algal biofuels hold significant potential through their enhanced carbon fixation efficiency, biomass productivity, and ability to survive in harsh or marginal conditions—such as arid and saline areas—based on non-portable water resources (Chisti, 2007; Chisti, 2008). In addition, algae have lower freshwater reliance than terrestrial crops and allow year-round production based on their non-seasonal patterns. The highly efficient photosynthetic machinery of algae, which is made up primarily of chlorophyll, permits ongoing energy capture and conversion into biomass. In contrast to land plants, which spend much energy on structural features such as roots and woody tissues, algae direct metabolic energy into biomass build-up. Some algal species are capable of doubling biomass within 24 hours and as quickly as 3.5 hours under ideal circumstances (Borowiak et al., 2022). Lipid content, a central characteristic for biodiesel production, is usually 30% of algal dry weight, but genetic improvements have yielded strains with lipid levels of up to 80%, well above the 5% usually found in palm oil (Masojídek et al., 2021; Singh and Dhar, 2011). Algae represent one of the highest biomass generating bioresource platforms per area and time (Heredia et al., 2021). It can produce an array of bioproducts such as renewable fuels, platform chemicals, livestock feed, nutraceuticals, fertilizers, and carbon sequestration products. Expanding the product portfolio of algal bioprocessing could demonstrate a potential way to decrease the costs of an algal biorefinery and improve its techno-economic viability. The high areal productivity, polysaccharides and lipid accumulation capabilities, and salt tolerant or wastewaters utilizing capabilities make algae an ideal substrate for thermochemical and biochemical biofuel conversion processes.

Macroalgae (seaweeds), or multicellular algal species, exhibit remarkable growth trajectories of up to 60 meters in length and are more productive than numerous terrestrial crops (McHugh, 2003). Brown seaweeds grow naturally with a yield of 3.3–11.3 kg dry weight/m^2 annually, whereas cultivated species can yield up to 13.1 kg/m^2 in under 7 months, even surpassing the yield of sugarcane (Kraan, 2010). Being mostly aquatic organisms, macroalgae do not occupy any terrestrial arable land and do not require chemical fertilization. Aside from their common uses in food and hydrocolloid processing, the high polysaccharide content present in these macroalgae makes them even a good source for bioethanol production (Wi et al., 2009). Different ways of converting algal biomass into ethanol can be taken up such as anaerobic digestion, pyrolysis, saccharification, and fermentation. Carlsson et al. (2007) showed that the best developed algal species for biofuel production have high volumetric productivity, cultivation and harvesting ease, ecological resistance, and scalability with minimal production expense in bulk systems. Recent research explores the double function of algae toward the production of biofuel and waste water bioremediation methods such as enzymatic esterification and anaerobic microbial conversion (Kraan, 2010). A vast diversity in morphology and physiology has been observed in this group of algae, and based on the coloration/pigmentation, these have been classified as brown (Phaeophyceae), red (Rhodophyceae), and green (Chlorophyceae) (Chan et al., 2006). Besides, few of the species are known to be farm-cultivated, while few contain fermentable sugars that can be utilized for conversion into bioethanol (Kraan, 2010). Furthermore, some of these also contain high amount of lipids, which have also been used for biodiesel production (Maceiras et al., 2011). Due to their inherent water content (80%–85%), macroalgae are more suitable for aqueous-phase conversion processes like fermentation than dry thermochemical conversion (Horn et al., 2000; Ross et al., 2008). Many algae belonging to genera like Sargassum, Gracilaria, Laminaria, Gelidium, etc. are composed of polysaccharides that can be hydrolyzed into monosaccharides through fermentation (Seo et al., 2010). Algae like *Kappaphycus alvarezii* and *Eucheuma* have even potential for the production of biofertilizer and bioethanol (Karunakaran and Gurusamy, 2011). Process optimizations by enzymatic hydrolysis and genome editing technology are under development to promote efficiency in sugar release and fermentability in strains such as *Saccharina latissima* (Adams et al., 2009; Uchida and Murata, 2004). In contrast to land-based lignocellulosic feedstocks, macroalgae do not contain lignin—an otherwise recalcitrant component—thereby making their saccharification and fermentation to ethanol easier (John et al., 2011). The occurrence of hydrocolloids such as agar, carrageenan, and galactans in red algae also enhances the fermentable content of sugars within them, placing them even stronger as viable substrates for bioethanol. Research today focus on enhancing ethanol biosynthesis efficiency and streamlining associated processes. Methods for increasing the productivity of ethanol from algal biomass include high-level genetic modification, growth parameter optimization, particularly through nutritional constraint, and further advancement of ethanol (Algenol Biofuels, 2011).

1.3 EMERGING ROLE OF HIGHER PLANTS IN SUSTAINABLE BIOFUEL PRODUCTION

Higher plants, especially corn (*Zea mays* L.), sugarcane (*Saccharum* spp.), soybean (*Glycine max*), and other oilseed crops, are critical biofuel feedstocks and are responsible for shaping the world renewable energy agenda (Table 1.2). With increasing global energy needs and the rapid progress of biotechnological interventions, research efforts have been focused on maximizing the efficiency of biofuel production, minimizing environmental effects, and maximizing economic feasibility. Of all types of biofuels, biodiesel—produced by using surplus food grade oils such as soybean, sunflower, and rapeseed—gets extremely well rooted in countries like the United States and all across Europe. This transition supports air pollution reduction, decreases geographically limited fossil fuel resources dependence, and serves as a hedge against crude oil price volatility (Agarwal, 2007; Berchmans and Hirata, 2008; Foidl et al., 1996; Ma and Hanna, 1999; Meher et al., 2006; Openshaw, 2000; Ranganathan et al., 2008; Sarin et al., 2007). Globally, the production of biodiesel is being led by a range of feedstocks, with castor, sunflower, palm, and jatropha oils being utilized in biodiesel plants across Europe, North America, South America, and several countries in Asia. Corn and sugarcane have also emerged as anchor crops for the production of bioethanol in India, the United States, France, China, Germany, and Australia. Approximately US$0.75 per liter was required to operate an ethanol plant with corn as a feedstock supply, taking into consideration installation, feedstock supply, production, and transportation (Ganesan et al., 2020). New approaches are also gaining popularity, such as the refining of biodiesel from vegetable oils, tallow, and waste cooking oil by Australian Renewable Fuels Limited. Simultaneously, production of bioethanol from lignocellulosic biomass like wheat straw, agrowastes, wood chips, and bagasse of sugarcane is also being promoted by companies like BlueFire Ethanol Fuels, Inc. (United States), Cosan (Brazil), and Coskata (United States) (Ganesan et al., 2020).

Sugarcane bagasse (SCB) has emerged as a promising lignocellulosic substrate for second-generation bioethanol production due to its favorable physicochemical properties. SCB typically

TABLE 1.2
Potential of Higher Plants for Biofuel Production

	Biofuels Yield			
Higher Plants	**Biogas**	**Biodiesel**	**Ethanol (L/ha)**	**Key References**
Barley	353–658 m^3 per ton of VS	0.130 g/L dry cell weight	1,100	Worldwatch Institute (2007)
Cassava	75,000 m^3	-	3,835	Kemausuor et al. (2015); Worldwatch Institute (2007)
Castor	-	393 L/ton	800	Johnston et al. (2009)
Coconut	833.33 to 976.67 mL	130 L/ton	2,151	Brown (2006); Johnston et al. (2009); Ramjani et al. (2023)
Corn	660–680 Nm^3 ton VS added	152 kg/ha/year	3,100	Bosnjakovic and Sinaga (2020); Liotta et al. (2012); Worldwatch Institute (2007)
Groundnut	290 mL CH4 g/VS added	309 L/ton	842	Brown (2006); Johnston et al. (2009); Olatunji and Madyira (2023)
Jatropha	220–250 L per kg of cake	656 kg/ha/year	-	Bosnjakovic and Sinaga (2020); Raheman and Mondal (2012)
Maize	560–650 Nm^3 ton VS added	-	2,500	Liotta et al. (2012); Worldwatch Institute (2007)
Oil palm	-	88 L/ha	4,500	Achinas et al. (2019); Worldwatch Institute (2007)
Rapeseed	-	1,570 L/ha	1,200	Achinas et al. (2019); Worldwatch Institute (2007)
Soybean	-	530 L/ha	600	Achinas et al. (2019); Worldwatch Institute (2007)
Sugarcane top portion	-	-	6,000	Niju et al. (2020)
Sugarcane bagasse	18.2 L/kg of biohydrogen dry biomass, 2.3 mol/mol glucose consumed	-	318–500 L/ton (or 10,000)	Corbin et al. (2015); Ivanova et al. (2009)
Sugar beet	730–770 Nm^3 ton VS added	-	5,000	Liotta et al. (2012); Worldwatch Institute (2007)
Sunflower oil	420–540 Nm^3 ton VS added	740 L/ha	1,000	Achinas et al. (2019); Liotta et al. (2012); Worldwatch Institute (2007)
Sweet sorghum	520–580 Nm^3 ton VS added	-	3,498	Brown (2006); Liotta et al. (2012)
Wheat	650–700 Nm^3 ton VS added	-	2,000	Liotta et al. (2012); Worldwatch Institute (2007)
Switch grass	-	-	10,790	Chaudhary et al. (2014)
Clover grass	290–390 m^3/ton	-	-	Feedstocks (2024)
Sudan grass	213–303 m^3/ton	-	-	Feedstocks (2024)
Corn stover	-	-	362–456 L/ton	Corbin et al. (2015)
Switchgrass	-	-	392–457 L/ton	Corbin et al. (2015)
Rice straw	-	-	416 L/ton	Corbin et al. (2015)

VS, Volatile solids.

comprises over 90% organic material on a dry weight basis and contains 20%–50% moisture on a wet weight basis. This provided abundant, renewable, and underutilized bioresource for making other valuable biofuel products (Dantas et al., 2013; Pan et al., 2022; Varshney et al., 2019). The conversion of this raw material into bioethanol involves pretreatment, enzymatic hydrolysis with cellulolytic enzymes, fermentation, and purification (Wani et al., 2023). Ethanol from SCB depends heavily upon pretreatment. For instance, Walter and Ensinas (2010) reported ethanol production of approximately 149.3 L per dry ton of bagasse after reaching 76% saccharification efficiency by enzymatic hydrolysis. Mesa (2011) also demonstrated a two-step pretreatment process involving dilute acid hydrolysis (1% H_2SO_4, 175°C, 40 minutes, solid to liquid ration 1:1) followed by organosolv enhanced ethanol production to 192 L per dry ton. A variation in modified protocol (120°C, 40 minutes, solid to liquid ratio) yielded 180 L per ton, demonstrating the critical role of thermochemical conditions to optimize sugar yield and conversion efficiency. Theoretical projections suggest that up to 300 L of ethanol can be derived from a single ton of SCB (Cerqueira-Leite et al., 2009). Moreover, the newer developments have come to identify the use of SCB in conventional ethanol production from sugarcane juice, following organosolv pretreatment with thermal treatment and dilute acid catalysis, thereby enhancing the yield and sustainability of the process (Singh et al., 2024). Globally, SCB supply is estimated to be around 243 million dry metric tonnes per year, yielding a considerable energy potential of approximately 4.3 EJ or nearly 6.8% of the global bioenergy potential of 63 EJ per year (Negrão et al., 2021).

Corn continues to be a preeminent feedstock for first-generation bioethanol, historically applying kernels with high starch content. Present-day strategies more and more confer value on the whole plant—including the stover—to maximize biomass conversion efficacy and reduce greenhouse emissions. As the most widely cultivated cereal crop globally, corn holds a central role in the bioenergy sector. The United States alone contributes around 40% of global corn production, followed by countries such as China, Brazil, Mexico, India, Indonesia, France, and Argentina. According to FAO (2009), global maize production in 2009 was approximately 817 million tonnes, surpassing both rice (678 million tonnes) and wheat (682 million tonnes). In the United States, corn stover, including stalks, leaves, husks, and cobs, has emerged as a promising lignocellulosic feedstock for cellulosic ethanol. Its high availability, fairly consistent composition, and advantageous logistical attributes support its potential for scalable biofuel use. Corn was grown on more than 34 million hectares in 41 states by 2010, with yield gains exhibiting a near-linear trend from the 1950s to 2010 due to improvements in hybrid breeding and agronomic management (Tan et al., 2012).

Sweet sorghum (*Sorghum bicolor* L. Moench), particularly its juice-rich stalks, is another ideal feedstock for bioethanol production. High content of cellulose, glucose, hemicellulose, and sucrose are the characteristics required for biofuel production (Dar et al., 2018; Kimm and Day, 2011). Besides these, several agronomic advantages such as rapid growth, high sugar accumulation, biomass productivity, drought and salinity tolerance, waterlogging resistance, wide adaptability, and early maturation under short photoperiods are also observed in this crop (Almodares and Hadi, 2009; Almodares et al., 2007; Almodares et al., 2008; Reddy et al., 2005; Tesso et al., 2005; Umakanth et al., 2019). Under optimal conditions, sweet sorghum can yield up to 13.2 tons/ha of total sugars, which is equivalent to 7,682 L/ha of ethanol (Murray et al., 2009). Ethanol yields have ranged from 3,000 to 8,000 L/ha depending on the region and processing efficiency (Bonin et al., 2016; Kimm and Day, 2011; Serna-Saldivar et al., 2012). In Brazil, where sweet sorghum can be added on sugarcane ethanol production, yields have been over 80 tons/ha, and pilot trials on large-scale with various stakeholders are ongoing (Rao et al., 2015; Renan et al., 2016). Techno-economic analysis, as in Central America, indicate that utilizing sweet sorghum during off-seasons of sugarcane can fetch substantial returns (Cutz and Santana, 2014). In the United States, feasibility reports that sweet sorghum is an inexpensive feedstock relative to corn, cotton, and soybean (Linton et al., 2011). Firms such as Delta BioRenewables and Bio Dimensions have experimented with incorporating it into current ethanol plants without extensive process modifications. Commercial-scale ethanol production from sweet sorghum commenced in China in 2012. The high feedstock process, which forms up to 80% of the cost of production, has been the economic hurdle (Liu et al., 2015). Nevertheless, companies, such as ZTE Energy, Fuxin Green BioEnergy, and Jilin Fuel Alcohol Company, are making progress toward commercialization (Rao et al., 2015). In Australia, operational trials have indicated that the sugar extraction efficiency of sweet sorghum can reach levels equivalent to sugarcane with mill modifications (Webster et al., 2004). Excellent yields of ethanol have been established through laboratory and pilot-scale fermentation experiments, with up to 9%–118% theoretical yield conversion efficiencies (Albertson et al., 2013). In India, the Indian Institute of Millets Research (IIMR) and the International Crops Research Institute for the Semi-Arid Tropics (ICRISAT) have led the research on sweet sorghum bioethanol. Partnerships with distilleries like Rusni Distilleries (world's first commercial sweet sorghum ethanol facility) and Tata Chemicals Ltd. have established ethanol production of 35–45 L/ton of crushed stalks (Basavaraj et al., 2013; Hunsigi et al., 2010; Vinutha et al., 2014). Although early success was recorded, low procurement prices and feedstock supply chain challenges impended long-term operations. Sweet sorghum has been identified as a crop of strategic importance for bioethanol. Latest initiatives to raise ethanol procurement prices to Rs. 48.50–49.50 per liter are meant to enhance blending targets (10% ethanol in petrol). Integrating sweet sorghum into sugar mills during sugarcane lean periods could enable year-round ethanol production, benefiting both industry and marginal farmers, and reducing fossil fuel imports.

Jatropha (*Jatropha curcas* L.) has emerged as a promising non-edible feedstock for biodiesel production due to its high oil yield and adaptability to marginal lands. Its seeds contain 30%–40% oil (Kamel et al., 2018), composed primarily of fatty acids such as arachidic, linoleic, oleic, palmitic, and stearic acids, which are suitable for conversion into biodiesel via transesterification. The extracted oil possesses desirable fuel attributes, including low acidity, moderate viscosity, high oxidative stability, and favorable cold flow behavior. Its high cetane number enhances combustion efficiency, further supporting its potential as a diesel alternative (Divakara et al., 2010; Jain and Sharma, 2010; Tapanes et al., 2008; Tian et al., 2009). One of the biggest strengths of *J. curcas* is that it can survive on dry, degraded land, without competing with food crops, thus increasing its sustainability profile. Nevertheless, to be suitable for automative use, the crude oil has to be processed by transesterification (Prueksakorn and Gheewala, 2008). In comparison with other vegetable oils, *Jatropha* oil provides better oxidative stability than soybean oil, lower viscosity compared to castor oil, and improved cold flow characteristics than palm oil. In addition, important fuel attributes like density, viscosity, and free fatty acid content are uniform during storage (Tapanes et al., 2008). With an oil content of 300–400 g/kg (Kamel et al., 2018), *J. curcas* is a promising and inexpensive feedstock for the synthesis of biodiesel. Its performance parameters are comparable to those of conventional diesel, making it a serious alternative fuel source (Endalew et al., 2011; Pandey et al., 2012; Rashid et al., 2010; Reddy et al., 2017).

Soybean, an oilseed crop in America and Asia, production has increased with surge in oil and protein demand by consumers. Though soybean is cultivated primarily for oil production, it even has potential to be utilized as biofuel feedstock as well. Soybean oil and protein processing yield large quantities of carbohydrate-rich byproducts, which are likely to be underutilized (Laman and Ju, 2016). Notably, soybean hulls, one of the important byproducts, contain sufficient carbohydrates to be employed as feedstock for the manufacture of ethanol. Even though such hulls find application in animal feed due to their protein composition, they also present a potential pathway for the manufacture of ethanol (Mielenz et al., 2009). While historically there has been a trend toward oil extraction, new methods now seek into integrate soybean biomass fully into the supply chain for biofuels to give higher energy returns and sustainability.

1.4 BIOFUEL AND SUSTAINABLE DEVELOPMENT GOAL: A GLOBAL PERSPECTIVE

Lora et al. (2011) utilized biomass productivity, GHG emissions, economic viability, water consumption and land use, and biodiversity effects for assessing the sustainability of biofuels. As earlier mentioned, second-generation biofuels utilize lignocellulosic biomass and agricultural residues, which contributes to various Sustainable Development Goals (SDG) like SDG 13 (Climate Action), SDG 9 (Industry, Innovation, and Infrastructure), and SDG 12 (Responsible

Consumption and Production) (Hassan et al., 2019; Mahmood et al., 2019). The environmental degradation can be easily implemented by these second-generation raw sources (Colla et al., 2019). Aside from their environmental benefits, biofuels from algal sources overcome the issue of competition for food sources, which can affect food security and biodiversity. In addition, they offer other advantages by the promotion of employment opportunities and earning potential, especially in communities dependent on seasonal industries (Doshi et al., 2016). Consequently, third- and fourth-generation biofuels are increasingly regarded as the means to achieve SDGs (Mat Arnon et al., 2020). The production of biofuels is inextricably linked with many of the SDGs, both directly and indirectly. For example, SDG 3, Good health and well-being, can be linked with reduced air pollution leading to cleaner energy sources, thus revealing the association between SDG 3 and SDG 7 (Raman et al., 2025).

With increasing global demand for energy and an urgent requirement to contain GHG emissions, biofuels have stepped into the center stage as a low-carbon fuel option capable of addressing climate and energy security goals simultaneously (Rial, 2024). Their usage finds special relevance within the transportation sector, which continues to rely upon fossil fuels in large numbers. Because the sector needs high energy density fuels to operate economically and efficiently, liquid biofuels will probably remain the center of activities in the foreseeable future (Aragon et al., 2022; Geissler et al., 2024). This has brought high interest in the manufacturing of liquid renewable fuels, particularly sustainable aviation fuels, which are critical for the decarbonization of air transport (Baral et al., 2019; Park et al., 2022; Perkis and Tyner, 2018). The initial phase of biofuel development was marked by first-generation fuels such as bioethanol from sugarcane or maize and biodiesel from vegetable oils, which set the stage for further innovation (Dharani et al., 2024). Technological advancements are now focusing the gaze on second- and third-generation biofuels produced from lignocellulosic biomass and algae, which do not compete with food and are more environment-friendly (Thakur and Singh, 2024).

Biofuel production also contributes to SDG 7, Affordable and Clean Energy, by utilizing biomass available on region basis. People of rural region can produce fuel and electricity for various purposes like household use, transport sector, small industries, which leads to lesser reliance on oil and fossil fuels (Acheampong et al., 2017).

Additionally, SDG 8 aims for inclusive and sustained economic growth alongside the promotion of decent work opportunities for all. Considering the aim of SDG 8, the production of biofuels through various sources has also opened up avenues for employment. While adoption of renewable energy has quickened in the developing world, including BRICS nations, due to domestic policy and international climate agreement, the implications for labor and economic stability are poorly explored (Ma and Wang, 2025). The majority of previous literature on these countries focuses on environmental impacts—lower emissions, cleaner air—and does not well incorporate the socioeconomic aspects, particularly green job creation (Han et al., 2024; Pata et al., 2024). With such high demand for jobs in these populous countries, renewable energy is an overdue opportunity for green job growth. This research seeks to examine how investment in clean energy affects both the creation of jobs and sustainable economic growth throughout the BRICS nation, and in doing so, better understand the relationship between renewable energy policy and inclusive growth among emerging markets (Hu and Li, 2024). Conversely, developed G-7 countries have witnessed in the last 10 years an increase in research on the employment-generating capacity of renewable technologies and other green industries. Li et al. (2025) proved that renewable energy investment generates more jobs per unit of energy compared to fossil fuels, reaffirming the worth of clean industries for employment in advanced economies such as United States and Germany. Job opportunities in this sector have also opened up across various disciplines from technical to operational (Stefek et al., 2022). Moreover, the production of biofuels contributes to broader expansion in the economical aspect, lowering energy costs and increasing reliance on energy-based products. Countries adapting the use of biofuel in various fields are at a higher beneficial stage with increased economic resilience and improved energy security. These relationships draw additional capital into research, innovation, and infrastructure, establishing a basis for long-term stability and competitiveness (Agbakwuru et al., 2024). The creation of employment is becoming a central factor in the formation of energy policy. In Germany's "Energiewende", employment, for instance, was found to be among the five essential issues driving national debate (Fischer et al., 2016). As the world's energy transition gains pace, the job effects of renewable technologies are coming under scrutiny from a broad range of stakeholders, ranging from academics and governments to private companies and civil society groups. Jobs trends differ between technologies and points along the energy value chain, including production, generation, storage, transmission, and flexibility solutions (Ram et al., 2022), which underscores the complexity of labor dynamics in the clean energy economy.

Besides, SDG 13 addresses the need for action on climate change and its adverse effects (Cohen et al., 2021). Biofuels, as renewable energy, act as an alternative to fossil fuels in power generation and transportation (Agbakwuru et al., 2024). Also, the role of biofuel in GHG emission has been emphasized by several studies (Agbakwuru et al., 2024; Raman et al., 2025). Additionally, biofuel production from waste and residues of agricultural crops and other non-food feedstocks plays significant role in decarbonization without compromising food security. The production of next-generation biofuels has also attained low carbon development. Furthermore, the association between SDG 13 and SDG 7 prioritizes the increase of low-emission energy technologies. Biofuel in this context bridges a link between these goals by serving as a sustainable energy alternative product while also curbing global warming (Raman et al., 2025).

SDG 12 encourages responsible production and consumption by the support of resource use efficiency as well as waste reduction. Biofuels drive the achievement of this goal by utilizing agricultural residues, food wastes, and municipal wastes to generate renewable fuels. The utilization of these raw materials for biofuel will also help in reducing landfills and promote the circular economy. This in turn extends the life cycle and utility of these raw materials. Basically, the biofuel sector supports sustainable consumption habits by converting waste into more efficient and sustainable resource economy (Raman et al., 2025).

Biofuels along with other benefits intersect with several SDGs. Achieving these advantages sustainable, however, requires an integrated approach that involves combining scientific progress, inclusive and adaptive policy environments, and higher international collaboration. With the guidance of good governance and global concerted efforts, biofuels can be a central driver in the creation of a low-emission, equitable, and climate resilient future (Chowdhury et al., 2025).

1.5 CHALLENGES OF BIOFUEL PRODUCTION IN LOWER (THIRD GENERATION) AND HIGHER (FIRST AND SECOND GENERATION) PLANTS

The production of biofuels from first-generation feedstocks, primarily derived from higher plants, has faced serious restraints, primarily the supply of feedstocks (Ahorsu et al., 2018). This generation is highly reliant on food crops such as maize and sugarcane, which are also required for human consumption. Currently, more than 70% of biodiesel is initially first-generation feedstocks like palm oil, and this further increases conflict with food stocks and raises ethical and environmental challenges (Patel et al., 2025). Due to the involvement of food crops and their availability for the production of these biofuels, it is obvious that the price of food and availability will increase, impacting low-income populations (Das, 2020). On the other hand, biofuels produced from waste and lignocellulose of plants show cleaner and more economic path. These sources, especially those regionally occurring at low prices, are used to produce bulk-scale biodiesel (Patel et al., 2025). Conversely, third- and fourth-generation biodiesel manufacture from algae and cyanobacteria also has its constraints, including the oil extraction technologies inefficiency and associated economic costs of advanced technologies. These technical and economical constraints have further shifted focus away from food-based feedstocks due to the food versus fuel issue. Nevertheless, the increased use of next-generation biofuels must also overcome numerous challenges. These include feedstock quality inconsistencies, high initial investment, advanced processing technology, and operation problems

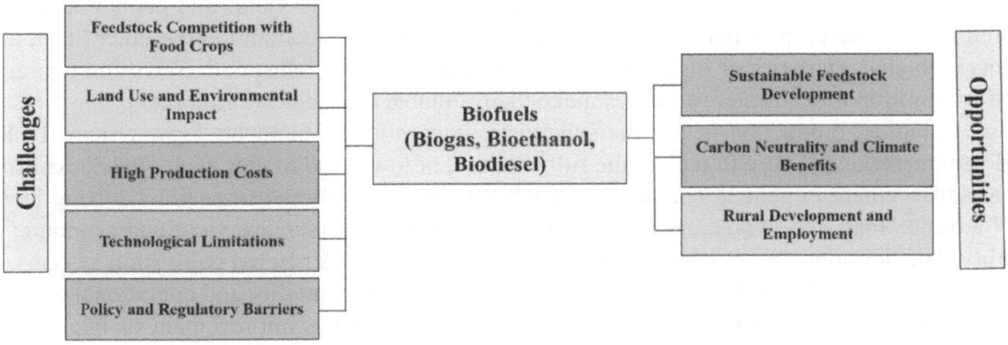

FIGURE 1.1 Challenges and opportunities of biofuels produced from plants.

such as reduced performance at low temperature and storage limitations. Aside from these, public perception, land use conflicts, and large resource requirements add additional barriers. Effective product management—particularly for feedstock like glycerol—and compatibility with existing fuel infrastructure are also critical to enabling the smooth integration of such alternative feedstocks into the biodiesel value chain (Malik et al., 2024).

The search for renewable energy has placed attention on plant-based biofuels, which can serve as a sustainable alternative to fossil fuels. As climate change, energy security, and environmental degradation become increasingly worrisome, biofuels offer many opportunities that can help deal with these challenges (Figure 1.1) on a global scale including sustainable feedstocks, carbon neutrality, rural development, technological advancement, and energy diversification (Rial, 2024; Sarma et al., 2024). As mentioned earlier, the use of sustainable feedstocks for biofuel production will also able to non-compete with food security (Das, 2020). In terms of environmental benefits, biofuels are generally considered carbon-neutral. Carbon neutrality refers to a balance being achieved in the amount of carbon dioxide, which is being released into the atmosphere and removing or off-setting, where carbon is neither released nor removed. This balance is aimed at mitigating climate change by preventing the increased levels of CO_2 being added to the atmosphere as a result of human activities (Chen, 2021). This carbon balance had driven plant-based biofuels that are a cleaner alternative for denominated fossil fuels, particularly in areas such as transportation, which are difficult to decarbonize (Rajendran et al., 2025). The incorporation of biofuels in national energy policies will help countries reduce greenhouse gas emissions and contribute to meeting climate targets in accordance with international agreements such as the Paris Accordance (Wuri et al., 2023). Advanced biofuels are becoming more widely understood to be necessary to achieve carbon neutrality targets. In alignment with its 2050 carbon neutrality goal, the European Union is actively promoting the generation and uptake of advanced biofuels under the Renewable Energy Directive 2.0 (Sharma et al., 2025). Another important opportunity offered by biofuels is the ability to promote rural development and jobs. The biofuel value chain, consisting of feedstock cultivation, harvesting, processing, and distribution, can also create significant livelihood opportunities (Tagwi and Chipfupa, 2023). In this context, farmers can also grow energy crops or provide agricultural waste, while processing plants and logistics can create new employment. Decentralized biofuel production systems can provide energy access to remote locations, enhancing local economies and energy independence (Mperejekumana et al., 2024).

1.6 CONCLUSION AND FUTURE PROSPECTS

With increasing global demand for biofuel usage, the production of biofuels from agricultural crops has now shifted to wastes and aquatic organisms like algae and bacteria. This shift has not only increased biofuel production but has also addressed the issue raised by crop utilization in the context of food security. Larger area of cultivation, high biomass yield, and economic value in biorefineries make crops a potential raw material for biofuel production. On the other hand, lower plants with characteristics of high lipid yields, increased growth rates, and growth on non-arable land with minimal freshwater resources make them suitable candidates for biofuel production. The future of biofuel production lies in the synergistic integration of biotechnology, systems biology, and sustainable agriculture to realize the full potential of lower and higher plant feedstocks. Lower plant advancement in genetic engineering, CRISPR-based genome editing, and synthetic biology will advance lipid and hydrocarbon biosynthesis continually to increase yield and economic feasibility. The development of low-cost photobioreactors, wastewater-based cultivation systems, and AI optimization models will be expected to overcome current scalability and operational efficiency constraints. Future research with higher plants will focus on the improvement of biomass conversion efficiency by lignin modification, augmented photosynthetic capacity, and stress-tolerant, high-yielding cultivars. The integration of lignocellulosic feedstocks with microbial consortia and advances in pretreatment will facilitate second-generation biofuel scalability. In addition, integration of carbon capture, precision agriculture, and circular bioeconomy models with biofuel production will enhance environmental and economic sustainability further. Global cooperation, policy support incentives, and inter-disciplinary research will be essential for creating resilient biofuel supply chains. Finally, a diversified, multi-feedstock strategy that combines the low-emission, carbon-neutral attributes of lower plants with the high biomass yield and versatility of higher plants will drive the next generation of low-emission, sustainable bioenergy system.

COMPETING INTERESTS

The authors declare that they have no competing interest.

DECLARATION OF GENERATIVE AI AND AI-ASSISTED TECHNOLOGIES IN THE WRITING PROCESS

During the preparation of this work, the book chapter author(s) have not taken any assistance from AI, and the book chapter author(s) reviewed and edited the content as needed and take(s) full responsibility for the content of the publication.

REFERENCES

Abusweireh, R. S., Rajamohan, N., Sonne, C. & Vasseghian, Y. 2023. Algae biogas production focusing on operating conditions and conversion mechanisms – A review. *Heliyon* 9(7): e17757.

Acheampong, M., Ertem, F. C., Kappler, B. & Neubauer, P. 2017. Is pursuit of sustainable development goal (SDG) number 7: Will biofuels be reliable? *Renewable and Sustainable Energy Reviews* 75: 927–937.

Achinas, S., Horjus, J., Achinas, V. & Euverink, G. J. 2019. A PESTLE analysis of biofuels energy industry in Europe. *Sustainability* 11(21): 5981.

Adams, J. M., Gallagher, J. A. & Donnison, I. S. 2009. Fermentation study on *Saccharina latissima* for bioethanol production considering variable pre-treatments. *Journal of Applied Phycology* 21(5): 569–574.

Agarwal, A. K. 2007. Biofuels (alcohols and biodiesel) applications as fuels for internal combustion engines. *Progress in Energy and Combustion Science* 33: 233–271.

Agbakwuru, V., Obidi, P. O., Salihu, O. & Ogwu, C. 2024. The role of renewable energy in achieving sustainable development goals. *International Journal of Engineering Research* 07(02): 013–027.

Ahorsu, R., Medina, F. & Constanti, M. 2018. Significance and challenges of biomass as a suitable feedstock for bioenergy and biochemical production: A review. *Energies* 11(12): 3366. https://doi.org/10.3390/en11123366.

Albertson, P., Wong, H. H., Moghaddam, L., Stolz, W., Harrison, M. D. & O'Hara, I. M. 2013. Energy products from sweet sorghum. In *Sweet sorghum* – opportunities for a new, renewable fuel and food industry in Australia. In: Australian Government-Rural Industries Research and Development Corporation (RIRDC) Publication No. 13/087, pp. 38–61.

Algenol Biofuels. 2011. https://www.algenolbiofuels.com (accessed December 20, 2021).

Almodares, A. & Hadi, M. 2009. Production of bioethanol from sweet sorghum: A review. *African Journal of Agricultural Research* 4: 772–780.

Almodares, A., Hadi, M. & Ahmadpour, H. 2008. Sorghum stem yield and soluble carbohydrates under different salinity levels. *African Journal of Biotechnology* 7: 4051–4055.

Almodares, A., Hadi, M. & Dosti, B. 2007. Effects of salt stress on germination percentage and seedling growth in sweet sorghum cultivars. *Journal of Biological Science* 7: 1492–1495.

Aragon, N. Z. U., Parker, N. C., VanLoocke, A., Bagley, J., Wang, M., Geprgescu, M. 2023. Sustainable land use and viability of biojet fuels. Nature Sustainability 6, 158–168.

Baral, N. R., Kavvada, O., Mendez-Perez, D., Mukhopadhyay, T., Lee, S., Simmons, B. A. & Scown, C. D. 2019. Technoeconomic analysis and life cycle greenhouse gas mitigation cost of five routes to biojet fuel blendstocks. *Energy Environmental Science* 12: 807–824. https://doi.org/10.1039/c8ee03266a.

Basavaraj, G., Rao, P. P., Basu, K., Reddy, C. R., Kumar, A. A., Rao, P. S. & Reddy, B. V. S. 2013. Assessing viability of bio-ethanol production from sweet sorghum in India. *Energy Policy* 56: 501–508.

Berchmans, H. J. & Hirata, S. 2008. Biodiesel production from crude *Jatropha curcas* L. seed oil with a high content of free fatty acids. *Bioresource Technology* 99: 1716–1721.

Bhattarai, K., Stalick, W. M., Mckay, S., Geme, G. & Bhattarai, N. 2011. Biofuel: An alternative to fossil fuel for alleviating world energy and economic crises. *Journal of Environmental Science and Health, Part A Toxic* 46: 1424–1442.

Bonin, C. L., Heaton, E. A., Cogdill, T. J. & Moore, K. J. 2016. Management of sweet sorghum for biomass production. *Sugar Technology* 18(2): 150–159.

Borowiak, D. & Krzywonos, M. 2022. Bioenergy, biofuels, lipids and pigments—research trends in the use of microalgae grown in photobioreactors. *Energies* 15: 5357.

Bosnjakovic, M. & Sinaga, N. 2020. The perspective of large-scale production of algae biodiesel. *Applied Sciences* 10(22): 8181.

BP. 2022. *BP statistical review of world energy*. https://www.bp.com/content/dam/bp/business-sites/en/global/corporate/pdfs/energy-economics/statistical-review/bp-stats-review-2022-full-report.pdf (accessed February 5, 2023).

Brown, L. R. 2006. Plan B 2.0: Rescuing a planet under stress and a civilization in trouble. New York: W Norton, p. 352.

Bruhn, A., Dahl, J., Nielsen, H. B., Nikolaisen, L., Rasmussen, M. B., Markager, S., Olesen, B., Arias, C. & Jensen, P. D. 2011. Bioenergy potential of *Ulva Lactuca*: Biomass yield methane production and combustion. *Bioresource Technology* 102(3): 2595–2604.

Carlsson, A. S., Van beilen, J. B., Moller, R. & Clayton, D. 2007. Micro and macro algae: Utility for industrial applications. Newbury, UK: CPL Press, p. 82.

Cerqueira-Leite, R. C., Leal, M. R. L. V. L., Cortez, L., Griffin, W. M. & Scandiffio, M. I. G. 2009. Can Brazil replace 5% of the 2025 gasoline world demand with ethanol? *Energy* 34: 655–661.

Chan, C. X., Ho, C. L. & Phang, S. M. 2006. Trends in seaweed research. *Trends in Plant Science* 11: 165–166.

Chaudhary, L., Pradhan, P., Soni, N., Singh, P. & Tiwari, A. 2014. Algae as a feedstock for bioethanol production: New entrance in biofuel world. *International Journal of ChemTech Research* 6(2): 1381–1389.

Chen, J. M. 2021. Carbon neutrality: Toward a sustainable future. *Innovation (Camb)* 2(3): 100127. https://doi.org/10.1016/j.xinn.2021.100127.

Chisti, Y. 2007. Biodiesel from microalgae. *Biotechnology Advances* 25: 294–306.

Chisti, Y. 2008. Biodiesel from microalgae beats bioethanol. *Trends in Biotechnology* 26: 126–131.

Chng, L. M., Teo, K. S., Chan, D. J., Lee, K. T. & Toh, P. Y. 2018. Fermentation of microalgae biomass through mild acid pretreatment for bioethanol production. *Journal of Energy Safety Technology* 1(2-2). https://doi.org/10.11113/jest.v1n2-2.27.

Chowdhury, P., Mahi, N. A., Yeassin, R., Chowdhury, N. U. R. & Farrok, O. 2025. Biomass to biofuel: Impacts and mitigation of environmental, health and socioeconomic challenges. *Energy Conversion and Management:* X 25: 1000889.

Cohen, B., Cowie, A., Babiker, M., Leip, A. & Smith, P. 2021. Co-benefits and trade-offs of climate change mitigation actions and the Sustainable Development Goals. *Sustainable Production and Consumption* 26: 805–813.

Colla, L. M., Rempel, A., Decesaro, A., Menegotto, A. L. L., Menegotto, E. C., Magro, F. G., Kreling, N. E. & Machado, T. S. 2019. Association of bioethanol and biogas for environmental impact mitigation. In Treichel, H., Alves Júnior, S. L., Fongaro, G. & Muller, C. (eds), *Ethanol as a green alternative fuel: Insight and perspectives*. Singapore: Springer, pp. 221–257.

Corbin, K. R., Byrt, C. S., Bauer, S., DeBolt, S., Chambers, D., Holtum, J. A. M., Karem, G., Henderson, M., Lahnstein, J., Beahen, C. T., Bacic, A., Fincher, G. B., Betts, N. S. & Burton, R. A. 2015. Prospecting for energy-rich renewable raw materials: Agave leaf case study. *PLoS One* 10: e0135382. https://doi.org/10.1371/journal.pone.0135382.

Cutz, L. & Santana, D. 2014. Techno-economic analysis of integrating sweet sorghum into sugar mills: The Central American case. *Biomass Bioenergy* 68: 195–214.

Dafnomilis, I., Hoefnagels, R., Pratama, Y. W., Schott, D. L., Lodewijks, G. & Junginger, M. 2017. Review of solid and liquid biofuel demand and supply in Northwest Europe towards 2030 – A comparison of national and regional projections. *Renewable and Sustainable Energy Reviews* 78: 31–45. https://doi.org/10.1016/j.rser.2017.04.108.

Dahman, Y., Syed, K., Begum, S., Roy, P. & Mohtasebi, B. 2019. 14-Biofuels: Their characteristics and analysis. In Verma, D., Fortunati, E., Jain, S. & Zhang, X. (eds), *Biomass, biopolymer-based materials, and bioenergy*. Woodhead Publishing Series in Composites Science and Engineering. Cambridge, UK: Woodhead Publishing, pp. 277–325.

Dantas, G. A., Mazzone, L. F. L. & Mazzone, A. 2013. Energy from sugarcane bagasse in Brazil: An assessment of the productivity and cost of different technological routes. *Renewable and Sustainable Energy Reviews* 21: 356–364.

Dar, R. A., Dar, E. A., Kaur, A. & Phutela, U. G. 2018. Sweet sorghum – A promising alternative feedstock for biofuel production. *Renewable and Sustainable Energy Reviews* 82: 4070–4090.

Das, S. 2020. The National Policy of biofuels of India - A perspective. *Energy Policy* 143: 111595. https://doi.org/10.1016/j.enpol.2020.111595.

Das, S. & Rosmann, M. 2023. *India: Biofuels annual* (GAIN Report, USDA FAS). Washington, DC.

Deora, P. S., Verma, Y., Muhal, R. A., Goswami, C. & Singh, T. 2022. Biofuels: An alternative to conventional fuel and energy source. *Materials Today: Proceedings* 48: 1178–1184.

Dharani, L., Umapriya, R., Arunkumar, N., Gokila, M. & Sakthi, R. 2024. Generations of biofuel. In Shah, M. & Deka, D. (eds), *Emerging sustainable technologies for biofuel production in generations of biofuel*. Springer, Cham, pp. 15–42.

Divakara, B., Upadhyaya, H., Wani, S. & Gowda, C. L. 2010. Biology and genetic improvement of *Jatropha curcas* L.: A review. *Applied Energy* 87: 732–742.

Doshi, A., Pascoe, S., Coglan, L. & Rainey, T. J. 2016. Economic and policy issues in the production of algae-based biofuels: A review. *Renewable and Sustainable Energy Reviews* 64: 329–337. https://doi.org/10.1016/j.rser.2016.06.027.

El-Araby, R. 2024. Biofuel production: Exploring renewable energy solutions for a greener future. *Biotechnology of Biofuels and Bioproducts* 17: 129.

Endalew, A. K., Kiros, Y. & Zanzi, R. 2011. Heterogeneous catalysis for biodiesel production from *Jatropha curcas* oil (JCO). *Energy* 36: 2693–2700.

FAO. 2009. *Statistics division maize, rice and wheat: Area harvested, production quantity, yield*. https://faostat.fao.org/site/567/DesktopDefault.aspx? PageID=567.

Feedstocks. 2024. *Anaerobic digestion*. https://www.biogas-info.co.uk/about/feedstocks/ (accessed December 26, 2024).

Fischer, W., Hake, J. F., Kuckshinrichs, W., Schröder, T. & Venghaus, S. 2016. German energy policy and the way to sustainability: Five controversial issues in the debate on the Energiewende. *Energy* 115: 1580–1591. https://doi.org/10.1016/j.energy.2016.05.069.

Foidl, N., Foidl, G., Sanchez, M., Mittelbach, M. & Hackel, S. 1996. *Jatropha curcas* L. as a resource for the production of biofuel in Nicaragua. *Bioresource Technology* 58: 77–82.

Folaranmi, J. 2013. Production of biodiesel (B100) from Jatropha oil using sodium hydroxide as catalyst. *Journal of Petroleum Engineering* 2013: 1–6.

Ganesan, R., Manigandan, S., Samuel, M. S., Shanmuganathan, R., Brindhadevi, K., Chi, N. T. L., Duc, P. A. & Pugazhendhi, A. 2020. A review on prospective production of biofuel from microalgae. *Biotechnology Reports* 27: e00509.

Gautam, A., Pant, M., Pant, G. & Kumar, G. 2024. Second-generation biofuels: Concepts, applications, and challenges. In Karnwal, A. & Mohammad Said Al-Tawaha, A. R. (eds), *Microbial applications for environmental sustainability*. Singapore: Springer, pp. 277–304. https://doi.org/10.1007/978-981-97-0676-1_16.

Geissler, G. H., Ryu, J. & Maravellas, C. T. 2024. The future of biofuels in the United States Energy Reviews. *Renewable and Sustainable Energy Reviews* 192: 114276.

Global bioenergy statistics 2019. (2019). World Bioenergy Association.

González-Fernández, C., Sialve, B., Bernet, N. & Steyer, J. P. 2012. Thermal pretreatment to improve methane production of Scenedesmus biomass. *Biomass Bioenergy* 40: 105–111.

Griffiths, M. J., van Hille, R. P. & Harrison, S. T. L. 2012. Lipid productivity, settling potential and fatty acid profile of 11 microalgal species grown under nitrogen replete and limited conditions. *Journal of Applied Phycology* 24: 989–1001.

Guo, H., Daroch, M., Liu, L., Qiu, G., Geng, S. & Wang, G. 2013. Biochemical features and bioethanol production of microalgae from coastal waters of Pearl River Delta. *Bioresource Technology* 127: 422–428.

Han, Y., Li, Z., Feng, T., Qiu, S., Hu, J., Yadav, K. K. & Obaidullah, A. J. 2024. Unravelling the impact of digital transformation on green innovation through microdata and machine learning. *Journal of Environmental Management* 354: e120271. https://doi.org/10.1016/j.jenvman.2024.120271.

Harun R. & Danquah, M. K. 2011. Influence of acid pre-treatment on microalgal biomass for bioethanol production. *Process Biochemistry* 46(1): 304–309. https://doi.org/10.1016/j.procbio.2010.08.027

Harun, R., Danquah, M. K. & Forde, G. M. 2010. Microalgal biomass as a fermentation feedstock for bioethanol production. *Journal of Chemical Technology & Biotechnology* 85: 199–203.

Hassan, S. S., Williams, G. A. & Jaiswal, A. K. 2019. Moving towards the second generation of lignocellulosic biorefineries in the EU: Drivers, challenges, and opportunities. *Renewable and Sustainable Energy Reviews* 101: 590–599. https://doi.org/10.1016/j.rser.2018.11.041.

Heredia, V., Gonçalves, O., Marchal, L. & Pruvost, J. 2021. Producing energy-rich microalgae biomass for liquid biofuels: Influence of strain selection and culture conditions. *Energies* 14: 1246.

Hernández, D., Riaño, B., Coca, M. & García-González, M. C. 2015. Saccharification of carbohydrates in microalgal biomass by physical, chemical and enzymatic pre-treatments as a previous step for bioethanol production. *Chemical Engineering Journal* 262: 939–945.

Hirani, A. H., Javed, N., Asif, M., Basu, S. K. & Kumar, A. 2018. A review on first- and second-generation biofuel productions. In Kumar, A., Ogita, S. & Yau, Y.-Y. (eds), *Biofuels: Greenhouse gas mitigation and global warming: Next generation biofuels and role of biotechnology*. New Delhi: Springer, pp. 141–154.

Horn, S. J., Aasen, I. M. & Ostgaard, K. 2000. Ethanol production from seaweed extract. *Journal of Industrial Microbiology and Biotechnology* 25: 249–254.

Hu, J. & Li, Q. 2024. From bits to green: Unraveling the digital economy's influence on natural resource efficiency. *Journal of Environmental Management* 365: e121203. https://doi.org/10.1016/j.jenvman.2024.121203.

Hunsigi, G., Yekkeli, N. R. & Kongawad, B. Y. 2010. Sweet stalk sorghum: An alternative sugar crop for ethanol production. *Sugar Technology* 12: 79–80. https://doi.org/10.1007/s12355-010-0017-5.

IEA. 2019. *Data & statistics – IEA 2019*. https://www.iea.org/data-and-statistics?country=WORLD&fuel=Energy supply&indicator=TPESbySource (accessed October 11, 2020).

IRENA. 2016. *Innovation technology outlook for advanced liquid biofuels*.

Ismail, M. M., Ismail, G. A. & El-Sheekh, M. M. 2020. Potential assessment of some micro- and macroalgae species for bioethanol and biodiesel production. *Energy Sources: Part A: Recovery, Utilization and Environmental Effects* 46(1): 1–17.

Ivanova, G., Rákhely, G. & Kovacs, K. L. 2009. Thermophilic biohydrogen production from energy plants by *Caldicellulosiruptor saccharolyticus* and comparison with related studies. *International Journal of Hydrogen Energy* 34: 3659–3670. https://doi.org/10.1016/j.ijhydene.2009.02.082.

Jain, S. & Sharma, M. 2010. Prospects of biodiesel from Jatropha in India: A review. *Renewable and Sustainable Energy Reviews* 14: 763–771.

John, R. P., Anisha, G. S., Nampoothiri, K. M. & Pandey, A. 2011. Micro and macroalgal biomass: A renewable source for bioethanol. *Bioresource Technology* 102: 186–193.

Johnston, M., Foley, J. A., Holloway, T., Kucharik, C. J. & Monfreda, C. 2009. Resetting global expectations from agriculture biofuels. *Environmental Research Letters* 4(1): 014004.

Kamel, D. A., Farag, H. A., Amin, N. K., Zatout, A. A. & Ali, R. M. 2018. Smart utilization of Jatropha (*Jatropha curcas* L.) seeds for biodiesel production: Optimization and mechanism. *Industrial Crops and Production* 111: 407–413.

Kandimalla, P., Desi, S. & Vurimindi, H. 2016. Mixotrophic cultivation of microalgae using industrial flue gases for biodiesel production. *Environmental Science and Pollution Research International* 23: 9345–9354.

Karunakaran, S. & Gurusamy, R. 2011. Bioethanol production as renewable biofuel from rhodophytes feedstock. *International Journal of Biol. Biotechnology* 2(2): 94–99.

Kemausuor, F., Addo, A. & Darkwah, L. 2015. Technical and socioeconomic potential of biogas from Cassava waste in Ghana. *Biotechnology Research International* 2015(1): 828576.

Kim, H. M., Oh, C. H. & Bae, H. J. 2017. Comparison of red microalgae (*Porphyridium cruentum*) culture conditions for bioethanol production. *Bioresource Technology* 233: 44–50.

Kimm, M. & Day, D. F. 2011. Composition of sugar cane, energy cane, and sweet sorghum suitable for ethanol production at Louisiana sugar mills. *Journal of Indian Microbiol Biotechnology* 38(7): 803–807.

Kraan, S. 2010. Mass cultivation of carbohydrate rich macroalgae, a possible solution for sustainable biofuel production. *Mitigation and Adaptation Strategies for Global Change* 18(1): 1–20. https://doi.org/10.1007/s11027-010-9275-5.

Kumar, S., Gupta, R., Kumar, G., Sahoo, D. & Kuhad, R. C. 2013. Bioethanol production from *Gracilaria verrucosa*, a red alga in a biorefinery approach. *Bioresource Technology* 135: 150–156.

Kumar, Y., Yogeshwar, P., Bajpai, S., Jaiswal, P., Yadav, S., Pathak, D. P., Sonker, M. & Tiwary, S. K. 2021. Nanomaterials: Stimulants for biofuels and renewables, yield and energy optimization. *Materials Advances* 2: 5318–5343.

Laman, A. A. & Jum, L. K. 2016. Soyabean carbohydrate as fermentation feedstock for production of biofuels and value added chemicals. *Process Biochemistry* 51(8): 1046–1057.

Li, Y., Umair, M., Guliyeva, S. & Shakaraliyeva, Z. 2025. The extractive industries transparency initiative: Achieving disclosure, but falling short on corruption reduction. *The Extractive Industries and Society* 22: e101602. https://doi.org/10.1016/j.exis.2024.101602.

Liang, K. 2024. Second generation biofuels: Utilization of Agricultural waste and non food parts. *Journal of Energy Bioscience* 15(5): 277–288.

Linton, J. A., Miller, J. C., Little, R. D., Petrolia, D. R. & Coble, K. H. 2011. Economic feasibility of producing sweet sorghum as an ethanol feedstock in the south-eastern United States. *Biomass Bioenergy* 35: 3050–3057.

Liotta, G., Panico, A., Esposito, G., Pirozzi, F. & Frunzo, L. 2012. Biomethane potential tests to measure the biogas production from the digestion and co-digestion of coAlbertsonmplex organic substrates. *The Open Environmental Engineering Journal* 5: 1–8.

Liu, H., Ren, L., Spiertz, H., Zhu, Y. & Xie, G. H. 2015. An economic analysis of sweet sorghum cultivation for ethanol production in North China. *GCB Bioenergy* 7: 1176–1184. https://doi.org/10.1111/gcbb.12222.

Liu, X., Ying, K., Chen, G., Zhou, C., Zhang, W., Zhang, X., Cai, Z., Holmes, T. & Tao, Y. 2017. Growth of *Chlorella vulgaris* and nutrient removal in the wastewater in response to intermittent carbon dioxide. *Chemosphere* 186: 977–985.

Lora, E. E. S., Palacio, J. C. E., Rocha, M. H., Renó, M. L. G., Venturini, O. J. & Del Olmo, O. A. 2011. Issues to consider, existing tools and constraints in biofuels sustainability assessments. *Energy* 36(4): 2097–2110. https://doi.org/10.1016/j.energy.2010.06.012.

Ma, B. & Wang, A. 2025. Exploring the role of renewable energy in green job creation and sustainable economic development: An empirical approach. *Energy Strategy Reviews* 28: 101642.

Ma, F. & Hanna, M. A. 1999. Biodiesel production: A review. *Bioresource Technology* 70: 1–15.

Maceiras, R., Rodriguez, M., Cancela, A., Urrejola, S. & Sanchez, A. 2011. Macroalgae: Raw material for biodiesel production. *Applied Energy* 88: 3318–3323.

Magneschi, L., Catalanotti, C., Subramanian, V., Dubini, A.; Yang, W., Mus, F., Posewitz, M. C., Seibert, M., Perata, P. & Grossman, A. R. 2012. A mutantin the ADH1 gene of *Chlamydomonas reinhardtii* elicits metabolic restructuring during anaerobiosis. *Plant Physiology* 158: 1293–1305.

Mahmood, H., Moniruzzaman, M., Iqbal, T. & Khan, M. J. 2019. Recent advances in the pretreatment of lignocellulosic biomass for biofuels and value-added products. *Current Opinion in Green and Sustainable Chemistry* 20: 18–24. https://doi.org/10.1016/j.cogsc.2019.08.001.

Malibari, R., Sayegh, F., Elazzazy, A. M., Baeshen, M. N., Dourou, M. & Aggelis, G. 2018. Reuse of shrimp farm wastewater as growth medium for marine microalgae isolated from red sea—Jeddah. *Journal of Clean. Production* 198: 160–169.

Malik, M. A. I., Zeeshan, S., Khubaib, M., Ikram, A., Hussain, F., Yassin, H. & Qazi, A. 2024. A review of major trends, opportunities, and technical challenges in biodiesel production from waste sources. *Energy Conversion and Management:* X 23: 1000675.

Malode, S. J., Prabhu, K. K., Mascarenhas, R. J., Shetti, N. P. & Aminabhavi, T. M. 2021. Recent advances and viability in biofuel production. *Energy Conversation and Management* X 10: 100070.

Masojídek, J., Ranglová, K., Lakatos, G. E., Silva Benavides, A. M. & Torzillo, G. 2021. Variables governing photosynthesis and growth in microalgae mass cultures. *Processes* 9: 820.

Mat Aron, N. S., Khoo, K. S., Chew, K. W., Show, P. L., Chen, W. H. & Nguyen, T. H. P. 2020. Sustainability of the four generations of biofuels – A review. *International Journal of Energy Research* 44(12): 9266–9282. https://doi.org/10.1002/er.5557.

McHugh, D. J. 2003. *A guide to the seaweed industry*. FAO Fisheries Technical Paper No. 441, 105.

Meher, L. C., Sagar, D. V. & Naik, S. N. 2006. Technical aspects of biodiesel production by transesterification - A review. *Renewable Sustainable Energy Review* 10: 248–268.

Mendez, L., Mahdy, A., Demuez, M., Ballesteros, M. & González-Fernández, C. 2014. Effect of high pressure thermal pretreatment on *Chlorella vulgaris* biomass: Organic matter solubilisation and biochemical methane potential. *Fuel* 117: 674–679.

Mesa, L., González, E., Romero, I., Ruiz, E., Cara, C. & Castro, E. 2011. Comparison of process configurations for ethanol production from two-step pretreated sugarcane bagasse. *Chemical Engineering Journal* 175: 185–191.

Mielenz, J. R. 2009. Biofuels and Biotechnology. In Walker, J. M. & Rapley, R. (eds), *Molecular biology and biotechnology*, 5th Edition. Cambridge: Royal Society of Chemistry, pp. 548–584.

Möllers, K. B., Cannella, D., Jørgensen, H. & Frigaard, N. U. 2014. Cyanobacterial biomass as carbohydrate and nutrient feedstock for bioethanol production by yeast fermentation. *Biotechnology Biofuels* 7(1): 1.

Mperejekumana, P., Shen, L., Zhang, S., Gaballah, M. S. & Muhirwa, F. 2024. Exploring the potential of decentralized renewable energy conversion systems on water, energy, and food security in Africa. *Energy Conversion and Management* 315: 118757.

Murray, S. C., Rooney, W. L., Hamblin, M. T., Mitchell, S. E. & Kresovich, S. 2009. Sweet sorghum genetic diversity and association mapping for brix and height. *The Plant Genome* 2: 48–62.

Mussgnugm, J. H., Klassen, V., Schlüter, A. & Kruse, O. 2010. Microalgae as substrates for fermentative biogas production in a combined biorefinery concept. *Journal of Biotechnology* 150(1): 51–56.

Negrão, D. R., Grandis, A., Buckeridge, M. S., Rocha, G. J., Leal, M. R. I. & Driemeier, C. 2021. Inorganics in sugarcane bagasse and straw and their impacts for bioenergy and biorefining: A review. *Renewable and Sustainable Energy Review* 148: 111268.

Nijum, S., Nishanthini, T. & Balaji, M. 2020. Alkaline hydrogen peroxide-pretreated sugarcane tops for bioethanol production-a process optimization study. *Biomass Conversion Biorefinery* 10(1): 149–165.

Olatunji, K. O. & Madyira, D. M. 2023. Anaerobic co-digestion of alkali pretreated groundnut shells and duck waste for methane yield optimization and sustainable environment. *E3S Web of Conferences* 433: 02005.

Openshaw, K. 2000. A review of *Jatropha curcas*: An oil plant of unfulfilled promise. *Biomass Energy* 19: 1–15.

Osman, M. E., Abo-Shady, A. M., Elshobary, M. E., Abd El-Ghafar, M. O. & Abomohra, A. E. F. 2020. Screening of seaweeds for sustainable biofuel recovery through sequential biodiesel and bioethanol production. *Environmental Science and Pollution Research* 27: 32481–32493. https://doi.org/10.1007/s11356-020-09534-1.

Pan, S., Zabed, H. M., Wei. Y. & Qi, X. 2022. Technoeconomic and environmental perspectives of biofuel production from sugarcane bagasse: Current status, challenges and future outlook. *Industrial Crops and Products* 188(B): 115684.

Pandey, V. C., Singh, K., Singh, J. S., Kumar, A., Singh, B. & Singh, R. P. 2012. *Jatropha curcas*: A potential biofuel plant for sustainable environmental development. *Renewable and Sustainable Energy Reviews* 16: 2870–2883.

Panwar, P., Singh, A. P., Chauhan, A. & Arora, R. 2025. Optimization of macro-algae-based biodiesel supply chain: A multi-objective approach. *Discover Sustainability* 6: 248.

Pardilho, S., Boaventura, R., Almeida, M. & Dias, J. M. 2022. Marine macroalgae waste: A potential feedstock for biogas production. *Journal of Environmental Management* 304: 114309.

Park, H., Chae, H. J., Suh, Y. W., Chung, Y. M. & Park, M. J. 2022. Techno-economic analysis and CO_2 emissions of the bioethanol-to-jet fuel process. *ACS Sustainable Chemical Engineering* 10: 12016. https://doi.org/10.1021/acssuschemeng.2c03853.

Park, J., Park, B. S., Wang, P., Patidar, S. K., Kim, J. H., Kim, S.-H. & Han, M.-S. 2017. Phycospheric native bacteria *Pelagibaca bermudensis* and *Stappia* sp. ameliorate biomass productivity of *Tetraselmis striata* (KCTC1432BP) in co-cultivation system through mutualistic interaction. *Frontier in Plant Sciences* 8: 289.

Passos, F., Solé, M., García, J. & Ferrer, I. 2013. Biogas production from microalgae grown in wastewater: Effect of microwave pretreatment. *Applied Energy* 108: 168–175.

Pata, U. K., Kartal, M. T. & Depren, S. K. 2024. The role of information and communication technologies and energy-related research and development investments in energy transition: Evidence from the United States of America by machine learning Algorithm. *Energy Technology* 12(5): e2301199. https://doi.org/10.1002/ente.202301199.

Patel, M. G. C., Ajith, B. S., Jagdish, Shettigar, A. K. & Samuel, O. D. 2025. Introduction to renewable energy resources and sustainable feedstocks for biodiesel. In Patel, M. G. C., Ajith, B. S., Jagdish, Shettigar, A. K. & Samuel, O. D. (eds), *Biofuel production, performance, and emission optimization: A comprehensive approach to modelling and optimization*, Vol. 215. Cham: Springer, pp. 1–21. https://doi.org/10.1007/978-3-03-84806-3_1

Perkis, D. F. & Tyner, W. E. 2018. Developing a cellulosic aviation biofuel industry in Indiana: A market and logistics analysis. *Energy* 142: 793–802. https://doi.org/10.1016/j.energy.2017.10.022.

Prajapati, S. K., Kaushik, P., Malik, A. & Vijay, V. K. 2013. Phytoremediation and biogas potential of native algal isolates from soil and wastewater. *Bioresource Technology* 135: 232–238.

Prasertsri, P. 2023. *Thailand: Biofuels annual* (GAIN Report, USDA FAS). Washington, DC.

Prueksakorn, K. & Gheewala, S. H. 2008. Full chain energy analysis of biodiesel from *Jatropha curcas* L. in Thailand. *Environ Science Technology* 42: 3388–3393.

Raheman, H. & Mondal, S. 2012. Biogas production potential of jatropha seed cake. *Biomass and Bioenergy* 37: 25–30.

Rajendran, S., Al-Samydai, A., Palani, G., Trilaksana, H., Sathish, T., Saravanan, R., Lalvani, J. I. R. & Nasri, F. 2025. Replacement of petroleum-based products with plant-based materials, green and sustainable energy - A review. *Engineering Reports* 7: e70108.

Ram, M., Osorio-Aravena, J. C., Aghahosseini, A., Bogdanov, D. & Breyer, C. 2022. Job creation during a climate complaint global energy transition across the power, heat, transport, and desalination sectors by 2050. *Energy* 238(A): 121690.

Raman, R., Sreenivasan, A., Kulkarni, N. V., Suresh, M. & Nedungadi, P. 2025. Analyzing the contributions of biofuels, biomass, and bioenergy to sustainable development goals. *iScience* 28: 112157.

Ramjani, S. A., Gunasekar, J. J. & Ramachandran, J. 2023. Evaluation of biogas production potential from waste coconut water. *International Journal of Environment and Climate Change* 13(12): 1383–1391.

Ranganathan, S. V., Narasimhan, S. L., & Muthukumar, K. 2008. An overview of enzymatic production of biodiesel. *Bioresource Technology*, 99(10): 3975–3981.

Rao, P. S., Prakasham, R. S., Parthasarathy Rao, P., Chopra, S. & Jose, S. 2015. Sorghum as a sustainable feedstock for biofuels. In: Jose, S. & Bhaskar, T. (eds), *Biomass and Biofuels - Advanced Biorefineries for Sustainable Production and Distribution*. Boca Raton, FL: CRC Press, Taylor & Francis Group, pp. 27–47.

Rashid, U., Anwar, F., Jamil, A. & Bhatti, H. N. 2010. Jatropha curcas seed oil as a viable source for biodiesel. *Pakistan Journal of Botany* 42: 575–582.

Reddy, A., Saleh, A., Islam, M. & Hamdan, S. 2017. Active razor shell CaO catalyst synthesis for *Jatropha* methyl ester production via optimized two-step transesterification. *Journal of Chemistry* 2017(1): 1489218. https://doi.org/10.1155/2017/1489218.

Reddy, B. V., Ramesh, S., Reddy, P. S., Ramaiah, B., Salimath, M. & Kachapur, R. 2005. *Sweet sorghuma* potential alternate raw material for bio-ethanol and bio-energy. *The International Sorghum and Millets Newsletter* 46: 79–86.

Renan, S., da Costa Parrella, R. A., de Souza, V. F., & Parrella, N. 2016. Maturation curves of sweet sorghum genotypes. *Ciência e Agrotecnologiaienc* 40(1): 46–56.

Reyimu, Z. & Ozcimen, D. (2017). Batch cultivation of marine microalgae *Nannochloropsis oculate* and *Tetraselmis suecica* in treated municipal water towards bioethanol production. *Journal of Cleaner Production* 150: 40–46.

Rial, R. C. 2024. Biofuels versus climate change: Exploring potentials and challenges in the energy transition. *Renewable and Sustainable Energy Reviews* 196: 114369.

Ross, A., Jones, J. M., Kubacki, M. L. & Bridgeman, T. G. 2008. Classification of macroalgae as fuel and its thermochemical behavior. *Bioresource Technology* 99: 6494–6504.

Ruatpuia, J. V. L., Halder, G., Vanlalchhhandama, M., Lalsangpuii, Boddula, R., Al-Qahtani, N., Niju, S., Mathimani, T. & Rokhum, S. L. 2024. *Jatropha curcas* oil a potential feedstock for biodiesel production: A critical review. *Fuel* 370: 131829.

Sandesh, K. & Ujwal, P. 2021. Trends and perspectives of liquid biofuel - Process and industrial viability. *Energy Conservation and Management:* X 10: 100075.

Saravanan, A. P., Pugazhendhi, A. & Mathimani, T. 2020. A comprehensive assessment of biofuel policies in the BRICS nations: Implementation, blending target and gaps. *Fuel* 272: 117635. https://doi.org/10.1016/J.FUEL.2020.117635.

Sarin, R., Sharma, M., Sinharay, S. & Malhotra, R. K. 2007. Jatropha-Palm biodiesel blends: An optimum mix for Asia. *Fuel* 86: 1365–1371.

Sarma, H. H., Paul, A., Sonowol, S., Medhi, M., Devi, O. R., Dutta, S. K., Borkotoky, B. & Talukdar, N. 2024. Biofuel: A sustainable approach to clean energy solutions, transition, carbon neutrality and green economy. *Plant Archives* 24(2): 92–98.

Sasaki, Y., Takagi, T., Motone, K., Shibata, T., Kuroda, K. & Ueda, M. 2018. Direct bioethanol production from brown macroalgae by co-culture of two engineered Saccharomyces cerevisiae strains. *Bioscience, Biotechnology, and Biochemistry* 82: 1459–1462. https://doi.org/10.1080/09168451.2018.1467262.

Schenk, P. M., Thomas-Hall, S. R., Stephens, E., Marx, U. C., Mussgnug, J. H., Posten, C., Kruse, O. & Hankamer, B. 2008. Second generation biofuels: High efficiency microalgae for biodiesel production. *Bioenergy Research* 1: 20–43.

Seo, Y. B., Lee, Y. W., Lee, C. H. & You, H. C. 2010. Red algae and their use in paper making. *Bioresource Technology* 101: 2549–2553.

Serna-Saldivar, S. O., Cristina, C., Esther, P. & Erick, H. 2012. Sorghum as a multifunctional crop for the production of fuel ethanol: Current status and future trends. In Lima, M. A. P. (ed.), *Bioethanol*. https://www.intechopen.com/books/bioethanol/sorghum-as-a-multifunctional-crop-forthe-production-offuel-ethanol-current-status-and-future-trend.

Sharma, T., Chauhan, P. S., Patel, M., Singh, A., Kaur, M., Chauhan, G., Rana, B. B., Kumar, N. & Walia, A. 2025. Carbon negative biofuels: A step ahead of carbon neutrality. *Biofuels*. https://doi.org/10.1080/17597269.2025.2450156.

Silva, A. M. B., Ranglová, K., Malapascua, J. R., Masojídek, J. & Torzillo, G. 2017. Diurnal changes of photosynthesis and growth of *Arthrospira Platensis* cultured in a thin-layer cascade and an open pond. *Algal Research* 28: 48–56.

Silva, J. F. L., Cantarella, H., Nogueira, L. A. H., Rossetto, R., Filho, R. M. & Souza, G. M. 2024. Biofuels in emerging markets of Africa and Asia: An overview of costs and greenhouse gas savings. *IEA Bioenergy Technology Collaboration Programme, Task* 39: 30.

Singh, A., Sadhvi, K., Saxena, M. & Sharma, N. K. 2024. Production process of sugarcane bagasse for the generation of bio-products and biofuel: An overview. *Archives of Current Research International* 24(12): 231–240.

Singh, N. K. & Dhar, D. W. 2011. Microalgae as second generation biofuel. A review. *Agronomy for Sustainable Development* 31: 605–629.

Soliman, R. M., Younis, S. A., El-Gendy, N. S., Mostafa, S. S. M., El-Temtamy, S. A. & Hashim, A. I. 2018. Batch bioethanol production via the biological and chemical saccharification of some Egyptian marine macroalgae. *Journal of Applied Microbiology* 125: 422–440. https://doi.org/10.1111/jam.13886.

Stefek, J., Constant, C., Clark, C., Tinnesand, H., Christol, C. & Baranowski, R. 2022. US Offshore Wind Workforce Assessment (No. NREL/TP-5000-81798). National Renewable Energy Lab. (NREL), Golden, CO (United States).

Stephenson, A. L., Dennis, J. S., Howe, C. J., Scott, S. A. & Smith, A. G. 2010. Influence of nitrogen-limitation regime on the production by *Chlorella vulgaris* of lipids for biodiesel feedstocks. *Biofuels* 1: 47–58.

Syafiuddin, A., Chong, J. H., Yuniarto, A. & Hadibarata, T. 2020. The current scenario and challenges of biodiesel production in Asian countries: A review. *Bioresource Technology Reports* 12: 100608. https://doi.org/10.1016/J.BITEB.2020.100608.

Tagwi, A. & Chipfupa, U. 2023. Participation of smallholder farmers in modern bioenergy value chains in Africa: Opportunities and constraints. *Bioenergy Research* 16: 248–262. https://doi.org/10.1007/s12155-022-10451-z.

Tahir, F., Ashfaq, H., Khan, A. Z., Amin, M., Akbar, I., Malik, H. A., Abdullah, M., Alessa, A. H., Alsaigh, A. A., Ralph, P. J., Mehmood, M. A. & Malik, S. 2024. Emerging trends in algae farming on no-arable lands for resource reclamation, recycling and mitigation of climate change driven food security challenges. *Reviews in Environmental Science and Biotechnology* 23: 869–896.

Tan, Z., Liu, S., Bliss, N. & Tieszen, L. L. 2012. Current and potential sustainable corn stover feedstock for biofuel production in the United States. *Biomass and Bioenergy* 47: 372–386.

Tapanes, N. C. O., Aranda, D. A. G., de Mesquita Carneiro, J. W. & Antunes. O. A. C. 2008. Transesterification of Jatropha curcas oil glycerides: Theoretical and experimental studies of biodiesel reaction. *Fuel* 87: 2286–2295.

Tesso, T. T., Claflin, L. E. & Tuinstra, M. R. 2005. Analysis of stalk rot resistance and genetic diversity among drought tolerant sorghum genotypes. *Crop Science* 45: 645–652.

Thakur, P. & Singh, A. 2024. Third generation biofuels: A promising approach for environmental sustainability. In Karnwal, A. & Mohammad Said Al-Tawaha, A. R. (eds), *Microbial applications for environmental sustainability*. Singapore: Springer, pp. 305–327.

Tian, Y., Zhao, L., Meng, H., Sun, L. & Yan, J. 2009. Estimation of un-used land potential for biofuels development in (the) People's Republic of China. *Applied Energy* 86: S77–S85.

Uchida, M. & Murata, M. 2004. Isolation of a lactic acid bacterium and yeast consortium from a fermented material of Ulva spp. (Chlorophyta). *Journal of Applied Microbiology* 97: 1297–1310.

Umakanth, A. V., Kumar, A. A., Vermerris, W. & Tonapi, V. 2019. Sweet sorghum for biofuel industry. In Aruna, C., Visarada, K. B. R. S., Bhat, B. V. & Tonapi, V. A. (eds), *Breeding sorghum for diverse end uses*. Woodhead Publishing Series in Food Science, Technology and Nutrition. Cambridge: Woodhead Publishing, pp. 255–270.

USAID. 2009. *Biofuels in Asia: An analysis of sustainability options*. Washington, DC: USAID.

Varshney, D., Mandade, P. & Shastri. Y. 2019. Multiobjective optimization of sugarcane bagasse utilization in an Indian sugar mill. *Sustainable Production and Consumption* 18: 96–114.

Vatsha, P. & Alam, M. R. 2025. Environmental aspects of biofuels used as sustainable resources: Current situation and future trends. In *Biofuel and sustainability: Life cycle assessments, system biology, policies and emerging technologies*, Woodhead Series in Bioenergy. Cambridge: Woodhead Publishing, pp. 19–31.

Velazquez-Lucio, J., Rodríguez-Jasso, R. M., Colla, L. M., Sáenz-Galindo, A., Cervantes-Cisneros, D. E., Aguilar, C. N., Fernandes, B. D. & Ruiz, H. A. 2018. Microalgal biomass pretreatment for bioethanol production: A review. *Biofuel Research Journal* 5: 780–791.

Vinutha, K. S., Rayaprolu, L., Yadagiri, K., Umakanth, A. V. & Srinivasa Rao, P. 2014. Sweet sorghum research and development in India: Status and prospects. *Sugar Technology* 16(2): 133–143.

Walter, A. & Ensinas, A. V. 2010. Combined production of second-generation biofuels and electricity from sugarcane residues. *Energy* 35: 874–879.

Wang, M., Ye, X., Bi, H. & Shen, Z. 2024. Microalgae biofuels: Illuminating the path to a sustainable future amidst challenges and opportunities. *Biotechnology for Biofuels and Bioproducts* 17: 10.

Wani, A. K., Rahayu, F., Fauziah, L. & Suhara, C. 2023. Advances in safe processing of sugarcane and bagasse for the generation of biofuels and bioactive compounds. *Journal of Agriculture and Food Research* 12: 100549.

Webster, A. J., Hoare, C. P., Sutherland, R. F. & Keating B. A. 2004. Observations of the harvesting, transporting and trial crushing of sweet sorghum in a sugar mill. *Proceedings of the Australian Society of Sugar Cane Technologist* 26: 1–10.

Wi, S. G, Kim, H. J, Mahadevan, S. A, Yang, D. & Bae, H. 2009. The potential value of the seaweed Ceylon moss (*Gelidium amansii*) as an alternative bioenergy resource. *Bioresource Technology* 100: 6658–6660.

Worldwatch Institute. 2007. *Biofuels for transportation: Global potential and implications for sustainable agriculture and energy in the 21st century*. Washington, DC: The Worldwatch Institute, p. 480.

Wuri, J., Rahayu, C. W. E., Hardanti, Y. R. & Kristianti, N. K. A. 2023. Assessing the emission reduction policies on global value chains: The renewable energy policy framework. *Energies* 17(23): 6031.

Yerizam, M., Jannah, A. M. & Aprianti, N. 2023. Bioethanol production from Chlorella pyrenoidosa by using enzymatic hydrolysis and fermentation method. *Journal of Ecological Engineering* 24(1): 34–40. https://doi.org/10.12911/22998993/156000.

Zabed, H., Sahu, J. N., Suely, A., Boyce, A. N. & Faruq, G. 2017. Bioethanol production from renewable sources: Current perspectives and technological progress. *Renewable and Sustainable Energy Reviews* 71: 475–501.

2 Cutting-Edge Approaches of Macro- and Microalgae as a Sustainable Energy Source for Biofuel Production

Huiying Zhang and Huu Hao Ngo

2.1 INTRODUCTION

Under the dual pressures of fossil fuel depletion and the escalating climate crisis, biofuels—as renewable energy sources—are reshaping the global energy landscape. From the early use of first-generation food-based fuels to the emerging promise of fourth-generation carbon-cycle biofuels, the evolution of biofuel technology represents more than a shift in feedstocks and production processes—it reflects a fundamental rethinking of energy paradigms. Biofuels, derived from organic matter, not only carry the strategic imperative of reducing dependence on fossil fuels but also foster a circular economy model of "resources–products–regenerated resources" (Rial, 2024).

The progression of biofuel technology exhibits a clear generational trajectory, with each stage addressing limitations of its predecessor while introducing new opportunities and challenges. These advances have resulted in the coexistence of four distinct technological systems:

- First-generation biofuels, such as corn ethanol and rapeseed oil biodiesel, leveraged the industrial use of agricultural commodities and achieved commercial-scale deployment in the early 21st century. However, their reliance on food crops has led to intense debate over "food versus fuel," given their competition for land and water resources and their implications for global food security (Malode et al., 2021).
- Second-generation biofuels addressed these concerns by utilizing non-edible biomass—such as agricultural residues (e.g., straw), forestry by products (e.g., sawdust), and food waste (e.g., fruit peels). Technologies like enzymatic hydrolysis, particularly using cellulase, enabled the conversion of lignocellulosic biomass into fermentable sugars. While this generation significantly mitigated food security concerns, high enzyme costs continue to hinder commercial viability (Binod et al., 2019).
- Third-generation biofuels rely primarily on microalgae, organisms known for their high oil content, rapid growth, and efficient carbon dioxide absorption. With greater photosynthetic efficiency and biomass productivity than terrestrial plants, microalgae are ideal for biodiesel production. However, the high production costs—often double those of petroleum-based fuels—remain a critical barrier to large-scale adoption (Hoang et al., 2023).
- Fourth-generation biofuels integrate synthetic biology with carbon capture technologies. Through the genetic engineering of microorganisms, they enable targeted evolution for

DOI: 10.1201/9781003600169-2

efficient production of biofuels and value-added biochemicals. This approach offers the potential for a synergistic solution to energy generation and environmental remediation (Pfleger et al., 2023).

Among the feedstocks for third- and fourth-generation biofuels, microalgae stand out for their remarkable adaptability. They can thrive in diverse environments—including saline-alkali soils, seawater, and wastewater—thus avoiding competition with food crops and offering co-benefits such as wastewater treatment and resource recovery (Wang et al., 2024).

Despite their potential, microalgal biofuels face significant economic and technical challenges. Harvesting alone accounts for 30%–40% of total production costs. Macroalgae, though slower growing, are easier to harvest due to their multicellular structure. Species such as kelp, nori, kombu, and sea lettuce undergo pretreatment processes—physical (e.g., grinding, pressing), chemical (e.g., acid/alkali treatment), and biological (e.g., enzymatic digestion)—to break down cell walls and release fermentable sugars. These are then converted via:

- Anaerobic fermentation into gaseous fuels (biohydrogen, biomethane).
- Aerobic fermentation into liquid fuels (bioethanol) (Kumar et al., 2021).

Applications include fuel cells and hydrogen-powered vehicles (biohydrogen), natural gas replacement for power generation (biomethane), and low-emission transport fuels (bioethanol). However, the industry still confronts critical obstacles: high collection and logistics costs, inefficient biomass pretreatment, and seasonal variability in algal productivity (Liu et al., 2022).

Common microalgal species used for biofuel production include *Chlorella*, *Dunaliella*, and various diatoms, valued for their high lipid content, short growth cycles, and strong carbon-fixation capacity. Cultivation methods such as open pond systems offer low-cost solutions but are vulnerable to contamination, while photobioreactors provide greater efficiency at higher operational costs (Siddiki et al., 2022).

Post-cultivation, multiple conversion pathways are employed: mechanical pressing, solvent extraction, supercritical fluid extraction, and ultrasound-assisted methods extract algal oils, which are then processed into biodiesel via esterification or transesterification. Fermentation techniques convert carbohydrates into bioethanol. Key biofuels derived from algae include (Figure 2.1).:

- Biodiesel, suitable for land and aviation transport.
- Bioethanol, used as a gasoline extender or substitute.
- Biohydrogen, powering fuel cells and hydrogen vehicles (Grira et al., 2023; Arora et al., 2021).

Nonetheless, commercialization remains constrained by high costs, variable product quality, and limited integration across technological stages.

To address these challenges, the integration of machine learning (ML) and life cycle assessment (LCA) is emerging as a transformative solution. ML can accelerate strain selection, optimize growth and lipid productivity models, and enable real-time adjustment of cultivation parameters by analyzing environmental and nutritional data (Syed et al., 2024). Meanwhile, LCA provides a comprehensive framework to evaluate environmental impacts, energy use, and resource efficiency across the biofuel production life cycle. It identifies high-impact stages, quantifies carbon reduction benefits, and assesses economic feasibility by analyzing material inputs, energy demands, and capital investment (Bradley et al., 2023).

The development of integrated ML–LCA models offers a robust, data-driven platform for decision-making, balancing technical, environmental, and economic dimensions. Looking ahead, emerging technologies—such as edge computing, digital twins, and automated control systems—will further enhance the precision and scalability of algal biofuel systems, catalyzing breakthroughs in the marine biomass economy.

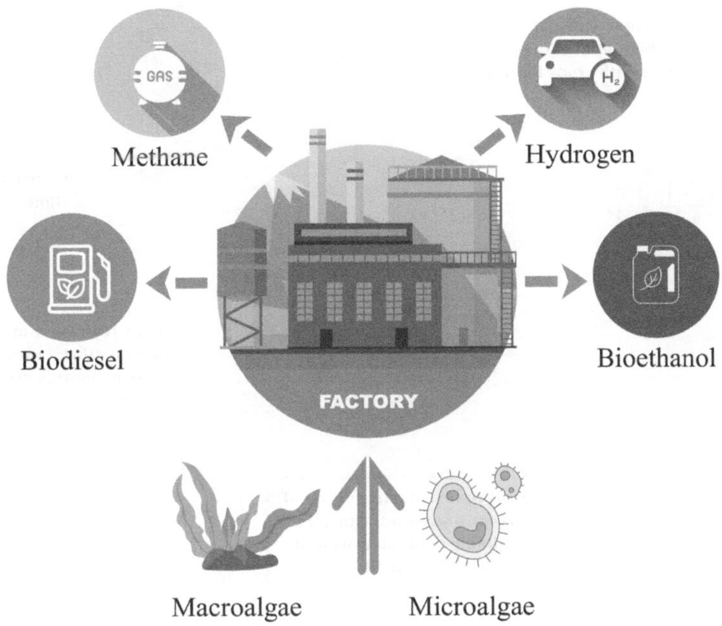

FIGURE 2.1 Multi-path conversion processes for biofuel production from algae post-cultivation.

2.2 EXPLORATION OF ENHANCING ALGAL BIOFUELS THROUGH EXTERNAL ENVIRONMENTAL FACTORS

2.2.1 Optimizing Cultivation Conditions

Environmental factors such as pH, temperature, light intensity, and light quality play pivotal roles in regulating microalgal growth and metabolic activity. Each microalgal species exhibits specific environmental preferences. For instance, *Spirulina* thrives under alkaline conditions, particularly within a pH range of 8.5–11.0, where its oil content and physiological traits are significantly enhanced (Monaselidze et al., 2002). Similarly, *Chlorella vulgaris* demonstrates optimal biomass and cell density when cultivated at pH 7 under 2,000 lux illumination and a 14 hours/10 hours light/dark photoperiod (Rajanren et al., 2017).

Temperature is equally critical, as it directly affects intracellular enzymatic activity, thereby influencing key metabolic processes including photosynthesis, respiration, and lipid synthesis. Furthermore, in macroalgae, pyrolysis temperature (typically ranging from 200°C to 600°C) determines the thermal degradation pathways of sugars and lipids, significantly impacting the composition of ethanol precursors such as monosaccharides and organic acids in the resulting bio-oil.

Light intensity and spectral quality are additional crucial parameters. While moderate light levels stimulate growth and lipid accumulation, excessive illumination may lead to photoinhibition. Spectral composition also exerts profound effects: red and blue wavelengths are particularly effective in promoting photosynthetic activity and lipid biosynthesis across a variety of microalgal species (Torres et al., 2025). In macroalgae such as *Laminaria japonica*, red light enhances callus formation and morphological development, whereas blue light stimulates photosynthetic pigment synthesis but can suppress growth rates (Mizuta et al., 2007). These findings underscore the species-specific responses to light quality and the need for tailored lighting strategies.

Optimizing cultivation parameters—such as dynamic pH control, precise temperature regulation, and the use of tailored LED lighting—can significantly enhance photosynthesis and oil

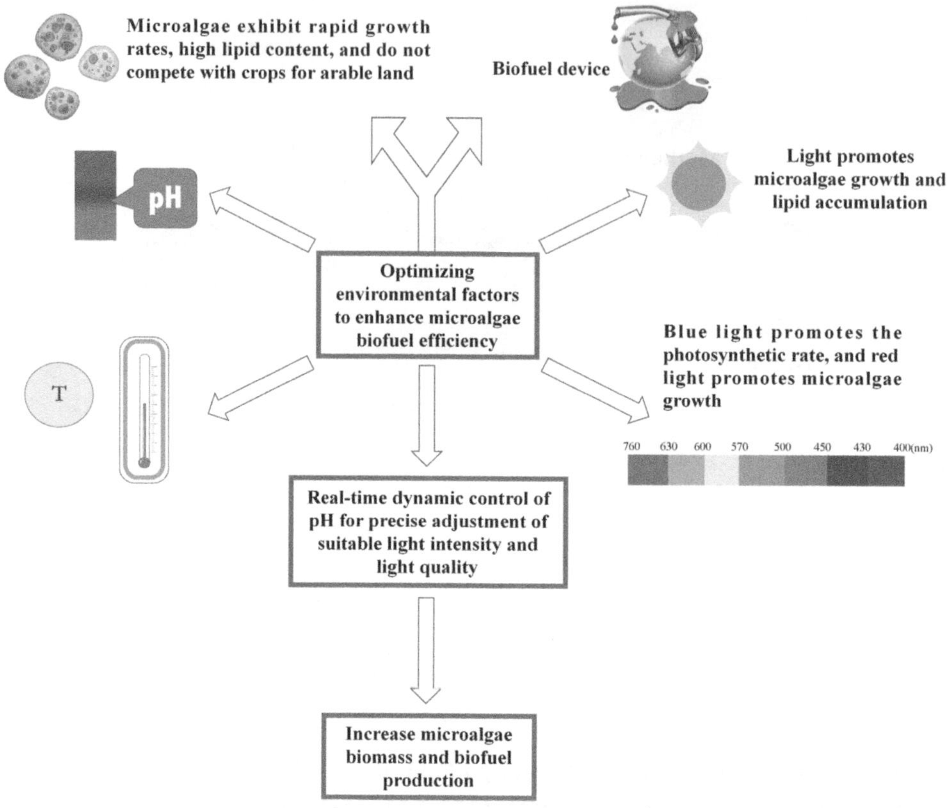

FIGURE 2.2 Effects of environmental factors on the growth and metabolism of microalgae.

synthesis, thereby improving biofuel productivity (Figure 2.2). Nevertheless, the intricate interplay among environmental factors requires further investigation to enable more precise and efficient cultivation strategies.

2.2.2 Modifying Culture Medium Composition

Manipulating the composition of the culture medium—particularly by limiting key nutrients such as nitrogen (N), phosphorus (P), and sulfur (S)—is a well-established strategy for enhancing lipid accumulation in algae (Figure 2.3). Under nitrogen-depleted conditions, many algal species redirect metabolic fluxes toward lipid biosynthesis as a survival response. For example, in the freshwater green alga UTEX 4-2219, lipid production is significantly elevated under combined nitrogen starvation, salt stress, and osmotic stress, with the most pronounced effect observed under dual nitrogen and osmotic stress (Chang et al., 2016).

Phosphorus, an essential component of nucleic acids and phospholipids, also plays a key role in algal physiology. In *Chlamydomonas reinhardtii*, phosphorus deficiency—similar to sulfur deprivation—reduces the oxygen-evolving capacity of photosystem II, creating anaerobic conditions that facilitate hydrogen production (Batyrova et al., 2015). *Laminaria abyssalis* has demonstrated efficient nitrogen uptake under low-nutrient conditions, with a particular preference for ammonium (NH_4^+) over nitrate (NO_3^-), enabling sustained growth even in nutrient-poor marine environments (Braga et al., 1996).

Under sulfur-limited conditions, certain algae activate alternative metabolic pathways conducive to biofuel production. For instance, *C. reinhardtii* strain KU204 exhibits robust light-induced hydrogen generation when subjected to concurrent nitrogen and sulfur deprivation (Pongpadung et al., 2015).

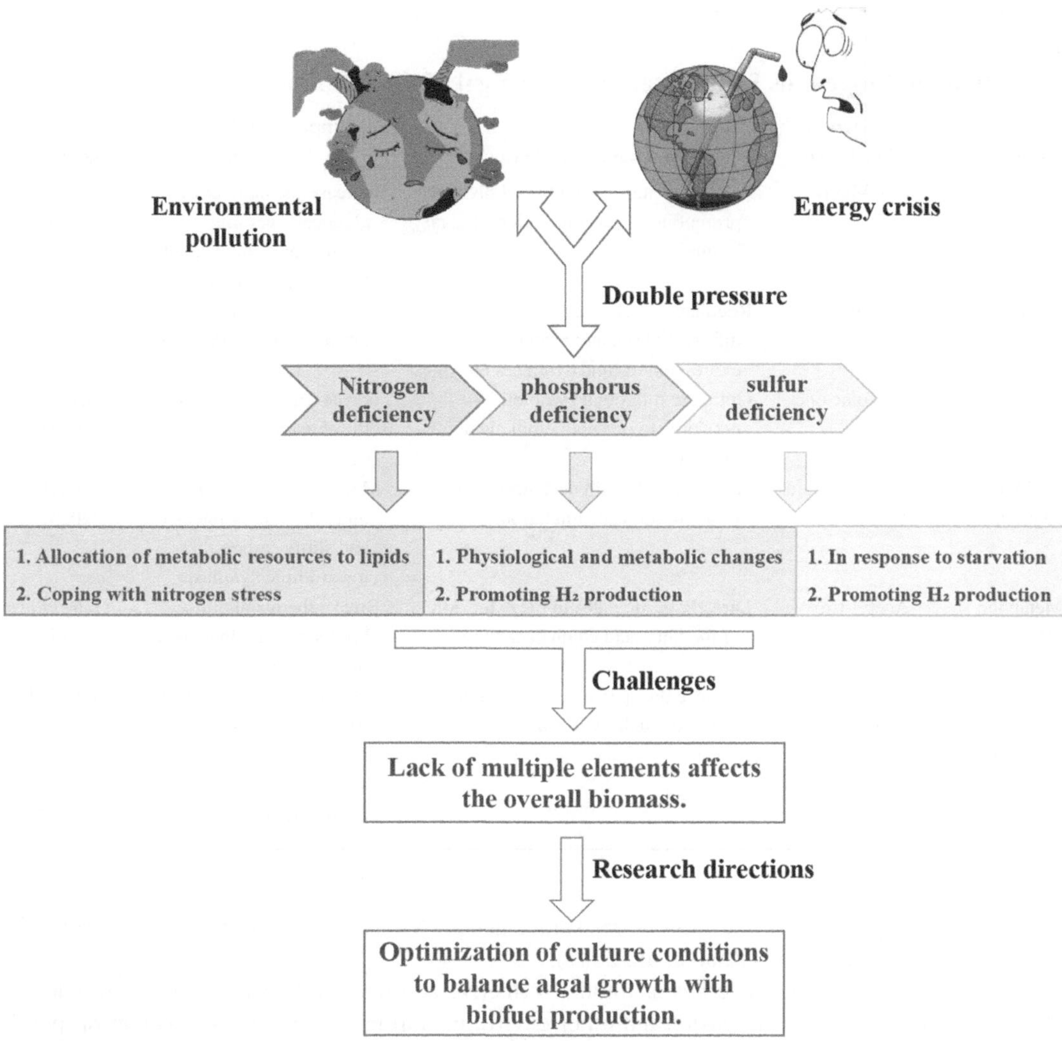

FIGURE 2.3 Regulating the nutrient composition of the culture medium enables a balance between microalgal growth and bioenergy accumulation.

Overall, nutrient stress-induced lipid enhancement is a relatively low-cost and scalable strategy for biofuel production. However, excessive nutrient limitation can inhibit algal growth and thus reduce overall productivity (Whitton et al., 2016). Therefore, a balanced approach to nutrient manipulation is essential to maximize both biomass yield and lipid content.

2.2.3 Supplementing with Exogenous Additives

The application of exogenous additives—including plant hormones, antioxidants, metabolic precursors, and gene expression inducers—offers a promising avenue to boost algal biofuel production (Table 2.1). Plant hormones such as cytokinins and auxins modulate algal cell growth and differentiation. Cytokinins promote cell division and biomass accumulation, while auxins enhance cell elongation and direct growth patterns, both of which contribute to increased lipid content.

Strigolactones, such as GR24, have also shown potential in modulating algal-fungal-bacterial symbiotic systems. For instance, the addition of GR24 not only enhances the degradation of pollutants—including antibiotics such as doxycycline, ciprofloxacin, and sulfamethoxazole—but

TABLE 2.1
The Impact of Exogenous Substances on Microalgal Biofuel Production

Category	Exogenous Substances	Mechanism of Action	Impact on Biofuel Production	References
	Cytokinins	Activate genes related to cell division, promoting the accumulation of algal biomass	Increase the number of algal cells per unit volume and enhance total biomass production	Singh et al. (2020)
Plant Hormones	Auxins	Regulate cell elongation and differentiation, and promote the expression of lipid synthesis enzymes	Increase lipid content and optimize cell morphology	Wei et al (2025)
	Strigolactones	Optimize microbial symbiotic networks to promote rapid algal growth	In a co-cultivation system, the abundance of ARGs decreases	Zhao et al. (2025)
Antioxidant System	Ascorbic Acid	Activates fatty acid desaturases to promote lipid accumulation	Maintains cell activity under high light intensity and shifts carbon flux toward lipid synthesis	Motlagh et al (2023)
Metabolic Precursors	Acetyl-CoA	Directly participates in the extension of the fatty acid chain backbone	Breaks the metabolic bottleneck and increases lipid synthesis flux	Gong et al (2024)
Gene Expression Regulators	Inducible Promoter Activators	Enhance the specific expression of key enzymes such as ACCase	Coordinate the metabolic pathways of sugar metabolism and lipid synthesis, reducing metabolic flux loss	Shin et al (2024)

also reduces the prevalence of antibiotic resistance genes (ARGs), while simultaneously promoting algal biomass accumulation (Zhao et al., 2025).

Vitamin C (ascorbic acid) serves as another effective additive, protecting algae from oxidative stress caused by high light intensity and other environmental pressures. Its antioxidant properties enhance algal viability and stimulate lipid biosynthesis by directing more carbon flux into fatty acid synthesis pathways.

In addition to regulatory compounds, direct supplementation with metabolic precursors such as acetyl-CoA can enhance fatty acid synthesis. Similarly, transcriptional activators that upregulate genes involved in lipid metabolism can further elevate biofuel yields.

In summary, the strategic use of exogenous additives represents an effective method to augment algal lipid production. However, further optimization of dosage, application timing, and compound combinations is required to ensure cost-effectiveness and scalability for industrial biofuel production.

2.3 THE APPLICATION OF SYNTHETIC BIOLOGY IN ALGAL BIOFUELS

2.3.1 Synthetic Biology Tools and Platforms

Synthetic biology, a cutting-edge interdisciplinary field of 21st-century life sciences, integrates principles from engineering, computer science, molecular biology, and systems biology. It aims to design, modify, or construct biological systems through standardization, modularization, and engineering approaches to create new organisms or systems with desired functions. In the context of algal biofuel development, synthetic biology plays a pivotal role in engineering algal strains to enhance lipid accumulation, stress tolerance, and growth rates.

By providing a suite of powerful tools and platforms, synthetic biology enables the rational design, construction, and optimization of biological systems (Mehrshahi et al., 2020), thereby offering novel methodologies for biofuel innovation (Figure 2.4).

2.3.1.1 CRISPR/Cas9 and Advanced Genome Editing Technologies

The CRISPR/Cas9 system has revolutionized algal genetic engineering by enabling precise genome editing with high efficiency. Innovations such as base editors now allow for single-nucleotide alterations without inducing double-stranded DNA breaks, minimizing off-target effects in model algae like *C. reinhardtii* and *Nannochloropsis* (Nguyen et al., 2020).

CRISPRa and CRISPRi systems enable the fine-tuned regulation of lipid biosynthesis pathways; for example, upregulating acetyl-CoA carboxylase (ACCase) enhances fatty acid synthesis (Arshad et al., 2025). Multiplex genome editing, employing guide RNA arrays, allows the simultaneous modification of multiple genes—such as knocking out starch synthesis pathways while upregulating lipid biosynthetic enzymes—thereby streamlining metabolic rewiring for enhanced lipid yields.

2.3.1.2 Biological Design Automation Platforms

Biological design automation platforms integrate robotics, artificial intelligence, and synthetic biology to streamline iterative processes from gene circuit design to strain validation. These platforms enable systematic resolution of technical challenges such as pathway optimization, editing efficiency,

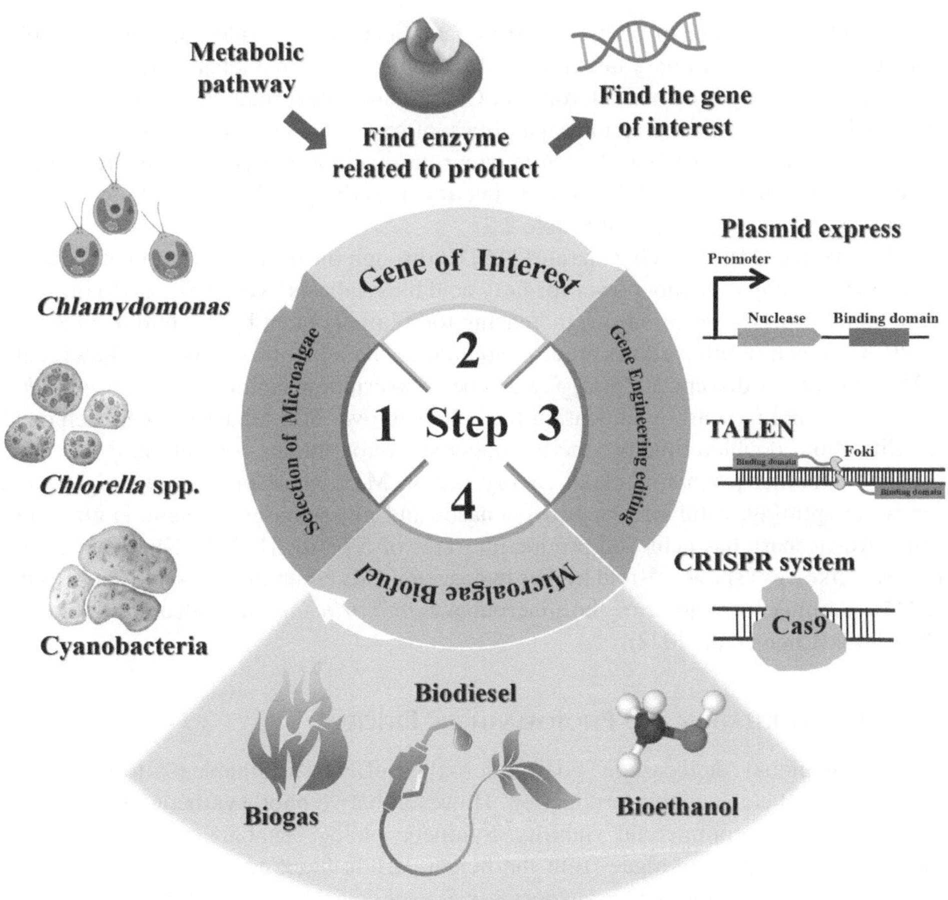

FIGURE 2.4 Engineering microalgae via synthetic biology technologies for enhanced bioenergy production.

and high-throughput phenotyping. For instance, Duman-Özdamar et al. (2025) implemented the Design-Build-Test-Learn (DBTL) cycle to enhance lipid production in *Yarrowia lipolytica*, demonstrating the effectiveness of automated platforms in improving biofuel-relevant traits.

2.3.1.3 Multiomics Integration Platforms

Multiomics platforms integrate genomic, transcriptomic, proteomic, and metabolomic datasets to construct dynamic biological network models, facilitating comprehensive insight into algal metabolic regulation. Tools such as the AlgaePath database provide interactive visualization of expression profiles and functional pathways, enabling the identification of regulatory bottlenecks in lipid accumulation under stress conditions (Zheng et al., 2014). Such integrative approaches accelerate hypothesis-driven research, guiding targeted interventions in metabolic engineering.

2.3.1.4 High-Throughput Screening (HTS) Platforms

High-throughput screening (HTS) platforms facilitate rapid optimization of algal cultivation parameters and robust prediction of large-scale productivity via automated, miniaturized experiments. These technologies are instrumental in overcoming industrialization barriers. For example, Park et al. (2024) developed a two-stage HTS system combining environmental control with phenotype analytics, significantly enhancing microalgal biomass and lipid optimization under heterotrophic conditions.

2.3.2 Enhanced Lipid and Hydrocarbon Production

Microalgae synthesize a range of lipids, including fatty acids, triglycerides, and phospholipids, with fatty acids representing the primary target for biofuel applications. In photosynthetic organisms, fatty acid biosynthesis primarily occurs in chloroplasts. Chloroplast transformation, owing to its precise gene integration and robust protein expression capabilities, offers advantages over nuclear transformation.

Li et al. (2018) engineered *Phaeodactylum tricornutum* chloroplasts to overexpress the ACCase gene (PtACC2), resulting in a 3.3-fold increase in enzyme activity and a 1.77-fold increase in neutral lipids—reaching 40.8% of dry weight (Table 2.2).

The malic enzyme (ME), which generates NADH through the decarboxylation of malate, plays a critical role in providing reducing power for fatty acid biosynthesis. Xue et al. (2015) introduced the *PtME* gene into *Chlorella pyrenoidosa*, increasing total lipid content by 2.5-fold to 57.8%.

CRISPR-based gene editing allows precise modulation of metabolic pathways. Ajjawi et al. (2017) used CRISPR/Cas9 to disrupt a $Zn(II)_2Cys_6$-type transcription factor gene in *Nannochloropsis gaditana*, increasing lipid carbon allocation from 20% to over 50% under nutrient-rich conditions. These modifications doubled lipid production under semi-continuous cultivation.

Hydrogen is another promising clean energy vector. Microalgae produce hydrogen via photosynthetic water-splitting, catalyzed by hydrogenases and nitrogenases. Genetic engineering of *C. reinhardtii* hydrogenase has achieved production rates of 5.77 mL H_2/L/h (Khetkorn et al., 2017). Mutant strains like *pgr5* show 2.5-fold higher hydrogen yields compared to wild-type strains (Nagy et al., 2024). Gamma irradiation of the macroalga *Laminaria japonica* also enhanced hydrogen output by 71.4% (Chen et al., 2022).

2.3.3 Carbon Utilization and Photosynthetic Efficiency

In alignment with global "dual carbon" goals, microalgae offer a sustainable platform for simultaneous CO_2 fixation and clean energy production. However, native photosynthetic efficiencies remain low (5%–8%), limiting commercial viability. Synthetic biology enables reengineering of photosynthetic systems to transition algae from inefficient carbon fixers to intelligent biofactories. For instance, overexpression of the BicA transporter in *Synechocystis* sp. PCC6803 doubled bicarbonate uptake, enhancing carbon assimilation (Kamennaya et al., 2015). Similarly, CRISPR-based knockout of *PhoA* in *P. tricornutum* reduced DOP hydrolase activity but significantly boosted cytochrome

TABLE 2.2
Instances of Synthetic Biology in Microalgal Biofuel Applications

Microalgae Strain	Gene/Target Site	Approach	Results	References
Phaeodactylum tricornutum (PtACC2)	ACCase	Foreign gene expression	The relative neutral lipid content increased substantially by 1.77-fold	Li et al. (2018)
Scenedesmus sp. MC-1	ACCase	Foreign gene expression	The intracellular lipid content has increased by 28.6%	Ma et al. (2019)
Chlorella pyrenoidosa	*Phaeodactylum tricornutum* ME (PtME)	Foreign gene expression	The total lipid content has significantly increased by 2.5 times	Xue et al. (2015)
Chlamydomonas reinhardtii PTS42	Malic enzyme isoform 2 (ME2)	Overexpression	The lipid synthesis rate has increased to 23.4%	Kim et al. (2019)
Phaeodactylum tricornutum	Glycerol-3-phosphate acyltransferase 2 (GPAT2)	Foreign gene expression and Overexpression	TAG content was notably increased by 2.9-fold than phospho- and glycolipids	Wang et al. (2020)
Chlamydomonas reinhardtii mutant strain L159I-N230Y	*PSBA*	Amino acid deletions in D1 protein	A hydrogen production rate of 5.77 mL $H_2 \cdot L^{-1} h^{-1}$ has been achieved	Khetkorn et al. (2017)
Synechococcus elongatus PCC 7942	Ethylene forming enzyme (efe)	Foreign gene expression and overexpression	The ethylene production rate remains stable at 140 µL/L/h/OD750	Carbonell et al. (2019)
Phaeodactylum tricornutum	UDP-glucose pyrophosphorylase	TALEN (knockout)	Developed a high-lipid-producing strain (45-fold increase in triacylglycerol accumulation)	Daboussi et al. (2014)
Nannochloropsis gaditana	Zn(ii)2Cys6	CRISPR-Cas9 (knockout)	Producing twice as much lipid as that in the wild type	Ajjawi et al. (2017)
Chlamydomonas reinhardtii	*Phospholipase* A_2	CRISPR-Cas9 (knockout)	The total lipid productivity increased by up to 64.25%	Shin et al. (2019)

levels, lipid content, and growth rates (Zhang et al., 2021). These advances not only improve biofuel yield but also pave the way for scalable CO_2 capture and circular biomanufacturing systems.

2.3.4 Stress Tolerance and Containment

Large-scale algal cultivation faces challenges from fluctuating environmental conditions (e.g., temperature, salinity, pollutants) and the risk of engineered gene escape. Synthetic biology addresses these issues by constructing robust and controllable strains. For instance, heterologous expression of the *ChbZIP1* transcription factor from alkaliphilic *Chlorella* sp. BLD in *C. reinhardtii* conferred tolerance to pH 9–9.5 conditions, enabling resilient growth in alkaline environments (Qu et al., 2021). Enhancing stress resilience reduces cultivation costs and extends applications to environmental remediation.

2.3.5 Computational and Systems Biology Approaches

The convergence of synthetic biology with artificial intelligence is transforming algal metabolic engineering by replacing traditional trial-and-error methods with predictive and integrative strategies.

Genome-scale models, such as iCZ843, combined with ^{13}C flux analysis, have enabled dynamic tracking of carbon flux in *Chlorella* under nitrogen deprivation, with prediction errors below 8% (Zuñiga et al., 2016). ML algorithms such as BioOptimus screened 10^7 enzyme combinations, identifying optimal pairs (e.g., KCS6-FATB3) to increase C16:0 fatty acid content from 35% to 58% (Markus et al., 2023).

LCA studies project that biofuel from engineered algae may achieve economic competitiveness at crude oil prices of $2/L, while reducing environmental impacts and conserving freshwater resources (Beal et al., 2015). These developments mark a paradigm shift from isolated pathway engineering to global, intelligent system design.

2.4 CONCLUSION AND FUTURE REMARKS

Contemporary biofuel technologies encompass four coexisting generations, ranging from first-generation (Gen-1) biofuels derived from food crops to emerging fourth-generation (Gen-4) carbon-neutral fuels centered on carbon cycling. Despite ongoing advances in feedstock innovation and process optimization, each generation continues to face persistent challenges. Gen-1 biofuels, in particular, have sparked the longstanding "food security versus energy demand" debate, while high costs remain a barrier across the entire biofuel value chain.

Microalgae and macroalgae (seaweed), as strategic feedstocks for Gen-3 and Gen-4 biofuels, offer notable advantages, including rapid growth cycles, high carbon capture efficiency, and non-competition with arable land. However, their industrial deployment is constrained by bottlenecks in large-scale cultivation, metabolic regulation, and product separation.

To overcome these challenges and enable next-generation biofuel commercialization, technological breakthroughs must target three key areas:

1. **Innovation in Microalgae Cultivation Technologies:**
   Advancements in genetic engineering are essential for improving genetic stability and enhancing photosynthetic efficiency. The optimization of photobioreactor design and scale-up processes will be critical to overcoming existing technical barriers. Additionally, establishing full-lifecycle regulatory frameworks is crucial to ensuring environmental sustainability and long-term viability. Commercial deployment is anticipated to follow a phased, three-stage industrial trajectory.
2. **Genetic Editing and Synthetic Biology Empowerment:**
   The integration of modular synthetic biology systems and standardized biological component libraries enables the design and optimization of algal cells for targeted functions. By incorporating genetic circuits, metabolic modules, and control elements, researchers can accelerate the rational engineering of algal metabolism. These advances allow for precise metabolic pathway construction and customized biosynthesis, significantly enhancing production efficiency. As synthetic biology ecosystems mature, algal biotechnology will become a central platform for innovation across the energy, environmental, and biomedical sectors.
3. **Intelligent Process Control and Bioeconomic Circularity:**
   The fusion of IoT, AI, and metabolic engineering paves the way for intelligent process control in algal biomanufacturing. Real-time monitoring of cultivation parameters, dynamic metabolic regulation, and precision synthesis of high-value products enhance overall system responsiveness and productivity. Coupled with multi-stage separation and electrochemical purification, this approach facilitates the conversion of byproducts into bioenergy and functional materials, forming an integrated "cultivation–extraction–conversion–reuse" polygeneration system. Blockchain and digital twin technologies further enable transparent tracking of carbon footprints and adaptive alignment with market demand, pushing resource utilization above 60% and advancing carbon mitigation efforts. This fosters a flexible, interconnected model of bioeconomic value co-creation.

Data-centric approaches are pivotal in breaking current industry impasses. ML accelerates the discovery of algal strains with high light use efficiency and resistance to contamination, while predictive maintenance reduces operational risks. Embedding LCA frameworks allows for comprehensive quantification of carbon emissions, water usage, and ecological impact, thereby informing continuous process improvement. Together, these tools create a closed-loop feedback system that propels algal biorefineries toward carbon-negative production models.

Moreover, the valorization of byproducts—such as natural pigments, phycobiliproteins, and functional polysaccharides—enhances material cycling efficiency and strengthens the economic viability of bio-based systems. This integrated "energy–resources–environment" solution provides transformative momentum for the marine biomass sector and accelerates the global transition toward a clean, low-carbon, and sustainable energy future.

COMPETING INTERESTS

The authors declare that they have no competing interest.

DECLARATION OF GENERATIVE AI AND AI-ASSISTED TECHNOLOGIES IN THE WRITING PROCESS

During the preparation of this work, the book chapter author(s) have not taken any assistance from AI, and the book chapter author(s) reviewed and edited the content as needed and take(s) full responsibility for the content of the publication.

REFERENCES

Ajjawi, I., Verruto, J., Aqui, M., Soriaga, L. B., Coppersmith, J., Kwok, K., Peach, L., Orchard, E., Kalb, R., Xu, W., Carlson, T. J., Francis, K., Konigsfeld, K., Bartalis, J., Schultz, A., Lambert, W., Schwartz, A. S., Brown, R., & Moellering, E. R. (2017). Lipid production in Nannochloropsis gaditana is doubled by decreasing expression of a single transcriptional regulator. *Nature Biotechnology*, 35(7), 647–652.

Arora, K., Kumar, P., Bose, D., Li, X., & Kulshrestha, S. (2021). Potential applications of algae in biochemical and bioenergy sector. *3 Biotech*, 11(6), 296.

Arshad, S., Qadir, M. L., Hussain, N., Ali, Q., Han, S., & Ali, D. (2025). Advances in CRISPR/Cas9 technology: Shaping the future of photosynthetic microorganisms for biofuel production. *Functional Plant Biology*, 52(2), FP24255.

Batyrova, K., Gavrisheva, A., Ivanova, E., Liu, J., & Tsygankov, A. (2015). Sustainable hydrogen photoproduction by phosphorus-deprived marine green microalgae Chlorella sp. *International Journal of Molecular Sciences*, 16(2), 2705–2716.

Beal, C. M., Gerber, L. N., Sills, D. L., Huntley, M. E., Machesky, S. C., Walsh, M. J., Tester, J. W., Archibald, I., Granados, J., & Greene, C. H. (2015). Algal biofuel production for fuels and feed in a 100-ha facility: A comprehensive techno-economic analysis and life cycle assessment. *Algal Research*, 10, 266–279.

Binod, P., Gnansounou, E., Sindhu, R., & Pandey, A. (2019). Enzymes for second generation biofuels: Recent developments and future perspectives. *Bioresource Technology Reports*, 5, 317–325.

Bradley, T., Rajaeifar, M. A., Kenny, A., Hainsworth, C., del Pino, V., del Valle Inclán, Y., Povoa, I., Mendonca, P., Brown, L., Smallbone, A., Roskilly, A. P., Joyce, S., & Heidrich, O. (2023). Life cycle assessment of microalgae-derived biodiesel. *The International Journal of Life Cycle Assessment*, 28(5), 590–609.

Braga, A. da C., & Yoneshigue-Valentin, Y. (1996). Nitrogen and phosphorus uptake by the Brazilian kelp Laminaria abyssalis (Phaeophyta) in culture. *Hydrobiologia*, 326–327(1), 445–450.

Brar, A., Kumar, M., Soni, T., Vivekanand, V., & Pareek, N. (2021). Insights into the genetic and metabolic engineering approaches to enhance the competence of microalgae as biofuel resource: A review. *Bioresource Technology*, 339, 125597.

Chang, W.-C., Zheng, H.-Q., & Chen, C.-N. N. (2016). Comparative transcriptome analysis reveals a potential photosynthate partitioning mechanism between lipid and starch biosynthetic pathways in green microalgae. *Botanical Studies*, 52(3), 305–312.

Chen, Y., Yin, Y., & Wang, J. (2022). Biohydrogen production using macroalgal biomass of Laminaria japonica pretreated by gamma irradiation as substrate. *Fuel*, 309, 122179.

Cheng, D. L., Ngo, H. H., Guo, W. S., Chang, S. W., Nguyen, D. D., & Kumar, S. M. (2019). Microalgae biomass from swine wastewater and its conversion to bioenergy. *Bioresource Technology*, 275, 109–122.

Choi, H. I., Lee, J. S., Choi, J. W., Shin, Y. S., Sung, Y. J., Hong, M. E., Kwak, H. S., Kim, C. Y., & Sim, S. J. (2019). Performance and potential appraisal of various microalgae as direct combustion fuel. *Bioresource Technology*, 273, 341–349.

Duman-Özdamar, Z. E., Julsing, M. K., Martins Dos Santos, V. A. P., Hugenholtz, J., & Suarez-Diez, M. (2025). Model-driven engineering of Yarrowia lipolytica for improved microbial oil production. *Microbial Biotechnology*, 18(3), 70089.

Gong, G., Liu, L., Wu, B., Li, J., He, M., & Hu, G. (2024). Simultaneous production of algal biomass and lipid by heterotrophic cultivation of linoleic acid-rich oleaginous microalga Chlorella sorokiniana using high acetate dosage. *Bioresource Technology*, 399, 130566.

Grira, S., Khalifeh, H. A., Alkhedher, M., & Ramadan, M. (2023). The conventional microalgal biofuel production process and the alternative milking pathway: A review. *Energy*, 277, 127547.

Hoang, A. T., Sirohi, R., Pandey, A., Nižetić, S., Lam, S. S., Chen, W. H., Luque, R., Thomas, S., Arıcı, M., & Pham, V. V. (2023). Biofuel production from microalgae: Challenges and chances. *Phytochemistry Reviews*, 22(4), 1089–1126.

Kamennaya, N. A., Ahn, S., Park, H., Bartal, R., Sasaki, K. A., Holman, H.-Y., & Jansson, C. (2015). Installing extra bicarbonate transporters in the cyanobacterium Synechocystis sp. *PCC6803 enhances biomass production. Metabolic Engineering*, 29, 76–85.

Khetkorn, W., Rastogi, R. P., Incharoensakdi, A., Lindblad, P., Madamwar, D., Pandey, A., & Larroche, C. (2017). Microalgal hydrogen production – A review. *Bioresource Technology*, 243, 1194–1206.

Kumar, D., Pugazhendi, A., Bajhaiya, A. K., & Gugulothu, P. (2021). Biofuel production from Macroalgae: Present scenario and future scope. *Bioengineered*, 12(2), 9216.

Li, D.-W., Xie, W.-H., Hao, T.-B., Cai, J.-X., Zhou, T.-B., Balamurugan, S., Yang, W.-D., Liu, J.-S., & Li, H.-Y. (2018). Constitutive and chloroplast targeted expression of acetyl-CoA carboxylase in oleaginous microalgae elevates fatty acid biosynthesis. *Marine Biotechnology*, 20, 566–572.

Liu, J., Zhou, F., Abed, A. M., Le, B. N., Dai, L., Ali, H. E., Khadimallah, M. A., & Zhang, G. (2023). Macroalgae as a potential source of biomass for generation of biofuel: Artificial intelligence, challenges, and future insights towards a sustainable environment. *Fuel*, 336, 126826.

Liu, J. J., Dickson, R., Niaz, H., Van Hal, J. W., Dijkstra, J. W., & Fasahati, P. (2022). Production of fuels and chemicals from macroalgal biomass: Current status, potentials, challenges, and prospects. *Renewable and Sustainable Energy Reviews*, 169, 1129.

Malode, S. J., Prabhu, K. K., Mascarenhas, R. J., Shetti, N. P., & Aminabhavi, T. M. (2021). Recent advances and viability in biofuel production. *Energy Conversion and Management:* X, 10, 100070.

Markus, B., C, G. C., Andreas, K., Arkadij, K., Stefan, L., Gustav, O., Elina, S., & Radka, S. (2023). Accelerating biocatalysis discovery with machine learning: A paradigm shift in enzyme engineering, discovery, and design. *ACS Catalysis*, 13(21), 14454–14469.

Mehrshahi, P., Nguyen, G. T. D. T., Rovira, A. G., Sayer, A., Llavero-Pasquina, M., Sin, M. L. H., Medcalf, E. J., Mendoza-Ochoa, G. I., Scaife, M. A., & Smith, A. G. (2020). Development of novel riboswitches for synthetic biology in the green alga Chlamydomonas. *ACS Synthetic Biology*, 9(6), 1406–1417.

Mizuta, H., Kai, T., Tabuchi, K., & Yasui, H. (2007). Effects of light quality on the reproduction and morphology of sporophytes of Laminaria japonica (Phaeophyceae). *Aquaculture Research*, 38(12), 1323–1329.

Monaselidze, J., Barbakadze, Sh., Kvirikashvili, Sh., Majagaladze, G., Khachidze, D., & Topchishvili, L. (2002). Thermal characteristics of Spirulina platensis cells under nongrowing conditions at various values of pH medium. *Biomacromolecules*, 3(4), 783–786.

Motlagh, Z. N., Mahdavi, M. A., & Gheshlaghi, R. (2023). Naphthalene acetic acid and scorbic acid significantly improve neutral lipid accumulation in Chlorella species under nutrient limitations. *Algal Research*, 76, 103311.

Nagy, V., Dabosi, Z., Kuntam, S., Csankó, K., Kovács, L., & Tóth, S. Z. (2024). Photoautotrophic and sustained H_2 production by the pgr5 mutant of Chlamydomonas reinhardtii in simulated daily light conditions. *International Journal of Hydrogen Energy*, 53, 760–769.

Nguyen, T. H. T., Park, S., Jeong, J., Shin, Y. S., Sim, S. J., & Jin, E. (2020). Enhancing lipid productivity by modulating lipid catabolism using the CRISPR-Cas9 system in Chlamydomonas. *Journal of Applied Phycology*, 32(5), 2829–2840.

Park, S.-B., Lee, Y. R., Yun, J.-H., Choi, H. I., Sim, E. J., Choi, D.-Y., Cho, D.-H., Kim, H.-S., & Lee, Y. J. (2024). Towards maximizing biomass and lipid productivity: High-throughput screening assay for prospecting heterotrophic growth for new microalgal isolates. *Microbial Cell Factories*, 23, 299.

Pfleger, B. F., & Takors, R. (2023). Recent progress in the synthesis of advanced biofuel and bioproducts. *Current Opinion in Biotechnology*, 80, 102913.

Pongpadung, P., Liu, J., Yokthongwattana, K., Techapinyawat, S., & Juntawong, N. (2015). Screening for hydrogen-producing strains of green microalgae in phosphorus or sulphur deprived medium under nitrogen limitation. *ScienceAsia*, 41(2), 97.

Qu, D., Show, P.-L., & Miao, X. (2021). Transcription factor ChbZIP1 from alkaliphilic microalgae Chlorella sp. BLD enhancing alkaline tolerance in transgenic Arabidopsis thaliana. *International Journal of Molecular Sciences*, 22(5), 2387.

Rajanren, J. R., & Ismail, H. M. (2017). Investigation of Chlorella vulgaris microalgae as a source for renewable fuel. *Biofuels*, 8(1), 37–47.

Rial, R. C. (2024). Biofuels versus climate change: Exploring potentials and challenges in the energy transition. *Renewable and Sustainable Energy Reviews*, 196, 114369.

Singh, J., Jain, D., Agarwal, P., & Singh, R. P. (2020). Auxin and cytokinin synergism augmenting biomass and lipid production in microalgae Desmodesmus sp. JS07. *Process Biochemistry*, 95, 223–234.

Shin, Y. S., Choi, H. I., Choi, J. W., Lee, J. S., Sung, Y. J., & Sim, S. J. (2018). Multilateral approach on enhancing economic viability of lipid production from microalgae: A review. *Bioresource Technology*, 258, 335–344.

Shin, S.-E., Koh, H. G., Park, K., Park, S.-H., Chang, Y. K., & Kang, N. K. (2024). Increasing lipid production in Chlamydomonas reinhardtii through genetic introduction for the overexpression of glyceraldehyde-3-phosphate dehydrogenase. *Frontiers in Bioengineering and Biotechnology*, 12, 1396127.

Siddiki, S. Y. A., Mofijur, M., Kumar, P. S., Ahmed, S. F., Inayat, A., Kusumo, F., Badruddin, I. A., Khan, T. Y., Nghiem, L. D., Ong, H. C., & Mahlia, T. M. I. (2022). Microalgae biomass as a sustainable source for biofuel, biochemical and biobased value-added products: An integrated biorefinery concept. *Fuel*, 307, 121782.

Syed, T., Krujatz, F., Ihadjadene, Y., Mühlstädt, G., Hamedi, H., Mädler, J., & Urbas, L. (2024). A review on machine learning approaches for microalgae cultivation systems. *Computers in Biology and Medicine*, 172, 108248.

Torres-Martínez, J. J., Martínez-Constantino, A., González-Ortega, O., Dibildox-Alvarado, E., & Soria-Guerra, R. E. (2025). Impact of blue and red light on lipid accumulation in Chlorella vulgaris and Scenedesmus acutus. *Journal of Applied Phycology*, 37(4), 2301–2311.

Wang, M., Ye, X., Bi, H., & Shen, Z. (2024). Microalgae biofuels: Illuminating the path to a sustainable future amidst challenges and opportunities. *Biotechnology for Biofuels and Bioproducts*, 17(1), 10.

Wei, Q., Zhao, D., Wang, M., Wang, C., Pang, F., & Ma, X. (2025). Enhanced lipid production in Chlorella vulgaris via indole-3-acetic acid salt stress in a two-stage culture for biofuels. *Biochemical Engineering Journal*, 221, 109795.

Whitton, R., Le Mével, A., Pidou, M., Ometto, F., Villa, R., & Jefferson, B. (2016). Influence of microalgal N and P composition on wastewater nutrient remediation. *Water Research*, 91, 371–378.

Xue, J., Niu, Y.-F., Huang, T., Yang, W.-D., Liu, J.-S., & Li, H.-Y. (2015). Genetic improvement of the microalga Phaeodactylum tricornutum for boosting neutral lipid accumulation. *Metabolic Engineering*, 27, 1–9.

Zhang, K., Li, J., Zhou, Z., Huang, R., & Lin, S. (2021). Roles of alkaline phosphatase PhoA in algal metabolic regulation under phosphorus-replete conditions. *Journal of Phycology*, 57(3), 703–707.

Zhao, C., Lu, B., Wang, Z., Wei, J., Zhao, Y., & Wang, S. (2025). Enhanced antibiotics and antibiotics resistance genes removal from aquaculture wastewater by microalgae-based system induced with plant hormones. *International Biodeterioration & Biodegradation*, 200, 106045.

Zheng, H.-Q., Chiang-Hsieh, Y.-F., Chien, C.-H., Hsu, B.-K. J., Liu, T.-L., Chen, C.-N. Nathan, & Chang, W.-C. (2014). AlgaePath: Comprehensive analysis of metabolic pathways using transcript abundance data from next-generation sequencing in green algae. *BMC Genomics*, 15, 196.

Zuñiga, C., Li, C.-T., Huelsman, T., Levering, J., Zielinski, D. C., McConnell, B. O., Long, C. P., Knoshaug, E. P., Guarnieri, M. T., Antoniewicz, M. R., Betenbaugh, M. J., & Zengler, K. (2016). Genome-scale metabolic model for the green alga Chlorella vulgaris UTEX 395 accurately predicts phenotypes under autotrophic, heterotrophic, and mixotrophic growth conditions. *Plant Physiology*, 172(1), 589–602.

3 Cutting-Edge Approaches of Vegetable Oil Fuels as Fuel for Biofuel Production

Nadia Riaz, Madiha Rashid, Afifa Younas, Sajjad Hyder, Aimen Razzaq Khan, Anam Ilyas, Saba Mazhar, and Noreen Kareem

3.1 INTRODUCTION

Vegetable oil (VO) is a naturally occurring hydrocarbon that can be found in a wide variety of natural plants, seeds, and fruit components (Farooq & Ngaini, 2024). Bioactive substances found in VOs, such as fatty acids, vitamins, phytosterols, carotenoids, and phenolic compounds, have significant positive effects on human health (Arab et al., 2022; Alsafadi et al., 2023; Baccouri et al., 2023; Guedes et al., 2023; Singh & Kumar, 2023; Tian et al., 2023). Oil is primarily produced from plant seeds, though it can be extracted from other sections of the plant as well (Calugar et al., 2024). VOs have so many uses in a variety of industries, such as biofuels, food, cosmetics, textiles, paint and coating, and plastic sectors, and they have become extremely popular in this new period (Farooq & Ngaini, 2024). VOs are an inexpensive food for undernourished people who require more energy intake (Meijaard et al., 2022), while in high-income environments, they supply 25%–30% of daily energy (Mannucci et al., 2023). Since the 1980s, VOs have become more widely used in a variety of consumer and industrial sectors worldwide (Mannucci et al., 2023). More than 80% of VOs are produced from palm, sunflower, soybean, and rapeseed oil, with most of the remaining sources coming from cotton, groundnut, coconut, and olive oil (Meijaard et al., 2020). In 2020, the gross output value of VOs worldwide was projected to be USD 335 billion (Mannucci et al., 2023). VOs have gained international attention for several reasons, including: (i) increased demand for VOs across the globe because of increased wealth; (ii) increased awareness of health issues associated with trans fatty acids; and (iii) increased use of VOs as a feedstock supply for biofuel production. To understand the demand features of VOs considering these conditions, it is crucial to examine the worldwide food and industrial demand for VOs (Parcell et al., 2018).

3.1.1 Physical Characteristics of Oil Products

The physical characteristics of VOs vary according to the oil type, composition, and temperature.

Density: During the cooling phase, density has an impact on the mass transfer rate and drainage rate, which are key factors that determine oil absorption. The density decreased linearly as the temperature increased (Sahasrabudhe et al., 2017; Mitrea et al., 2022).

Viscosity: One of the primary variables influencing oil drainage and absorption is viscosity. The rate of oil drainage decreases with increasing oil viscosity (Sahasrabudhe et al., 2017). Both oil type and quality affect oil viscosity. The viscosity dropped exponentially as the temperature increased. For instance, the composition of olive oil makes it more viscous than sunflower oil (Hlaváč et al., 2019).

DOI: 10.1201/9781003600169-3

Surface Tension: Surface tension is crucial in systems of oil absorption that involve steam and water. As temperature increases, surface tension decreases linearly (Sahasrabudhe et al., 2017).

Smoke Point: The temperature at which oil begins to emit smoke is known as the smoke point, and it is greatly influenced by the fatty acid level (Guillaume et al., 2018). High smoke points (210°C–232°C) are recommended by the USDA. For example, Palm oil has a smoke point at 223°C, rapeseed oil at 218°C, sunflower oil at 209°C, soybean oil at 213°C, peanut oil at 207°C, olive oil at 207°C, and coconut oil at 194°C (Guillaume et al., 2018; Ranilović et al., 2024).

3.1.2 Food Applications

VOs are an essential and common lipid source in the human diet, because they give food texture, color, and flavor. Its uses for food purposes are growing daily (Memon et al., 2024). Between 2011–2012 and 2015–2016, the global consumption of VOs increases from 151.68 to 177.16 million metric tons, according to the US Department of Agriculture. The oil used for food applications accounts for 73.4% of the total production. In the food application, VOs can be used in their natural state, emulsified, or even as cooking, frying, salad dressing (SD), marinades, baking, and flavor enhancing media (Roychowdhury et al., 2024). Due to its affordability, convenience of use, and versatility, frying is one of the most used techniques in food application (Timilsena et al., 2017; Pereira et al., 2019; Correa et al., 2024). It produces extremely appealing foods by evoking highly valued sensory qualities in fried food, such as golden color, crispy surface texture, juicy interior, and distinctive flavor and aroma (Sahasrabudhe et al., 2017; Abrante-Pascual et al., 2024). Because of their excellent oxidative stability and increased nutritional value compared to saturated fatty acid (SFA)-rich oil, monounsaturated fatty acid (MUFA)-rich oils are strongly advised for frying (Abrante-Pascual et al., 2024). The semisolid emulsion known as SD (Song & McClements, 2021) is a common condiment in the United States, where it is purchased by domestic customers for about two billion dollars (Bakhtavoryan, 2017; Yin et al., 2022). SD is a thick, creamy sauce or condiment made using oil, egg yolks, vinegar or lemon juice, and seasonings (Vishwakarma et al., 2022). VOs, which are mostly composed of fatty acids, calcium, iron, phosphorus, potassium, vitamins, and other minerals, are a vital source of SD (Bakhtavoryan, 2017; Gautron et al., 2022). The application of various VOs in food, along with burning point and culinary uses, is mentioned in Table 3.1.

TABLE 3.1
Application of Various VOs in Culinary Uses

Oil Type	Culinary Uses	Burning Point °C	Key References
Peanut oil	Cooking & salads	207	Calugar et al. (2024); Ranilović et al. (2024)
Walnut oil	Salads	160	
Coconut oil	Preserves the aroma of foods & frying	177	
Soybean oil	Salads & frying	213	
Palm oil	Frying	235	
Sunflower oil	Cooking	209	
Rapeseed oil	Frying	218	
Olive oil	Frying	210	de Carvalho et al. (2022); García-González et al. (2023)

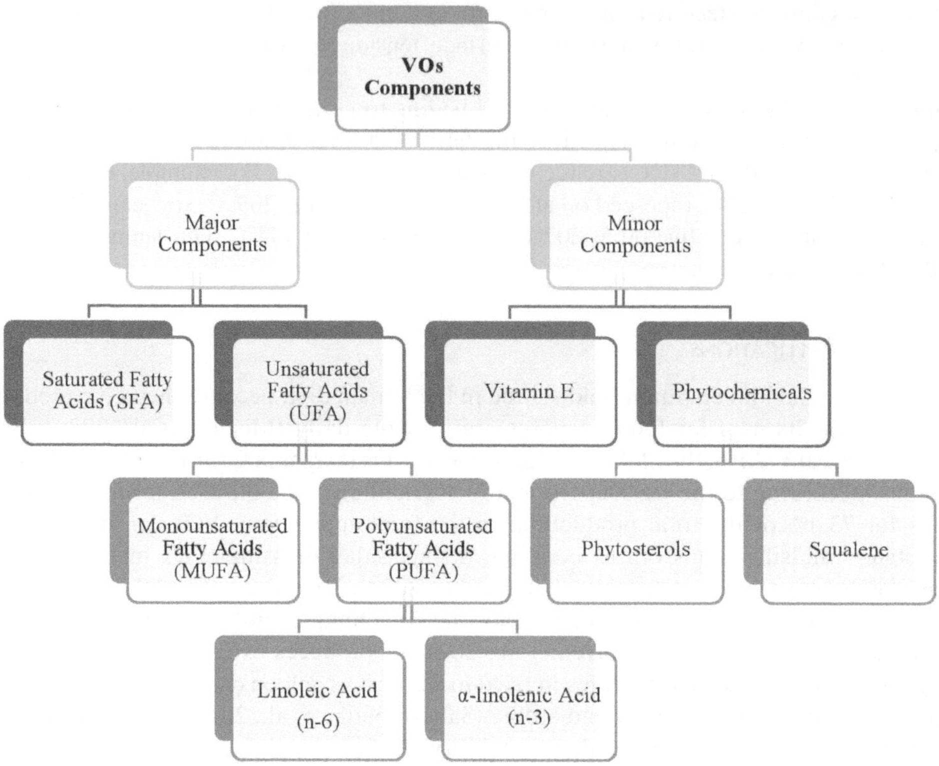

FIGURE 3.1 General overview of chemical composition of VOs.

3.1.3 Nutritional Value

The main components of VOs, triacylglycerol (TG) (95%–98%) and other trace compounds (2%–5%), make up a significant portion of the human diet, as shown in Figure 3.1, and can supply most of the fatty acids needed by the body for physiological processes (Abrante-Pascual et al., 2024). As precursors of different regulatory lipids, unsaturated fatty acids play a crucial role in cell metabolism, cell-to-cell interaction, signaling, and the preservation of cell membrane fluidity and structure. Fatty acids affect people's health, well-being, and risk of disease because of these impacts (Correa et al., 2024). However, by blocking its receptor activation, SFAs in the diet can raise serum LDL-C levels, causing cholesterol to accumulate inside arterial blood vessels and making the body more vulnerable to different cardiovascular diseases (Piernas et al., 2019). Fat-soluble vitamin E, a naturally occurring antioxidant and immunomodulator, is mostly obtained from VOs (Zhao et al., 2021). VOs rich in MUFA (olive oil, rapeseed oil, and peanut oil), Polyunsaturated fatty acids (PUFA) (sunflower oil, corn oil, soybean oil, and cottonseed oil), and SFA (palm oil and coconut oil) (Abrante-Pascual et al., 2024). The SFA, MUFA, and PUFA contain VOs vary in type, content, and ratio, as shown in Table 3.2 (Figure 3.1).

3.2 VEGETABLE OILS SUPPLY AND DEMAND

The VOs market is steadily expanding. The global market value of palm oil, olive oil, sunflower oil, soybean oil, and rapeseed oil in 2022 was USD 212.6 billion, considering their respective production quantities. It is anticipated to reach USD 268.9 billion by 2027, growing at a compound annual growth rate (CAGR) of 4.8%. Global VOs use is expected to be 249 million metric tons, with the food industry consuming most of these VOs (Ranilović et al., 2024). 60% of the world's total VOs

TABLE 3.2
Nutritional Quality of Different VOs

Oil Type	SFA (%)	MUFA (%)	Linoleic Acid (*n*-6) (%)	α-Linolenic Acid (*n*-3) (%)	Key References
Peanut oil	8.7–27.7	35–80	4.0–43.0	nd–0.5	Abrante-Pascual et al. (2024)
Rapeseed oil	3.5–12.3	51–70	15.0–30.0	5.0–14.0	
Soybean oil	10.1–21	17–30	48–59	4.5–11	
Olive oil	8.0–26.2	55–83	3.5–21.0	nd–1.5	García-González et al. (2023)
Cottonseed	21.4–26.4	14.7–21.7	46.7–58.2	nd–0.4	Bhattacharya (2023); Abrante-Pascual et al. (2024)
Coconut oil	93.0–96.3	4.2–5.8	1.0–2.5	nd–0.2	Abrante-Pascual et al. (2024); Calugar et al. (2024)
Palm oil	39.9–56.3	36.3–47.3	9.0–12.0	nd–0.5	
Corn oil	12.0–15.2	25.5–40.7	34.0–65.6	nd–2.0	
Sunflower oil	8.1–16.9	14–43	45.4–74.0	nd–0.3	Rauf et al. (2017); Abrante-Pascual et al. (2024)

SFA, saturated fatty acids; MUFA, monounsaturated fatty acids; PUFA, polyunsaturated fatty acids; nd, not detectable (≤0.05%).

production is reportedly derived from the production of soybean (Zhang et al., 2023). Furthermore, 30% of global production is palm oil, and during the past few decades, its consumption has grown significantly, especially in China and India (Tian et al., 2023). Global demand for VOs is substantial, reaching 212.82 million tons in 2022, a 1.25-fold increase from 2014, and its production, import, and export volumes essentially continue their upward trend (Tian et al., 2023). Due to the impact of COVID-19, the scarcity of arable land, and the challenge of growing the production of oilseed crop, China's refined VOs production fell to 48.819 million tons in 2022, a decline of 912,000 tons from the previous year. Concurrently, VOs self-sufficiency has been declining annually due to the fast development in consumption, which has left them heavily dependent on the external market (Tian et al., 2023). The supply and demand balance for VOs consumption would be further upset by population expansion, rising living standards, longer life expectancy, and the incidence of chronic diseases. VOs supplementation is crucial for maintaining social stability (Tian et al., 2023).

3.3 INTERNATIONAL MARKET FOR MAJOR VEGETABLE OILS AS BIOFUELS

The growing need for sustainable energy and encouraging government policies are driving the robust expansion of the global market for key VOs used as biofuels, especially palm, soybean, and canola oils.

3.3.1 Palm Oil

Palm oil is utilized by almost every home and manufacturing sector globally, in conjunction with agricultural appraisal (Suroso & Ramadhan, 2014). Indonesia and Malaysia are the world's largest producers of palm oil, making up over 80% of global production, according to the Food and Agriculture Organization of the United Nations (FAO). Currently, Indonesia produces 25 billion liters of palm oil annually, making it the world's largest producer (Mukherjee & Sovacool, 2014). Indonesia ranked first in palm oil production about 55.33%, Malaysia ranked second with 30.38% of palm oil production, Thailand is the third highest palm oil producer with 3.70% of palm oil production, and Colombia and Nigeria ranked fourth and fifth with 2.20% and 1.56%, respectively (Tandra et al., 2022). Figure 3.2 demonstrates the percentage of palm oil production in the top 10 countries worldwide.

The global palm oil trade in 2019 involved 111 exporters and 153 importers. With a combined palm oil export of USD 14.7 billion, USD 8.3 billion, USD 922 million, and USD 390 million,

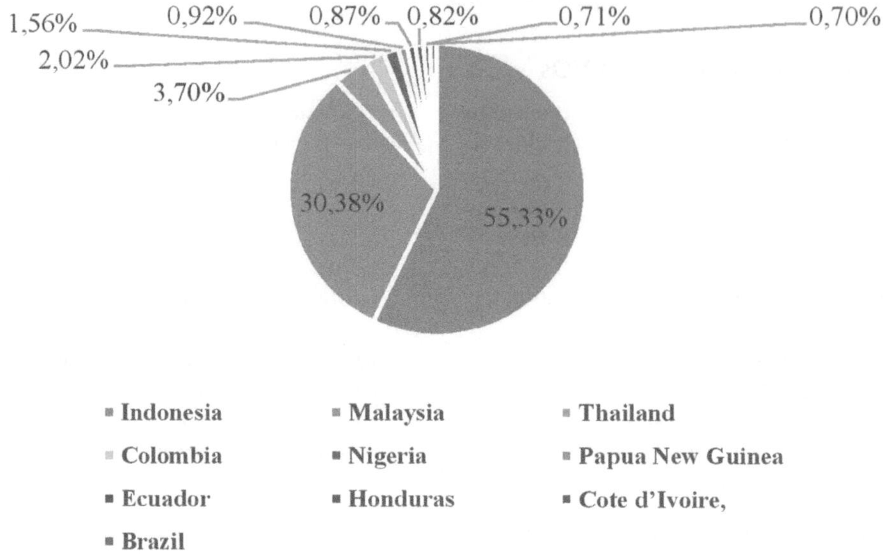

FIGURE 3.2 Global production percentage of palm oil.

Indonesia was the largest exporter, followed by Malaysia, the Netherlands, and Guatemala. Furthermore, Asian nations accounted for the majority of palm oil imports, namely India (USD 5.4 billion), China (USD 5.4 billion), and Pakistan (USD 1.7 billion). The United States and European Union (EU) nations (Netherlands, Spain, and Italy) came in second and third, respectively. But this commodity is linked to sustainability problems, particularly in the social, economic, and environmental spheres (Suroso et al., 2021). Consequently, a spike in agricultural trade could result from circumstances like the global financial crisis and the drop in its prices (Anderson & Nelgen, 2012). Tariffs and non-tariff measures were put in place by different countries to make the palm oil trade less accessible (Annas & Nurmalina, 2020). In Southeast Asia, where government regulations like Indonesia's B40 biodiesel blending strategy have greatly raised domestic consumption for renewable energy, palm oil is the favored feedstock for biofuels (Dimawarnita et al., 2021). Global output of palm oil is anticipated to increase gradually due to growing demand for biofuels as well as for culinary and industrial applications (Mathur et al., 2023).

3.3.2 Soybean Oil

United States, Brazil, and Argentina are the world's top producers of soybeans. Argentina and Brazil are catching up with the United States, which was the biggest producer. Due to the conversion of land to maize for the then-more-profitable manufacturing of ethanol, soybean production in the United States witnessed a significant fall in 2007. However, US soybean production returned to its prior levels in 2008, suggesting that the impact was only short-lived. Animal feed has been the main application for soybean meal, but soybean oil has mainly been a byproduct. United states ranked first in soybean oil production about 32%, Argentina ranked second with 21% of soybean oil production, and Brazil and China ranked third and fourth with 19% and 13%, respectively (Rosillo-Calle et al., 2009). Figure 3.3 demonstrates the percentage of Soybean oil production in the top countries worldwide.

3.3.3 Canola Oil

Canola oilseed species rank third in the world's VO supply, after soybean and palm (McVetty et al., 2015). Canola oil has been utilized for the generation of biofuel and as a dielectric fluid in electrical transformers. It has been utilized as lubricants and for specific chemical feedstock applications.

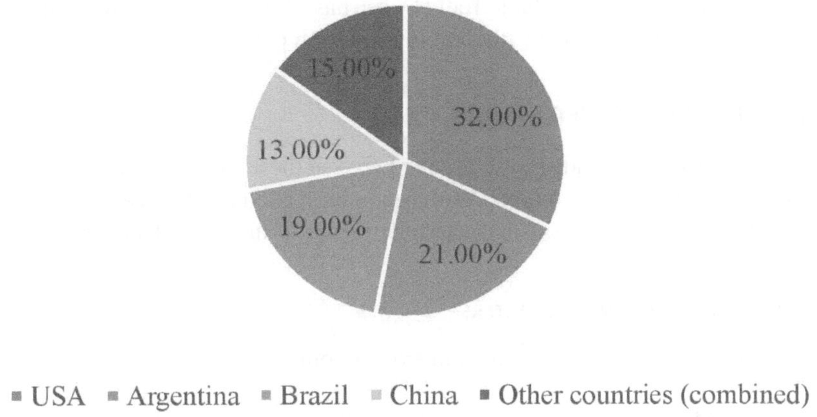

FIGURE 3.3 Global production percentage of soybean oil.

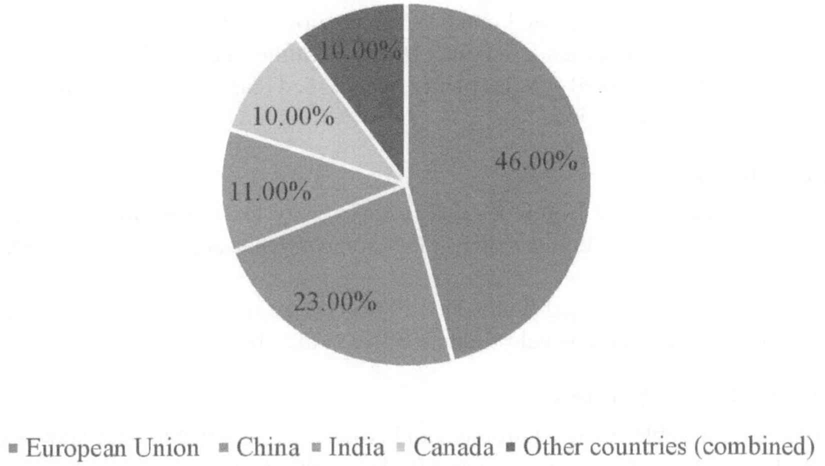

FIGURE 3.4 Global production percentage of canola oil.

The EU ranked first in canola oil production about 46%, China ranked second with 23% of Canola oil production, and India and Canada ranked third and fourth with 11% and 10%, respectively (Rosillo-Calle et al., 2009). Figure 3.4 demonstrates the percentage of Canola oil production in the top countries worldwide.

3.4 WORLD PRODUCTION AND POTENTIAL OF BIOFUELS (BIODIESEL AND BIOGAS)

Biofuels, particularly biogas and biodiesel, have gathered substantial attention as vital components in increasing dependence on sustainable energy. These fuels play an essential role in mitigating greenhouse gas (GHG) emissions, stimulating rural economies, and minimizing reliance on fossil fuels, which are key to advancing sustainable energy objectives worldwide. This section will give a light on current trends in global production and predict the prospective development of biodiesel and biogas, highlighting their significance within the broader bioenergy site. The process of transesterification is used for the conversion of waste oil, animal fats, or VO into biofuels as an ecofriendly and alternative biodegradable fuel (Demirbas, 2009). It can conveniently be used with classical diesel engines, either alone or in combination with petroleum diesel.

3.5 GLOBAL TRADE OF VEGETABLE OILS

VOs sit at the heart of the worldwide biofuel scene because these oils serve as the main raw material for making biodiesel and other green fuels. With rising food needs and a steady move toward cleaner energy, the global market for these oils has expanded sharply in recent decades.

3.5.1 Major Producers and Exporters

Only a few countries now drive production and export volumes:

- Indonesia and Malaysia together supply almost 85% of the world's palm oil, with a large share heading to biofuel plants in the EU and to local blending programs (Mukherjee & Sovacool, 2014).
- In South America and an ever-growing Asia, Argentina and Brazil remain the leading sources of soybean oil for biodiesel (Masiero, 2011).
- Ukraine and Russia once supplied much of the sunflower oil market, though ongoing conflict has thrown those shipments into uncertainty (Kumar et al., 2025).
- Within Europe, the EU itself presses plenty of rapeseed (canola) oil, most of which stays at home for biodiesel production.

3.5.1.1 Major Importers

- India remains the biggest buyer of edible oils, mostly to feed its kitchens, yet its early moves into biofuel plans could slowly push some of that appetite toward clean energy (Singh & Gu, 2010).
- China, the EU, and the United States also import on large scale to meet the dual demand on food consumption and biofuel production (Su et al., 2015).

3.5.1.2 Potential Market Trends

Mainly, Europe's Renewable Energy Directive and America's Renewable Fuel Standard keep pushing up the need for VOs as clean fuel. As a result, the market now sees:

- an intense clash between cooking oil and biofuel users, which makes prices jump around;
- a wider mix of raw materials, with governments tapping waste fats, used frying oil and non-food crops to stay green and secure energy; and
- new shipping lanes that shift after every political jolt, such as the Russia-Ukraine war, which cut off sunflower oil.

In analysis of global production, current results showed that biodiesel output surpassed 50 billion liters annually. The main producer countries include United States, Indonesia, Brazil, Germany, and Argentina (Figure 3.5).

The upward chart of biodiesel production is closely tied to exact national blending commands that edict fixed proportions of biodiesel using conventional diesel, often labeled as B10, B20, or B30. A notable shift in feedstock trends can be seen. Biodiesel production is now taking place of conventional edible oil as more sustainable sources that include:

1. Non-edible oils, such as jatropha, karanja, and castor.
2. Used cooking oil (UCO).
3. Algal oils, which are emerging as innovative feedstocks.

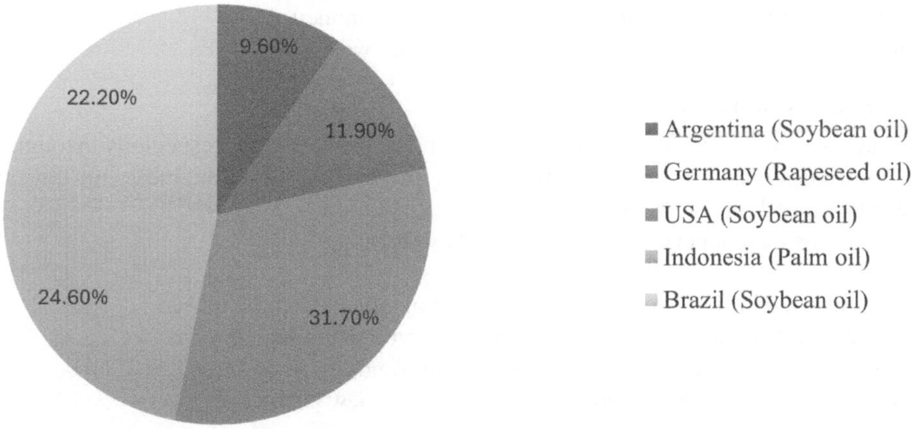

FIGURE 3.5 Global major producing countries of biodiesel (%).

Biogas—a mix of methane (CH_4) and carbon dioxide (CO_2)—forms when organic matter breaks down without air. Wastes from farms, manure, sewage, and leftover food feed this process. Today, the world makes about 35–40 billion cubic meters of biogas each year, with major output coming from:

1. Europe, especially Germany, Italy, and France, home to the smallest digesters.
2. China, where thousands of on-farm and city plants sprang up as part of clean-waste efforts.
3. India, where compact units power village grids and cook stoves alike.
4. The United States, which tends to build big digesters that handle trailer loads of farm and city scraps.

A big step forward is cleaning that raw gas into biomethane, a near-pure methane that acts like pipeline gas. It slips right into gas grids or fuel trucks under the name bio-CNG, closing the circle on renewable energy. Room to grow in biodiesel and biogas output is still huge, fueled by climate worries, cost-saving opportunity, and friendly policy.

The International Renewable Energy Agency (IRENA) even estimates that worldwide biodiesel use might top 136 billion liters in 2019 to 378 billion liters by 2030. Large regions of negligible or wasteland can be utilized for non-edible oil crops to reduce competition with food production. Microalgal biodiesel production may transform the industry by offering greater yields with less land use.

Biogas also shows huge promise. The International Energy Agency (IEA) believes that sustainably produced biogas could cover nearly 15% of the world's gas needs by 2040. If every bit of agricultural residue was captured, farmers alone could deliver more than 100 billion cubic meters of gas each year. City-run waste-to-energy plants add another route, turning everyday rubbish into clean fuel.

3.5.2 Stimulating Factors and Obstacles

3.5.2.1 Stimulators

- Climate policies and targets that demand lower emissions.
- The quest for diverse energy supplies and rural jobs.
- Government rules and payments that boost output.
- New equipment that speeds up digestion and improves transesterification.

3.5.2.2 Obstacles

- The price and supply of feedstock can swing wildly.
- Growing nations often lack roads, plants, or grids to handle production.

- Edible oils still compete with biofuels for land and money.
- Methane leaks from production sites raise climate worries.

3.5.2.3 Key Position and Growth Trajectory

Biofuels have become a key piece in the puzzle for hitting net-zero goals, especially in tough-to-clean sectors like transport and farming. Biodiesel slips into current diesel engines with little fuss, and biogas can be tapped for electricity, heat, or even vehicle fuel.

Success in this field depends on a handful of key steps:

- Building reliable supply chains for waste and plants that can be reused.
- Funding fresh ideas and experiments to make production leaner.
- Crafting smart policies and backing them with funds.
- Sharing know-how across borders and welcoming new partners.

Overall, demand for cleaner and renewable fuel keeps pushing up the worlds output of biodiesel and biogas. Much room to grow still exists, particularly in countries rich in crops and discarded matter. With steady backing from governments and sharp-eyed inventors, biofuels can play a major part in the shift toward greener energy grids.

3.6 FUTURE PREDICTIONS OF VEGETABLE OILS FOR BIOFUELS

As countries intensify efforts toward cleaner energy transitions, VO is emerging as a central component in the biofuel sector. With new environmental targets, concerns about energy security, and government blending policies, demand for biodiesel mainly derived from these vegetable sources is projected to rise sharply (Marzouk, 2025). However, future expansion strategies must address complex issues such as land competition, food security, and environmental durability. According to the International Energy Agency, global biodiesel production is expected to grow by over 20% between 2024 and 2030, primarily due to stricter biofuel regulations in the EU, United States, Brazil, and various Asian countries. Fortune business insights estimate that the market value will reach USD 65 billion by 2,030 feedstocks (Golden et al., 2024). Common edible oils such as soybean, palm, rapeseed, and sunflower are anticipated to maintain dominance as primary feedstocks.

Future biofuel strategies increasingly will promote non-edible oils, such as Jatropha, Pongamia, Camelina, and Castor, produced on marginal land without contributing to food production (Atabani et al., 2013). The growth of biodiesel from UCO is expected to be 6.8% CAGR due to circular economy factors. The EU and US regulators are increasingly adopting sustainability criteria that promote the use of waste feedstocks over food (Crossley, 2024). Crops biotechnology (Gene editing) will be an important driver in improving oil yields and pest resistance. Enhanced oil yields, drought resistance, and pest protection will all involve using CRISPR and other gene editing technology of crops such as soybean, sunflower, and canola (Zhang et al., 2023). A series of tools within smart farming, e.g. drones, AI-based monitoring, and satellite mapping, will optimize oil crop irrigation, fertilization, and harvesting, ultimately lowering energy impact and increasing yield (FAO, 2022).

The outlook for VO toward biofuel is promising, but conditional. Biodiesel production is expected to meaningfully increase before 2030; however, this may have long-term consequences if VO biofuels are to be part of it.

- The transition away from edible and waste feedstocks.
- The efficiency gains made in agricultural and processing.
- The diversification of sustainability criteria.
- Avoiding competition with food supply.

Emerging economies are pumping more money into green energy, and that push is sparking a big rise in trade among themselves. As the world is moving toward low-carbon energy, VOs will remain an essential component of the biofuel landscape. However, maintaining market stability and environmental sustainability while harmonizing their usage of food and fuel will be a significant challenge, impacting the dynamics of future trade.

Developments in global policy frameworks are all moving toward developing advanced biofuels. Worries about palm oils toll on forests and wildlife have prompted many European nations to tighten import rules. These measures now include:

- Requirements like the Roundtable on Sustainable Palm Oil (RSPO) badge.
- Tariffs or red tape tied to a products carbon score or deforestation risk.
- Talks of carbon border taxes on shipments that fail the climate test.

Increasingly, the EU renewable energy directive II and the US renewable fuel standard stand to prefer low-ILUC-risk (Indirect Land Use Change) and waste feedstocks such as UCO over food-based oils (Bennett & Buckley, 2022; Crossley, 2024).

Carbon intensity scoring and certification Carbon intensity scoring will govern the future of VO biofuels. Certification will be essential and taken into consideration for market access (especially in North America and Europe).

Volatility in VO markets may also occur if demand for both food and fuel increases. For example, there was a supply disruption of sunflower oil stemming from the 2022 Russia-Ukraine war that had implications for both the food and fuel markets. In the future, the markets will need a way to manage the risk of volatility and ensure food security. Countries including Brazil, India, and Indonesia are poised to scale up their domestic biodiesel programs while promoting both food security and energy independence. These countries are investing in local oilseed crop development and developing decentralized biodiesel units.

In the longer term, biofuels from VOs will become one component of a more expansive bio-based economy, where oils may be used for energy transitions and bioplastics, lubricants, and green chemicals. The integration of VO biofuels with algal oils, synthetic biology, and biorefineries may also contribute to the biorefineries' bioeconomy value chain (Bennett & Buckley, 2022).

A viable alternative may include liquid biofuels such as biodiesel and bioethanol. These biofuels contribute to a reduction in reliance on external sources for energy, support regional engineering, enhance research and development, reduce contributions from power generation and transformation, enhance the quality of service provided to rural populations, create jobs, and so forth (Meijaard et al., 2022).

3.7 SUSTAINABILITY OF VEGETABLE OIL BIOFUELS

3.7.1 Renewable Energy Source

Biofuels from VO are renewable because they utilize crops like palm, soybean, and rapeseed, and these crops can be produced on a yearly basis compared to fossil fuels that are finite (Kumar et al., 2022).

3.7.2 Reducing Greenhouse Gases

Because VO biofuels can be produced sustainably, they can also produce lower net GHG emissions. For example, palm oil biodiesel can net up to 60% GHG reductions compared to fossil diesel if there is no land use change (Jeswani et al., 2020).

3.7.3 Biodegradability

VO biodiesel is significantly safer with lower toxicity and, in the case of an accidental spill, a reduced potential ecological impact because of the increased biodegradability (Kumar et al., 2022).

3.7.4 Land Use Change and Deforestation

Sustainability issues include land use change and deforestation. One of the main issues from developing oil crops is deforestation on a monumental scale, with consequences for carbon emissions, particularly in relation to palm oil in Malaysia and Indonesia. As of 1996, over 3.8 million hectares of forest have been converted to oil palm plantation in Indonesia (Kumar et al., 2022; Raza et al., 2024).

3.7.5 Loss of Biodiversity

The monoculture plantations required for vegetable oil crops will almost undoubtedly displace native species and degrade habitat, leading to biodiversity loss. This biodiversity loss may be most pronounced in tropical countries after destruction of rainforest (Kumar et al., 2022).

3.7.6 High Input Requirements

Edible oil crops have the potential to change food prices and food access particularly in food insecure conditions. High input requirement oil crops, such as soy and palm, require significant amounts of water, fertilizers, and pesticides that can affect water quality, as well as soil (Kumar et al., 2022).

3.7.7 Opportunities

There are many opportunities in the biodiesel feedstock market due to the worldwide movement toward renewable energy and sustainability. As climate change continues to affect the way we live and the contributions of fossil fuel consumption, the biodiesel feedstock market will continue to expand and develop. While there are many opportunities for biodiesel feedstocks, the market will face difficulties that may hinder its growth. The competition for feedstocks with the food industry presents a clear threat, especially in terms of VOs such as soybean and rapeseed oils, as a point-in-time competition with demand could cause regulatory problems and impact the economy for biodiesel production. Further, agricultural yield fluctuations could result in a lack of supply, price volatility, and therefore economic implications, as prices negatively impact the economic viability of biodiesel production. Many feedstocks present some very clear environmental concerns in relation to sustainability such as palm oils, which may create a situation of conflict and regulatory requirements for the current market of biodiesel feedstocks. Ultimately, this raises the question in an emerging market of feed stocks, continuous use, as well as the innovation of feedstocks may somewhat resolve these long-term problems.

3.8 CONCLUSION

VOs are primarily made up of essential nutrients like fatty acids, calcium, iron, phosphorus, potassium, vitamins, and other minerals. However, our ability to produce enough VOs has been decreasing each year. This is due to rapid increase in consumption, making us heavily reliant on outside markets to meet demand. Several factors are expected to further imbalance the supply and demand for VOs, including population growth, improved lining standards, longer life expectancies, and the rise of chronic diseases. Because of this, ensuring a sufficient supply of VOs is crucial for maintaining social stability. Simultaneously, biofuels, especially biogas and biodiesel, are gaining significant attention as vital contributors to increasing our reliance on sustainable energy. These fuels are essential for reducing GHG emission, boosting rural economies, and lessening our dependence on fossil fuels. All of these efforts are key to achieving global sustainable energy goals. With new environmental targets, growing concerns about energy security, and government policies promoting blending, demand for biodiesel (which is mainly derived from these vegetable sources) is expected to rise sharply.

COMPETING INTERESTS

The authors declare that they have no competing interest.

DECLARATION OF GENERATIVE AI AND AI-ASSISTED TECHNOLOGIES IN THE WRITING PROCESS

During the preparation of this work, the book chapter author(s) have not taken any assistance from AI, and the book chapter author(s) reviewed and edited the content as needed and take(s) full responsibility for the content of the publication.

REFERENCES

Abrante-Pascual, S., Nieva-Echevarría, B., & Goicoechea-Oses, E. (2024). Vegetable oils and their use for frying: A review of their compositional differences and degradation. *Foods*, 13(24), 4186.

Alsafadi, D., Alhesan, J. A., Mansour, A., & Oqdeh, S. (2023). Extraction and quantification of bioactive phenolic compounds in olive oil by acid hydrolysis method. *Food Analytical Methods*, 16(3), 581–595.

Anderson, K., & Nelgen, S. (2012). Agricultural trade distortions during the global financial crisis. *Oxford Review of Economic Policy*, 28, 235–260.

Annas, A., & Nurmalina, R. (2020). The effect of the European Union biomass regulation and export taxation on palm oil export. *Jurnal Manajemen & Agribisnis*, 17(1), 1.

Arab, R., Casal, S., Pinho, T., Cruz, R., Freidja, M. L., Lorenzo, J. M., Hano, C., Madani, K., & Boulekbache-Makhlouf, L. (2022). Effects of seed roasting temperature on sesame oil fatty acid composition, lignan, sterol and tocopherol contents, oxidative stability and antioxidant potential for food applications. *Molecules*, 27(14), 4508.

Atabani, A. E., Silitonga, A. S., Badruddin, I. A., Mahlia, T. M. I., Masjuki, H. H., & Mekhilef, S. (2013). A comprehensive review on biodiesel as an alternative energy resource and its characteristics. *Renewable and Sustainable Energy Reviews,* 18, 211–245. https://doi.org/10.1016/j.rser.2012.10.016

Baccouri, B., Sieren, T., Rajhi, I., & Willenberg, I. (2023). Characterization of the fingerprint profile of bioactive constituents of extra virgin olive oils from Peninsula Tunisian Cap Bon with regard to altitude. *European Food Research and Technology*, 249(2), 497–509.

Bakhtavoryan, R. (2017). Brand-level analysis of demand for mayonnaise in Northeast Texas. *Texas Journal of Agriculture and Natural Resources*, 30, 53–62.

Bhattacharya, S. (2023). *Chapter 10—Fats and Oils* (pp. 251–281). Cambridge, MA: Academic Press.

Calugar, L., Grozea, I., & Butnariu, M. (2024). Vegetable oils in maintaining health. *Journal of Medical and Clinical Nursing Studies*, 2(2), 1–7.

Correa, K. L., de Carvalho-Guimarães, F. B., Mourão, E. S., Oliveira Santos, H. C., da Costa Sanches, S. C., Lamarão, M. L. N., Pereira, R. R., Barbosa, W. L. R., Ribeiro-Costa, R. M., Converti, A., & Silva-Júnior, J. O. C. (2024). Physicochemical and nutritional properties of vegetable oils from Brazil diversity and their applications in the food industry. *Foods*, *13*(10), 1565.

Crossley, P. (2024). From the climate change to war: The evolving role of renewable energy law and policy in meeting the EU's energy security challenges. In *Research Handbook on EU Energy Law and Policy* (pp. 617–641). London: Edward Elgar Publishing.

de Carvalho, A. G. A., Olmo-García, L., Gaspar, B. R. A., Carrasco-Pancorbo, A., Castelo-Branco, V. N., & Torres, A. G. (2022). Evolution of the metabolic profile of virgin olive oil during deep-frying: Assessing the transfer of bioactive compounds to the fried food. *Food Chemistry*, 380, 132205.

Demirbas, M. F. (2009). Biorefineries for biofuel upgrading: a critical review. *Applied Energy*, *86*, S151–S161.

Dimawarnita, F., Kartika, I. A., & Hambali, E. (2021, May). Sustainability of biodiesel B30, B40, and B50 in Indonesia with addition of emulsifier. In *IOP Conference Series: Earth and Environmental Science* (Vol. 749, No. 1, p. 012026). Bristol: IOP Publishing.

FAO. (2022). *Precision Agriculture and Sustainability*. Food and Agriculture Organization. https://www.fao.org

Farooq, S., & Ngaini, Z. (2024). Introduction to vegetable oils. In *Vegetable Oil-Based Composites: Processing, Properties and Applications* (pp. 1–20). Singapore: Springer Nature Singapore.

García-González, A., Quintero-Flórez, A., Ruiz-Méndez, M. V., & Perona, J. S. (2023). Virgin olive oil ranks first in a new nutritional quality score due to its compositional profile. *Nutrients*, 15(9), 2127.

Gautron, J., Dombre, C., Nau, F., Feidt, C., & Guillier, L. (2022). Production factors affecting the quality of chicken table eggs and egg products in Europe. *Animal*, 16, 100425.

Golden, J. S., Handfield, R. B., Daystar, J., & Pires, S. (2024). *An Economic Impact Analysis of the US Biobased Products Industry: 2023, USA.*

Guedes, A. M. M., Wilhelm, A. E., Moura, J. I. L., Lopes, R., Faria-Machado, A. F. D., & Antoniassi, R. (2023). Bioactive compounds of fractionated palm oil with a higher content of oleic acid. *Revista Brasileira de Fruticultura*, 45, e–555.

Guillaume, C., De Alzaa, F., & Ravetti, L. (2018). Evaluation of chemical and physical changes in different commercial oils during heating. *Acta Scientific Nutritional Health*, 2(6), 2–11.

Hlaváč, P., Božiková, M., & Petrović, A. (2019). Selected physical properties assessment of sunflower and olive oils. *Acta Technologica* Agriculturae, 3(2), 86–91.

Jeswani, H. K., Chilvers, A., & Azapagic, A. (2020). Environmental sustainability of biofuels: A review. *Proceedings of the Royal Society A,* 476(2243), 20200351.

Kumar, P., Kumar, A., Patil, M., Hussain, S., Sharma, S., Singh, S., Yadav, R., Tokas, D., & Singh, A. N. (2022). Biodiversity resources and ecosystem services in the context of the industrialized world. In *Biodiversity* (pp. 23–39). England: Taylor & Francis Group.

Kumar, D., Thakur, G., Singh, P., Dutt, S., Mangal, V., Kumar, D., & Singh, B. (2025). Biofuel production from starchy crops: advanced technology and current perspectives. *Archives of Microbiology*, *207*(9), 220.

Mannucci, P. M., Jolliet, O., Meijaard, E., Slavin, J., Rasetti, M., Aleta, A., Moreno, Y., & Agostoni, C. (2023). Sustainable nutrition and the case of vegetable oils to match present and future dietary needs. *Frontiers in Public Health*, 11, 1106083.

Marzouk, O. A. (2025). Summary of the 2023 report of TCEP (Tracking Clean Energy Progress) by the International Energy Agency (IEA), and proposed process for computing a single aggregate rating. In *E3S Web of Conferences* (Vol. 601, p. 00048). Les Ulis: EDP Sciences.

Masiero, G. (2011). Developments of biofuels in Brazil and East Asia: experiences and challenges. *Revista Brasileira de Política Internacional*, *54*(2), 97–117.

Mathur, R. K., Sujatha, M., Bera, S. K., Rai, P. K., Babu, B. K., Suresh, K., Sudhakara Babu, S. N., Sharma, M., Rani, K., Ajay, B. C., Sharma, H. K., & Singh, V. V. (2023). Oilseeds and oil palm. In *Trajectory of 75 Years of Indian Agriculture After Independence* (pp. 231–264). Singapore: Springer Nature Singapore.

McVetty, P. B., & Duncan, R. W. (2015). Canola, rapeseed, and mustard: For biofuels and bioproducts. In *Industrial Crops: Breeding for BioEnergy and Bioproducts* (pp. 133–156). New York, NY: Springer New York.

Meijaard, E., Abrams, J. F., Slavin, J. L., & Sheil, D. (2022). Dietary fats, human nutrition and the environment: Balance and sustainability. *Frontiers in Nutrition*, 9, 878644.

Meijaard, E., Brooks, T. M., Carlson, K. M., Slade, E. M., Garcia-Ulloa, J., Gaveau, D. L., Lee, J. S. H., Santika, T., Juffe-Bignoli, D., Struebig, M. J., & Sheil, D. (2020). The environmental impacts of palm oil in context. *Nature Plants*, 6(12), 1418–1426.

Memon, H. D., Mahesar, S. A., Kara, H., Sherazi, S. T. H., & Talpur, M. Y. (2024). A review: Health benefits and physicochemical characteristics of blended vegetable oils. *Grain & Oil Science and Technology*, 7(2), 113–123.

Mitrea, L., Teleky, B. E., Leopold, L. F., Nemes, S. A., Plamada, D., Dulf, F. V., Pop, I. D., & Vodnar, D. C. (2022). The physicochemical properties of five vegetable oils exposed at high temperature for a short-time-interval. *Journal of Food Composition and Analysis*, 106, 104305.

Mukherjee, I., & Sovacool, B. K. (2014). Palm oil-based biofuels and sustainability in Southeast Asia: A review of Indonesia, Malaysia, and Thailand. *Renewable and Sustainable Energy Reviews*, 37, 1–12.

Parcell, J., Kojima, Y., Roach, A., & Cain, W. (2018). Global edible vegetable oil market trends. *Biomedical Journal of Scientific & Technical Research*, 2(1), 2282–2291.

Pereira, E., Ferreira, M. C., Sampaio, K. A., Grimaldi, R., de Almeida Meirelles, A. J., & Maximo, G. J. (2019). Physical properties of Amazonian fats and oils and their blends. *Food Chemistry*, 278, 208–215.

Piernas, C., Tsiountsioura, M., Astbury, N. M., Madigan, C., Aveyard, P., & Jebb, S. A. (2019). Primary Care SHOPping intervention for cardiovascular disease prevention (PC-SHOP): Protocol for a randomised controlled trial to reduce saturated fat intake. *BMJ Open*, 9(4), e027035.

Ranilović, J., Cvetković, T., Kušević, T., Gajari, D., Tomić-Obrdalj, H., Aladić, K., & Jokić, S. (2024). The impact of smoke point and aroma profile of sweet red pepper seed oil on culinary application and consumer acceptance. *Croatian Journal of Food Science and Technology*, 16(1), 173–187.

Rauf, S., Jamil, N., Tariq, S. A., Khan, M., Kausar, M., & Kaya, Y. (2017). Progress in modification of sunflower oil to expand its industrial value. *Journal of the Science of Food and Agriculture*, 97(7), 1997–2006.

Raza, A., Shahid, M. A., Safdar, M., Zaman, M., Sabir, R. M., Muzammal, H., & Ahmed, M. M. (2024). Impact of land use and land cover change on agricultural production in District Bahawalnagar, Pakistan. *Environmental Sciences Proceedings,* 29(1), 46.

Rosillo-Calle, F., Pelkmans, L., & Walter, A. (2009). *A global overview of vegetable oils, with reference to biodiesel.* A report for the IEA Bioenergy Task, 40.

Roychowdhury, P., Chatterjee, M., Bhattacharjya, A., & Lahiri, S. (2024). Smoke points: A crucial factor in cooking oil selection for public health. *Current Functional Foods*, *2*(2), E041223224179.

Sahasrabudhe, S. N., Rodriguez-Martinez, V., O'Meara, M., & Farkas, B. E. (2017). Density, viscosity, and surface tension of five vegetable oils at elevated temperatures: Measurement and modeling. *International Journal of Food Properties*, 20(sup2), 1965–1981.

Singh, A., & Kumar, V. (2023). Phyto-chemical and bioactive compounds of pumpkin seed oil as affected by different extraction methods. *Food Chemistry Advances*, 2, 100211.

Singh, J., & Gu, S. (2010). Commercialization potential of microalgae for biofuels production. *Renewable and Sustainable Energy Reviews*, *14*(9), 2596–2610.

Song, H. Y., & McClements, D. J. (2021). Nano-enabled-fortification of salad dressings with curcumin: Impact of nanoemulsion-based delivery systems on physicochemical properties. *LWT*, 145, 111299.

Su, Y., Zhang, P., & Su, Y. (2015). An overview of biofuels policies and industrialization in the major biofuel producing countries. *Renewable and Sustainable Energy Reviews*, *50*, 991–1003.

Suroso, A. I., Pahan, I., & Tandra, H. (2021). Triple bottom line in Indonesia commercial palm oil mill business: Analytical network process approach. *International Journal of Sustainable Development and Planning*, 16(5), 965–972.

Suroso, A. I., & Ramadhan, A. (2014). Decision support system for agricultural appraisal in dryland areas. *Advanced Science Letters*, *20*(10–11), 1980–1986.

Tandra, H., Suroso, A. I., Syaukat, Y., & Najib, M. (2022). The determinants of competitiveness in global palm oil trade. *Economies*, 10(6), 132.

Tian, M., Bai, Y., Tian, H., & Zhao, X. (2023). The chemical composition and health-promoting benefits of vegetable oils—A review. *Molecules*, 28(17), 6393.

Timilsena, Y. P., Vongsvivut, J., Adhikari, R., & Adhikari, B. (2017). Physicochemical and thermal characteristics of Australian chia seed oil. *Food Chemistry*, 228, 394–402.

Vishwakarma, S., Dalbhagat, C. G., Mandliya, S., & Mishra, H. N. (2022). Investigation of natural food fortificants for improving various properties of fortified foods: A review. *Food Research International*, 156, 111186.

Yin, M., Chen, M., Yanagisawa, T., Matsuoka, R., Xi, Y., Tao, N., & Wang, X. (2022). Physical properties, chemical composition, and nutritional evaluation of common salad dressings. *Frontiers in Nutrition*, 9, 978648.

Zhang, H., Zhou, G., Varshney, R. K., & Wan, S. (2023). Advances in plant cultivation and physiology of oilseed crops. *Frontiers in Plant Science,* 14, 1280392.

Zhang, M., Wang, O., Cai, S., Zhao, L., & Zhao, L. (2023). Composition, functional properties, health benefits and applications of oilseed proteins: A systematic review. *Food Research International*, 171, 113061.

Zhao, X., Xiang, X., Huang, J., Ma, Y., Sun, J., & Zhu, D. (2021). Studying the evaluation model of the nutritional quality of edible vegetable oil based on dietary nutrient reference intake. *Acs Omega*, 6(10), 6691–6698.

4 Cutting-Edge Approaches of Edible Crop Plants for Biofuel Production

Rabeea Zafar, Sahar Saleem, Ayesha Ashfaq, and Nazish Iftikhar

4.1 INTRODUCTION

The accessible oil reserves of the world are gradually becoming exhausted because of an increasing demand for fossil fuels. Furthermore, fossil fuel extraction and utilization have resulted in massive environmental pollution through carbon dioxide emissions, which lead to the greenhouse effect. To this end, a need to explore alternative energy sources that can sufficiently replace fossil fuels has emerged with urgency (Pickering et al., 2022). In the last few years, liquid and gaseous biofuels have gained popularity due to their beneficial implications for the environment (Cavelius et al., 2023). Therefore, the scientific community has invested considerable effort in utilizing various sources and innovative technologies to create renewable fuels that are more cost-effective, efficient, and sustainable, while producing lower emissions (Algarni et al., 2023). Biofuels have gained serious attention due to a range of benefits in contrast to fossil fuels (Pickering et al., 2022; Cavelius et al., 2023). Historically, biofuels have been derived from both edible and non-edible biomass; however, edible crops such as sugarcane, maize, wheat, soybean, and cassava have become prominent feedstocks to produce biofuels due to their high carbohydrate and oil content (Algarni et al., 2023).

Biofuels are usually divided into two broad categories: bioethanol and biodiesel (Mizik & Gyarmati, 2021), as illustrated in Figure 4.1. These liquid biofuels are an economical alternative to traditional transport fuels, accounting for approximately 18% of primary energy consumption (Wang et al., 2007; Demirbas et al., 2016). Bioethanol currently constitutes approximately 80% of the total liquid biofuel production, while biodiesel constitutes the remaining portion (Joyia et al., 2024). Unlike petroleum-based fuels, which contain no oxygen, biofuels are notable for their oxygen content, ranging from 10% to 45%. Another advantage of biofuels is their reduced sulfur and nitrogen levels relative to petroleum sources (Topare et al., 2022). Nevertheless, promoting the large-scale adoption of biofuels requires urgent attention to challenges involving cost (which exceeds that of other renewable sources), efficiency, and land use (Whitaker et al., 2018).

To address these challenges, biotechnological methods present an exciting opportunity to increase plant yields and enhance biofuel production while minimizing land use and environmental consequences. In the 21st century, biotechnology emerged as a cutting-edge solution for meeting our energy needs and reducing environmental risks, especially in agricultural biotechnology. Research in this domain spans from fundamental to adaptive studies, all aiming to enhance crop characteristics through genetic manipulation and selective breeding. By enhancing crop characteristics and refining biochemical conversion processes, biotechnology can significantly lower the costs tied to biofuel production. Through genetic engineering, a promising solution is to raise crops yield per area by either increasing stress resistance or selecting varieties ideal for bioenergy use (Lynd et al., 2008). Moreover, biotechnological tools also aid in optimizing bioenergy production by improving microorganisms and enzymes, creating efficient bioprocesses, and refining manufacturing methods.

DOI: 10.1201/9781003600169-4

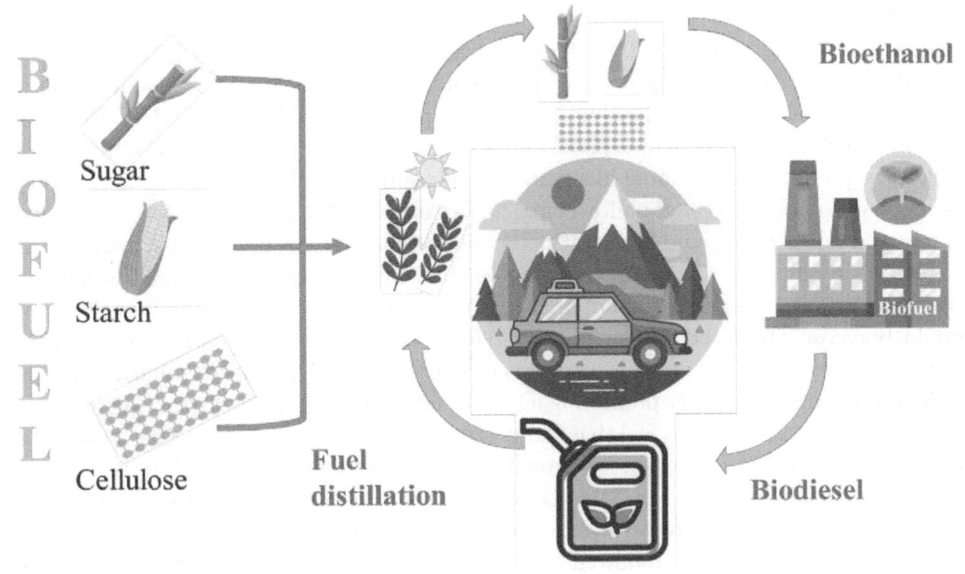

FIGURE 4.1 Schematic representation of biofuel production from edible feedstock.

This chapter explains biofuel production from edible crops, focusing on key types like bioethanol and biodiesel. It dives into the cutting-edge biotechnological techniques that help improve biofuel yields, such as genetic modifications, enzyme-based conversions, and innovative microbial processes. This chapter also touches on crucial policies, as well as socioeconomic and environmental factors that play a role in the viability and sustainability of biofuels made from food crops. Through a variety of case studies and success stories, practical applications and valuable insights from different national and regional efforts are depicted. Finally, the chapter outlines future opportunities and research directions, emphasizing the need for sustainable solutions, integrated systems, and continuous innovation in the biofuel sector.

4.1.1 Biofuel Types Derived from Edible Crops

The use of edible biomass including starch and sugars from crops like potato, sugarcane, corn, wheat, barley, and sugar beet holds great potential for biofuel generation and reducing fossil fuel use. These crops also play an significant role in capturing CO_2 from the atmosphere as they grow (Rodionova et al., 2017). Currently, around 50 billion liters of bioethanol, biodiesel, and biogas are produced annually on a commercial scale. Among these, biogas, which results from the anaerobic digestion of biomass, has limited application in the transportation sector (Naik et al., 2010). The performance of biofuels is measured by how well they mix with petroleum fuels for engine use and their compatibility with both natural gas and flexible fuel vehicles (Naik et al., 2010). Importantly, producing biofuel from edible biomass requires careful consideration of various factors. The chemical composition of biomass is of primary importance. Additionally, aspects such as energy balance, availability of cropland, and impact on biodiversity must be considered. Competition with food supplies, cultivation techniques, and pollutant emission are vital factors, alongside the impact on water and soil systems, with a specific emphasis on mineral uptake. Furthermore, the use of pesticides, transportation costs, and the risk of soil erosion are fundamental concerns. Lastly, variables such as employability and water availability are also essential in evaluating the sustainability of biofuel production (Singh et al., 2013).

4.1.1.1 Bioethanol

The fermentation of feedstocks like sugar beet, corn, and wheat yields bioethanol, a liquid alcohol commonly used to power vehicles either on its own or combined with gasoline. The blended form involves mixing liquid ethyl alcohol with gasoline, a practice commonly used in flex-fuel vehicles. One such blend is E10, which contains 10% bioethanol and 90% gasoline. Additionally, ethyl tertiary butyl ether (ETBE) can be produced by blending bioethanol with gasoline to enhance its oxygen content, thereby helping in reducing pollution (Norkobilov et al., 2017). Around 1894, bioethanol was used in industries in countries such as Germany and France (Chelf et al., 1993). By 1952, it was being used in Brazil as a vehicle fuel. Its use also became common in regions like United States (USA) and Europe during the early 1900s. However, bioethanol was not seen as a practical option during World War II because of its high production cost. This trend continued prior to the oil crisis in the period of the 1970s, after which bioethanol began receiving attention as an alternative vehicle fuel. Developed countries such as Brazil and USA have since been actively promoting its domestic production (Packer et al., 2016).

Bioethanol, being a renewable fuel, has the potential to reduce atmospheric CO_2 concentration by curbing the use of fossil fuel. Further, the growth of feedstocks for bioethanol absorbs CO_2, adding to greenhouse gas (GHG) reductions. In 2014, the world produced more than 93 billion liters of bioethanol reflecting a nearly fourfold rise from 10 years prior (Li et al., 2017). The major crop sources to produce biofuels are maize and sugarcane, followed by wheat and sorghum. Yet these plants are food staples for approximately 200 million individuals, these crops face conflicting demands for food consumption and fuel production (Rulli et al., 2016). In contrast to wheat and maize, sugarcane contributes the highest percentage to bioethanol production among edible biomass feedstocks (Rulli et al., 2016). The USA contributes around 47% of global bioethanol production, being at the forefront of its development (Balat & Balat, 2009). When producing high-quality bioethanol from crops, two important factors should be considered, particularly regarding GHG benefits. These include practices that reduce carbon concentration in soil and lower nitrous oxide emissions (Alkimim & Clarke, 2018; Huang et al., 2018). Furthermore, the process can be made more cost-effective by focusing on the production and utilization of byproducts.

The fermentation of carbohydrate-rich crops produces bioethanol. These feedstocks fall under two categories. The first include sugar-rich wheat, palm juice, sugarcane, and various fruits. All of these are sugar-containing crops. The second category includes starch-rich crops such as rice, corn, barley, sorghum, wheat, and other grains. These are all starch-containing crops. However, due to their primary structural component glucose in the long chain polymers, it is not possible to convert starch directly into bioethanol using conventional fermentation technologies (Zabed et al., 2016). First, the starch macromolecular structure is broken down into smaller glucose molecules. To do this, starch-containing feedstocks (15%–20% starch content) are ground first into a fine mash. The starch after grinding is then blended with water and subsequently cooked at or higher than its boiling point. During this process, two enzymes are used. Amylase, the starter enzyme, breaks down starch molecules into two components: dextrins and oligosaccharides. Following this, the saccharification process is carried out, in which the components are hydrolyzed using enzymes such as glucoamylase and pullulanase. During this step, dextrins are further converted into maltose, glucose, and isomaltose. In the next step, the mesh is cooled to 30°C, after which yeast is added to initiate fermentation (Cereda, 2024).

Corn can be utilized to produce bioethanol through two primary processing techniques: wet milling and dry milling. Wet milling has the benefit of higher capacity production than dry milling. Also, it produces useful co-products such as organic acids, solvents, drugs, and nutraceuticals (De Corato et al., 2018). Wet milling corn-based bioethanol production is shown in Figure 4.2. Dry milling is also an effective method, as it not only produces but also generates co-products such as dried grains and soluble distillers. These products are valuable as livestock feed due to the rich content of proteins, fats, and carbohydrates (Kurambhatti, 2021).

4.1.1.2 Biodiesel

Diesel fuel, which is widely used in transportation, has a chemical formula that typically ranges from $C_{10}H_{20}$ to $C_{15}H_{28}$. Biodiesel can be produced from various feedstocks by employing multiple technological

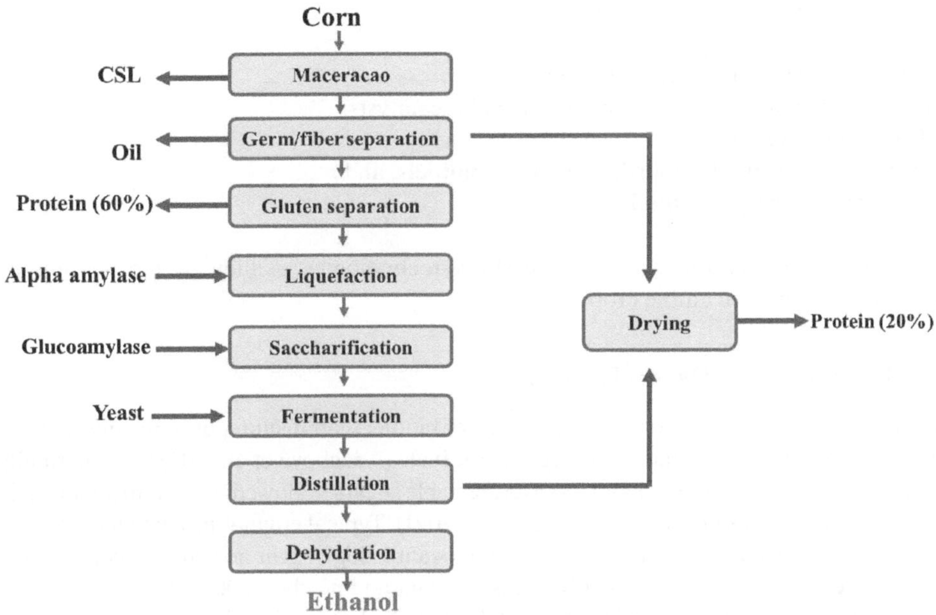

FIGURE 4.2 Flow chart of wet milling corn-based bioethanol production. (Cortez et al., 2019.)

approaches (Alalwan et al., 2019). It is worth noting that up to 75% production cost of biodiesel from oil crops comes from the cost of feedstock (Mahmudul et al., 2017). For biodiesel production, approximately 350 oil-bearing crops have been identified as promising feedstock, including both edible and non-edible varieties (Sajjadi et al., 2016; Alalwan et al., 2019). Common edible sources among these include castor, linseed rapeseed, sunflower, rice barn, coconut, milkweed seed, peanut, palm, safflower, soybean, and palm. To obtain high-quality biodiesel, several technologies are employed, including direct use and blending transesterification, microemulsion, and pyrolysis of vegetable oil.

One of the challenges of using pure vegetable oil is its high viscosity, which can cause engine issues. This drawback can be mitigated by blending vegetable oil with biodiesel fuel for direct use. However, the use of vegetable oil can lead to several engine-related issues, such as carbon deposition, gelling and thickening of lubricants, and the buildup of deposits in the engine resembling trumpet shape (Mahmudul et al., 2017; Alalwan et al., 2019). The use of biodiesel can effectively overcome the problems associated with using vegetable oil.

4.2 ADVANCED BIOTECHNOLOGICAL APPROACHES FOR BIOFUEL PRODUCTION

Over the past decade, biotechnology has revolutionized biofuel production by leveraging engineered microorganisms, nanoscale biocatalysts, algal cultivation, and cutting-edge genetic tools. Modern biofuel strategies focus on maximizing yield and sustainability: for example, baker's yeast *Saccharomyces cerevisiae* can convert sugars to ethanol with >90% of theoretical yield. Even modest improvements in microbial yield can translate into huge economic gains. By exploiting metabolic and evolutionary engineering, researchers have enhanced yeast and bacterial strains for first- and second-generation bioethanol and advanced liquid fuels. Similarly, leveraging nanobiotechnology (e.g., enzyme immobilization on nanoparticles) improves catalyst stability and reusability for biodiesel and other fuels (Topaloğlu et al., 2023). Third- and fourth-generation approaches use microalgae and photosynthetic systems to generate oils, hydrogen, and even synthetic hydrocarbons without competing for arable land (Narala et al., 2016). Concurrently, anaerobic processes – including methane generation via anaerobic digestion and biohydrogen via dark fermentation – convert waste biomass into gas fuels under mild conditions (Subbarao et al., 2023; Albuquerque et al., 2024).

Five advanced biotechnological approaches:

1. Microorganism-driven fermentation,
2. nanobiotechnology via nano-immobilized biocatalysts,
3. algal biotechnology,
4. genetic engineering for fourth-generation biofuels, and
5. anaerobic bioenergy methods.

In this chapter, we will be focusing on the first two technologies, as they are most commonly used in biofuel production form edible crops.

4.2.1 Microorganism-Driven Fermentation

Microorganism-based fermentation lies at the heart of biofuel manufacture, utilizing microbial metabolism to transform organic substrates to energy-dense fuels (Adegboye et al., 2021). It is normally based on the hydrolysis of lignocellulosic biomass to fermentable sugars, followed by their metabolism into biofuels with certain enzymes or microbes (Braga et al., 2021). Typical ethanol-producing microorganisms, such as *S. cerevisiae* and *Zymomonas mobilis*, are resistant to high sugar and ethanol concentrations and achieve high yields. *S. cerevisiae* is highly stress-resistant and reaches >90% theoretical ethanol yield (~0.51 g/g glucose), whereas *Z. mobilis*, operated through the Entner–Doudoroff pathway, ferments well despite high substrate concentrations (~400 g/L sugars, ~160 g/L ethanol). Metabolic engineering has been used to increase yields by lowering byproducts such as glycerol (Topaloğlu et al., 2023). Aside from ethanol, the ABE (acetone–butanol–ethanol) process with *Clostridium* sp. produces butanol, a less volatile, higher-energy fuel. Yet toxicity from butanol and gas stripping continue to be issues (see Table 4.1). Co-cultures and batch-wise fermentations, e.g., yeast with cellulolytic bacteria or algal consortia, maximize sugar consumption and ethanol yield. Developments in metabolic engineering, adaptive evolution, and genome editing increasingly enhance yield, tolerance, and substrate range in microbial systems.

4.2.2 Nano-immobilized Biocatalyst Approach

Nanobiotechnology offers new ways to immobilize enzymes and cells, enhancing biocatalyst performance for biofuel synthesis. Immobilization on nanoparticles or nanostructures can dramatically increase enzyme stability, loading, and reusability. Nanocarriers (e.g., magnetic Fe_3O_4 particles, silica nanoparticles, carbon nanotubes, and nanofibers) provide a high surface area-to-volume ratio, allowing more enzyme molecules per volume and improved substrate access (Kim et al., 2018). Critically, magnetic nanoparticles permit easy recovery of the biocatalyst with an external field. For

TABLE 4.1
Representative Microorganisms for Biofuel Fermentation

Microorganism	Biofuel Type	Key Advantages	Challenges
Saccharomyces cerevisiae	Bioethanol	High ethanol tolerance, well-established fermentation (Topaloğlu et al., 2023)	Limited substrate range
Escherichia coli	Biodiesel/bioethanol	Genetic tractability, rapid growth	Requires extensive engineering
Clostridium species	Biobutanol	Natural solvent production (Ganeshan and Tülbek, 2025)	Strict anaerobic conditions
Zymomonas mobilis	Bioethanol	High ethanol yield, efficient fermentation (Braga et al., 2021)	Narrow substrate utilization
Yarrowia lipolytica	Biodiesel	High lipid accumulation	Complex cultivation requirements

These organisms have been genetically and process-optimized to enhance biofuel yields.

TABLE 4.2
Examples of Nanomaterials Used to Immobilize Lipases or Whole-Cell Biocatalysts for Biofuel (Biodiesel) Production

Nanocarrier Types	Biocatalyst (Example)	Reported Benefits
Magnetic nanoparticles (Fe_3O_4)	Lipase (*Candida antarctica*, *Burkholderia* sp.)	Enhanced thermal/operational stability; facile magnetic recovery and reuse
Silica/mesoporous nanoparticle	Lipase (Porcine, *Pseudomonas*)	High surface area for enzyme loading; improved activity retention
Carbon nanotubes (SWNT/MWNT)	Lipase (*Burkholderia*, *Rhizopus*)	Elevated enzyme loading and activity retention; good electron/heat conduction
Nanofibers (e.g. PAN nanofibers)	Lipase (*Candida rugosa*, *Candida antarctica*)	High enzyme loading and stability; low mass transfer limitation

Nanocarriers improve enzyme stability, loading, and reusability

instance, lipases attached to superparamagnetic Fe_3O_4 allow complete removal of the catalyst after biodiesel synthesis, enabling dozens of reuse cycles (Hajareh Haghighi et al., 2024).

In biodiesel production, nano-immobilized lipases are widely studied. Lipase enzymes (e.g. from *Candida antarctica* or *Burkholderia cepacia*) catalyze the transesterification of oils to fatty acid methyl esters. When immobilized on nanomaterials, their activity and thermal stability increase. For example, *C. antarctica* lipase on carbon nanotubes retained nearly 100% of activity over 10 cycles. Similarly, lipase on nanofibers (acrylonitrile or chitosan fibers) achieved high loading and very low leaching (Kim et al., 2018). Table 4.2 summarizes key nanocarriers and their biocatalytic benefits.

By combining various biotechnologies, modern biofuel technology is steadily overcoming the limitations of first-generation fuels; continued integration of these strategies promises truly sustainable, next-generation biofuels.

4.3 WASTE VALORIZATION AND CIRCULAR BIOECONOMY

Integrating circular economy concepts into the bioeconomy, the circular bioeconomy prioritizes maximizing biomass utilization and valorizing waste streams. It seeks to replace fossil inputs with renewable biomass feedstock in a regenerative system that minimizes waste, closes nutrient loops, and maximizes resource efficiency. This approach is expected to improve ecoefficiency, lower GHG footprints, and reduce reliance on finite resources (Leong et al., 2021). A key strategy is waste biorefining, where agricultural and food-processing residues are converted into fuels, chemicals, and materials, thereby "closing the loop" on raw materials and reducing environmental impacts. For example, circular bioeconomy practices interconnect biomass supply chains with renewable energy, hydrogen, carbon, and water networks, enabling synergies that policymakers can use to amplify benefits across sectors (Schipfer et al., 2024).

4.3.1 Global Availability and Current Use of Crop Residues

Agricultural residues (stems, husks, stalks, etc.) greatly exceed harvested yields and represent a large untapped biomass pool. For instance, cereal residues alone (wheat straw, rice straw, etc.) embody about 15% of global primary energy consumption. Globally, crop residues provide ~19% of livestock feed and constitute roughly 73% of all crop-derived biomass residues. Yet about 10% of residues are open burned in the field, causing pollution and forfeiting their energy/nutrient value. Typically, only a fraction of residues is returned to soil as an organic matter; much is underutilized or wasted (Bogdanski et al., 2010; Smerald et al., 2023).

4.3.2 Valorization Pathways and Technologies

A suite of biological and thermochemical processes can convert crop residues and byproducts into valuable outputs. Anaerobic digestion (AD) co-digests crop residues (often with manure) to produce biogas (methane for heat/electricity) and nutrient-rich digestate for fertilizers. AD is widely deployed (e.g., on farms) and exemplifies waste-to-energy valorization (Schipfer et al., 2024). Bioethanol and biogas production from lignocellulose (cellulose, hemicellulose) is also advancing; second-generation bioethanol (using straw or corn stover) is commercially proven in Brazil/USA, though techno-economic hurdles (pretreatment costs) remain. Combustion or gasification can co-fire residues for heat or power (as in many sugar mills and power plants). For instance, sugarcane bagasse (fiber waste) is routinely burned or gasified on-site in Brazil and India to supply process heat and power. In one modeling study, sugarcane byproducts (bagasse, vinasse, filter cake, etc.) in Cuba and Peru were projected to yield up to 15.5 billion Nm³ of biogas and 7–13 GWh of electricity, with bagasse accounting for ~60% of the energy output (Castillo Alvarez et al., 2025).

Other valorization routes include composting/soil amendment (returning residue nutrients to fields), biochar production (pyrolysis to stabilize carbon and filter soils), and extraction of biochemicals (e.g. lignin-derived phenolics, dietary fibers). Agro-industrial byproducts (molasses, fruit peels, oilseed meals) likewise feed into biorefineries: they can yield feed supplements, enzymes, edible oils, or precursors for bioplastics. Table 4.3 summarizes key residues and their valorization applications globally.

4.3.3 Integration of Food and Energy Production

Synergizing food and energy systems entails structuring agriculture to incorporate energy production and recycling of resources. Agroforestry, farm waste-run biogas units, and agrivoltaics solar panels integrated with crops for augmenting land productivity by 30%–70% (Padilla et al., 2022).

TABLE 4.3
Examples of Crop Residues and Byproducts and Their Circular Bioeconomy Valorization

Residue/Byproduct	Applications	Valorization Products/ Benefits	Region/Example
Sugarcane bagasse (fiber)	Cogeneration, fermentation	Electricity/heat, ethanol, paper pulp	Brazil sugar mills (cogeneration, ethanol) (Castillo Alvarez et al., 2025)
Rice straw/husk	Anaerobic digestion, compost, pellets	Biogas, fertilizer, animal bedding, biochar	South/Southeast Asia (biogas co-digestion; soil amendment) (Smerald et al., 2023)
Wheat straw	2G ethanol (cellulosic), AD, pellet heating	Bioethanol, methane (heat/electricity), animal feed, biochar	EU and China (pilot bioethanol plants; pellet boilers) (Bogdanski et al., 2010; Smerald et al., 2023)
Corn stover	Bioethanol, AD, gasification	Ethanol, biogas, syngas	USA (cellulosic ethanol demo plants), China (biochar projects)
Fruit/veg wastes	Anaerobic digestion, animal feed, compost	Biogas, feed concentrate, soil conditioner	Global (food industry waste; household organics)
Palm oil residues (palm fronds, shells)	Pellet fuel, AD, pyrolysis	Solid biofuel, biodiesel feedstock, bio-oil	Malaysia/Indonesia (palm mills)
Animal manures	Anaerobic digestion (co-digest with crops)	Biogas, fertilizer/nutrient recycling	Mixed crop-livestock farms (global) (Schipfer et al., 2024)

Technological advances in pretreatment (steam explosion, enzymes) and in thermo-chemical conversion (hydrothermal liquefaction, fast pyrolysis) are improving yields of biofuels and chemicals from residues, advancing the circular bioeconomy.

Intercropping for energy grasses or utilizing fruit trees for both food and firewood are further methods (Bogdanski et al., 2010). At greater scales, bioenergy systems substitute for artificial inputs such as employing biogas or CO_2 for fertilizers and greenhouse agriculture. Circular bioeconomy models seek to connect agriculture, forestry, aquaculture, and energy, increasing the level of sustainability and ensuring food, energy, and water security.

4.4 POLICY, SOCIOECONOMIC, AND ENVIRONMENTAL ASPECTS OF BIOFUEL

4.4.1 Biofuel Policies Across the Globe

A few nations have promoted biofuel uptake by robust policy backing (McCarl et al., 2017). Brazil is a world leader, with a 27% ethanol and 10% biodiesel blend being required and 100% hydrous ethanol being sold at every fuel outlet (Cardoso et al., 2019). China implemented major biofuel standards between 2001 and 2010, with emphasis on ethanol rather than biodiesel because of the dependence on imported vegetable oils (Chen et al., 2020). The USA started out with environmental objectives but widened biofuel policy to cover wider economic and social interests (Soratana et al., 2014; Sajid et al., 2021). India initiated the National Policy on Biofuels in 2008, revised in 2018, which included monetary incentives and subsidies to promote the use of biofuels (Das, 2020; Saravanan et al., 2021). Worldwide, biofuel markets are policy-oriented, with effective implementation depending on robust regulation (Sinha et al., 2019). Lower- and middle-income countries (LMICs) may leverage dual-policy models that integrate technology-push (for Research and Development) and market-pull (for adoption) approaches (Ebadian et al., 2020; Roy & Abedin, 2022).

4.4.2 Environmental Implications of Biofuels

Biodiesel has significant environmental advantages over petroleum-based fossil fuels, mostly through a notable decrease in GHG emissions. The CO_2 emitted when combusted is considerably compensated for by the CO_2 fixed by feedstock plants through photosynthesis, leading to a reduced net atmospheric CO_2 increment (Chozhavendhan et al., 2020). Biodiesel is also lower in harmful emissions like carbon monoxide, sulfuric compounds, and particulates than regular diesel, thus enhancing air quality and minimizing health hazards (Chozhavendhan et al., 2020). Its biodegradable nature also reduces the chances of long-term soil and water contamination from spills (Ahmad et al., 2023). Latest purification technologies such as membrane separation and adsorption improve the quality of biodiesel while reducing water consumption and wastewater emission, increasing sustainability in production (Gupta & Sharma, 2023). Furthermore, biodiesel's increased combustion efficiency translates to less unburned hydrocarbons and particulate matter (Ali & Akbar, 2020). Overall, biodiesel promotes cleaner energy systems with decreased environmental and health burdens.

4.4.2.1 Biofuels and Impact on Biodiversity

Biodiversity is essential in sustaining ecosystem processes and functions and, therefore, supporting the delivery of an array of ecosystem services critical to human welfare (Quijas et al., 2010). Hence, any danger posed to biodiversity automatically risks human livelihoods (Pyšek et al., 2017). Of all the factors implicated in the production of biofuels, land use change is largely identified as the leading cause of biodiversity loss (Tudge et al., 2021). This is most apparent in instances where large tracts of natural forests, grasslands, shrublands, or woodland are cleared for biofuel feedstock farming (Pyšek et al., 2017). Land conversion has been found to impact species richness, habitat availability, abundance of species, and community composition negatively (Tudge et al., 2021). In addition, some second-generation biofuel crops, especially those that are invasive, will also pose an additional risk to indigenous biodiversity (Quijas et al., 2010; Pyšek et al., 2017). Nonetheless, overall, the influence of biofuel production on biodiversity is complicated and hard to measure because it differs with spatial scales, production systems, geographical locations, and the use of specific biodiversity indicators in measurement (Tudge et al., 2021).

4.4.2.2 Biofuels and Food Security

Biofuel production negatively impacts the availability of food, particularly when resources like land, labor, and capital are removed from food crop production to go into feedstock production (Kgathi et al., 2017). This is notably important for first-generation biofuels, which use food crops like maize and sugarcane as feedstocks. Because cereals are a staple component of diets worldwide, their utilization in biofuels has the potential to undermine food availability (Naylor & Higgins, 2018). Variations in supply and price of these staple foods thus have considerable implications for food security (Naylor & Higgins, 2018). In addition, biofuel production can decrease access to food indirectly by contributing to rising food commodity prices (Renzaho et al., 2017). Increasing consensus exists that the increase in biofuel production has contributed to rising global food prices, thus worsening food insecurity, especially for developing nations (Naylor & Higgins, 2018).

In order to reduce these dangers, there is a need to formulate integrated national policies and have in place legal and regulatory frameworks that integrate biofuel development with sustainable food security goals (Fischer et al., 2009). For instance, where land available for arable purposes is scarce, policies can stipulate the ratio of agricultural land that is acceptable to be allocated for the production of biofuel crops, so that there is a balance between food and energy requirements (Kgathi et al., 2017).

4.4.2.3 Biofuels and Sustainable Environment

Biofuels are promoted as green substitutes for fossil fuels, with CO_2 released on combustion approximately balanced by carbon sequestered through feedstock crops (Koch et al., 2023). Large-scale production of biofuels, however, can be detrimental to the environment. The overuse of agrochemicals undermines biodiversity and accumulates heavy metals in soils, which can find their way into the food chain (Kashe et al., 2021; Koch et al., 2023). Fertilizer runoff can lead to eutrophication and toxic algal blooms, posing risks to aquatic life. Nevertheless, biodiesel biodegradability minimizes long-term pollution hazards. Therefore, whereas biofuels have climate advantages, their environmental effects need to be managed with utmost care (Tudge et al., 2021).

4.4.2.4 Biofuels and Water Resources

The rapid expansion of biofuel production is increasing water demand, especially in agriculture, jeopardizing water-scarce regions (Ambaye et al., 2021; Malode et al., 2021). Agriculture alone consumes 80%–90% of freshwater worldwide, much of it through irrigation (Ahmed et al., 2022). Enhanced production of biofuel crops can also strain water resources, especially in arid areas. Simulation models indicate that irrigation demand for biofuel crops can be 1,481–3,880 km^3/year (Stenzel et al., 2019). In Botswana, *Jatropha curcas* is more water hungry than maize or sorghum (Moseki et al., 2019), while for the US Southern Great Plains, sorghum biomass is more water efficient on rainfed (Yimam et al., 2014). Large-scale irrigation can drain aquifers and change river flow. Water use differs based on crop type, land area, and management. Biofuel production needs to be sustainable, minimizing the water footprint and preventing invasive species import (Quijas et al., 2010; Tudge et al., 2021).

4.4.3 Economical Implication of Biofuels

Biodiesel has major operational, economic, and environmental benefits over conventional diesel. High lubricity enhances engine performance, minimizes wear, and lowers maintenance requirements (Ogunkunle & Ahmed, 2019). It is fully compatible with current diesel engines, without the requirement for any change, and is a simple renewable alternative for transport and fleet operations.

Mass manufacturing decreases costs by 20%–30%, and byproduct recovery, e.g., glycerol, enhances process economics by as much as 10% (Fasaei et al., 2018; Chilakamarry et al., 2021). Promotional measures can increase capacity by up to 15%–20% (Gielen et al., 2019). Clean combustion biodiesel emits lower emissions and minimizes engine wear, leading to a longer engine life. Its biodegradability and high combustion quality lead to cleaner air and lower GHG emissions. With energy security, environmental benefits, and cost savings, biodiesel facilitates the shift to global sustainable energy systems and lowers fuel dependence on fossil fuels.

4.4.4 Limitations of Biofuels

Transesterification, the most critical process of producing biodiesel, reacts methanol and a catalyst with triglycerides to give biodiesel and glycerol (Salaheldeen et al., 2021). The resulting biodiesel must comply with international standards such as EN 14214, ASTM D6751, and MS 2008:2014 (Meher et al., 2006). Glycerol, which is hygroscopic and of high viscosity, can compromise fuel stability and engine performance. It must be less than 0.02% by weight to prevent injector deposits, engine knocking, and maintenance issues (Chilakamarry et al., 2021). Purification is therefore essential. Processes include wet washing, dry washing, distillation, membrane separation, adsorption, and solvent-assisted crystallization (SAC) (Veljković et al., 2015). Wet washing is excellent but generates wastewater, dry washing is environment-friendly but inferior, distillation yields purity at high energy expense, membranes foul easily, and SAC requires strict control. Purification technology development is essential for making biodiesel sustainable and commercially competitive (Osman et al., 2024).

4.5 CASE STUDIES AND SUCCESS STORIES OF BIOFUELS

Biofuels present a renewable energy source reducing fossil fuel reliance and decreasing CO_2 emissions in the transportation sector (Koch et al., 2023). Bioethanol can be used in gasoline engines with satisfactory efficiency at 5%–10%, and biodiesel blends in diesel engines can sustain 2%–20% without significant overhaul (Ogunkunle & Ahmed, 2019). The USA and Brazil produced 62% of worldwide biofuels in 2019, employing corn and sugarcane as main feedstocks (Mahmud et al., 2022). See Table 4.4 for success stories. World biodiesel production was 699,000 barrels of oil equivalent in 2019 and is expected to rise to 41.4 billion litres by 2025 (Mahmud et al., 2022). In Asia, palm oil-based production is dominant in Indonesia, Malaysia, and Thailand (Syafiuddin et al., 2020). Biofuels release less particulate matter, making the air cleaner (Chanthon & Assabumrungrat, 2022). They are divided into four generations: 1G (food crops), 2G (crop residues), 3G (microalgae, waste oils), and 4G (solar fuels, electro-fuels) (Syafiuddin et al., 2020).

TABLE 4.4
Success Story of Biofuel Production

Case Study	Technology Used	Product	Benefits	Key Reference
Novozymes (Denmark)	Enzymatic transesterification	Biodiesel	• Reduce catalyst waste (50 times reusability) • Lower production cost • Energy efficient • Few impurities • Greenhouse gas emissions reduced by 40%	Norjannah et al. (2016)
Japan	Microwave-assisted transesterification; membrane purification	Biodiesel	• Time efficient • Effective removal of impurities • Energy efficient (40% less energy than conventional methods)	Chanthon et al. (2022)

Bioethanol is obtained from sugar-based (molasses, sugarcane, bagasse), starch-based (corn, wheat), and lignocellulosic (rice straw, bagasse) feedstocks through distillation, fermentation, and sugar extraction (Susmozas et al., 2020; Zandoná Filho et al., 2022). Biodiesel is produced from edible and non-edible oils, animal fats, and waste cooking oil through microemulsion, pyrolysis, and predominantly transesterification, the most commonly adopted industrial process (Susmozas et al., 2020; Roy & Abedin, 2022; Osman et al., 2024).

4.6 FUTURE PROSPECTS AND RESEARCH DIRECTIONS

This chapter discussed the technical viability of the production of biodiesel from food crops. Major obstacles to biofuel diffusion such as technical, social, economic, market, and policy problems were highlighted. Pilot projects and case studies are required to address these problems so that the impediments to large-scale deployment of biofuels are understood better. In addition, finding effective mechanisms and business models is critical for effective diffusion, particularly in highly populated LMICs such as Bangladesh and Pakistan, where market growth and environmental sustainability need to go hand in hand. Current research on biodiesel manufacturing involves improving purification processes for increased efficiency, cost-effectiveness, and sustainability. Apart from minimizing fossil fuel reliance, biodiesel promotes waste management using waste oils and animal fats. Its low sulfur content also ensures cleaner air. Nevertheless, there is a need for constant innovation to enhance both production and purification. Regardless of feedstock, catalyst, or conditions of the process, all raw biodiesel have to be refined to comply with international regulations like EN 14214 and ASTM D6751 because they contain impurities such as glycerol, soap, water, and free fatty acids.

4.7 CONCLUSION

Finally, this chapter stresses the pivotal position of purification and synthesis of biodiesel for making biofuel production more sustainable. It underscores continuous innovation to minimize effects on the environment, maximize efficiency, and decrease costs. Transesterification is critically evaluated in addition to its economic and environmental outputs, particularly when edible feedstocks are used. Purification methods like wet washing, distillation, adsorption, membrane separation, and SAC are assessed in terms of efficacy and sustainability. Technologies such as enzyme catalysts (e.g., Novozymes) and tandem processes (e.g., in situ transesterification with membrane separation) improve energy efficiency. Green technologies including solvent-free approaches and integration with renewable energy enhance sustainability. The contributions of AI, IoT, and machine learning to the optimization of processes need to be explored and integrated. Policy and economic analyses highlight the importance of supportive policies and cost-reduction measures. Overall, the chapter emphasizes technological and interdisciplinary collaboration as essential to a cleaner, more efficient biofuel industry.

COMPETING INTERESTS

The authors declare that they have no competing interest.

DECLARATION OF GENERATIVE AI AND AI-ASSISTED TECHNOLOGIES IN THE WRITING PROCESS

During the preparation of this work, the book chapter author(s) have not taken any assistance from AI, and the book chapter author(s) reviewed and edited the content as needed and take(s) full responsibility for the content of the publication.

REFERENCES

Adegboye, M. F., Ojuederie, O. B., Talia, P. M., & Babalola, O. O. (2021). Bioprospecting of microbial strains for biofuel production: Metabolic engineering, applications, and challenges. *Biotechnology for Biofuels*, 14(1), 1–21. https://doi.org/10.1186/s13068-020-01853-2

Ahmad, A., Yadav, A. K., & Singh, A. (2023). An environmental impact assessment and optimisation study of biodiesel production from microalgae. *International Journal of Global Warming*, 31(3), 294–313.

Ahmed, W., Safdar, U., Ali, A., Haider, K., Tahir, N., Sajid, S., Ahmad, M., Khalid, M. N., & Sattar, M. T. (2022). Sustainable water use in agriculture: A review of worldwide research. *International Journal of Agriculture and Biosciences*, *11*(4), 246–250.

Alalwan, H. A., Alminshid, A. H., & Aljaafari, H. A. (2019). Promising evolution of biofuel generations. Subject review. *Renewable Energy Focus*, 28, 127–139.

Albuquerque, M. M., Sartor, G. de B., Martinez-Burgos, W. J., Scapini, T., Edwiges, T., Soccol, C. R., & Medeiros, A. B. P. (2024). Biohydrogen produced via dark fermentation: A review. *Methane*, 3(3), 500–532. https://doi.org/10.3390/methane3030029

Algarni, S., Tirth, V., Alqahtani, T., Alshehery, S., & Kshirsagar, P. (2023). Contribution of renewable energy sources to the environmental impacts and economic benefits for sustainable development. *Sustainable Energy Technologies and Assessments*, 56, 103098.

Ali, M., & Akbar, N. (2020). Biofuel is a renewable environment friendly alternate energy source for the future. *Modeling Earth Systems and Environment*, 6, 557–565.

Alkimim, A., & Clarke, K. C. (2018). Land use change and the carbon debt for sugarcane ethanol production in Brazil. *Land Use Policy*, 72, 65–73.

Ambaye, T. G., Vaccari, M., Bonilla-Petriciolet, A., Prasad, S., van Hullebusch, E. D., & Rtimi, S. (2021). Emerging technologies for biofuel production: A critical review on recent progress, challenges and perspectives. *Journal of Environmental Management*, 290, 112627.

Balat, M., & Balat, H. (2009). Recent trends in global production and utilization of bio-ethanol fuel. *Applied Energy*, 86(11), 2273–2282.

Bogdanski, A., Dubois, O., Jamieson, C., & Krell, R. (2010). *Making Integrated Systems work for People and Climate*. https://www.fao.org/docrep/013/i2044e/i2044e.pdf

Braga, A., Gomes, D., Rainha, J., Amorim, C., Cardoso, B. B., Gudiña, E. J., Silvério, S. C., Rodrigues, J. L., & Rodrigues, L. R. (2021). *Zymomonas mobilis* as an emerging biotechnological chassis for the production of industrially relevant compounds. *Bioresources and Bioprocessing*, 8(1). https://doi.org/10.1186/s40643-021-00483-2

Cardoso, C. M. M., Zavarize, D. G., & Vieira, G. E. G. (2019). Transesterification of Pequi (*Caryocar brasiliensis* Camb.) bio-oil via heterogeneous acid catalysis: Catalyst preparation, process optimization and kinetics. *Industrial Crops and Products*, 139, 111485.

Castillo Alvarez, Y., Jiménez Borges, R., Alfonso-Francia, G., Rodríguez Pérez, B., Patiño Vidal, C. D., Iturralde Carrera, L. A., & Rodríguez-Reséndiz, J. (2025). Transition to a circular bioeconomy in the sugar agro-industry: Predictive modeling to estimate the energy potential of by-products. *Technologies*, 13(6), 238. https://doi.org/10.3390/technologies13060238

Cavelius, P., Engelhart-Straub, S., Mehlmer, N., Lercher, J., Awad, D., & Brück, T. (2023). The potential of biofuels from first to fourth generation. *PLoS Biology*, 21(3), e3002063.

Cereda, M. P. (2024). Starch hydrolysis: Physical, acid, and enzymatic processes. In Cereda, M. P., & Vilpoux, O. F. (eds.), *Starch Industries: Processes and Innovative Products in Food and Non-Food Uses* (pp. 75–113). Cambridge, MA: Academic Press.

Chanthon, N., Petchsoongsakul, N., Ngaosuwan, K., Kiatkittipong, W., Wongsawaeng, D., Appamana, W., & Assabumrungrat, S. (2022). Mainstream strategies for biodiesel production. In Rokhum, S. L., Halder, G., Assabumrungrat, S., & Ngaosuwan, K. (eds.), *Biodiesel Production: Feedstocks, Catalysts, and Technologies* (pp. 311–330). Hoboken, NJ: John Wiley and Sons Ltd.

Chelf, P., Brown, L. M., & Wyman, C. E. (1993). Aquatic biomass resources and carbon dioxide trapping. *Biomass and Bioenergy*, 4(3), 175–183.

Chen, H., Ding, M., Li, Y., Xu, H., Li, Y., & Wei, Z. (2020). Feedstocks, environmental effects and development suggestions for biodiesel in China. *Journal of Traffic and Transportation Engineering (English Edition)*, 7(6), 791–807.

Chilakamarry, C. R., Sakinah, A. M., Zularisam, A. W., & Pandey, A. (2021). Glycerol waste to value added products and its potential applications. *Systems Microbiology and Biomanufacturing*, 1(4), 378–396.

Chozhavendhan, S., Singh, M. V. P., Fransila, B., Kumar, R. P., & Devi, G. K. (2020). A review on influencing parameters of biodiesel production and purification processes. *Current Research in Green and Sustainable Chemistry*, 1, 1–6.

Das, S. (2020). The national policy of biofuels of India – A perspective. *Energy Policy*, 143, 111595.

De Corato, U., De Bari, I., Viola, E., & Pugliese, M. (2018). Assessing the main opportunities of integrated biorefining from agro-bioenergy co/by-products and agroindustrial residues into high-value added products associated to some emerging markets: A review. *Renewable and Sustainable Energy Reviews*, 88, 326–346.

Demirbas, A., Bafail, A., Ahmad, W., & Sheikh, M. (2016). Biodiesel production from non-edible plant oils. *Energy Exploration & Exploitation*, 34(2), 290–318.

Ebadian, M., van Dyk, S., McMillan, J. D., & Saddler, J. (2020). Biofuels policies that have encouraged their production and use: An international perspective. *Energy Policy*, 147, 111906.

Fasaei, F., Bitter, J. H., Slegers, P. M., & Van Boxtel, A. J. (2018). Techno-economic evaluation of microalgae harvesting and dewatering systems. *Algal Research*, 31, 347–362.

Fischer, G., Hizsnyik, E., Prieler, S., Shah, M., & Van Velthuizen, H. T. (2009). *Biofuels and Food Security*. Final Report to Sponsor: The OPEC Fund for International Development (OFID), Vienna, Austria.

Ganeshan, S., & Tülbek, M. Ç. (2025). Fermentative butanol production—perspectives and scale-up challenges. *Encyclopedia*, *5*(2), 50.

Gielen, D., Boshell, F., Saygin, D., Bazilian, M. D., Wagner, N., & Gorini, R. (2019). The role of renewable energy in the global energy transformation. *Energy Strategy Reviews*, 24, 38–50.

Gupta, S., & Sharma, M. P. (2023). Impact of binary blends of biodiesels on fuel quality, engine performance and emission characteristics. *Clean Energy*, 7(2), 417–425.

Hajareh Haghighi, F., Binaymotlagh, R., Palocci, C., & Chronopoulou, L. (2024). Magnetic iron oxide nanomaterials for lipase immobilization: Promising industrial catalysts for biodiesel production. *Catalysts*, 14(6). https://doi.org/10.3390/catal14060336

Huang, C., Ragauskas, A. J., Wu, X., Huang, Y., Zhou, X., He, J., Huang, C., Lai, C., Li, X., & Yong, Q. (2018). Co-production of bio-ethanol, xylonic acid and slow-release nitrogen fertilizer from low-cost straw pulping solid residue. *Bioresource Technology*, 250, 365–373.

Joyia, M. A. K., Ahmad, M., Chen, Y. F., Mustaqeem, M., Ali, A., Abbas, A., & Gondal, M. A. (2024). Trends and advances in sustainable bioethanol production technologies from first to fourth generation: A critical review. *Energy Conversion and Management*, 321, 119037.

Kashe, K., Kgathi, D. L., & Teketay, D. (2021). Invasiveness of biofuel crops: Implications for energy research and policy in Botswana. *South African Geographical Journal*, 103(2), 259–281.

Kgathi, D. L., Mmopelwa, G., Chanda, R., Kashe, K., & Murray-Hudson, M. (2017). A review of the sustainability of Jatropha cultivation projects for biodiesel production in southern Africa: Implications for energy policy in Botswana. *Agriculture, Ecosystems & Environment*, 246, 314–324.

Kim, K. H., Lee, O. K., & Lee, E. Y. (2018). Nano-immobilized biocatalysts for biodiesel production from renewable and sustainable resources. *Catalysts*, 8(2). https://doi.org/10.3390/catal8020068

Koch, M., Spierling, S., Venkatachalam, V., Endres, H. J., Owsianiak, M., Vea, E. B., Daffert, C., Neureiter, M., & Fritz, I. (2023). Comparative assessment of environmental impacts of 1st generation (corn feedstock) and 3rd generation (carbon dioxide feedstock) PHA production pathways using life cycle assessment. *Science of the Total Environment*, 863, 160991.

Kurambhatti, C. (2021). *Technical and Economic Viability of Value Added Product Recovery in the Corn Dry Grind Ethanol Process* (Doctoral dissertation, University of Illinois at Urbana-Champaign).

Leong, H. Y., Chang, C. K., Khoo, K. S., Chew, K. W., Chia, S. R., Lim, J. W., Chang, J. S., & Show, P. L. (2021). Waste biorefinery towards a sustainable circular bioeconomy: A solution to global issues. *Biotechnology for Biofuels*, 14(1), 1–15. https://doi.org/10.1186/s13068-021-01939-5

Li, Z., Wang, D., & Shi, Y. C. (2017). Effects of nitrogen source on ethanol production in very high gravity fermentation of corn starch. *Journal of the Taiwan Institute of Chemical Engineers*, 70, 229–235.

Lynd, L. R., Laser, M. S., Bransby, D., Dale, B. E., Davison, B., Hamilton, R., Himmel, M., Keller, M., McMillan, J. D., Sheehan, J., & Wyman, C. E. (2008). How biotech can transform biofuels. *Nature Biotechnology*, 26(2), 169–172.

Mahmud, S., Haider, A. R., Shahriar, S. T., Salehin, S., Hasan, A. M., & Johansson, M. T. (2022). Bioethanol and biodiesel blended fuels—Feasibility analysis of biofuel feedstocks in Bangladesh. *Energy Reports*, 8, 1741–1756.

Mahmudul, H. M., Hagos, F. Y., Mamat, R., Adam, A. A., Ishak, W. F. W., & Alenezi, R. (2017). Production, characterization and performance of biodiesel as an alternative fuel in diesel engines – A review. *Renewable and Sustainable Energy Reviews*, 72, 497–509.

Malode, S. J., Prabhu, K. K., Mascarenhas, R. J., Shetti, N. P., & Aminabhavi, T. M. (2021). Recent advances and viability in biofuel production. *Energy Conversion and Management:* X, 10, 100070.

McCarl, B. A., Jones, J. P. H., Kapilakanchana, M., Wang, M. Z., Zhao, G., Riazi, M. R., & Chiaramonti, D. (2017). Economical assessment of biofuel production. In Riazi, M. R., & Chiaramonti, D. (eds.), *Biofuels Production and Processing Technology* (pp. 613–640). Boca Raton, FL: CRC Press.

Meher, L. C., Sagar, D. V., & Naik, S. N. (2006). Technical aspects of biodiesel production by transesterification—A review. *Renewable and Sustainable Energy Reviews*, 10(3), 248–268.

Mizik, T., & Gyarmati, G. (2021). Economic and sustainability of biodiesel production—A systematic literature review. *Clean Technologies*, 3(1), 19–36.

Moseki, O., Murray-Hudson, M., & Kashe, K. (2019). Crop water and irrigation requirements of *Jatropha curcas* L. in semi-arid conditions of Botswana: Applying the CROPWAT model. *Agricultural Water Management*, 225, 105754.

Naik, S. N., Goud, V. V., Rout, P. K., & Dalai, A. K. (2010). Production of first and second generation biofuels: A comprehensive review. *Renewable and Sustainable Energy Reviews*, 14(2), 578–597.

Narala, R. R., Garg, S., Sharma, K. K., Thomas-Hall, S. R., Deme, M., Li, Y., & Schenk, P. M. (2016). Comparison of microalgae cultivation in photobioreactor, open raceway pond, and a two-stage hybrid system. *Frontiers in Energy Research*, 4(August), 1–10. https://doi.org/10.3389/fenrg.2016.00029

Naylor, R. L., & Higgins, M. M. (2018). The rise in global biodiesel production: Implications for food security. *Global Food Security*, 16, 75–84.

Norjannah, B., Ong, H. C., Masjuki, H. H., Juan, J. C., & Chong, W. T. (2016). Enzymatic transesterification for biodiesel production: A comprehensive review. *RSC Advances*, 6(65), 60034–60055.

Norkobilov, A., Gorri, D., & Ortiz, I. (2017). Process flowsheet analysis of pervaporation-based hybrid processes in the production of ethyl tert-butyl ether. *Journal of Chemical Technology & Biotechnology*, 92(6), 1167–1177.

Ogunkunle, O., & Ahmed, N. A. (2019). A review of global current scenario of biodiesel adoption and combustion in vehicular diesel engines. *Energy Reports*, 5, 1560–1579.

Osman, W. N. A. W., Rosli, M. H., Mazli, W. N. A., & Samsuri, S. (2024). Comparative review of biodiesel production and purification. *Carbon Capture Science & Technology*, 13, 100264.

Packer, M. A., Harris, G. C., & Adams, S. L. (2016). Food and feed applications of algae. In Bux, F., & Chisti, Y. (eds.), *Algae Biotechnology: Products and Processes* (pp. 217–247). Cham: Springer International Publishing.

Padilla, J., Toledo, C., & Abad, J. (2022). Enovoltaics: Symbiotic integration of photovoltaics in vineyards. *Frontiers in Energy Research*, 10(September), 1–11. https://doi.org/10.3389/fenrg.2022.1007383

Pickering, B., Lombardi, F., & Pfenninger, S. (2022). Diversity of options to eliminate fossil fuels and reach carbon neutrality across the entire European energy system. *Joule*, 6(6), 1253–1276.

Pyšek, P., Pergl, J., Essl, F., Lenzner, B., Dawson, W., Kreft, H., Weigelt, P., Winter, M., Kartesz, J., Nishino, M., & Antonova, L. A. (2017). Naturalized alien flora of the world: Species diversity, taxonomic and phylogenetic patterns, geographic distribution and global hotspots of plant invasion. *Preslia*, *89*(3), 203–274.

Quijas, S., Schmid, B., & Balvanera, P. (2010). Plant diversity enhances provision of ecosystem services: A new synthesis. *Basic and Applied Ecology*, 11(7), 582–593.

Renzaho, A. M., Kamara, J. K., & Toole, M. (2017). Biofuel production and its impact on food security in low and middle income countries: Implications for the post-2015 sustainable development goals. *Renewable and Sustainable Energy Reviews*, 78, 503–516.

Rodionova, M. V., Poudyal, R. S., Tiwari, I., Voloshin, R. A., Zharmukhamedov, S. K., Nam, H. G., Zayadan, B. K., Bruce, B. D., Hou, H. J. M., & Allakhverdiev, S. I. (2017). Biofuel production: Challenges and opportunities. *International Journal of Hydrogen Energy*, 42(12), 8450–8461.

Roy, D. K., & Abedin, M. Z. (2022). Potentiality of biodiesel and bioethanol production from feedstock in Bangladesh: A review. *Heliyon*, *8*(11), e11213.

Rulli, M. C., Bellomi, D., Cazzoli, A., De Carolis, G., & D'Odorico, P. (2016). The water-land-food nexus of first-generation biofuels. *Scientific Reports*, 6(1), 22521.

Sajid, Z., da Silva, M. A. B., & Danial, S. N. (2021). Historical analysis of the role of governance systems in the sustainable development of biofuels in Brazil and the United States of America (USA). *Sustainability*, 13(12), 6881.

Sajjadi, B., Raman, A. A. A., & Arandiyan, H. (2016). A comprehensive review on properties of edible and non-edible vegetable oil-based biodiesel: Composition, specifications and prediction models. *Renewable and Sustainable Energy Reviews*, 63, 62–92.

Salaheldeen, M., Mariod, A. A., Aroua, M. K., Rahman, S. A., Soudagar, M. E. M., & Fattah, I. R. (2021). Current state and perspectives on transesterification of triglycerides for biodiesel production. *Catalysts*, *11*(9), 1121.

Saravanan, A., Kumar, P. S., Vo, D. V. N., Yaashikaa, P. R., Karishma, S., Jeevanantham, S., Gayathri, B., & Bharathi, V. D. (2021). Photocatalysis for removal of environmental pollutants and fuel production: A review. *Environmental Chemistry Letters*, 19, 441–463.

Schipfer, F., Burli, P., Fritsche, U., Hennig, C., Stricker, F., Wirth, M., Proskurina, S., & Serna-Loaiza, S. (2024). The circular bioeconomy: A driver for system integration. *Energy, Sustainability and Society*, 14(1), 1–19. https://doi.org/10.1186/s13705-024-00461-4

Singh, A., Olsen, S. I., & Pant, D. (2013). Importance of life cycle assessment of renewable energy sources. In *Life Cycle Assessment of Renewable Energy Sources* (pp. 1–11). London: Springer London.

Sinha, S. K., Subramanian, K. A., Singh, H. M., Tyagi, V. V., & Mishra, A. (2019). Progressive trends in bio-fuel policies in India: Targets and implementation strategy. *Biofuels*, 10(1), 155–166.

Smerald, A., Rahimi, J., & Scheer, C. (2023). A global dataset for the production and usage of cereal residues in the period 1997–2021. *Scientific Data*, 10(1), 1–9. https://doi.org/10.1038/s41597-023-02587-0

Soratana, K., Harden, C. L., Zaimes, G. G., Rasutis, D., Antaya, C. L., Khanna, V., & Landis, A. E. (2014). The role of sustainability and life cycle thinking in US biofuels policies. *Energy Policy*, 75, 316–326.

Stenzel, F., Gerten, D., Werner, C., & Jägermeyr, J. (2019). Freshwater requirements of large-scale bioenergy plantations for limiting global warming to 1.5C. *Environmental Research Letters*, 14(8), 084001.

Subbarao, P. M. V., D' Silva, T. C., Adlak, K., Kumar, S., Chandra, R., & Vijay, V. K. (2023). Anaerobic digestion as a sustainable technology for efficiently utilizing biomass in the context of carbon neutrality and circular economy. *Environmental Research*, 234(May), 116286. https://doi.org/10.1016/j.envres.2023.116286

Susmozas, A., Martín-Sampedro, R., Ibarra, D., Eugenio, M. E., Iglesias, R., Manzanares, P., & Moreno, A. D. (2020). Process strategies for the transition of 1G to advanced bioethanol production. *Processes*, 8(10), 1310.

Syafiuddin, A., Chong, J. H., Yuniarto, A., & Hadibarata, T. (2020). The current scenario and challenges of biodiesel production in Asian countries: A review. *Bioresource Technology Reports*, 12, 100608.

Topaloğlu, A., Esen, Ö., Turanlı-Yıldız, B., Arslan, M., & Çakar, Z. P. (2023). From *Saccharomyces cerevisiae* to ethanol: Unlocking the power of evolutionary engineering in metabolic engineering applications. *Journal of Fungi*, 9(10). https://doi.org/10.3390/jof9100984

Topare, N. S., Jogdand, R. I., Shinde, H. P., More, R. S., Khan, A., & Asiri, A. M. (2022). A short review on approach for biodiesel production: Feedstock's, properties, process parameters and environmental sustainability. *Materials Today:* Proceedings, 57, 1605–1612.

Tudge, S. J., Purvis, A., & De Palma, A. (2021). The impacts of biofuel crops on local biodiversity: A global synthesis. *Biodiversity and Conservation*, 30(11), 2863–2883.

Veljković, V. B., Banković-Ilić, I. B., & Stamenković, O. S. (2015). Purification of crude biodiesel obtained by heterogeneously-catalyzed transesterification. *Renewable and Sustainable Energy Reviews*, 49, 500–516.

Wang, Y., Ou, S., Liu, P., & Zhang, Z. (2007). Preparation of biodiesel from waste cooking oil via two-step catalyzed process. *Energy Conversion and Management*, 48(1), 184–188.

Whitaker, J., Field, J. L., Bernacchi, C. J., Cerri, C. E., Ceulemans, R., Davies, C. A., Delucia, E. H., Donnison, I. S., McCalmont, J. P., Paustian, K., Rowe, R. L., Smith, P., Thornley, P., & McNamara, N. P. (2018). Consensus, uncertainties and challenges for perennial bioenergy crops and land use. *GCB Bioenergy*, 10(3), 150–164.

Yimam, Y. T., Ochsner, T. E., Kakani, V. G., & Warren, J. G. (2014). Soil water dynamics and evapotranspiration under annual and perennial bioenergy crops. *Soil Science Society of America Journal*, 78(5), 1584–1593.

Zabed, H., Sahu, J. N., Boyce, A. N., & Faruq, G. (2016). Fuel ethanol production from lignocellulosic biomass: An overview on feedstocks and technological approaches. *Renewable and Sustainable Energy Reviews*, 66, 751–774.

Zandoná Filho, A., Lorenci Woiciechowski, A., Letti, L. A. J., Zevallos Torres, L. A., Valladares-Diestra, K. K., & Soccol, C. R. (2022). Feedstocks for first-generation bioethanol production. In *Liquid Biofuels: Bioethanol* (pp. 13–27). Cham: Springer International Publishing.

5 Cutting-Edge Approaches of Non-Edible Crop Plants for Biodiesel Production

Sadia Abdullah, Mamoona Munir, Wasim Akhtar, and Mushtaq Ahmed

5.1 INTRODUCTION

In recent decades, the world has faced a worsening energy crisis, largely due to the rapid exhaustion of fossil fuel supplies and the rising energy requirements of both developing and industrialized countries. The global reliance on fuels derived from petroleum has resulted in numerous problems. It has increased the exhaustion of finite natural resources and rendered economies globally highly susceptible to price changes, geopolitical conflicts, and persistent supply chain disruptions (K. Malik et al., 2024). The International Energy Agency states that global energy consumption is expected to increase by over 25% by 2040, largely driven by significant input from developing nations in Asia and Africa. The continuous rise in industrial output, urban expansion, and transportation demands has increased the demand for energy, thereby accelerating the use of fossil fuels (Nassar & Khaleel, 2024). The ongoing environmental effects of fossil fuel use, including greenhouse gas (GHG) emissions, air pollution, and climate change, have compelled global leaders and researchers to swiftly seekout alternative, sustainable energy solutions (Hassan et al., 2024). The 2023 Intergovernmental Panel on Climate Change (IPCC) report emphasized that to maintain global warming below 1.5°C, a rapid transition away from fossil fuels is crucial, combined with significant funding for renewable energy infrastructures. Simultaneously, studies show that fossil fuel combustion accounts for approximately 75% of global CO_2 emissions highlighting the urgent need for decarbonization in the energy sector (Bilgili et al., 2024). Figure 5.1 suggests that the world's total energy consumption is expected to continue growing over next few decades with projected increase of around 140% from 1990 to 2040. This trend is likely to be driven by certain factors i.e. population growth, economic development and increasing energy demands.

Consequently, the emphasis has transitioned increasingly toward renewable and sustainable energy sources such as biofuels, solar energy, wind power, and hydropower. Among these, biofuels, particularly those derived from non-edible oilseed crops and algal biomass, are gaining popularity due to their compatibility with existing combustion engines, biodegradability, and capacity to lower GHG emissions (Gayen et al., 2024). Recent advancements in biotechnology and genetic engineering have enhanced the effectiveness and feasibility of biofuel production, allowing for the development of high-yield, stress-resistant energy crops that thrive in marginal areas with minimal input (Patel & Singh, 2024). These innovations offer an optimistic pathway to reduce dependence on fossil fuels, promote environmental sustainability, and stimulate economic development in rural areas (Nath, 2024).

Biodiesel, a sustainable and biodegradable fuel mainly derived from biological sources such as vegetable oils and animal fats, has garnered significant attention as a potential substitute for diesel produced from petroleum (Randive et al., 2024). Its ability to reduce emissions, enhance energy security, and promote rural development makes it an appealing option for nations seeking

DOI: 10.1201/9781003600169-5

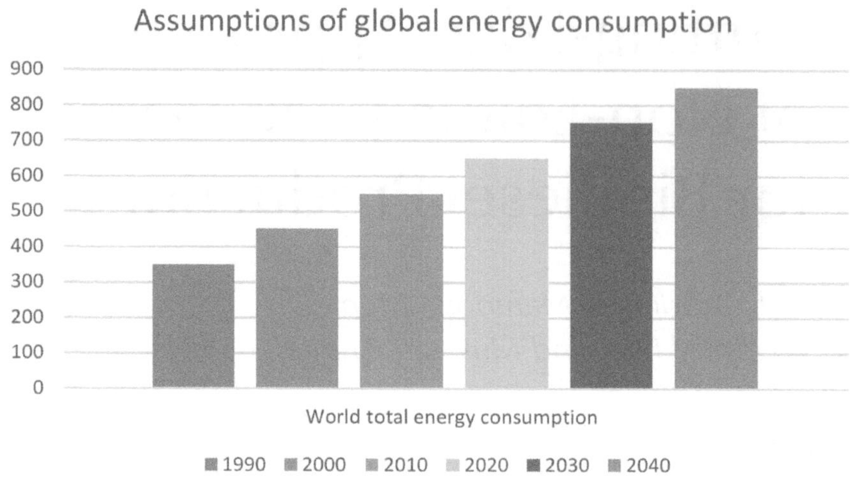

FIGURE 5.1 Energy consumption worldwide.

to diversify their energy sources. Despite its environmental benefits, conventional biodiesel production, mainly dependent on food oil crops like soybean, rapeseed, and palm, has raised considerable ethical and economic concerns (Zabermawi et al., 2022). The extensive rerouting of food crops for fuel generation not only drives up food prices but also intensifies food shortages in at-risk communities, weakening global food security initiatives (Snow, 2023).

The biodiesel industry has grown considerably in recent times. It will rise from \$57.91 billion in 2024 to \$61.52 billion in 2025, reflecting a compound annual growth rate (CAGR) of 6.2%. The rise during the historical period can be associated with rising crude oil costs, environmental regulations, variations in fossil fuel availability, energy security concerns, consumer awareness and preferences, and demand from the transportation industry. The biodiesel market is expected to see considerable expansion in the years ahead. It is projected to attain \$77.94 billion by 2029, with a CAGR of 6.1%. The growth throughout the projected period can be associated with the global surge in biodiesel production, focusing on sustainable aviation fuels, requirements for biodiesel blending, enhance-ments in biodiesel distribution, and advancements in sustainable agriculture (Kumari et al., 2025).

To ensure the role of biodiesel in a sustainable energy future, it is vital to develop production methods that do not threaten food supply or ecosystem health. This requires innovative approaches that emphasize the use of overlooked resources, particularly those that thrive in less arable areas, require minimal inputs, and do not vie with food crops (Brahma et al., 2022). The continuous debate over "food versus fuel" has persistently created difficulties for the biofuel sector, particularly in regions already struggling with hunger and malnutrition. When edible oil crops are prioritized for energy production, they directly compete with food resources, leading to reduced availability and increased prices for vital products (Løkke et al., 2021). This issue is especially pressing in developing countries, where a significant portion of the population spends the majority of their income on food. In such cases, reallocating agricultural land and water resources for biofuel production can exacerbate poverty, increase social tensions, and obstruct efforts to achieve Sustainable Development Goals (SDGs) aimed at eradicating hunger and reducing inequalities (Agholor, 2021).

Non-food crops provide a practical and ethical solution to this issue. Plants such as *Jatropha curcas, Pongamia pinnata, Calophyllum inophyllum,* and *Camelina sativa* can flourish on degraded or marginal lands unsuitable for food production, requiring minimal water, fertilizers, and pesticides. Their capacity to flourish in harsh environments and need little maintenance makes them outstanding candidates for biodiesel production in regions with limited agricultural resources. Utilizing these various feedstocks allows for biodiesel production without affecting food supplies, offering a sustainable path to clean energy while preserving food sovereignty and aiding rural communities (Anuragi et al., 2025). Cultivating non-consumable crops can support land restoration, improve soil quality, and generate rural employment, thereby boosting their significance in comprehensive sustainable development strategies (Thomas, 2016).

5.2 OVERVIEW OF NON-EDIBLE OILSEED CROPS

Table 5.1 offers an extensive summary of different non-edible oilseed crops along with their associated oil content and biodiesel yield. These plants, such as Jatropha, Castor, Neem, Karanja, Mahua, and more, have received considerable focus due to their capabilities in sustainable biodiesel creation. This comparative study highlights the significance of choosing crops that are both productive in oil yield and suitable for marginal lands, where energy needs are increasing and cultivable land is scarce.

5.2.1 *Jatropha curcas*

J. curcas is a perennial shrub or small tree that originated in Central America, but it is now widely cultivated in tropical and subtropical regions around the world (Anuragi et al., 2025). It belongs to the Euphorbiaceae family and can grow up to a height of 6 m. The plant is characterized by its sturdy, gray stem, lobed leaves, and small greenish-yellow flowers. The oil of Jatropha primarily comes from its seeds, which are toxic and unfit for consumption but highly sought after for producing biofuels. Jatropha thrives in semi-arid conditions and is drought-resistant, making it ideal for cultivation on marginal lands unsuitable for food crops (Agrawal et al., 2023). It requires minimal inputs such as water and fertilizers, and it can flourish in degraded or saline soils. The plant has a long productive life of 30–50 years and, once established, provides annual yields with little maintenance. It starts bearing fruit in 2–3 years and reaches full maturity in about 5 years (Ruatpuia et al., 2024).

TABLE 5.1
Non-Edible Oilseed Crops

Plants	Scientific Name	Biodiesel Yield/Oil Content	Key References
Soapnut	*Sapindus mukorossi*		I. U. Khan et al. (2021)
Jatropha	*Jatropha curcas*	~97 % conversion; ~27%–40% seed oil content	Almady et al. (2024)
Castor bean	*Ricinus communis*	45%–50% oil content; yield optimized with NaOH/KOH catalysts	Baioni e Silva et al. (2023)
Thlaspi (pennycress)	*Thlaspi arvense*	High erucic; considered for biodiesel (jet fuel capable)	J. Liu et al. (2023)
Crambe	*Crambe abyssinica*	High suitability for biofuels (industrial)	Cavalheiro et al. (2023)
Hibiscus/Ambadi	*Hibiscus cannabinus*	~13 000 t oil/year; 18%–22% oil content; used as biodiesel	Kukana and Jakhar (2022)
Pilu	*Salvadora persica*	32%–44% oil content; potential industrial/biodiesel use	Thakur et al. (2025)
Misc non-edibles	Multiple species	Croton, Prunus, rubber seed, wild Brassica, etc.	Devale and Mahajan (2024)

The seeds of *J. curcas* are prized for their high oil content, typically between 30% and 40%, depending on the variety and growing conditions. This oil cannot be consumed due to harmful compounds, yet it holds significant promise as a renewable energy resource (Riayatsyah et al., 2022). Jatropha oil, when processed and purified, can be converted into biodiesel through a transesterification process, producing a sustainable, clean-burning alternative to conventional fossil fuels. Biodiesel derived from Jatropha oil is biodegradable, generates reduced pollutant emissions such as carbon monoxide and sulfur oxides, and helps lessen dependence on imported oil products (Aigba et al., 2021).

5.2.2 *Pongamia pinnata*

P. pinnata is a medium-sized, fast-growing legume tree that comes from the Indian subcontinent and Southeast Asia. Belonging to the Fabaceae family, it can grow to a height of 15–25 m (Islam et al., 2021). The tree is easily identifiable by its feathered leaves, fragrant purple flowers, and flat, bean-shaped pods containing oil-rich seeds. Pongamia has historically been used in agroforestry and for offering shade along roadways and boulevards. The tree is highly adaptable and flourishes in tropical and subtropical regions, including saline and alkaline soils (Fugare et al., 2021). Being a legume, it improves the soil by capturing nitrogen from the air, benefiting sustainable agriculture. Once established, Pongamia requires minimal water and is resistant to pests and diseases. It typically begins to yield pods within 4–5 years and can continue to produce seeds for numerous years (Degani et al., 2022).

It improves the soil by capturing nitrogen from the air, beneficial for sustainable agriculture. Once established, Pongamia requires minimal water and is tolerant to pests and diseases. It typically begins to yield pods in 4–5 years and can continue to create seeds for numerous years (Sambasivam et al., 2023). Despite being unsafe for human consumption due to toxic compounds like karanjin and pongamol, the oil holds considerable value in industrial applications, especially within the bioenergy sector. Recently, Pongamia has gained increased attention as a promising renewable energy option because of its low cultivation costs, minimal input requirements, and high oil yield per hectare relative to other biofuel crops (Islam et al., 2021). Biodiesel derived from Pongamia oil meets international fuel standards and can be blended with conventional diesel for engine usage, reducing reliance on fossil fuels. Moreover, cultivating Pongamia enhances environmental sustainability through carbon sequestration, since fully grown trees can take in significant amounts of carbon dioxide from the atmosphere during their lifespan. This trait positions Pongamia as an essential crop in the fight against climate change (Goel & Kumar, 2025).

5.2.3 *Calophyllum inophyllum* (Tamanu or *Alexandrian laurel*)

C. inophyllum is a tall, perennial tree located in tropical areas, originally from the coastal regions of Southeast Asia and the Pacific Islands. It is part of the Calophyllaceae family and can reach heights of 20–25 m. The tree is notable for its wide, glossy, dark green leaves and aromatic white blossoms (Sreekumar, 2023). The round, green fruits hold a seed, which is where its oil, widely referred to as Tamanu oil, originates. This tree flourishes in coastal and saline conditions and can develop effectively in sandy, rocky, or poor soils, making it ideal for erosion management and reforestation efforts. It endures salt and wind, commonly used as a windbreak or decorative tree. *C. inophyllum* requires approximately 7–10 years to grow and commence fruit production, producing around 20–100 kg of fruit each year for each mature tree (Farida et al., 2024; Hasnah et al., 2025).

Tamanu oil has emerged as a potential source for biodiesel production in recent years, owing to its significant oil content and favorable fatty acid composition. The oil-rich seed of the Calophyllum fruit contains 40%–60% oil, which can be efficiently extracted and transformed into biodiesel

(Yuniastuti et al., 2021). This establishes the tree as a viable renewable energy choice, especially in tropical regions where conventional oilseed crops may have difficulty thriving (Adenuga et al., 2021). Unlike edible oils that often raise ethical concerns as fuel, the non-edible nature of Tamanu oil makes it an eco-friendly and socially accountable choice for energy production. Moreover, cultivating Calophyllum in degraded or coastal regions enhances the effective utilization of underutilized lands, thus supporting the green economy (Aneeque et al., 2021).

5.2.4 *Camelina sativa* (False Flax)

C. sativa is an annual herbaceous plant that is part of the Brassicaceae (mustard) family, native to Europe and Central Asia. It attains a height of 30–120 cm and has branching stems, narrow leaves, and small yellow flowers (Frizen et al., 2021). The plant produces small, round seeds rich in oil that are not safe for consumption due to the presence of glucosinolates and other anti-nutritional elements (Karamac et al., 2020). Camelina is a hardy plant with a short growing cycle (85–100 days), which makes it ideal for rotation or cover cropping. It endures cold, arid environments and poor soil, requiring little fertilizers and pesticides. It can be cultivated on less fertile lands without competing with food plants. It is similarly resilient to a range of common pests and diseases affecting oilseeds, enhancing its sustainability profile (Blume et al., 2024).

C. sativa seeds are notable for their high oil content, ranging from 30% to 40%, making them an excellent feedstock for biofuel manufacturing. Camelina is becoming known as a next-generation bioenergy crop due to its low environmental footprint. It requires minimal water, fertilizer, and chemical inputs, making it a more sustainable choice than conventional oilseeds (Dida, 2024). Non-consumable oilseed crops offer several agronomic advantages, which make them especially attractive for cultivation, particularly in resource-limited settings. A key feature of these crops is their low input requirement; they can flourish with minimal dependence on irrigation, synthetic fertilizers, and chemical pesticides (Shahee et al., 2023).

A significant advantage of non-edible oilseed crops is their remarkable capacity to flourish on marginal, degraded, and wastelands usually deemed unsuitable for traditional agriculture. These regions could experience various environmental challenges, including high salinity, low soil nutrients, rocky or shallow terrain, insufficient organic matter, and limited water availability (Carrino et al., 2020; Dinesha, 2023; Shah et al., 2023). Table 5.2 shows the comparison of different crop oils and their free fatty acid contents.

5.3 CUTTING EDGE GENETIC AND BIOTECHNOLOGICAL APPROACHES

In recent years, advances in genetic and biotechnological tools have revolutionized the improvement of nonedible oilseed crops for biodiesel production. These approaches aim to enhance the oil yield, stress tolerance, and specific fatty acid profiles to make biodiesel production more efficient and sustainable (Msanne et al., 2020). Tools like CRISPR/Cas9, genetic transformation, omics platforms, and synthetic biology are being employed to unlock the genetic potential of underutilized oilseed species such as *J. curcas, P. pinnata*, and *C. sativa*. These crops are particularly promising due to their adaptability to marginal lands and non-competition with food crops (Shabhanani et al., 2025). In Pakistan, the dependence on imported fossil fuels presents economic and ecological difficulties, making the use of these cutting-edge biotechnologies highly promising (Abbasi et al., 2022; Kanwal et al., 2022) (Figure 5.2).

5.3.1 CRISPR/Cas9 and Gene Editing for Trait Improvement

CRISPR/Cas9 gene editing technology has become a revolutionary instrument in plant biotechnology (Abd-Elsalam et al., 2023). Its capacity to execute exact modifications in the genome has allowed researchers to enhance oil quality, boost resistance to biotic and abiotic stressors, and adjust

TABLE 5.2
Comparison of Oil Content and Fatty Acid Profiles

Crop Name	Oil Content (%)	Major Fatty Acids	Notable Features
Jatropha curcas	30–40	Oleic, linoleic, palmitic	Non-edible, drought-tolerant, biodiesel use
Camelina sativa	35–40	α-linolenic, linoleic, oleic	Cold-tolerant, short life cycle
Castor bean	45–55	Ricinoleic acid (~90%)	Industrial oil, high viscosity
Pongamia pinnata	25–40	Oleic, linoleic, palmitic	Hardy, suited for marginal lands
Soybean (edible)	18–20	Linoleic, oleic, palmitic	Major food oil, less suitable for biodiesel
Neem (*Azadirachta indica*)	30–40	Oleic, stearic, palmitic	Biopesticidal properties, medicinal uses, non-edible
Mahua (*Madhuca indica*)	35–40	Oleic, palmitic, stearic	Used in rural lighting and soap, non-edible
Tobacco seed	30–40	Linoleic, oleic	Toxic, used in biodiesel
Rubber seed	40–50	Linoleic, oleic, linolenic	Industrial oil, non-edible due to cyanogenic compounds
Cottonseed (gossypol-rich)	15–20	Linoleic, palmitic, oleic	Toxic (gossypol), industrial applications
Euphorbia lathyris	40–50	Linoleic, oleic	Hydrocarbon-rich latex, potential biofuel crop
Azadirachta indica (Neem)	30–40	Oleic, palmitic, stearic	Anti-fungal, pesticide, medicinal uses

metabolic pathways in biodiesel plants (E. Ali & Zhang, 2023). Researchers have effectively utilized CRISPR/Cas9 in *C. sativa* to disable the FAD2 gene, leading to increased oleic acid levels, a sought-after characteristic for biodiesel quality. Likewise, *J. curcas* has experienced genome editing initiatives aiming at improving drought resistance and altering oil composition (J. Ali et al., 2020; Kawall, 2021).

5.3.2 Genetic Transformation and Metabolic Engineering

Genetic transformation entails the incorporation of new genes into a plant's genetic material to manifest preferred characteristics like improved oil production or resistance to pests (Kumawat et al., 2025). Using Agrobacterium-mediated methods or gene gun techniques, researchers have effectively modified Jatropha to enhance the expression of DGAT1 and WRINKLED1, which are essential genes in the lipid biosynthesis pathway (Al-Khayri et al., 2022). These metabolic engineering strategies have resulted in increases of up to 40% in triacylglycerol (TAG) accumulation, significantly enhancing oil yields per hectare (Al-Khayri et al., 2022). Significant advancements have occurred in the genetic modification of crops such as canola and cotton (Almas et al., 2023).

Omics technologies offer extensive understanding of plant biology through system-wide examination of genes (genomics), transcripts (transcriptomics), proteins (proteomics), and metabolites (metabolomics). These instruments assist in pinpointing crucial regulatory genes and metabolic constraints in oil production and stress reactions (Winck et al., 2022). Transcriptomic analysis of *P. pinnata* has shown varying gene expressions in response to salt and drought stress, which may be utilized for genetic enhancement (Rajarajan et al., 2022; Abd-Elsalam & Ahmad, 2023).

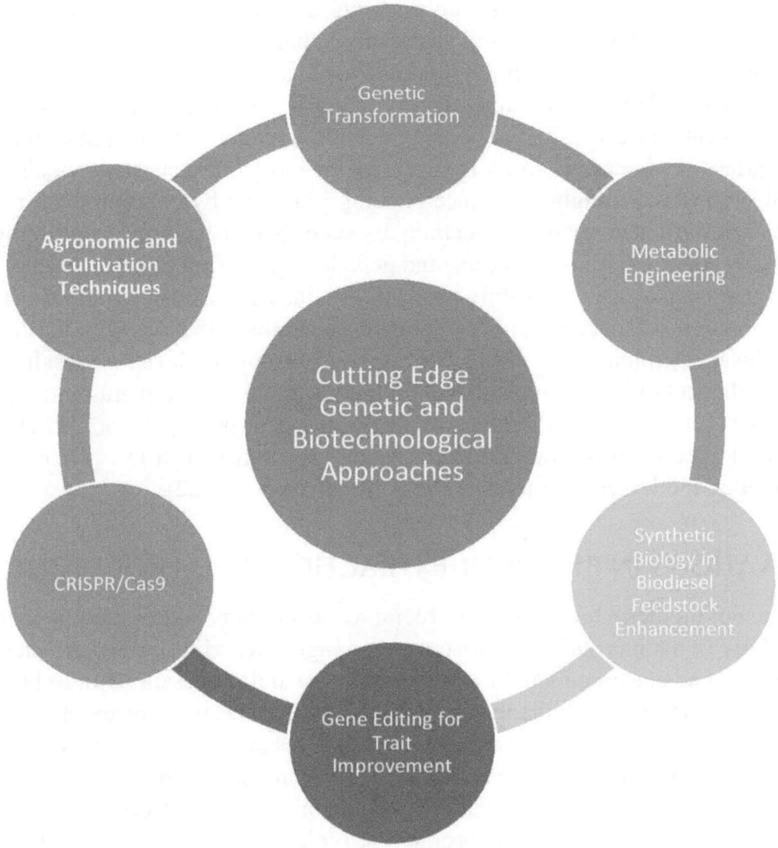

FIGURE 5.2 Cutting edge genetic and biotechnological approaches.

5.3.3 Synthetic Biology in Biodiesel Feedstock Enhancement

Synthetic biology provides an elevated form of genetic engineering, enabling the creation of tailored metabolic pathways and even artificial genomes. Through the construction of innovative biosynthetic circuits, researchers can enhance oil yield, customize fatty acid compositions, and incorporate new characteristics into non-model plants (Makepa & Chihobo, 2025). For example, incorporating synthetic routes for acetyl-CoA recycling in Camelina has considerably enhanced its oil production and biomass, increasing its competitiveness as a biodiesel crop (Stamenković et al., 2023).

5.3.4 Advanced Agronomic and Cultivation Techniques

Precision agriculture involves utilizing advanced technologies like GPS, GIS, remote sensing, and data analysis to oversee and improve the management of crops at the field level. Farmers can make data-driven choices to improve productivity and resource efficiency by gathering real-time information on soil characteristics, crop health, moisture content, and nutrient levels (Surendran et al., 2024). Devices such as drones, satellite images, and IoT sensors enable targeted interventions instead of generalized input applications, leading to less waste and enhanced sustainability. Smart farming enhances precision agriculture through the incorporation of automation, artificial intelligence (AI), and machine learning into farming methods. Examples encompass self-driving tractors, intelligent irrigation systems, and applications that help identify pests and diseases. These technologies enhance efficiency, minimize manual work, and offer predictive analysis of crop performance, thus boosting both quality and quantity of yield. The collaboration of digital tools aids in reducing environmental effects while enhancing economic gains (Yazdinejad et al., 2021).

Enhancing stress resilience in crops is essential for ensuring consistent yields in challenging environmental situations. Drought resistance is being improved by creating deep-rooted varieties, implementing water-efficient irrigation techniques such as drip systems, and applying mulch to minimize water evaporation. Plant breeding and genetic engineering also play a role by introducing characteristics that enable crops to sustain productivity during periods of water shortage (Gonzalez et al., 2022). Another key aspect is salinity tolerance, particularly in areas with ineffective irrigation methods and elevated groundwater salinity. Strategies encompass developing salt-resistant varieties, using soil enhancers such as gypsum, and integrating halophytic plants in areas impacted by salinity (H. Verma et al., 2021). Practices of integrated pest management (IPM), which include biological control agents, crop rotation, and resistant crop types, aim to decrease dependence on chemical pesticides and lessen pest-related stress (Rivero et al., 2022).

5.4 INNOVATIONS IN BIODIESEL EXTRACTION AND PROCESSING

The development of biodiesel technology is crucial for sustainably addressing the increasing energy needs, particularly in nations such as Pakistan, where agricultural resources are plentiful (S. Khan et al., 2022). Recent advancements in biodiesel extraction and processing aim to boost yield, minimize environmental effects, and enhance economic viability. These advancements not only enhance the conversion process but also support worldwide initiatives aiming at more sustainable and greener biofuel production (Innocenzi & Prisciandaro, 2021). A major advancement in biodiesel production is the transition to green solvents and environmentally friendly transesterification techniques (Yan et al., 2024). Traditional biodiesel production typically relies on methanol and strong acids or bases, which can be hazardous and detrimental to human health and the environment (Ameen et al., 2022). In contrast, environmentally friendly alternatives such as ionic liquids, deep eutectic solvents (DES), and bio-based solvents offer a sustainable choice. These solvents display low volatility, significant thermal stability, and outstanding solvation characteristics, rendering them perfect for sustainable processing (Iqbal et al., 2021). Sustainable transesterification techniques decrease the reliance on aggressive chemicals, reduce energy usage, and limit wastewater production. These innovations lead to a more sustainable production process, enhancing the environmental appeal of biodiesel (Orege et al., 2022).

The application of enzymatic and microbial catalysis signifies a fundamental change in biodiesel production. Lipases, especially those sourced from fungi and bacteria, have demonstrated effectiveness as biocatalysts in transesterification and esterification reactions (Ramos et al., 2022). Enzymatic catalysis functions at milder conditions (room temperature and pressure), greatly lowering energy requirements and avoiding thermal breakdown of delicate compounds (Pasha et al., 2021). Microbial systems, comprising genetically modified strains, have demonstrated the capability to transform waste oils and non-edible feedstocks into biodiesel with high specificity and output (Mandari & Devarai, 2022). These biocatalytic processes improve sustainability while enabling continuous production systems, since enzymes and microbes can be immobilized and reused across multiple cycles, which reduces long-term operational costs (R. Singh et al., 2022).

Nanotechnology has become a revolutionary instrument in biodiesel production, providing answers for catalysis as well as feedstock preparation. Nanocatalysts, including metal oxides (like ZnO, TiO_2, and Fe_3O_4) and mesoporous materials, offer extensive surface areas, adjustable pore sizes, and distinct electronic characteristics that greatly improve catalytic activity and selectivity (Bin Rashid, 2023). These nanomaterials hasten the transesterification process, shorten reaction time, and frequently remove the necessity for elevated temperatures and pressures. Moreover, magnetic nanoparticles enable straightforward separation and retrieval of catalysts, enhancing process efficiency (Arya et al., 2021).

The integrated biorefinery model embodies a comprehensive method for biodiesel production, generating various products from one biomass source. This model not only enhances resource efficiency but also boosts the overall financial viability of biodiesel manufacturing (Bin Rashid, 2023). In a combined system, oilseed plants or algal biomass may be processed to yield oils for biodiesel, whereas the leftover biomass is transformed into useful co-products like bioethanol, biogas, bioplastics, fertilizers, and animal feed (Qiao et al., 2022). This waste-free method improves energy efficiency and financial gains. Connecting with waste management systems, such as using municipal or agricultural waste as feedstock, enhances the sustainability component. In nations such as Pakistan, where biomass assets are abundant yet not fully utilized, establishing integrated biorefineries can generate fresh prospects for rural growth, energy independence, and ecological preservation (Goswami et al., 2021).

5.5 ENVIRONMENTAL AND ECONOMIC ASSESSMENTS

5.5.1 Life Cycle Assessment (LCA) of Non-Edible Crop Biodiesel:

Life cycle assessment (LCA) has become an essential method for assessing the environmental sustainability of biodiesel produced from non-edible oilseed crops like *J. curcas*, *P. pinnata,* and *Ricinus communis* (castor bean) (S. Kumar & Pal, 2023). LCA takes into account every phase of biodiesel production, including land preparation, crop growth, oil extraction, transesterification, and final combustion in engines. Research indicates that biodiesel derived from crops that are not meant for human consumption typically produces fewer GHG emissions than fossil diesel (Hosseinzadeh-Bandbafha et al., 2022). This decrease, however, relies significantly on aspects like fertilizer application, irrigation methods, and alterations in land use. Utilizing marginal or degraded lands for growing non-edible oil crops greatly enhances the environmental impact by preventing deforestation and reducing emissions from indirect land use changes (Abreu et al., 2022).

Recent improvements in modeling and remote sensing technologies have increased the precision of LCA studies by allowing for more localized evaluations (Yang et al., 2024). In nations such as India, Brazil, and Pakistan, customized LCAs have shown that using indigenous biomass for biodiesel generation can result in a 60%–70% decrease in CO_2-equivalent emissions. A significant challenge persists regarding the energy requirements in oil extraction and conversion processes. The integration of renewable energy during these phases, such as employing solar dryers or biomass-driven extraction systems, can enhance the life cycle emissions profile even more (Ingrao et al., 2021; Zubair et al., 2024). Additionally, LCA studies are starting to encompass more environmental impact categories apart from GHG emissions, including water consumption, soil quality, and biodiversity decline (Xing et al., 2022). Non-consumable plants such as Pongamia have demonstrated the ability to boost soil fertility via nitrogen fixation and demand considerably less water compared to conventional food crops, promoting greater environmental sustainability (Hasselsteen et al., 2025). Integrating these wider metrics into LCA can assist policymakers and investors in selecting the most suitable feedstocks and production systems for biodiesel advancement (Esmaeili, 2022).

5.5.2 Economic Viability and Cost-Benefit Analysis

The economic feasibility of biodiesel derived from non-edible crops depends on various factors, such as feedstock supply, yield per hectare, input expenditures, processing effectiveness, and the market rates for diesel and biodiesel (Leary, 2021; Lal et al., 2022). Current economic models suggest that biodiesel from Jatropha and Castor can compete with petroleum diesel when crude oil prices surpass $70 per barrel, particularly when policy incentives like tax credits and subsidies are available. In numerous developing countries, state-sponsored initiatives that promote the growing and processing of non-consumable oil crops have greatly lessened the financial investment load and operating expenses (Sharvini et al., 2022).

A thorough cost-benefit assessment should consider both direct and indirect financial impacts. In terms of costs, the biggest expenses continue to be seed procurement, land preparation, pest control, and transesterification (Arshad et al., 2025). On the advantage side, non-edible biodiesel provides significant benefits from reduced spending on imported fossil fuels, possible carbon credit earnings, and valuable byproducts such as glycerin and oilcake. Moreover, decentralized biodiesel facilities can lower transportation expenses and provide localized energy alternatives, particularly in isolated and off-grid regions (Y. Liu et al., 2024). Current trends also emphasize the importance of technological progress in enhancing economic gains. Enzyme-driven transesterification and supercritical fluid techniques are minimizing reaction durations and chemical consumption, which in turn decreases operational expenses. Automated harvesting and high-yield crop types, complemented by precision farming technologies such as drones and IoT devices, are greatly improving productivity per hectare. Collectively, these advancements are reinforcing the argument for biodiesel as a financially feasible alternative energy option (Arshad et al., 2025).

5.5.3 Socioeconomic Impacts in Rural Development

The economic feasibility of biodiesel derived from non-food crops depends on several factors such as the availability of feedstock, yield per hectare, costs of inputs, efficiency of processing, and the market prices of both diesel and biodiesel (Sikiru et al., 2024). Current economic models suggest that biodiesel produced from Jatropha and Castor could rival petroleum diesel when crude oil prices rise above $70 per barrel, particularly when supportive policies like tax incentives and subsidies are implemented. In numerous developing countries, state-sponsored initiatives that promote the growing and processing of non-food oil plants have considerably lessened the financial investment load and operational expenses (Shweta et al., 2024). The encouragement of non-edible oil crops for biodiesel production is increasingly acknowledged as a means for rural development, especially in areas with underutilized or degraded land (Kichonge et al., 2024). Job creation is one of the most immediate socioeconomic effects. Farming, gathering, oil production, and biodiesel transformation provide various employment possibilities, ranging from agricultural work to specialized positions in processing facilities. In nations such as India and Kenya, extensive Jatropha farms (Miftah & Mutta, 2024) have offered temporary jobs for numerous rural families, aiding in poverty reduction and diversifying livelihoods (Das et al., 2024; Kumari et al., 2025).

Besides job creation, the creation of local biodiesel industries promotes the enhancement of rural infrastructure, including roads, storage units, and power sources. This infrastructure supports biodiesel operations and simultaneously boosts the wider economic activities in these regions (Okhota et al., 2024). Additionally, the availability of locally made biodiesel can lessen reliance on expensive diesel imports and ensure a steady fuel supply for agricultural equipment, irrigation systems, and small generators, ultimately enhancing agricultural efficiency and food security (M. Kumar & Sharma, 2024).

5.6 POLICY, REGULATION, AND GLOBAL PERSPECTIVES

5.6.1 National and International Biofuel Policies

In recent years, national and international biofuel policies have gained renewed momentum as countries seek sustainable energy alternatives to mitigate climate change, reduce the reliance on fossil fuels, and meet their carbon neutrality commitments (Ebadian et al., 2020). The International Energy Agency (IEA) report that over 65 countries now have dedicated bioenergy strategies integrated into their National Energy and Climate Plans (NECPs). These policies emphasize second- and third-generation biofuels, including biodiesel derived from non-edible oilseed crops, algae, and waste (Saleh & Hassan, 2024). In United States, the Renewable Fuel Standards (RFS) continues to drive biofuel production by mandating annual blending targets, while the European Union's Renewable Energy Directive III (RED III), implemented in 2023, raises the bar by requiring at least 29% of transport fuels to come from renewables, with a sustainable share from advanced biofuels. Asian and African countries are also advancing the biofuel policy frameworks. India's National Policy on Biofuels 2018, amended in 2022, targets 20% ethanol blend by 2025 and encourages biodiesel production from nonedible feedstocks such as Jatropha, Karanja, and Pongamia. China under its 14th five-year plan (2021–2025) (Parvez et al., 2024) supports biofuel development through provincial pilot programs focusing on the marginal lands and rural revitalization. Meanwhile, Brazil continues to lead globally in biodiesel blending mandates, maintaining a minimum B13 blend (13% biodiesel in diesel fuel) and exploring oil seed crop diversification to avoid competition with food crops (Caprace et al., 2025).

The international policy coherence remains a challenge, with inconsistencies in trade rules, sustainability criteria, and certification schemes hindering the global harmonization of biofuel markets. Organizations like the Global Bioenergy Partnership (GBEP) and the International Renewable Energy Agency (IRENA) are actively working to bridge these gaps by promoting voluntary guidelines, knowledge exchange, and multilateral cooperation (Souza, 2025). Efforts are also being made to align biofuel policies with United Nations SDGs, particularly SDG 7 (affordable and clean energy), SDG 13 (climate action), and SDG2 (zero hunger) to ensure that energy transitions do not compromise food security or environmental integrity (N. Verma et al., 2024).

5.6.2 Incentives and Barriers to Non-Edible Biodiesel Crops

Non-edible biodiesel crops such as *J. curcas, P. pinnata, C. sativa,* and castor bean offer significant potential due to their adaptability to marginal lands and minimal competition with food crops (Anuragi et al., 2025). Many governments have introduced incentives to encourage their cultivation and use in biodiesel production. These include tax rebates, subsidies for plantation and processing infrastructure, guaranteed procurement prices, and research grants (Carlucci et al., 2025). The U.S. Department of Agriculture has included Camelina in its Biomass Crop Assistance Program (BCAP), supporting growers through establishment maintenance payments (Gregg et al., 2024).

Despite these initiatives, several barriers limit the widespread adoption of non-edible biodiesel crops. Agronomic challenges such as low initial yields, lack of high-quality seed varieties, and pest susceptibility hinder crop productivity (Yadav et al., 2022). Moreover, a lack of supply chain infrastructure ranging from collection to processing discourages private sector investment. Financial institutions are often reluctant to lend to biodiesel ventures due to perceived risks and uncertain returns, further stifling growth (Begna, 2020). A lack of unified land use policies results in clashes with forestry and conservation initiatives, especially in nations where land tenure systems are fragile or disputed.

Environmental and societal issues also serve as obstacles. Although non-edible crops are frequently advocated as sustainable, their farming may still cause unexpected ecological effects if implemented in delicate ecosystems (Dharmasena & Kadupitiya, 2021). In certain instances, extensive plantations have taken over lands used by smallholder farmers or indigenous groups, prompting concerns regarding fair land access and social equity. To address these challenges, an all-encompassing framework is needed that harmonizes economic feasibility with environmental sustainability and social equity. This encompasses focused research, training for farmers, collaborative land use planning, and robust regulatory oversight to guarantee accountable expansion (Durán-Díaz, 2023).

Numerous nations offer compelling illustrations of how non-food biodiesel crop initiatives can be effectively executed with the appropriate mix of policy, funding, and community involvement. In Brazil, the Social Fuel Seal initiative, introduced in 2004, has been crucial for encouraging biodiesels sourced from plants such as castor and babassu. The program mandates biodiesel producers to source feedstock from smallholder farmers and offer tax incentives for compliance (Everton et al., 2023). This has led to significant rural employment, improved farmer incomes, and a diversified energy portfolio without impinging on food resources. Brazil's success underscores the importance of integrating social welfare objectives into biofuel strategies (Everton et al., 2023).

In Africa, Mali's Jatropha program stands out as a grassroots model for community-based energy production. Initiated by local NGOs and cooperatives, the program trained farmers to grow Jatropha on degraded lands, enabling them to produce biodiesel for local consumption (El Bassam, 2021). By linking energy access with rural development, Mali was able to achieve localized energy security and reduce dependence on imported diesel. Although yields have fluctuated, adaptive management, local capacity-building, and decentralized processing facilities have ensured resilience and long-term impact (Coulibaly, 2021). Italy has emerged as a leader in cultivating *C. sativa* under the EU's advanced biofuel mandate. Supported by both public funding and private industry partnerships, Italian farmers have adopted Camelina as a rotational crop, improving soil health and reducing GHG emissions (Zanetti et al., 2021). Integrated biorefineries have been set up to process Camelina into biodiesel and jet fuel, aligning with the aviation sector's decarbonization goals.

5.7 CHALLENGES AND FUTURE PROSPECTS

Non-edible crop plants such as *J. curcas*, *P. pinnata*, *R. communis*, and *Azadirachta indica* (neem) have emerged as promising candidates for biodiesel production due to their adaptability to marginal lands and non-competition with food crops. Their existing yield capacity continues to be a significant obstacle (Yadav et al., 2022). Numerous plants display variable oil content based on climate factors, soil health, and farming methods. For instance, although Jatropha was previously celebrated as a miraculous biodiesel source, it has actually demonstrated inconsistent and low yields in various locations due to genetic diversity and insufficient cultivation efforts. Scalability presents another considerable challenge (Mohan et al., 2021). The cultivation of non-edible oil crops on a large scale demands significant land resources, infrastructure, and long-term investment, which are often impractical in nations where land is already strained for food production. The majority of non-edible oil crops require 3–5 years to mature, posing an economic bottleneck in biodiesel supply chains. Farmers are reluctant to switch from traditional crops to biodiesel plants without dependable markets or government backing. Monoculture plantations can adversely affect biodiversity and ecosystem stability if they are not managed in a sustainable manner (Rajpoot et al., 2022).

Logistics, encompassing harvesting, oil extraction, transportation, and conversion methods, continue to be underdeveloped for non-edible feedstocks (Wacal et al., 2021). The high free fatty acid (FFA) content in certain oils necessitate more sophisticated transesterification techniques, raising the expenses of biodiesel production (Salaheldeen et al., 2021). These constraints collectively impede the market potential and wider acceptance of biodiesel sourced from non-food plant materials. A significant research gap exists in the genetic improvement of non-food oil crops to boost their productivity, oil quality, and resilience to environmental stresses. Numerous species stay semi-domesticated and are less studied in relation to primary food crops (Ghosh et al., 2023). Progress in genomic selection, CRISPR/Cas9 gene editing, and transcriptomics presents exciting opportunities to create superior cultivars featuring greater oil yields, reduced maturation durations, and resilience to salinity, drought, and pest characteristics critical for effective farming in challenging environments (V. S. Malik, 2025).

Future initiatives ought to emphasize the combination of cutting-edge biotechnologies with innovations in large-scale industrial processes. Investigating microbial fermentation, enzymatic transesterification, and the valorization of byproducts (like seed cakes for biofertilizers or biopesticides) can enhance the cost-effectiveness and circularity of the biodiesel production chain (Hu et al., 2024). Creating affordable, decentralized biodiesel processing facilities in rural regions can lower logistics expenses and generate local job opportunities, promoting rural growth while ensuring sustainable energy generation. The future of non-edible crop-based biodiesel hinges on robust interdisciplinary collaboration across plant science, biotechnology, agronomy, chemical engineering, environmental science, and socioeconomics. Breeding high-yielding, stress-resilient cultivars requires the combined expertise of molecular biologists, plant physiologists, and geneticists. Simultaneously, input from agronomists is essential to translate lab findings into scalable field practices, while chemical engineers play a critical role in optimizing conversion technologies tailored to high-FFA feedstocks (Okhota et al., 2024). Environmental scientists and economists need to collaborate to assess the long-term ecological and economic impacts of biodiesel initiatives (Batra, 2023).

5.8 CONCLUSION

The field of biodiesel production has undergone a notable shift, moving from feedstocks obtained from food to sustainable non-food crop sources. Recent developments have enabled the practical utilization of robust, underutilized species such as *J. curcas, P. pinnata, R. communis,* and *C. sativa,* which thrive in low-quality soils with minimal agricultural inputs. Advancements in plant genomics, metabolic engineering, and synthetic biology have opened new avenues for boosting oil content, stress tolerance, and yield traits in these crops, making them more suitable for commercial biodiesel production. These developments are crucial for countries like Pakistan, India, and various African nations, where concerns regarding food security limit the use of edible oil crops for energy. Cutting-edge technologies such as CRISPR-Cas9 gene editing and transcriptome-guided breeding are presently employed to enhance lipid biosynthesis pathways in non-food oilseed plants. These techniques enable precise, targeted modification of crucial genes that regulate fatty acid synthesis and storage. For example, successfully modifying the DGAT1, WRINKLED1, and FAD2 genes in non-food crops has led to significant improvements in yield and oil quality. Molecular techniques, rapid phenotyping, and AI-based trait forecasting are enhancing the breeding process, enabling the quick development of improved, high-oil cultivars. Integrating these advanced strategies into a cohesive framework is essential for unlocking the full potential of non-edible biodiesel crops. With precision agriculture, remote sensing, and climate-resilient farming methods, the future of biodiesel production transcends traditional constraints. This cohesive, systems-focused approach seeks to align the economic, environmental, and social dimensions of biofuel creation, delivering a strong and flexible solution to global energy demands.

A future involving sustainable biodiesel from non-edible crops requires a strategically coordinated approach encompassing scientific, economic, and policy dimensions. Investment in R&D should mainly continue to focus on improving crop genetics through advanced molecular methods. Public entities and private companies should support joint research initiatives focusing on high-oil, low-input, and climate-resilient cultivars. Establishing international gene banks and digital breeding systems will improve the exchange of germplasm and the development of cultivars in diverse agro-ecological areas. Improving infrastructure for large-scale cultivation and processing of non-food oilseed crops is essential. This includes setting up decentralized biodiesel production plants, especially in rural or semi-arid regions apt for growing these crops. Public–private partnerships (PPPs) should be encouraged to develop robust value chains, spanning from seed supply to fuel delivery. Promoting the cultivation of non-food crops on lower-yield lands through subsidies, training initiatives, and supply contracts will ensure a steady supply of raw materials while maintaining food production. The shift toward biodiesel sustainability needs to be guided by encouraging policies and international collaboration. Quality standards for biodiesel, sustainability certifications for raw materials, and trade regulations are essential to boost market trust. Aligning biodiesel programs with national climate goals and energy security plans will ensure ongoing sustainability. As climate change continues to intensify, effectively employing non-food biodiesel crops allows for the diversification of energy sources, reduction of GHG emissions, and promotion of rural development, thus attaining multiple SDGs simultaneously.

COMPETING INTERESTS

The authors declare that they have no competing interest.

DECLARATION OF GENERATIVE AI AND AI-ASSISTED TECHNOLOGIES IN THE WRITING PROCESS

During the preparation of this work, the book chapter author(s) have not taken any assistance from AI, and the book chapter author(s) reviewed and edited the content as needed and take(s) full responsibility for the content of the publication.

REFERENCES

Abbasi, K. R., Shahbaz, M., Zhang, J., Irfan, M., & Alvarado, R. (2022). Analyze the environmental sustainability factors of China: The role of fossil fuel energy and renewable energy. *Renewable Energy,* 187, 390–402.

Abd-Elsalam, K. A., & Ahmad, A. (2023). *Global regulatory outlook for CRISPRized plants*: Elsevier.

Abreu, M., Silva, L., Ribeiro, B., Ferreira, A., Alves, L., Paixão, S. M., . . . Duarte, L. C. (2022). Low indirect land use change (ILUC) energy crops to bioenergy and biofuels—A review. *Energies,* 15(12), 4348.

Adenuga, A. A., Oyekunle, J. A. O., & Idowu, O. O. (2021). Pathway to reduce free fatty acid formation in Calophyllum inophyllum kernel oil: A renewable feedstock for biodiesel production. *Journal of Cleaner Production,* 316, 128222.

Agholor, A. I. (2021). Examining biofuel sustainability as an alternative energy source and implication for food security in Sub-Saharan Africa. *Journal of Critical Reviews, 8*(2), 2021.

Agrawal, A., Jain, S., & Gupta, A. (2023). Effectiveness of *Jatropha curcas* as biodiesel and antiviral: A review. *International Journal of Newgen Research in Pharmacy & Healthcare, 1*(2), 32–41.

Aigba, P., Anyadiegwu, F., & Ogoke, J. (2021). Characterization of jatropha oil and its biodiesel. *Advance Environmental Studies,* 5(1), 376–381.

Al-Khayri, J. M., Sudheer, W. N., Preetha, T. R., Nagella, P., Rezk, A. A., & Shehata, W. F. (2022). Biotechnological research progress in Jatropha, a biodiesel-yielding plant. *Plants,* 11(10), 1292.

Ali, E., & Zhang, K. (2023). CRISPR-mediated technology for seed oil improvement in rapeseed: Challenges and future perspectives. *Frontiers in Plant Science,* 14, 1086847.

Ali, J., Mahender, A., Prahalada, G., Sevilla, M. A. L., Galang, A., Asis, E. J. D., . . . Vinarao, R. (2020). Genomics-assisted breeding of climate-smart inbred and hybrid rice varieties. In C. Kole (Ed.), *Genomic designing of climate-smart cereal crops* (pp. 1–43). Springer.

Almady, S. S., Moussa, A. I., Deef, M. M., Zayed, M. F., Al-Sager, S. M., & Aboukarima, A. M. (2024). Biodiesel production through the transesterification of non-edible plant oils using glycerol separation technique with ac high voltage. *Sustainability,* 16(7), 2896.

Almas, H. I., Azhar, M. T., Atif, R. M., Iqbal, M. A., Khan, A. I., & Rana, I. A. (2023). Adaptation of genetically modified crops in Pakistan. In M. A. Nawaz & G. Chung (Eds.), *GMOs and political stance* (pp. 93–114): Elsevier.

Ameen, M., Ahmad, M., Zafar, M., Munir, M., Abbas, M. M., Sultana, S., . . . Kalam, M. (2022). Prospects of catalysis for process sustainability of eco-green biodiesel synthesis via transesterification: A state-of-the-art review. *Sustainability,* 14(12), 7032.

Aneeque, M., Alshahrani, S., Kareemullah, M., Afzal, A., Saleel, C. A., Soudagar, M. E. M., . . . Ahmed, M. H. (2021). The combined effect of alcohols and *Calophyllum inophyllum* biodiesel using response surface methodology optimization. *Sustainability,* 13(13), 7345.

Anuragi, H., Rajarajan, K., Taria, S., Bharati, A., Yadav, S., Ram, A., . . . Arunachalam, A. (2025). Non-edible tree-based oilseeds as sustainable source of biodiesel and bioenergy production. In P. Kumar, A. Singh, V. D. Rajput, T. Minkina, & A. K. Singh (Eds.),*Smart technologies in sustainable agriculture* (pp. 107–121): Apple Academic Press.

Arshad, M. U., Archer, D., Wasonga, D., Namoi, N., Boe, A., Mitchell, R., . . . Lee, D. (2025). Comparative economic analysis between bioenergy and forage types of switchgrass for sustainable biofuel feedstock production: A data envelopment analysis and cost–benefit analysis approach. *GCB Bioenergy,* 17(2), e70020.

Arya, I., Poona, A., Dikshit, P. K., Pandit, S., Kumar, J., Singh, H. N., . . . Kumar, S. (2021). Current trends and future prospects of nanotechnology in biofuel production. *Catalysts,* 11(11), 1308.

Baioni e Silva, G., Manicardi, T., Longati, A. A., Lora, E. E., & Milessi, T. S. (2023). Parametric comparison of biodiesel transesterification processes using non-edible feedstocks: Castor bean and jatropha oils. *Biofuels, Bioproducts and Biorefining,* 17(2), 297–311.

Batra, G. (2023). Renewable energy economics: Achieving harmony between environmental protection and economic goals. *Social Science Chronicle,* 2(2), 1–32.

Begna, T. (2020). Major challenging constraints to crop production farming system and possible breeding to overcome the constraints. *International Journal of Research Studies in Agricultural Sciences (IJRSAS),* 6(7), 27–46.

Bilgili, M., Tumse, S., & Nar, S. (2024). Comprehensive overview on the present state and evolution of global warming, climate change, greenhouse gasses and renewable energy. *Arabian Journal for Science and Engineering,* 49(11), 14503–14531.

Bin Rashid, A. (2023). Utilization of nanotechnology and nanomaterials in biodiesel production and property enhancement. *Journal of Nanomaterials, 2023*(1), 7054045.

Blume, R., Sakharova, V., Rabokon, A., Shumilova, A., Shiyan, N., Mosyakin, S., & Blume, Y. (2024). Distribution and infraspecific diversity of little-pod false flax (*Camelina microcarpa*, Brassicaceae) in Ukraine. *Ukrainian Botanical Journal,* 81(1), 52–62.

Brahma, S., Nath, B., Basumatary, B., Das, B., Saikia, P., Patir, K., & Basumatary, S. (2022). Biodiesel production from mixed oils: A sustainable approach towards industrial biofuel production. *Chemical Engineering Journal Advances,* 10, 100284.

Caprace, J.-D., Marques, C. H., Assis, L. F., Lucchesi, A., & Pereda, P. (2025). Sustainable shipping: Modeling technological pathways toward zero-emission in maritime transport (part I). *Sustainability Journal, 17*(8).

Carrino, L., Visconti, D., Fiorentino, N., & Fagnano, M. (2020). Biofuel production with castor bean: A win–win strategy for marginal land. *Agronomy,* 10(11), 1690.

Cavalheiro, L. F., Prado, E. R. L., de Freitas, O. N., Nazário, C. E. D., Rial, R. C., & Viana, L. H. (2023). Biofuels obtained from the crambe (*Crambe abyssinica*) oil. *Journal of Analytical and Applied Pyrolysis,* 175, 106214.

Das, P., Jha, C. K., Saxena, S., & Ghosh, R. K. (2024). Can biofuels help achieve sustainable development goals in India? A systematic review. *Renewable and Sustainable Energy Reviews,* 192, 114246.

Degani, E., Prasad, M., Paradkar, A., Pena, R., Soltangheisi, A., Ullah, I., . . . Tibbett, M. (2022). A critical review of *Pongamia pinnata* multiple applications: From land remediation and carbon sequestration to socioeconomic benefits. *Journal of Environmental Management,* 324, 116297.

Devale, R. R., & Mahajan, Y. S. (2024). Transesterification reactions as a means to produce esters: A critical review. *The Canadian Journal of Chemical Engineering,* 102(11), 3787–3802.

Dharmasena, P., & Kadupitiya, H. (2021). *UNEP-GEF project: Managing agricultural landscapes in socio-ecologically sensitive areas to promote food security, wellbeing and ecosystem health in Sri Lanka.*

Dida, G. (2024). Biotechnology towards energy crops. *CABI Agriculture and Bioscience,* 5(1), 45.

Dinesha, S. (2023). Short-rotation tree-based biofuel production in India from agroforestry and marginal lands without compromising food security. *Journal of Agricultural Technology,* 10(2), 58–71.

Durán-Díaz, P. (2023). Sustainable land governance for water–energy–food systems: A framework for rural and peri-urban revitalisation. *Land,* 12(10), 1828.

Ebadian, M., van Dyk, S., McMillan, J. D., & Saddler, J. (2020). Biofuels policies that have encouraged their production and use: An international perspective. *Energy Policy,* 147, 111906.

El Bassam, N. (2021). Current distributed renewable energy in rural and urban communities. In N. El Bassam (Ed.), *Distributed renewable energies for off-grid communities* (2nd Ed., pp. 297–326). Elsevier.

Esmaeili, H. (2022). A critical review on the economic aspects and life cycle assessment of biodiesel production using heterogeneous nanocatalysts. *Fuel Processing Technology,* 230, 107224.

Everton, S. S., Sousa, I., da Silva Dutra, L., Cipolatti, E. P., Aguieiras, E. C. G., Manoel, E. A., . . . Pinto, J. C. (2023). The role of Brazil in the advancement of enzymatic biodiesel production. *Brazilian Journal of Chemical Engineering,* 40(1), 67–80.

Farida, S., Jenie, R. I., & Fakhrudin, N. (2024). *Calophyllum inophyllum*: A comprehensive analysis of its ethnobotanical, phytochemical, and pharmacological properties. *Majalah Obat Tradisional,* 29(2), 121–142.

Fugare, A. G., Shete, R. V., Adak, V. S., & Murthy G, K. (2021). A review on *Pongamia pinnata* (L.): Traditional uses, phytochemistry and pharmacological properties. *Journal of Drug Delivery & Therapeutics,* 11, 207.

Gayen, D., Chatterjee, R., & Roy, S. (2024). A review on environmental impacts of renewable energy for sustainable development. *International Journal of Environmental Science and Technology,* 21(5), 5285–5310.

Ghosh, N., Rhithuparna, D., Rokhum, S. L., & Halder, G. (2023). Ethical issues pertaining to sustainable biodiesel synthesis over trans/esterification process. *Sustainable Chemistry and Pharmacy,* 33, 101123.

Goel, V., & Kumar, S. (2025). Optimization of process parameters and biodiesel extraction of *Pogamia pinnata* via response surface methodology. *International Journal of Applied and Computational Mathematics,* 11(3), 1–20.

Gonzalez Guzman, M., Cellini, F., Fotopoulos, V., Balestrini, R., & Arbona, V. (2022). New approaches to improve crop tolerance to biotic and abiotic stresses. *Physiologia Plantarum,* 174(1), e13547.

Goswami, R. K., Agrawal, K., & Verma, P. (2021). Microalgae-based biofuel-integrated biorefinery approach as sustainable feedstock for resolving energy crisis. In M. Srivastava, N. Srivastava, & R. Singh (Eds.), *Bioenergy research: Commercial opportunities & challenges* (pp. 267–293). Springer.

Gregg, S., Gesch, R. W., & Garcia y Garcia, A. (2024). Nitrogen uptake and use efficiency in winter camelina with applied N. *Nitrogen,* 5(2), 509–517.

Hasnah, T. M., Hanudin, E., Leksono, B., Rahman, S. A., Purwanto, B. H., Windyarini, E., . . . Baral, H. (2025). Carbon stock in *Calophyllum inophyllum* provenances from eight islands in Indonesia: Associate soil physicochemical properties and litter fiber content. *Forest Science and Technology,* 21(1), 51–63.

Hassan, Q., Viktor, P., Al-Musawi, T. J., Ali, B. M., Algburi, S., Alzoubi, H. M., . . . Jaszczur, M. (2024). The renewable energy role in the global energy Transformations. *Renewable Energy Focus,* 48, 100545.

Hasselsteen, L., Stapel, E. B., Birgisdottír, H., Sørensen, C. G., & Kanafani, K. (2025). Evaluating the environmental impact of construction waste: A comprehensive analysis of end-of-life scenarios in environmental product declarations. *Building and Environment, 280,* 113159.

Hosseinzadeh-Bandbafha, H., Nizami, A.-S., Kalogirou, S. A., Gupta, V. K., Park, Y.-K., Fallahi, A., . . . Aghbashlo, M. (2022). Environmental life cycle assessment of biodiesel production from waste cooking oil: A systematic review. *Renewable and Sustainable Energy Reviews,* 161, 112411.

Hu, X., Xu, B., Chen, M., Li, K., Xiao, Y., Liang, S., . . . Song, H. (2024). Development and assessment of cutting-edge biotechnologies. *Journal of Biosafety and Biosecurity, 6*(1), 51–63.

Ingrao, C., Evola, R. S., Cantore, P., De Bernardi, P., Del Borghi, A., Vesce, E., & Beltramo, R. (2021). The contribution of sensor-based equipment to life cycle assessment through improvement of data collection in the industry. *Environmental Impact Assessment Review,* 88, 106569.

Innocenzi, V., & Prisciandaro, M. (2021). Technical feasibility of biodiesel production from virgin oil and waste cooking oil: Comparison between traditional and innovative process based on hydrodynamic cavitation. *Waste Management,* 122, 15–25.

Iqbal, N., Sharma, R., Hazra, D. K., Dubey, S., Kumar, N., Agrawal, A., & Kumar, J. (2021). Successful utilization of waste cooking oil in Neem oil based fungicide formulation as an economic and eco-friendly green solvent for sustainable waste management. *Journal of Cleaner Production,* 288, 125631.

Islam, A. K. M. A., Chakrabarty, S., Yaakob, Z., Ahiduzzaman, M., & Islam, A. K. M. M. (2021). Koroch (*Pongamia pinnata*): A promising unexploited resources for the tropics and subtropics. In A. C. Gonçalves (Ed.), *Forest biomass – From trees to energy*: IntechOpen.

Janssen, J. H., Spoelder, J., Koehorst, J. J., Schaap, P. J., Wijffels, R. H., & Barbosa, M. J. (2020). Time-dependent transcriptome profile of genes involved in triacylglycerol (TAG) and polyunsaturated fatty acid synthesis in *Nannochloropsis gaditana* during nitrogen starvation. *Journal of Applied Phycology,* 32, 1153–1164.

Jeevan Kumar, S., Avanthi, A., Chintagunta, A. D., Gupta, A., & Banerjee, R. (2020). Oleaginous lipid: A drive to synthesize and utilize as biodiesel. In M. Mitra & A. Nagchaudhuri (Eds.), *Practices and perspectives in sustainable bioenergy: A systems thinking approach* (pp. 105–129). Springer.

Kanwal, S., Mehran, M. T., Hassan, M., Anwar, M., Naqvi, S. R., & Khoja, A. H. (2022). An integrated future approach for the energy security of Pakistan: Replacement of fossil fuels with syngas for better environment and socio-economic development. *Renewable and Sustainable Energy Reviews,* 156, 111978.

Karamac, M., Gai, F., & Peiretti, P. G. (2020). Effect of the growth stage of false flax (*Camelina sativa* L.) on the phenolic compound content and antioxidant potential of the aerial part of the plant. *Polish Journal of Food and Nutrition Sciences,* 70(2), 189–198.

Kawall, K. (2021). Genome-edited *Camelina sativa* with a unique fatty acid content and its potential impact on ecosystems. *Environmental Sciences Europe,* 33(1), 38.

Khan, I. U., Chen, H., Yan, Z., & Chen, J. (2021). Extraction and quality evaluation of biodiesel from six familiar non-edible plants seeds. *Processes,* 9(5), 840.

Khan, S., Naushad, M., Iqbal, J., Bathula, C., & Al-Muhtaseb, A. a. H. (2022). Challenges and perspectives on innovative technologies for biofuel production and sustainable environmental management. *Fuel,* 325, 124845.

Kichonge, B., & Kivevele, T. (2024). Sustainable biofuel production in Sub-Saharan Africa: Exploring transesterification process, nonedible feedstocks, and policy implications. *Wiley Interdisciplinary Reviews: Energy and Environment,* 13(3), e519.

Kukana, R., & Jakhar, O. P. (2022). Performance, combustion and emission characteristics of a diesel engine using composite biodiesel from waste cooking oil-*Hibiscus cannabinus* oil. *Journal of Cleaner Production,* 372, 133503.

Kumar, M., & Sharma, S. (2024). Renewable energy and sustainable transportation. In R. C. Sobti (Ed.), *Role of science and technology for sustainable future: Volume 1: Sustainable development: A primary goal* (pp. 375–414): Springer.

Kumar, S., & Pal, A. (2023). Life cycle analysis of biodiesel derived from fresh water microalgae and Karanja. *Sustainable Energy Technologies and Assessments,* 56, 103082. ISSN 2213-1388, Elsevier.

Kumari, A., Kumar, D., & Kumar, S. (2025). Societal impact of biofuels industry. In A. Kuila & D. Kumar (Eds.), *Optimizing biofuel production with artificial intelligence* (pp. 329–346). John Wiley & Sons.

Kumawat, S., Pragati, K., Garkoti, P., Singh, A., & Senthilkumar, V. (2025). Next-gen seeds: Harnessing genetic engineering for improving seed quality traits. In B. Roy, G. Shukla, V. Dunna, P. Sharma, & P. S. Shukla (Eds.), *Advances in seed quality evaluation and improvement* (pp. 273–307): Springer.

Lal, R., Kumar, S., & Naidu, S. (2022). A cost-benefit analysis of small biofuel projects in Fiji: Lessons and implications. *Journal of Cleaner Production,* 340, 130812.

Leary, M. (2021). Economic feasibility and cost-benefit analysis. In I. Yadroitsev, I. Yadroitsava, A. du Plessis, & E. MacDonald (Eds.), *Fundamentals of laser powder bed fusion of metals* (pp. 597–620): Elsevier.

Liu, J., Zhang, Z., Tang, S., Yu, Z., Zhang, Y., & Zheng, B. (2023). Ultrasound-assisted production of biodiesel from field pennycress (*Thlaspi arvense* L.) seeds: Process optimization and quality evaluation. *Industrial Crops and Products,* 203, 117224.

Liu, Y., Zhu, Z., Zhang, R., & Zhao, X. (2024). Life cycle assessment and life cycle cost analysis of Jatropha biodiesel production in China. *Biomass Conversion and Biorefinery,* 14(22), 28635–28660.

Løkke, S., Aramendia, E., & Malskær, J. (2021). A review of public opinion on liquid biofuels in the EU: Current knowledge and future challenges. *Biomass and Bioenergy,* 150, 106094.

Makepa, D. C., & Chihobo, C. H. (2025). Synthetic biology-enhanced microalgae for biofuel production: A perspective. *Biofuels, Bioproducts and Biorefining,* 19(2), 277–287.

Malik, K., Capareda, S. C., Kamboj, B. R., Malik, S., Singh, K., Arya, S., & Bishnoi, D. K. (2024). Biofuels production: A review on sustainable alternatives to traditional fuels and energy sources. *Fuels,* 5(2), 157–175.

Malik, V. S. (2025). Enhancing oil quality and yield in Brassica juncea via CRISPR/Cas9 technology. *Plant Physiology Reports,* 30(1), 11–17.

Mandari, V., & Devarai, S. K. (2022). Biodiesel production using homogeneous, heterogeneous, and enzyme catalysts via transesterification and esterification reactions: A critical review. *BioEnergy Research,* 15(2), 935–961.

Miftah, F. K., & Mutta, D. (2024). Potential markets and policies for sustainable liquid biofuel production with emphasis to Eastern Africa countries: A review. *Energy, Sustainability and Society,* 14(1), 1.

Miranda, A. M., Hernandez-Tenorio, F., Villalta, F., Vargas, G. J., & Sáez, A. A. (2024). Advances in the development of biofertilizers and biostimulants from microalgae. *Biology,* 13(3), 199.

Mohan, C., Srivastva, D. K., Verma, R., Kumar, V., Singh, V., Chaudhary, S., & Saini, P. N. (2021). Current status, challenges and future prospects of vegetable seed system in India: Vegetable seed system in India. *Journal of AgriSearch,* 8(3), 188–200.

Msanne, J., Kim, H., & Cahoon, E. B. (2020). Biotechnology tools and applications for development of oilseed crops with healthy vegetable oils. *Biochimie,* 178, 4–14.

Nasar, S., Muhammad, M., Siddiqi, E. H., Majeed, M., Ameen, N., Ullah, S., . . . Abdullah, M. (2024). OMICS-based knowledge for achieving food and nutritional security. In S. Fiaz & C. S. Prakash (Eds.), *OMICs-based techniques for global food security* (pp. 67–90). Wiley.

Nassar, Y., & Khaleel, M. (2024). Sustainable development and the surge in electricity demand across emerging economies. *International Journal of Electrical Engineering and Sustainability*, *2*(1), 51–60.

Nath, S. (2024). Biotechnology and biofuels: Paving the way towards a sustainable and equitable energy for the future. *Discover Energy,* 4(1), 8.

Okhota, Y., Titov, D., & Dotsiuk, S. (2024). Mechanisms for ensuring the economic development of rural areas through the formation of their energy dependence based on the intensification of biofuel production. *Baltic Journal of Economic Studies,* 10(5), 282–292.

Orege, J. I., Oderinde, O., Kifle, G. A., Ibikunle, A. A., Raheem, S. A., Ejeromedoghene, O., . . . Fagbohun, E. O. (2022). Recent advances in heterogeneous catalysis for green biodiesel production by transesterification. *Energy Conversion and Management,* 258, 115406.

Parvez, M., Lal, S., Khan, O., Ahmad, M., Lal Meena, S., & Salvi, B. L. (2024). Sustainable regulatory framework for ethanol and biodiesel blending with petro-fuels in India to meet fuel demand: A review of biofuel policies. *Biofuels*, *16*, 1–14.

Pasha, M. K., Dai, L., Liu, D., Du, W., & Guo, M. (2021). Biodiesel production with enzymatic technology: Progress and perspectives. *Biofuels, Bioproducts and Biorefining,* 15(5), 1526–1548.

Patel, K., & Singh, S. (2024). Sustainable biodiesel from used cooking oil: A comparative life cycle, energy, and uncertainty analysis. *Environment, Development and Sustainability*, *27*(6), 1–26.

Qiao, J., Cui, H., Wang, M., Fu, X., Wang, X., Li, X., & Huang, H. (2022). Integrated biorefinery approaches for the industrialization of cellulosic ethanol fuel. *Bioresource Technology,* 360, 127516.

Rajarajan, K., Sakshi, S., Taria, S., Prathima, P., Radhakrishna, A., Anuragi, H., . . . Arunachalam, A. (2022). Whole plant response of *Pongamia pinnata* to drought stress tolerance revealed by morpho-physiological, biochemical and transcriptome analysis. *Molecular Biology Reports,* 49(10), 9453–9463.

Rajpoot, L., Tagade, A., Deshpande, G., Verma, K., Geed, S. R., Patle, D. S., & Sawarkar, A. N. (2022). An overview of pyrolysis of de-oiled cakes for the production of biochar, bio-oil, and pyro-gas: Current status, challenges, and future perspective. *Bioresource Technology Reports,* 19, 101205.

Ramos, M. D. N., Milessi, T. S., Candido, R. G., Mendes, A. A., & Aguiar, A. (2022). Enzymatic catalysis as a tool in biofuels production in Brazil: Current status and perspectives. *Energy for Sustainable Development,* 68, 103–119.

Randive, A. B., Gaikwad, S. K., Khadake, S. B., & Mallad, H. (2024). Biodiesel: A renewable source of fuel. *International Journal of Advanced Research in Science, Communication and Technology (IJARSCT)*, *4*(3).

Rangan, P., Maurya, R., & Singh, S. (2022). Can omic tools help generate alternative newer sources of edible seed oil? *Plant Direct,* 6(6), e399.

Riayatsyah, T., Sebayang, A., Silitonga, A., Padli, Y., Fattah, I., Kusumo, F., . . . Mahlia, T. (2022). Current progress of *Jatropha curcas* commoditisation as biodiesel feedstock: A comprehensive review. *Frontiers in Energy Research,* 9, 815416.

Rivero, R. M., Mittler, R., Blumwald, E., & Zandalinas, S. I. (2022). Developing climate-resilient crops: Improving plant tolerance to stress combination. *The Plant Journal,* 109(2), 373–389.

Ruatpuia, J. V., Halder, G., Vanlalchhandama, M., Lalsangpuii, F., Boddula, R., Al-Qahtani, N., . . . Rokhum, S. L. (2024). *Jatropha curcas* oil a potential feedstock for biodiesel production: A critical review. *Fuel,* 370, 131829.

Salaheldeen, M., Mariod, A. A., Aroua, M. K., Rahman, S. A., Soudagar, M. E. M., & Fattah, I. R. (2021). Current state and perspectives on transesterification of triglycerides for biodiesel production. *Catalysts,* 11(9), 1121.

Saleh, H. M., & Hassan, A. I. (2024). The challenges of sustainable energy transition: A focus on renewable energy. *Applied Chemical Engineering,* 7(2), 2084.

Sambasivam, K. M., Kuppan, P., Laila, L. S., Shashirekha, V., Tamilarasan, K., & Abinandan, S. (2023). Kernel-based biodiesel production from non-edible oil seeds: Techniques, optimization, and environmental implications. *Energies,* 16(22), 7589.

Shabhanani, H. R., Jog, S. M., Pundir, P., Vekariya, P., & Murugan, A. C. (2025). Enhancing oilseed crop productivity and quality with CRISPR/Cas gene-editing technologies. In Y. Suryawanshi, M. Shahnawaz, J. Parihar, R. A. Dar, S. Samyal, A. B. Ade (Eds.), *Oilseed Crops* (pp. 357–416). Wiley-Scrivener.

Shah, T. M., Khan, A. H., Nicholls, C., Sohoo, I., & Otterpohl, R. (2023). Using landfill sites and marginal lands for socio-economically sustainable biomass production through cultivation of non-food energy crops: An analysis focused on South Asia and Europe. *Sustainability,* 15(6), 4923.

Sharvini, S., Noor, Z., Stringer, L., Afionis, S., & Chong, C. (2022). Energy generation from palm oil mill effluent: A life cycle cost-benefit analysis and policy insights. *Renewable and Sustainable Energy Reviews,* 156, 111990.

Shweta, Capareda, S. C., Kamboj, B. R., Malik, K., Singh, K., Bhisnoi, D. K., & Arya, S. (2024). Biomass resources and biofuel technologies: A focus on Indian development. *Energies,* 17(2), 382.

Sikiru, S., Abioye, K. J., Adedayo, H. B., Adebukola, S. Y., Soleimani, H., & Anar, M. (2024). Technology projection in biofuel production using agricultural waste materials as a source of energy sustainability: A comprehensive review. *Renewable and Sustainable Energy Reviews,* 200, 114535.

Singh, R., Langyan, S., Rohtagi, B., Darjee, S., Khandelwal, A., Shrivastava, M., . . . Kaur, J. (2022). Production of biofuels options by contribution of effective and suitable enzymes: Technological developments and challenges. *Materials Science for Energy Technologies,* 5, 294–310.

Snow, M. S. (2023). The invasion of Ukraine and the global food crisis of 2022: Responding to food supply shocks. In A. Farhadi, M. Grzegorzewski, A. J. Masys (Eds.), *The great power competition volume 5: The Russian invasion of Ukraine and implications for the central region* (pp. 351–387): Springer.

Souza, M. (2025). The bioeconomy revolution: How trade fuels the green transition. *Global Trade and Customs Journal,* 20(5), 321–331.

Sreekumar, R. (2023). *Beauty Leaf Tree (Calophyllum inophyllum L.)—Its biogeographic variability for salinity tolerance and kernel oil content as determined by FTIR/FTNIR technologies.* CQUniversity,

Stamenković, O. S., Gautam, K., Singla-Pareek, S. L., Dhankher, O. P., Djalović, I. G., Kostić, M. D., . . . Veljković, V. B. (2023). Biodiesel production from camelina oil: Present status and future perspectives. *Food and Energy Security,* 12(1), e340.

Surendran, U., Nagakumar, K. C. V., & Samuel, M. P. (2024). Remote sensing in precision agriculture. In P. M. Priyadarshan, S. M. Jain, S. Penna, J. M. Al-Khayri (Eds.), *Digital agriculture: A solution for sustainable food and nutritional security* (pp. 201–223): Springer.

Thakur, P., Ahmed, M., Wani, A. W., Rahim, A., Thakur, V., Kumar, S., . . . Zimare, S. B. (2025). Use of non-edible oilseeds crops for the production of biofuels. In Y. Suryawanshi, M. Shahnawaz, J. Parihar, R.A. Dar, S. Samyal, & A.B. Ade *Oilseed Crops* (pp. 439–454). https://doi.org/10.1002/9781394186426.ch17

Thomas, I. (2016). *An assessment of the feasibility of using vegetable oil fuels in light of the impending post fossil fuel dilemma.* RMIT University,

Verma, H., Kumar, D., Kumar, V., Kumari, M., Singh, S. K., Sharma, V. K., . . . Kumar, A. (2021). The potential application of endophytes in management of stress from drought and salinity in crop plants. *Microorganisms,* 9(8), 1729.

Verma, N., Talwar, P., Upadhyay, A., Singh, R., Lindenberger, C., Pareek, N., . . . Vivekanand, V. (2024). Food-energy-water nexus in compliance with sustainable development goals for integrating and managing the core environmental verticals for sustainable energy and circular economy. *Science of the Total Environment,* 930, 172649.

Wacal, C., Basalirwa, D., Okello-Anyanga, W., Murongo, M. F., Namirembe, C., & Malingumu, R. (2021). Analysis of sesame seed production and export trends; challenges and strategies towards increasing production in Uganda. *OCL, 28,* 4.

Winck, F. V., Monteiro, L. D. F., & Souza, G. M. (2022). Introduction: Advances in plant omics and systems biology. In F. V Winck (Ed.), *Advances in plant omics and systems biology approaches* (pp. 1–9): Springer.

Xing, W., Tam, V. W., Le, K. N., Hao, J. L., & Wang, J. (2022). Life cycle assessment of recycled aggregate concrete on its environmental impacts: A critical review. *Construction and Building Materials,* 317, 125950.

Yadav, R., Kalia, S., Rangan, P., Pradheep, K., Rao, G. P., Kaur, V., . . . Langyan, S. (2022). Current research trends and prospects for yield and quality improvement in sesame, an important oilseed crop. *Frontiers in Plant Science,* 13, 863521.

Yan, M., Shen, Y., Wang, S., Zhu, Z., Cui, P., & Wang, Y. (2024). Green separation of azeotropes in dimethyl carbonate synthesis by transesterification. *Renewable and Sustainable Energy Reviews,* 202, 114687.

Yang, X., Dong, X., Bezama, A., & Liu, Y. (2024). Improving rapeseed carbon footprint evaluation via the integration of remote sensing technology into an LCA approach. *Science of the Total Environment,* 946, 174262.

Yazdinejad, A., Zolfaghari, B., Azmoodeh, A., Dehghantanha, A., Karimipour, H., Fraser, E., . . . Duncan, E. (2021). A review on security of smart farming and precision agriculture: Security aspects, attacks, threats and countermeasures. *Applied Sciences,* 11(16), 7518.

Zabermawi, N. M., Alsulaimany, F. A., El-Saadony, M. T., & El-Tarabily, K. A. (2022). New eco-friendly trends to produce biofuel and bioenergy from microorganisms: An updated review. *Saudi Journal of Biological Sciences, 31*(8), 103214.

Zanetti, F., Alberghini, B., Marjanović Jeromela, A., Grahovac, N., Rajković, D., Kiprovski, B., & Monti, A. (2021). Camelina, an ancient oilseed crop actively contributing to the rural renaissance in Europe. A review. *Agronomy for Sustainable Development,* 41, 1–18.

Zubair, M. U., Ali, M., Khan, M. A., Khan, A., Hassan, M. U., & Tanoli, W. A. (2024). BIM-and GIS-based life-cycle-assessment framework for enhancing eco efficiency and sustainability in the construction sector. *Buildings,* 14(2), 360.

6 Role of Essential Oil in Biofuel Production

Shifa Shaffique, Muhammad Saleem Khan, Anis Ali Shah, Muhammad Aaqil Khan, Javier A. Trujillo-Rugamas, Sang-Mo Kang, and In-Jung Lee

6.1 INTRODUCTION

6.1.1 What Are the Essential Oils?

Essential oils (EOs) are defined as the concentrated, volatile aromatic compounds that are extracted from plant sources. It is present in several parts of the plants, depending on the species. Mostly, it is abundant in the leaves, flowers, roots, seeds, and bark of the plants (Hanif et al., 2019; Tisserand & Young, 2013). It is naturally synthesized by the plants and stored in the glandular trichomes, secretory ducts, and oil cells of plants, which gives a unique essence to plants (Biswas et al., 2009; Tiwari, 2016). It can be extracted through several processes such as solvent extraction, cold pressing, and the steam distillation method (Aziz et al., 2018; El Asbahani et al., 2015). It is hydrophobic in nature, and mainly EOs are composed of terpenes, alcohol, aldehydes, etc. As it contains several bioactive compounds, it is used as an antimicrobial, an antioxidant, and an anti-inflammatory (Kalemba & Kunicka, 2003; Lopez-Romero et al., 2015). Interestingly, they have several other properties, such as energy-dense compounds, volatile in nature, flammable, and biologically active, which make them attractive to do more research and explore their other properties, such as biofuel production (Bakkali et al., 2008; Biberci, 2023; Jugreet et al., 2020).

6.2 WHY ARE ESSENTIAL OILS USED IN BIOFUEL PRODUCTION?

Biofuel can be defined as the renewable energy that is derived from organic matter, such as plants, algae, and animal waste, which can be used as fossil fuels like coal, petrol, and diesel (Rahman et al., 2019). There are several kinds of biofuels, such as bioethanol, biodiesel, biogas, bio-oil, and algal biofuel (Demirbas, 2010; Fernandes et al., 2007; Hassan & Kalam, 2013; Rodionova et al., 2017). The purpose of the production of biofuel from EO is the conversion of agricultural waste into energy. Moreover, it is biodegradable in nature and helps to clean up the environment. Several advanced countries are using the EO in the production of biofuel, and they promote the production of EO from plant-based materials (Anestis & Vasiliki, 2025; Kurnia et al., 2016). The properties are given in Figure 6.1.

6.3 ROLE OF ESSENTIAL OIL IN BIOFUEL PRODUCTION

EOs that are rich in oxygenated hydrocarbons and terpenes are being used as direct fuels or additives in biofuels. They have properties like fossil fuels, such as density, viscosity, volatile contents, cetane numbers, flash point, and sulfur contents. In some of the properties, it has more powerful characteristics, such as higher oxygen content, greater biodegradability, and lower sulfur content (Mewalal et al., 2017; Rahman et al., 2019).

DOI: 10.1201/9781003600169-6

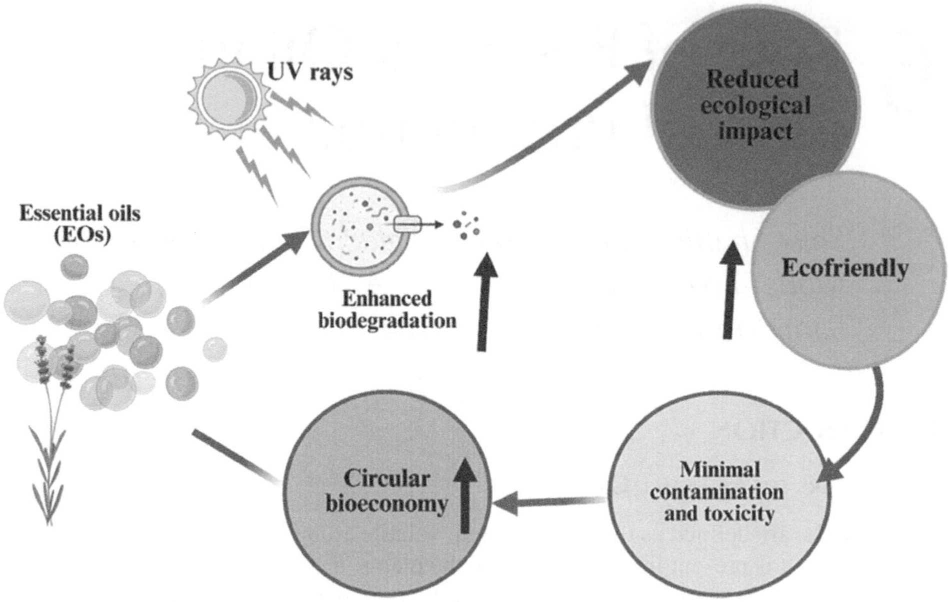

FIGURE 6.1 Essential oils (EOs) and their potential advantages to the ecosystem. The figure illustrates the potential application of essential oils (EOs) as biofuels, highlighting their eco-friendly nature, their ability to reduce environmental stress, and their role in enhancing the circular economy. The figure was created using the online software, www.biorender.com.

6.4 ROLE OF ESSENTIAL OILS IN BIOFUEL PRODUCTION

6.4.1 Global Warming and Greenhouse Gas Emissions

Global warming is a significant stress on sustainability. Due to global warming, there is reduced crop productivity and enhanced climatic change conditions. Global warming is escalating at an alarming rate. It is predicted that in 2050, the heat index will reach up to 50°C. Several factors are contributing to increasing global warming such as burning fossil fuels that release stored carbon from several million years into the atmosphere (Hosseini, 2022; Houghton, 2001; Jones & Wigley, 1990; Zecca & Chiari, 2010). In comparison to EOs that are derived from the grown biomass, the carbon dioxide CO_2 that is released from its combustion is directly absorbed by the plants during the process of photosynthesis. This emphasizes the importance of EOs in the production of biofuel. Moreover, its green gas emission is so less or net zero. Moreover, it contains the oxygenated compounds such as terpenes, ketones, and alcohols that enhance the combustion power and reduce the C emissions; for example, the usage of palm oil reduces the carbon-based emission of greenhouse gas (De Souza et al., 2010; Reijnders & Huijbregts, 2008).

6.5 SULFUR CONTENT

Conventional biofuel precursors, such as fossil fuels, contain sulfur compounds in the form of oxides, such as sulfur dioxide, which may become part of the environment and sometimes cause acidic rain (Alves et al., 2015; Kularathne et al., 2019; Smith et al., 2005). When we use it in the form of oil, it may cause the erosion of the engine (Matbouei, 2018). While EOs have minimal or no sulfur content and fewer contaminants, the environmental properties of EOs, such as having more oxygenate compounds and fewer sulfur contents, make them more efficient for combustion. Whether it's alone or blended with the conventional biofuel, 20% blended enhances combustion and efficiency of engine system (Doppalapudi et al., 2024; Rahman et al., 2018; Rahman et al., 2019).

6.6 BIODEGRADABLE AND ECO-FRIENDLY

The most important feature of EOs is that they are biodegradable and consist of natural organic compounds. Upon exposure to UV rays, soil microbes break it down quickly in a shorter duration. It has lower toxicity toward the environment, unlike other resources of biofuel production, which cause environmental contamination and enter the environment through subsequent leaching, causing environmental pollution (Adorjan & Buchbauer, 2010; Chellachamy & Rajadurai, 2019; Guddaraddi et al., 2023; Parakh et al., 2020). The production of biofuel with EOs, which is cultivated from the plants, is our sustainable approach. It is easy to handle and store EOs; moreover, it can use agricultural waste and produce EOs. Moreover, it's naturally derived from plant biomass and exhibits high levels of biodegradability, which makes it an environmentally favorable alternative to synthetic fossil fuels. They degrade rapidly in environmental conditions, including exposure to ultraviolet light, microbiome, and atmospheric oxygen, due to the presence of their molecular structures that consist mainly of terpenes, alcohols, ketones, and aldehydes (Bakkali et al., 2008; Dhifi et al., 2016). This method of natural degradation means that EO residues will not remain in the environment, so the chances are far less that EO residues will contaminate water or soil on a long-term basis (Gamayel & Priyadi, 2024; Nirmala et al., 2023; Sivalingam et al., 2023) (Figure 6.2).

Unlike fossil-based fuels, consisting of persistent hydrocarbons and toxic sulfur compounds, EOs have a lower ecological risk estimate since they are non-toxic and non-carcinogenic (Lak et al., 2024; Nazzaro et al., 2017; Powers et al., 2018). It was found that compounds of EO, such as limonene and linalool, are readily broken down and quickly adopted by microbial communities, contributing to soil fitness and reducing unacceptable bioaccumulation (Duetz et al., 2003). Further, the fact that many EOs have low aquatic toxicity adds value, as they can be used in sensitive, environmentally vulnerable regions, further contributing to their usefulness in sustainable energy systems (Khan et al., 2007; Svoboda & Hampson, 1999).

These features provide EOs with a strong incentive as an environmental safety tool in biofuels. Their biodegradability puts minimal ecological pressure and is aligned with global sustainability objectives in the context of reduced dependency on such persistent organic pollutants. As such, the use of EOs as a part of the biofuel industry will allow a reduction in the overall life cycle and make the supply of renewable energy forms more eco-safe in general (Ainane et al., 2020; Ramalingam et al., 2019).

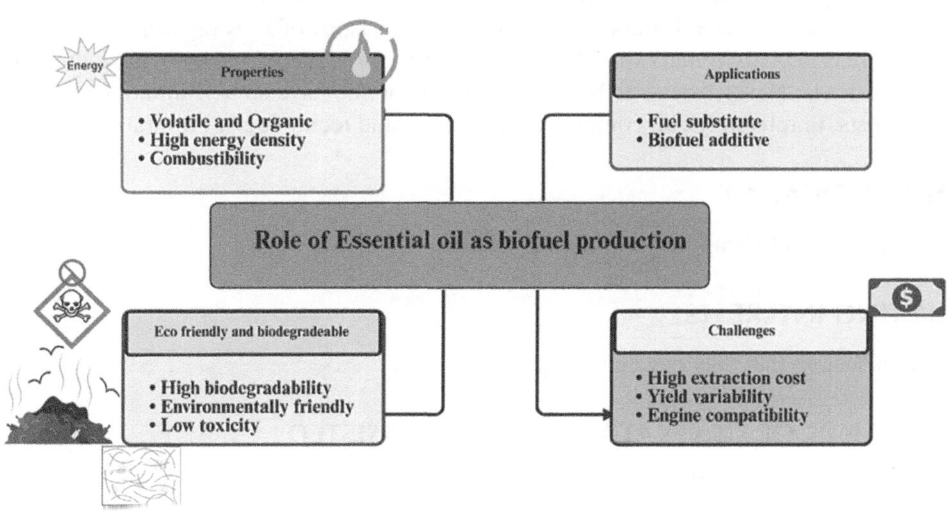

FIGURE 6.2 The role of essential oils (EOs) in biofuel production. The figure represents their properties, challenge and potential application. The figure was created using the online software, www.biorender.com.

6.6.1 Research and Development Needs

To fully realize the potential of EOs as viable biofuel components, significant research and development (R&D) efforts, hand-to-hand practices, and workshops are necessary across multiple platforms. The other main challenge is the improvement of the aromatic plants through genetic approach to enhance an increased yield of EO. This may be performed by breeding technique or genomic intervention, such as marker-assisted selection as well as genetic engineering, which are already proven fruitful in enhancing terpene production mechanism in plants (Daviet & Schalk, 2010; Hammouti et al., 2025; Kainer et al., 2015).

The other challenge is technological intervention in the extraction and refinement of EOs, which are of most importance to enhance efficiency and lowered cost of production. Conventional approaches to extraction, such as steam distillation and solvent extraction, are valuable but are energy-consuming and could lead to chemical breakdown of thermolabile compounds. New processes such as supercritical CO_2 extraction, microwave-assisted extraction, and optimization of hydrodistillation will provide more sustainable, energy-efficient, and high-yielding processes (Cardoso-Ugarte et al., 2013; Dhotre, 2025; Matinfard et al., 2025). Furthermore, to be compatible with current energy systems, there must be further research conducted regarding the nature of EO–fuel blend interactions in present-day combustion engines. Recent studies show that EO–diesel blends (maximum 20% blending) have the potential to enhance engine performance and decrease emissions, yet there is still much to be done to guarantee long-term compatibility of the blended engine, fuel blending stability, and emission standards. Lastly, research and development funds and policies are also important to continue the research. This covers the cost of growing EO crops as well as green fuel research grants and the regulatory mechanisms that recognize EO-based fuels as viable resources in national energy mixes (Demirbas, 2010; Dincer & Acar, 2015).

6.7 CONCLUSION

In conclusion, EOs represent a promising, sustainable alternative in the evolving landscape of renewable energy, particularly within biofuel production. It is composed of oxygen-containing, volatile organic substances, such as terpenes and alcohols, which are derived and produced by a variety of aromatic plants. It provides several distinctive characteristics that make it an alternative fuel, which include high energy density, flammability, and low sulfur content. They may be used in a broader scope than fuel substitution and can serve well as additives to biofuels, enabling better combustion and less harmful emissions. The characteristics of EOs regarding their ecological advantages are especially remarkable. They are naturally produced, extremely biodegradable, and, at low levels, toxic. Nevertheless, EOs, despite their benefits, have some limitations such as large extraction costs, unreliable yield, complex composition, and technological limitations.

ACKNOWLEDGMENT

We are very grateful to brain Korea BK214th phase.

COMPETING INTERESTS

The authors declare that they have no competing interest.

DECLARATION OF GENERATIVE AI AND AI-ASSISTED TECHNOLOGIES IN THE WRITING PROCESS

During the preparation of this work, the book chapter author(s) have not taken any assistance from AI, and the book chapter author(s) reviewed and edited the content as needed and take(s) full responsibility for the content of the publication.

REFERENCES

Adorjan, B., & Buchbauer, G. (2010). Biological properties of essential oils: An updated review. *Flavour and Fragrance Journal*, 25(6), 407–426.

Ainane, A., Mohamed Abdoul-Latif, F., Mohamed Abdoul-Latif, T., & Ainane, T. (2020). Evaluation of biological activities of two essential oils as a safe environmental bioinsecticides: Case of Eucalyptus globulus and Rosmarinus officinalis. *Przegląd Naukowy. Inżynieria i Kształtowanie Środowiska*, *29*(4), 544–556.

Alves, L., Paixão, S. M., Pacheco, R., Ferreira, A. F., & Silva, C. M. (2015). Biodesulphurization of fossil fuels: Energy, emissions and cost analysis. *RSC Advances*, 5(43), 34047–34057.

Anestis, P., & Vasiliki, K. (2025). Utilization of aromatic plants residual biomass after distillation mixed with wood in solid biofuels production. *Renewable Energy*, *248*, 123198.

Aziz, Z. A., Ahmad, A., Setapar, S. H. M., Karakucuk, A., Azim, M. M., Lokhat, D., Rafatullah, M., Ganash, M., Kamal, M. A., & Ashraf, G. M. (2018). Essential oils: Extraction techniques, pharmaceutical and therapeutic potential – A review. *Current Drug Metabolism*, 19(13), 1100–1110.

Bakkali, F., Averbeck, S., Averbeck, D., & Idaomar, M. (2008). Biological effects of essential oils – A review. *Food and Chemical Toxicology*, 46(2), 446–475.

Biberci, M. A. (2023). Bibliometric analysis of the use of biodiesel production from essential oils as biofuels. *Processes*, 11(4), 974.

Biswas, K. K., Foster, A. J., Aung, T., & Mahmoud, S. S. (2009). Essential oil production: Relationship with abundance of glandular trichomes in aerial surface of plants. *Acta Physiologiae Plantarum*, 31, 13–19.

Cardoso-Ugarte, G. A., Juárez-Becerra, G. P., SosaMorales, M. E., & López-Malo, A. (2013). Microwave-assisted extraction of essential oils from herbs. *Journal of Microwave Power and Electromagnetic Energy*, 47(1), 63–72.

Chellachamy, A., & Rajadurai, S. J. (2019). Evaluation of essential oils as antioxidant for biodiesel. *Journal of Biobased Materials and Bioenergy*, 13(3), 380–388.

Daviet, L., & Schalk, M. (2010). Biotechnology in plant essential oil production: Progress and perspective in metabolic engineering of the terpene pathway. *Flavour and Fragrance Journal*, 25(3), 123–127.

De Souza, S. P., Pacca, S., De Ávila, M. T., & Borges, J. L. B. (2010). Greenhouse gas emissions and energy balance of palm oil biofuel. *Renewable Energy*, 35(11), 2552–2561.

Demirbas, A. (2010). Use of algae as biofuel sources. *Energy Conversion and Management*, 51(12), 2738–2749.

Dhifi, W., Bellili, S., Jazi, S., Bahloul, N., & Mnif, W. (2016). Essential oils' chemical characterization and investigation of some biological activities: A critical review. *Medicines*, 3(4), 25.

Dhotre, I. (2025). A comprehensive review on progression and innovations in microwave assisted extraction technology for essential oils. *Journal of Chemical Technology & Biotechnology*, 100(5), 894–907.

Dincer, I., & Acar, C. (2015). Review and evaluation of hydrogen production methods for better sustainability. *International Journal of Hydrogen Energy*, *40*(34), 11094–11111.

Doppalapudi, A. T., Azad, A. K., & Khan, M. M. K. (2024). Exergy, energy, performance, and combustion analysis for biodiesel NOx reduction using new blends with alcohol, nanoparticle, and essential oil. *Journal of Cleaner Production*, 467, 142968.

Duetz, W., Bouwmeester, H., Van Beilen, J., & Witholt, B. (2003). Biotransformation of limonene by bacteria, fungi, yeasts, and plants. *Applied Microbiology and Biotechnology*, 61, 269–277.

El Asbahani, A., Miladi, K., Badri, W., Sala, M., Addi, E. A., Casabianca, H., El Mousadik, A., Hartmann, D., Jilale, A., & Renaud, F. (2015). Essential oils: From extraction to encapsulation. *International Journal of Pharmaceutics*, *483*(1–2), 220–243.

Fernandes, S. D., Trautmann, N. M., Streets, D. G., Roden, C. A., & Bond, T. C. (2007). Global biofuel use, 1850–2000. *Global Biogeochemical Cycles*, *21*(2), GB2019.1–GB2019.15.

Gamayel, A., & Priyadi, M. U. Z. (2024). Fuel properties and diesel engine exhaust gas temperature of biodiesel-*Calophyllum inophyllum*-essential oil mixture. *Jurnal Dinamika Vokasional Teknik Mesin*, *9*(1), 44–50.

Guddaraddi, A., Singh, A., Amrutha, G., Saikanth, D., Kurmi, R., Singh, G., Chowdhury, M., & Singh, B. V. (2023). Sustainable biofuel production from agricultural residues an eco-friendly approach: A review. *International Journal of Environment and Climate Change*, 13(10), 2905–2914.

Hammouti, Y., Marhri, A., & Addi, M. (2025). Integrated molecular strategies for enhanced production of terpenes, terpenoids, and essential oils from medicinal plants. In J.-T. Chen (Ed.), *Biotechnology, multiple omics, and precision breeding in medicinal plants* (pp. 333–340). CRC Press.

Hanif, M. A., Nisar, S., Khan, G. S., Mushtaq, Z., & Zubair, M. (2019). Essential oils. In S. Malik (Ed.), *Essential oil research: Trends in biosynthesis, analytics, industrial applications and biotechnological production* (pp. 3–17). Springer International Publishing.

Hassan, M. H., & Kalam, M. A. (2013). An overview of biofuel as a renewable energy source: Development and challenges. *Procedia Engineering*, 56, 39–53.

Hosseini, S. E. (2022). Fossil fuel crisis and global warming. In S. E. Hosseini (Ed.), *Fundamentals of low emission flameless combustion and its applications* (pp. 1–11). Elsevier.

Houghton, J. (2001). The science of global warming. *Interdisciplinary Science Reviews*, 26(4), 247–257.

Jones, P. D., & Wigley, T. M. (1990). Global warming trends. *Scientific American*, 263(2), 84–91.

Jugreet, B. S., Suroowan, S., Rengasamy, R. K., & Mahomoodally, M. F. (2020). Chemistry, bioactivities, mode of action and industrial applications of essential oils. *Trends in Food Science & Technology*, 101, 89–105.

Kainer, D., Lanfear, R., Foley, W. J., & Külheim, C. (2015). Genomic approaches to selection in outcrossing perennials: Focus on essential oil crops. *Theoretical and Applied Genetics*, 128, 2351–2365.

Kalemba, D., & Kunicka, A. (2003). Antibacterial and antifungal properties of essential oils. *Current Medicinal Chemistry*, 10(10), 813–829.

Khan, N., Warith, M. A., & Luk, G. (2007). A comparison of acute toxicity of biodiesel, biodiesel blends, and diesel on aquatic organisms. *Journal of the Air & Waste Management Association*, 57(3), 286–296.

Kularathne, I., Gunathilake, C., Rathneweera, A., Kalpage, C., & Rajapakse, S. (2019). The effect of use of biofuels on environmental pollution—A review. *International Journal of Renewable Energy Research*, 9(3), 1355–1367.

Kurnia, J. C., Jangam, S. V., Akhtar, S., Sasmito, A. P., & Mujumdar, A. S. (2016). Advances in biofuel production from oil palm and palm oil processing wastes: A review. *Biofuel Research Journal*, 3(1), 332–346.

Lak, S. Z., Rezaei, J., & Rahimpour, M. R. (2024). Health and pollution challenges of fossil fuels utilization. *Encyclopedia of Renewable Energy, Sustainability and the Environment*, 8, 155.

Lopez-Romero, J. C., González-Ríos, H., Borges, A., & Simões, M. (2015). Antibacterial effects and mode of action of selected essential oils components against *Escherichia coli* and *Staphylococcus aureus*. *Evidence-Based Complementary and Alternative Medicine*, *2015*(1), 795435.

Matbouei, M. (2018). *The effect of biofuel on the corrosion and wear of automotive engine components* (PhD thesis). University of Leicester.

Matinfard, S., Tavakolipour, H., Sodeifian, G., Sajadian, S. A., & Kalvandi, R. (2025). Supercritical CO_2 green solvent extraction of Nepeta crispa: Evaluation of process optimization, chemical analysis, and biological activity. *The Journal of Supercritical Fluids*, 216, 106451.

Mewalal, R., Rai, D. K., Kainer, D., Chen, F., Külheim, C., Peter, G. F., & Tuskan, G. A. (2017). Plant-derived terpenes: A feedstock for specialty biofuels. *Trends in Biotechnology*, 35(3), 227–240.

Nazzaro, F., Fratianni, F., Coppola, R., & De Feo, V. (2017). Essential oils and antifungal activity. *Pharmaceuticals*, 10(4), 86.

Nirmala, M. J., Sekar, P. C., Johnson, A., Kizhuveetil, U., Shora, S., & Nagarajan, R. (2023). A comprehensive review of nanoadditives in plant-based biodiesels with a special emphasis on essential oils. *Fuel*, 351, 128934.

Parakh, P. D., Nanda, S., & Kozinski, J. A. (2020). Eco-friendly transformation of waste biomass to biofuels. *Current Biochemical Engineering*, 6(2), 120–134.

Powers, C. N., Osier, J. L., McFeeters, R. L., Brazell, C. B., Olsen, E. L., Moriarity, D. M., Satyal, P., & Setzer, W. N. (2018). Antifungal and cytotoxic activities of sixty commercially-available essential oils. *Molecules*, 23(7), 1549.

Rahman, S. A., Nabi, M. N., Van, T. C., Suara, K., Jafari, M., Dowell, A., Islam, M. A., Marchese, A. J., Tryner, J., & Hossain, M. F. (2018). Performance and combustion characteristics analysis of multi-cylinder CI engine using essential oil blends. *Energies*, 11(4), 738.

Rahman, S. A., Van, T. C., Hossain, F., Jafari, M., Dowell, A., Islam, M., Nabi, M. N., Marchese, A., Tryner, J., & Rainey, T. (2019). Fuel properties and emission characteristics of essential oil blends in a compression ignition engine. *Fuel*, 238, 440–453.

Ramalingam, K., Kandasamy, A., & Joshua Stephen Chellakumar, P. J. T. (2019). Production of eco-friendly fuel with the help of steam distillation from new plant source and the investigation of its influence of fuel injection strategy in diesel engine. *Environmental Science and Pollution Research*, 26(15), 15467–15480. https://doi.org/10.1007/s11356-019-04773-3

Reijnders, L., & Huijbregts, M. A. (2008). Palm oil and the emission of carbon-based greenhouse gases. *Journal of Cleaner Production*, 16(4), 477–482.

Rodionova, M. V., Poudyal, R. S., Tiwari, I., Voloshin, R. A., Zharmukhamedov, S. K., Nam, H. G., Zayadan, B. K., Bruce, B. D., Hou, H. J., & Allakhverdiev, S. I. (2017). Biofuel production: Challenges and opportunities. *International Journal of Hydrogen Energy*, 42(12), 8450–8461.

Sivalingam, A., Perumal Venkatesan, E., Roberts, K. L., & Asif, M. (2023). Potential effect of lemon peel oil with novel eco-friendly and biodegradable emulsion in un-modified diesel engine. *ACS Omega*, 8(21), 18566–18581.

Smith, S. J., Pitcher, H., & Wigley, T. (2005). Future sulfur dioxide emissions. *Climatic Change*, 73(3), 267–318.

Svoboda, K. P., & Hampson, J. B. (1999). Bioactivity of essential oils of selected temperate aromatic plants: Antibacterial, antioxidant, antiinflammatory and other related pharmacological activities. *Plant Biology Department, SAC Auchincruive,* Ayr, Scotland, UK., KA6 5HW, 16, 1–7.

Tisserand, R., & Young, R. (2013). *Essential oil safety: A guide for health care professionals.* Elsevier Health Sciences.

Tiwari, P. (2016). Recent advances and challenges in trichome research and essential oil biosynthesis in *Mentha arvensis* L. *Industrial Crops and Products*, 82, 141–148.

Zecca, A., & Chiari, L. (2010). Fossil-fuel constraints on global warming. *Energy Policy*, 38(1), 1–3.

7 Role of Lignin in Biofuel Production

Current Prospects and Challenges

Tesfaye Kassaw Bedru, Workisa Bacha Garuma, Abayneh Getachew Demesa, Selvakumar Periyasamy, and Mani Jayakumar

7.1 INTRODUCTION

Lignin, the world's second most abundant biopolymer, following cellulose, has been underemphasized in favor of more easily accessed biomass components. Historically treated as a low-value by-product and primarily used for energy recovery in pulping industries, lignin is now gaining attention due to its rich aromatic framework, which offers substantial opportunities for the sustainable production of fuels and high-value chemicals (Lian & Zhou, 2024). Such a shift has been prompted by growing energy demands, climate change urgency, and worldwide transition toward circular bioeconomy models that necessitate utilization of the whole biomass, including lignin (Zhou et al., 2025).

Chemically, lignin is an amorphous, highly branched polymer of three dominant monolignol-oxidative coupling products: *p*-coumaryl, coniferyl, and sinapyl alcohols. These yield hydroxyphenyl (H), guaiacyl (G), and syringyl (S) units, the proportions of which differ according to plant species, with softwoods being G-dominant, hardwoods more S-G-dominant, and grasses containing all three in a balanced proportion (Vural, 2024). The structural diversity of lignin underlies its high variability and recalcitrance. Its complex network of β-O-4, β-5, and C–C linkages not only reinforces plant cell walls and enhances pathogen resistance but also hinders enzymatic and catalytic degradation (Obrzut et al., 2023) Despite its recalcitrance, lignin's high energy density of around 26–30 MJ/kg and availability in agricultural and forest residues position it as a strategic feedstock for the production of biofuels. More recent studies show that lignin can be processed into liquid fuels (e.g., syngas, bio-oil, and drop-in hydrocarbons) using thermochemical processes like pyrolysis, hydrothermal liquefaction (HTL), and gasification. In addition, new biological and catalytic routes enable selective transformation of lignin-derived aromatics to high-value chemicals such as vanillin, phenols, and muconic acid (Madigal et al., 2025).

New advances in computational modeling, deep eutectic solvents (DESs), and fungal pretreatment technologies increasingly enhance the effectiveness and selectivity of lignin valorization (Lian & Zhou, 2024; Madigal et al., 2025). These approaches enhance lignin depolymerization by disrupting the native plant cell wall architecture and selectively cleaving key chemical linkages. In parallel, genetic engineering of biomass crops to alter lignin content and composition is being increasingly employed to improve biomass saccharification and biofuel yields (Ntsomboh-Ntsefong et al., 2025). Table 7.1 shows an overview of lignin types, composition, and structural characteristics. Additionally, Figure 7.1 also shows the structure of lignin.

 DOI: 10.1201/9781003600169-7

TABLE 7.1
Overview of Lignin Types, Composition, and Structural Characteristics

Lignin Type	Typical Biomass Sources	Main Monolignol	Structural Characteristics	Key Reference
Guaiacyl	Softwoods	Coniferyl alcohol	High G unit content	Zhang et al. (2022)
Syringyl–Guaiacyl	Hardwoods	Sinapyl and coniferyl alcohols	Higher S/G ratio	Balk et al. (2023)
p-Hydroxyphenyl	Grasses and monocots	*p*-Coumaryl, coniferyl, and sinapyl alcohols	Mixed S/G/H units	Zhang et al. (2022)
Native (in situ) Lignin	Untreated plant cell walls	S, G, and H alcohols	Intact β-O-4, β–β, 5-5 linkages	Karmanov et al. (2024)
Technical Lignin	Industrial pulping residues	Variable; degraded monolignol fragments	Chemically modified; condensed structures	Bogolitsyn et al. (2021)

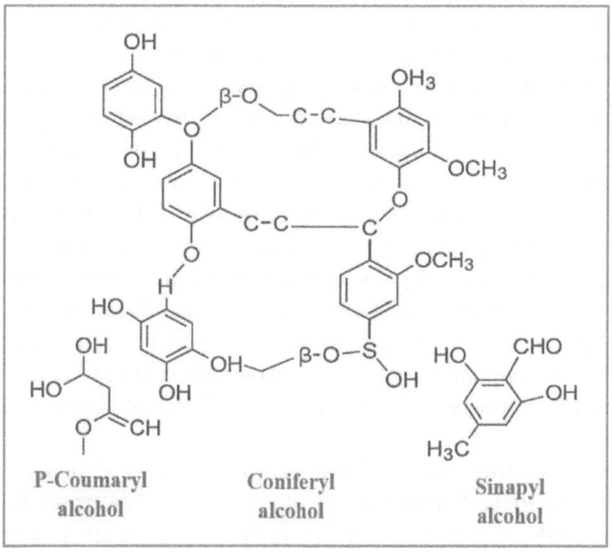

FIGURE 7.1 Primary mono-lignols of lignin.

7.2 PROPERTIES OF LIGNIN

7.2.1 Biological Properties of Lignin

Lignin is not only the structural material for the cell walls of the plant but also a biologically active macromolecule with multifunctional roles in plant physiology, microbial ecology, and medicine. Lignin provides the rigidity, hydrophobicity, and defens of cell walls in plant cells. It is primarily deposited in secondary cell walls in which it is mixed with hemicellulose and cellulose to form a matrix, adding mechanical strength and allowing efficient water transport in vascular tissues (Jaffur et al., 2021). Its cross-linked and hydrophobic phenolic chemistry makes it a solid physical barrier to pathogens' invasion and environmental stresses and hence plays a substantial role in plant immunity. It is complemented by the antioxidant properties of lignin, which enable it to entrap reactive oxygen species (ROS) and prevent oxidative damage to plant tissues (Marturano et al., 2023).

Ecologically, lignin plays an important role in soil organic matter formation and carbon sequestration. Lignin is one of the recalcitrant organic polymers; it slows down the decomposition of biomass and participates in long-term soil organic matter formation. Lignin degradation is regulated primarily by certain fungi and bacteria that are capable of producing lignin-lytic enzymes such as lignin peroxidase (LiP), manganese peroxidase (MnP), and laccases. Such enzymes oxidatively degrade lignin's complex aromatic polymer into more delicate phenolic intermediates that can feed into microbial metabolic processes (Saini et al., 2022). White-rot fungi *Phanerochaete chrysosporium* and *Trametes versicolor* are particularly effective lignin degraders and are actively studied today for biotechnological applications, including lignin valorization in biorefineries and bioremediation (Jaffur et al., 2021; Lesko et al., 2025).

Recent studies also concentrated on lignin's interactions with microbial communities, not only as a carbon source but also as a driver of microbial diversity in lignin-rich environments. Its decomposition products, such as ferulic acid, vanillic acid, and syringaldehyde, can influence microbial signaling pathways and serve as bioactive molecules with antimicrobial or quorum-sensing inhibitory activities (Martínková et al., 2023). Such activities are being exploited for soil conditioning, plant growth promotion, and microbiome engineering uses.

Apart from its ecological and microbiological roles, lignin also possesses remarkable bioactive characteristics that have been of interest in biomedical sciences. Nanoparticles and hydrogels derived from lignin proved to be outstanding in biocompatibility, antioxidant, and low cytotoxicity, which makes them a good option for wound healing, tissue engineering, and drug delivery systems (Lesko et al., 2025). For instance, lignin scaffolds have also found applications in regenerative dentistry since they are structurally similar to the constituents of extracellular matrix (ECM) and are capable of supporting cellular adhesion and differentiation. The phenolic groups of lignin also render it endowed with intrinsic antimicrobial activity against Gram-positive bacteria, making it suitable for use in antimicrobial packaging materials and coating (Marturano et al., 2023).

In agriculture, biologically active lignin derivatives are increasingly being viewed as natural alternatives to synthetic agrochemicals. Studies have indicated the possibility of lignosulfonates and alkali lignin derivatives in modulating plant growth regulators, enhancing nutrient uptake, and stimulating root development in abiotic stress conditions (Martínková et al., 2023). Therefore, this opens new possibilities for lignin-based biostimulants and soil conditioners in sustainable agriculture systems.

Finally, the biological significance of lignin is much deeper than its conventional role as a structural polymer. The ecological stability, enzyme resistance, antimicrobial capability, and biocompatibility of lignin make it a biologically multifunctional material with increasingly diverse applications in agriculture, medicine, biotechnology, and environmental engineering. Further examination of molecular interactions of lignin with living organisms will be essential to realize the integration of lignin into biorefinery, biomedical, and ecological innovations.

7.2.2 Chemical Properties of Lignin

Lignin is a highly complex, irregular, and heterogeneous biopolymer synthesized through oxidative radical coupling of three primary phenyl-propanoid alcohols, *p*-coumaryl, coniferyl, and sinapyl alcohol. Such precursors form hydroxyphenyl (H), guaiacyl (G), and syringyl (S) units, respectively, which are connected by an array of ether (β-O-4) and carbon–carbon (β-5, β–β, 5-5) bridges. The β-O-4 is the most common and chemically labile one and hence an ideal target for lignin depolymerization strategies (Shorey et al., 2024). However, the prevalence of recalcitrant C–C linkages, especially in softwood lignin, renders it resistant to chemical and enzymatic degradation. Such intrinsic complexity introduces processing challenges as well as the possibility of selective valorization.

Chemically, lignin is described by its aromatic-rich backbone and presence of diverse functional groups, including methoxyl, carboxyl, aliphatic and phenolic hydroxyl, and carbonyl groups.

These moieties have multiple reactive sites for chemical amendment and catalytic conversion. The presence of phenolic and aliphatic hydroxyl groups can facilitate esterification, etherification, oxidation, and polymer grafting. The above sensitivity is the foundation for intensified interest in lignin as a renewable biomass feedstock for functional polymers, adhesives, coatings, and carbon material (Hasan et al., 2023). Recent work suggests that variation of the methoxylation level of lignin monomers influences directly the polarity and solubility of the polymer, which can be used for designing lignin for specific industrial applications (Borrero-López et al., 2022).

One of the most attractive chemical properties of lignin is its high carbon content (60–65%) and heating value (typically 25–27 MJ/kg), which makes it an energy-dense feedstock susceptible to thermochemical conversion to biofuels such as bio-oil, syngas, and biochar. Pyrolysis and gasification processes fully exploit the aromatics and thermal stability of lignin. Catalytic depolymerization allows lignin to be selectively transformed into platform chemicals like vanillin, syringaldehyde, phenol, and benzene derivatives, especially under hydrogenolysis or oxidative cleavage mediated by metal catalysts or DESs (Zhou et al., 2025). The solvent systems have shown the ability to disrupt hydrogen bonding and selectively solubilize lignin for facilitating high-value product extraction from lignocellulosic biomass.

The chemical versatility of lignin also makes it a prime candidate to be integrated into polymeric materials. Functionalized lignins were used to synthesize polyurethane foam, epoxy resin, and thermoplastics with improved thermal stability and reduced fossil carbon usage (Shorey et al., 2024). In addition, lignin is a precursor material for carbon fiber synthesis due to its high aromatic concentration and graphitization potential upon thermal treatment. Chemical modification of lignin, such as demethylation, hydroxylation, or sulfonation, yet widens its range of applications by a favorable interaction with synthetic polymers or enhanced hydrophilicity for aqueous dispersion.

In short, chemical properties of lignin, such as functional diversity, aromaticity, and thermal reactivity, are the core of its increasing contribution to the bioeconomy. Emerging technologies in catalytic systems, solvent design, and analytical chemistry are triggering the efficient application of lignin as a leading building block for renewable chemicals, energy, and green materials.

7.2.3 Physical Properties of Lignin

Lignin possesses a distinctive pattern of physical properties that arise from its heterogeneous, highly branched, and irregular polymeric character. Its physical properties, including solubility and thermal stability, are governed by monomer composition (S/G/H ratio), degree of polymerization, molecular weight distribution, and the nature and abundance of interunit linkages (C–C and C–O bonds). As opposed to cellulose or hemicellulose, lignin is amorphous and non-crystalline, generating characteristic viscoelastic and thermal characteristics differing significantly based on plant origin as well as isolation procedure (Alam et al., 2024).

One of the significant physical properties of lignin is its heat stability, which makes it highly prospective for pyrolysis, combustion, and carbon material synthesis. Lignin tends to have generally a degradation onset of 200–300°C and maintains high carbon content (~60–65%), and therefore, it is an energy-dense polymer for thermochemical conversions. The glass transition temperature (Tg) of lignin is extremely prone to variation (90–180°C), which varies with the origin and purity of lignin, with lower Tg for organosolv lignins and higher Tg for Kraft lignins due to differences in crosslinking and availability of hydroxyls (Jasiukaitytė-Grojzdek et al., 2025). Thermal properties control the processibility of lignin in composite and thermoplastic applications.

Solubility is another crucial property that is governed by lignin's polarity, functional groups, and molecular weight. Native lignins are insoluble in water but are soluble in alkaline media and polar organic solvents like DMSO or acetone. Modified lignins, particularly those that are recovered by DESs or organosolv processes, are soluble due to reduced molecular weight and altered polarity (Lu et al., 2025). These differences in solubility are relevant to lignin application in functional inks, hydrogels, and adhesives. Novel studies demonstrate that fractionation by solvent allows for the

TABLE 7.2
Summary of Biological, Chemical, and Physical Properties of Lignin

Property Type	Key Characteristics	Functional Applications	Key References
Biological	Structural component in plant cell walls; involved in plant defense; recalcitrant to microbial degradation; influences microbial ecology	Antioxidant and antimicrobial; slows decomposition; substrate for ligninolytic enzymes (e.g., LiP, MnP, laccase); used in biomedical and agricultural applications	Karmanov et al. (2024), I. Ullah et al. (2022), Zheng et al. (2021)
Chemical	Aromatic-rich polymer with H, G, S units; contains hydroxyl, methoxyl, and carbonyl groups; dominated by β-O-4 and C–C linkages	Feedstock for bio-oil, vanillin, phenol, and bio-based resins; platform for polymer and nanocomposite synthesis	Karagoz et al. (2024), Karmanov et al. (2024), Singh et al. (2021)
Physical	Amorphous structure; thermally stable; glass transition varies (Tg: 90°C–180°C); solubility depends on polarity and molecular weight	Used in carbon materials, adhesives, nanofibers, and coatings; tunable for dispersion and composite blending	Bouhfid et al. (2025), Hachimi Alaoui et al. (2023), Mastouri et al. (2025)

separation of lignin fractions with narrow molecular weight distribution and more consistent physical performance in designed systems (Chen et al., 2023).

Morphology and dispersion in composite matrices determine lignin mechanical properties. Pure lignin is brittle but can enhance tensile strength, thermal stability, and UV resistance if it is blended with biopolymers like starch, PLA, or epoxy resins (Li et al., 2025). It was also revealed in studies that the nanostructure of lignin can be processed to develop lignin nanoparticles, which contribute to mechanical reinforcement and barrier properties in films and membranes. Morphologically, lignin is found to exist in a range of particle sizes and morphologies from disordered pieces in native biomass to spherical nanoparticles under controlled process conditions. Such structures are important to lignin functions in coatings, drug delivery, and emulsification. Surface area and porosity are also factors that influence its activity in catalysis and adsorbent systems.

The physical properties of lignin are amorphousity, limited crystallinity, a broad thermal window, and flexible solubility, making it a uniquely versatile material in bio-based platforms and energy products. With recent advances in solvent systems, thermal analysis, and nanostructuring, the physical performance of lignin can also be further developed to meet the circular bioeconomy platform. Table 7.2 shows the summary of biological, chemical, and physical properties of lignin.

7.3 SYNTHESIS OF LIGNIN

7.3.1 Biological Synthesis of Lignin from Lignocellulosic Biomass for Biofuel Production

Lignin, being an extremely aromatic biopolymer, is biosynthesized in plants through the phenylpropanoid pathway and is deposited in the secondary cell wall, where it adds rigidity, impermeability, and resistance to microbial invasion. This condensation is initiated in the cytoplasm, where L-phenylalanine is converted into cinnamic acid by phenylalanine ammonia-lyase (PAL), and a sequence of enzymatic reactions catalyzed by cinnamate 4-hydroxylase (C4H), 4-coumarate CoA ligase (4CL), caffeic acid O-methyl-transferase (COMT), and cinnamyl alcohol dehydrogenase (CAD).

These steps yield three principal monolignols, p-coumaryl, coniferyl, and sinapyl alcohols, which are transported to the cell wall and polymerized by the action of laccases and peroxidases to the lignin macromolecule, adding hydroxyphenyl (H), guaiacyl (G), and syringyl (S) units (Sulis et al., 2025). The ratio of these monomers decides lignin's structure and influences its reactivity in downstream biofuel conversion.

Although lignin biosynthesis performs critical physiological roles, it also accounts for biomass recalcitrance, a barrier in enzymatic hydrolysis of cellulose for bioethanol production. To circumvent this, recent biological strategies have focused on genetic engineering of lignin biosynthetic genes in plants to reduce the overall lignin content, alter monomer composition (e.g., increase S/G ratio), and introduce labile linkages. CRISPR/Cas9 and transcription factor control (MYB and NAC families, for example) have enabled specific manipulation of monolignol biosynthesis and deposition, thus- enhancing biomass digestibility (Adeleke et al., 2025; Sharma et al., 2025).

Concurrently, biological methods have been evolved to valorize and treat lignin post-harvesting, especially through microbial degradation and bioconversion. Fungi, such as *P. chrysosporium* and *T. versicolor*, and bacteria, such as *Pseudomonas putida* and *Rhodococcus jostii*, have been used because they are capable of secreting ligninolytic enzymes, i.e., LiPs, MnPs, and versatile peroxidases (VPs), which preferentially break the recalcitrant C–C and C–O bonds in lignin (Fu et al., 2025; Saxena et al., 2025).

These microorganisms are able to degrade lignin into aromatic monomers, which are routed through metabolic pathways toward biosynthesis of valuable products such as bioethanol, bioplastics, or organic acids. So far, more sophisticated synthetic biology platforms enable the development of engineered microbial strains for complete lignin valorization. For instance, *Pseudomonas putida* strains engineered metabolically can accumulate lignin-derived intermediates such as vanillin or p-hydroxybenzoic acid and convert them to polyhydroxyalkanoates (PHAs) or muconic acid, which are precursors to bioplastics and biofuels (Liang et al., 2025). In other cases, microbial consortia are designed to capitalize on synergistic breakdown, in which one strain breaks down the lignin and another microbe catabolizes its byproducts, increasing yield and process efficiency.

Furthermore, biological pretreatment processes such as white-rot fungal delignification, enzymatic hydrolysis, and anaerobic microbial digestion were effectively integrated in biorefineries for reducing chemical use and achieving maximal sustainability (Ali et al., 2024; Shakeel et al., 2025). These types of processes are becoming increasingly popular due to minimal environmental footprints, specificity, and amenability toward green bioprocessing.

7.3.2 Chemical Synthesis of Lignin from Lignocellulosic Biomass for Biofuel Production

Chemical depolymerization and extraction of lignin from lignocellulosic biomass are important steps in biorefineries for sustainable biofuel production (Asfaw et al., 2024; Jayakumar et al., 2022, 2023). Lignin exists as an extremely entangled blend with cellulose and hemicellulose within the plant cell wall, and it must be separated by breaking up these advanced matrices through various chemical pretreatment schemes. Among the most widely utilized processes are alkaline pretreatment (e.g., sodium hydroxide or ammonia), organosolv extraction, and more recently, favored DESs and ionic liquids (ILs). All these processes target ether and ester bond cleavage between lignin and hemicelluloses and cellulose, facilitating effective lignin solubilization and separation (Alam et al., 2024; Bedru et al., 2025; Melesse et al., 2022).

Alkaline pretreatment, particularly by NaOH or lime, cleaves ester and ether bonds and releases lignin pieces with high phenolic content. It is inexpensive and routine practice but produces lignin with broad molecular weight distributions and unknown solubility. Organosolv processes use organic solvents such as ethanol, methanol, or acetic acid (generally acidified) at elevated temperatures (140°C–200°C) to depolymerize the lignin and dissolve it in higher purity with lesser condensation. Organosolv lignin tends to be low in sulfur, high in β-O-4 linkages and thus more appropriate for catalytic upgrading to bio-oil intermediates (Mujtaba et al., 2023). Other than this, DESs and ILs also gained popularity because they possess the ability to selectively dissolve lignin at mild conditions, preserve native linkages, and save energy. For instance, choline chloride- and lactic acid-based DESs can solubilize 70% of lignin without losing its viability for subsequent use in biofuel or biopolymer production (Wang et al., 2025).

Following isolation, the lignin is chemically depolymerized to yield its value-added aromatic products such as vanillin, phenol, syringol, and other bio-oil precursors. These are achieved by methods such as oxidative cleavage (with O_2, H_2O_2, or metal oxides), hydrogenolysis (with metal catalysts such as Ru, Ni, or Pd), and acid/base hydrolysis. These processes are often conducted under high pressure and temperature to cleave the strongest C–C and C–O bonds. Upgrades in catalytic systems, such as bi-functional catalysts with combined acidic and redox capabilities, have hugely improved monomer yields and selectivity, making viability higher in incorporating lignin valorization into existing value chains in biofuel production (Alazaiza et al., 2025).

Particularly, lignin-first biorefinery approaches selectively extract and valorize lignin before cellulose processing is a new paradigm shift. The approach not only enhances the quality of lignin but also overall biomass utilization efficiency. Combining DES-based pretreatment with catalytic hydrogenolysis has been found to yield good results with high-quality lignin-derived bio-oils being achieved with reduced energy input (Ntunka et al., 2025). These technologies are essential for the fulfilment of circular economy goals by turning lignin, which is otherwise considered a low-value byproduct.

7.3.3 Thermochemical Synthesis of Lignin from Lignocellulosic Biomass for Biofuel Production

Thermochemical processing, particularly by pyrolysis, HTL, and gasification, is a prominent method for the thermocatalytic conversion of lignin-rich lignocellulosic biomass to energy-dense biofuels and platform chemicals. Compared to biological or chemical routes, thermochemical routes are conducted at elevated temperatures and often at reduced oxygen levels, which facilitates the breakdown of lignin's highly recalcitrant, cross-linked aromatic structure. Among them, fast pyrolysis is generally applied to lignin valorization, which typically operates between 400°C and 600°C to break down lignin into bio-oil, syngas, and biochar. Lignin-derived bio-oils are a mixture of complex phenolic components such as guaiacol, vanillin, syringol, and catechol, which can be applied as biofuel precursors, resins, and fine chemicals (Alazaiza et al., 2025; Ntunka et al., 2025).

HTL is another promising technique with subcritical water (200–350°C, 10–20 MPa) to liquefy lignin-containing wet biomass to bio crude oil without a costly drying step. HTL favors aqueous C–C and C–O bond cleavage with fewer char residues but more fuel-range compounds. It is especially well-suited for the treatment of wet or aqueous lignin streams, as commonly recovered from black liquor of pulping mills (Durak, 2025; Gundupalli et al., 2022). Gasification, usually at >700°C with the presence of limited oxygen or steam, converts lignin into syngas (a H_2–CO mixture), which in turn can be catalytically converted into liquid biofuels using the Fischer-Tropsch process.

Thermochemical lignin valorization has been greatly enhanced by advanced catalytic systems. Current reports highlight the utilization of metal-supported catalysts (e.g., Ni, Ru, Mo, and Fe) and bi-functional acidic catalysts to enhance the selectivity and yield of target compounds in lignin pyrolysis and catalytic depolymerization. For instance, catalysts like Ni/Al_2O_3 or Co/MCM-41 may be employed to enhance hydro-deoxygenation, removing oxygenates from pyrolysis vapor in an effort to increase hydrocarbon content in the final biofuel (Tuler et al., 2025). Moreover, improved microwave-assisted pyrolysis technology and reactor design, such as fluidized-bed reactors and co-feeding strategies, have increased energy efficiency as well as scalability.

Most notably, thermochemical processes are being integrated more and more into lignin-first biorefinery systems in which the lignin is targeted for extraction and valorization prior to cellulose conversion. This increases both the energy yield per unit of biomass as well as the economic viability of biofuel production pipelines. Despite these advances, challenges remain to be overcome, such as the heterogeneity of lignin feedstocks, deactivation of catalysts by formation of char, and upgrading crude bio-oils to transportation fuel level. Advances in recent times in reactive distillation, fractional condensation, and models of integrated biorefining are addressing these gaps and taking thermochemical lignin conversion commercial viable (Mujtaba et al., 2023; Vanisree et al., 2024). Table 7.3 shows a comparative overview of lignin synthesis routes for biofuel production.

TABLE 7.3
Comparative Overview of Lignin Synthesis Routes for Biofuel Production

Synthesis Route	Main Processes	Key Products	Advantages	Disadvantages	Key References
Biological synthesis	Microbial degradation; enzymatic depolymerization	Bioethanol, PHAs, muconic acid, monolignols, organic acids	Environmentally friendly; selective lignin breakdown; low-energy processes	Slow degradation; enzyme cost; strain optimization required	Ingle et al. (2025), Sulis et al. (2025)
Chemical synthesis	Alkaline/organosolv pretreatment; deep eutectic solvents (DES); oxidative cleavage; hydrogenolysis	Vanillin, phenol, syringol, bio-oils, aromatic monomers	High lignin purity; efficient monomer recovery; flexible for various lignin types	Harsh conditions; expensive solvents; complex purification;	Ali et al. (2024), Mujtaba et al. (2023)
Thermochemical synthesis	Fast pyrolysis, hydrothermal liquefaction (HTL), gasification, microwave-assisted depolymerization	Bio-oil, syngas, biochar, phenolic compounds	High energy efficiency; works with wet feedstocks; broad product spectrum	Catalyst deactivation; product upgrading needed; high process temperatures	Lago et al. (2021), Sethupathy et al. (2022)

7.4 ROLE OF LIGNIN IN BIOFUEL PRODUCTION

Lignin is a recalcitrant, structured biopolymer that has traditionally been considered an industrial byproduct and is now becoming a valuable resource in the biofuel industry. It constitutes 15%–30% of lignocellulosic biomass and contains high carbon content combined with aromatic molecular structure, resulting in high energetic and chemical potential. Previously, it was largely combusted to recover process heat or disposed of in biorefineries and contributed little to the economic value of biofuel operations. Technological advancements and environmental concerns more recently led to its reconsideration as a value-added feedstock in second-generation and integrative biorefineries (Lu & Gu, 2022).

More and more publications identify lignin as key to enhancing the overall efficiency and sustainability of biofuel systems. According to Sethupathy et al. (2022), lignin is now being converted directly into high-value biofuels such as liquid hydrocarbons, syngas, and biohydrogen (Sethupathy et al., 2022). Such conversion pathways not only enhance the carbon efficiency of the biorefinery process but also assist in reducing the carbon intensity of energy products. Furthermore, Bajwa et al. (2019) point out that life cycle analyses consistently show that the inclusion of lignin valorization processes results in a net reduction of greenhouse gas (GHG) emissions and, as such, lignin is a strategic contributor to climate change mitigation goals (Bajwa et al., 2019).

New valorization technologies, such as the Ligno-Boost and Ligno-Force process, and organosolv pretreatment are now enabling one to obtain lignin in high purity for catalytic upgrading to fuels. Kumar et al. (2025) note that these technologies, when implemented through conventional pulp and paper mills, can enable closed-loop operations where the lignin-containing wastes are removed from waste streams and sent back to energy generation. This circularity not only improves process economics but also conforms to general principles of resource efficiency and waste reduction (Kumar et al., 2024).

Economically, the valorization of lignin enhances rural development and industrial competitiveness. Giwa et al. (2023) demonstrate that the integration of lignin-derived bio-oil or biogas into the overall stream of biofuel production raises the yields of products and lowers the minimum selling price (MSP) of biofuels, thereby enhancing commercial viability. Black liquor from craft mills, which is one of the greatest sources of lignin, is now considered more of a renewable energy source than an industrial liability. This notion is substantiated by policy incentives that reward low-carbon technologies and promote the substitution of fossil resources with bio-based alternatives (Giwa et al., 2023).

Environmentally, lignin utilization addresses multiple sustainability objectives. Vuppaladadiyam et al. (2023 argue that incorporating lignin into energy production systems prevents the environmentally detrimental disposal practices like landfilling and open burning. In addition, lignin valorization mitigates the food-versus-fuel conflict by exploiting a non-edible biomass, thereby supporting more socially equitable and sustainable energy development (Vuppaladadiyam et al., 2023).

Lignin has evolved from a bottleneck to biomass conversion to a keystone to next-generation biofuel platforms. Valorization of lignin not only improves the environmental footprint and economic attractiveness of biorefineries but also enables system goals of carbon neutrality and integration into the circular economy. As the global energy systems transition to renewals, the triple utility of lignin as a fuel feedstock, carbon sink, and industrial feedstock reconfirms its inimitable value in the transition toward sustainable energy.

7.5 CHALLENGES

Despite the growing promise of lignin as a raw material for the production of green biofuel, its proper utilization is restricted by numerous technical, biological, and economic hurdles. Possibly the most long-standing obstacle is its inherent heterogeneity and complex composition of lignin, which shifts quite significantly between plant families and extraction methods. Such heterogeneity results in unpredictable reactivity and low depolymerization selectivity, excluding the possibility of regular production of biofuel intermediates (Ali et al., 2024). In addition, the stiff C–C and ether crosslinks of lignin render it chemically recalcitrant, only to be degraded under severe thermal or catalytic conditions, much of which is an energetically expensive and environmentally unfriendly process (Ullah et al., 2022).

Even with more effective depolymerization processes, such as catalytic pyrolysis or reductive fractionation, char yield, poor yield of monomers, and catalyst deactivation limit commercial applicability. Biological valorization is also impeded; enzymatic degradation and microbial funnelling show great promise, but most microbial strains are not inherently able to effectively metabolize aromatics derived from lignin, and engineered systems are hampered by toxicity and regulatory issues (Singhania et al., 2022). While some progress has been made in lignin-lytic enzyme systems, total biological conversion product yields remain modest, and bioconversion of lignin remains behind that of cellulose fermentation in terms of efficiency and industrial maturity (Hafez et al., 2024).

From an economic perspective, lignin valorization is still less competitive compared to carbohydrate biofuels because of partially missing market drivers and incomplete lignin product supply chains. Technical lignin from the majority of pulp and paper mills still serves as a fuel for low-grade process heat with very little incentive to divert it for fuel synthesis routes (Ali et al., 2024). This problem is added to the limited regulatory integration of lignin into renewable energy systems. Sustainability evaluation has revealed that without high-efficiency valorization, lignin processing can reduce net carbon savings within a biorefinery application (Sulis et al., 2025).

Furthermore, process integration challenges persist in existing biorefinery infrastructures. The majority of platforms are carbohydrate-optimized and lack equipment and flexibility for lignin valorization. High development expenditures for catalysts, complex solvent recovery systems, and downstream separations also represent operational challenges. New hybrid strategies and modular biorefinery architectures are promising to counter these challenges, but widespread adoption is limited by capital investment risk and low technology readiness levels (J. Li et al., 2024).

7.6 FUTURE PROSPECTS FOR THE ROLE OF LIGNIN IN BIOFUEL PRODUCTION

The prospects for lignin valorization toward biofuel production improve with a synergy of technological innovation, circular economy strategies, and global necessity to transition into net-zero carbon energy systems. Emerging trends in catalytic, microbial, and hybrid conversion platforms are overcoming traditional roadblocks posed by the heterogeneity of lignin, opening doors toward its application toward the production of bio-oil, renewable hydrocarbons, and even green aviation fuels (Wan et al., 2025). Progress in recent years' lignin depolymerization reactions such as oxidative cleavage, hydrogenolysis, and electrochemical conversion provides higher selectivity and better product yield, with the potential to make lignin an economically viable co-feedstock in integrated biorefineries. Furthermore, engineered microbial platforms and new biocatalysts also possess incredible potential for aromatics derived from lignin to high-value chemicals and fuels (Rocha Balbino et al., 2023).

From a strategic perspective, lignin plays a role in the expansion of the scope of biomass valorization beyond carbohydrate paradigms. With its carbon and energy density, it can serve as a precursor for biofuels and as a renewable alternative to fossil-based aromatics, thus representing the cornerstone for the development of low-emission, fossil-free fuel infrastructures. Lignin valorization is an essential driver to propel climate action objectives and hard-to-abate industry decarbonization, such as shipping and aviation. Policy agendas and country-level bioeconomy action plans are beginning to recognize the worth of lignin-rich waste streams to meet renewable energy targets as well as reduce life cycle emissions (Ning et al., 2021).

Besides, innovations in combining biorefinery and modularity allow lignin to be valorized in situ with cellulose and hemicellulose in real time, maximizing resource use and value. Simulation modeling and techno-economic analysis show that the optimization of lignin streams can decrease the minimum selling price of biofuels by over 15%, in addition to producing co-products such as biochar, biohydrogen, and biodegradable polymers (de Morais & de Carvalho, 2023). In addition, machine learning and artificial intelligence programs are being used more and more to optimize catalyst selection, reactor configurations, and reaction conditions for lignin conversion at a considerably higher rate for innovation cycles (Hu et al., 2024).

The future also includes policy-driven deployment of lignin valorization platforms in traditional industries such as pulp and paper, where black liquor is diverted from combustion into higher-value conversion streams. This not only enhances revenue diversification but also helps emission reduction in traditional biomass-intensive industries. As global sustainability standards change and carbon pricing mechanisms tighten, the economic and environmental attractiveness of lignin valorization improves. Given current technological and policy trends, lignin has the potential to emerge as a foundational resource for building a resilient, decentralized, and circular biofuel economy that simultaneously delivers climate, economic, and material security.

7.7 CONCLUSION

Lignin, once rated as a low-grade lignocellulosic biomass by-product, is now recognized as an economically viable resource toward the pursuit of economically and sustainably produced biofuels. With its aromatic character, energy content, and abundance, it is a strategic feedstock for forward-thinking biorefineries looking to maximize biomass conversion while minimizing fossil fuel dependency. This chapter has discussed the biological, chemical, and thermochemical pathways supporting lignin conversion to value-added biofuels and co-products and its increasing role in making bioenergy systems more profitable and sustainable. Despite current challenges, structural heterogeneity, limited conversion rates, and limited industrial integration, long-term developments in catalysis, microbial engineering, and circular biorefinery design are rapidly expanding lignin's horizon. With increased global effort toward carbon neutrality and resource recyclability, lignin is poised to become a redeveloped non-traditional waste stream into an economy pillar for a bio-based economy supporting cleaner energy futures and green industrial transformation.

COMPETING INTERESTS

The authors declare that they have no competing interest.

DECLARATION OF GENERATIVE AI AND AI-ASSISTED TECHNOLOGIES IN THE WRITING PROCESS

During the preparation of this work, the book chapter author(s) have not taken any assistance from AI, and the book chapter author(s) reviewed and edited the content as needed and take(s) full responsibility for the content of the publication.

REFERENCES

Adeleke, A. O., Babatunde, K. A., Kasavan, S., Omar, R. C., & Dele-Afolabi, T. T. (2025). Shell-based agricultural wastes biomass valorization for biofuel production: A bibliometric analysis into the circular bioeconomy. *Environmental Science and Pollution Research*, 32(3), 1095–1119. https://doi.org/10.1007/s11356-024-35766-6

Alam, M. M., Greco, A., Rajabimashhadi, Z., & Esposito Corcione, C. (2024). Efficient and environmentally friendly techniques for extracting lignin from lignocellulose biomass and subsequent uses: A review. *Cleaner Materials*, 13, 100253. https://doi.org/10.1016/j.clema.2024.100253

Alazaiza, M. Y. D., Alzghoul, T. M., Ramu, M. B., & Nassani, D. E. (2025). Catalysis in biofuel production and biomass valorization: Trends, challenges, and innovations through a bibliometric analysis. *Catalysts*, *15*(3). https://doi.org/10.3390/catal15030227

Ali, S., Rani, A., Dar, M. A., Qaisrani, M. M., Noman, M., Yoganathan, K., Asad, M., Berhanu, A., Barwant, M., & Zhu, D. (2024). Recent advances in characterization and valorization of lignin and its value-added products: Challenges and future perspectives. *Biomass*, 4(3), 947–977. https://doi.org/10.3390/biomass4030053

Asfaw, B. T., Gari, M. T., Jayakumar, M., & Baskar, G. (2024). Sustainable bioethanol production from the pretreated waste lignocellulosic feedstocks. In B. Gurunathan, S. L. Rokhum, A. Veeramuthu, & V. S. Moholkar (Eds.), *Circular bioeconomy perspectives in sustainable bioenergy production* (pp. 377–394). Springer Nature Singapore. https://doi.org/10.1007/978-981-97-2523-6_16

Bajwa, D. S., Pourhashem, G., Ullah, A. H., & Bajwa, S. G. (2019). A concise review of current lignin production, applications, products and their environmental impact. *Industrial Crops and Products*, 139, 111526. https://doi.org/10.1016/j.indcrop.2019.111526

Balk, M., Sofia, P., Neffe, A. T., & Tirelli, N. (2023). Lignin, the lignification process, and advanced, lignin-based materials. *International Journal of Molecular Sciences*, 24(14). https://doi.org/10.3390/ijms241411668

Bedru, T. K., Meshesha, B. T., Mohammed, S. A., Demesa, A. G., & Jayakumar, M. (2025). Efficient biomass fractionation via organosolv for sustainable bioenergy production: A comprehensive review. *International Journal of Chemical Engineering*, 2025(1), 3120449. https://doi.org/10.1155/ijce/3120449

Bhardwaj, P., & Kumar, P. (2025). Microbial enzymes acting on lignocellulosic biomass: An overview. In R. K. Kapoor & K. Rajan (Ed.), *Lignocellulosic biomass and enzymes: Fundamentals, emerging technologies and applications* (pp. 177–209). Springer Nature Singapore. https://doi.org/10.1007/978-981-96-3037-0_8

Bogolitsyn, K. G., Gusakova, M. A., & Krasikova, A. A. (2021). Molecular self-organization of wood lignin–carbohydrate matrix. *Planta*, 254(2), 30. https://doi.org/10.1007/s00425-021-03675-4

Borrero-López, A. M., Valencia, C., & Franco, J. M. (2022). Lignocellulosic materials for the production of biofuels, biochemicals and biomaterials and applications of lignocellulose-based polyurethanes: A review. *Polymers*, *14*(5). https://doi.org/10.3390/polym14050881

Bouhfid, R., Qaiss, A. E. K., & Raji, M. (2025). Morphological properties of lignin. In M. Jawaid, A. Ahmed, & A. Meraj (Eds.), *Handbook of lignin* (pp. 1–35). Springer Nature Singapore. https://doi.org/10.1007/978-981-97-2664-6_17-1

Chen, Y.-L., Lo, C.-C., Liu, Y.-L., & Sun, Y.-M. (2023). Lignin extraction and fractionation from rice straw biorefinery residues. *Separation and Purification Technology*, 326, 124778. https://doi.org/10.1016/j.seppur.2023.124778

de Morais, L. C., & de Carvalho, G. B. (2023). Potential applications of lignin. In S. Thomas, M. Hosur, D. Pasquini, & C. J. Chirayil (Eds.), *Handbook of biomass* (pp. 1–23). Springer Nature Singapore. https://doi.org/10.1007/978-981-19-6772-6_24-1

Durak, H. (2025). Comparative analysis of electrochemical and thermochemical hydrogenation of biomass-derived phenolics for sustainable biofuel and chemical production. *Processes*, *13*(5). https://doi.org/10.3390/pr13051581

Fu, N., Liu, R.-Y., Zhou, Y., Li, B.-Z., Yuan, Y.-J., & Liu, Z.-H. (2025). Technological advances in ligninolytic enzymes for the biological valorization of lignin. *Green Chemistry*, 27(16), 4016–4039. https://doi.org/10.1039/D4GC05724D

Giwa, T., Akbari, M., & Kumar, A. (2023). Techno-economic assessment of an integrated biorefinery producing bio-oil, ethanol, and hydrogen. *Fuel*, 332, 126022. https://doi.org/10.1016/j.fuel.2022.126022

Gundupalli, M. P., Gundupalli, S. P., Somavarapu Thomas, A. S., Jayakumar, M., Bhattacharyya, D., & Gurunathan, B. (2022). Chapter 4 – Hydrothermal liquefaction of lignocellulosic biomass for production of biooil and by-products: current state of the art and challenges. In B. Gurunathan, R. Sahadevan, & Z. A. Zakaria (Eds.), *Biofuels and bioenergy* (pp. 61–84). Elsevier. https://doi.org/10.1016/B978-0-323-85269-2.00001-0

Hachimi Alaoui, C., Réthoré, G., Weiss, P., & Fatimi, A. (2023). Sustainable biomass lignin-based hydrogels: A review on properties, formulation, and biomedical applications. *International Journal of Molecular Sciences*, 24(17). https://doi.org/10.3390/ijms241713493

Hafez, I., Yang, H.-S., Liew, F. J., Schilling, J. S., & Tze, W. T. Y. (2024). Fully biobased thermosetting adhesive from enzymatic saccharification residue. *ACS Sustainable Chemistry & Engineering*, 12(34), 12726–12735. https://doi.org/10.1021/acssuschemeng.3c07759

Hasan, Md. S., Sardar, Md. R., Shafin, A., Rahman, Md. S., Mahmud, Md., & Hossen, Md. M. (2023). A brief review on applications of lignin. *Journal of Chemical Reviews*, 5(1), 56–82. https://doi.org/10.22034/jcr.2023.359861.1186

Hu, M., Lv, X., Wang, Y., Ma, L., Zhang, Y., & Dai, H. (2024). Recent advance on lignin-containing nanocelluloses: The key role of lignin. *Carbohydrate Polymers*, 343, 122460. https://doi.org/10.1016/j.carbpol.2024.122460

Ingle, A. P., Saxena, S., Moharil, M. P., Rivaldi, J. D., Ramos, L., & Chandel, A. K. (2025). Production of biomaterials and biochemicals from lignocellulosic biomass through sustainable approaches: Current scenario and future perspectives. *Biotechnology for Sustainable Materials*, 2(1), 3. https://doi.org/10.1186/s44316-025-00025-2

Jaffur, N., Jeetah, P., & Kumar, G. (2021). A review on enzymes and pathways for manufacturing polyhydroxybutyrate from lignocellulosic materials. *3 Biotech*, *11*(11), 483. https://doi.org/10.1007/s13205-021-03009-x

Jasiukaitytė-Grojzdek, E., Ročnik Kozmelj, T., Tofani, G., Segers, B., Nimmegeers, P., Billen, P., Pogorevc, R., Likozar, B., & Grilc, M. (2025). Design of organosolv lignin fractionation: Influence of temperature, antisolvent, and source on molecular weight, structure, and functionality of lignin fragments. *ACS Sustainable Chemistry & Engineering*, 13(9), 3452–3466. https://doi.org/10.1021/acssuschemeng.4c08125

Jayakumar, M., Gindaba, G. T., Gebeyehu, K. B., Periyasamy, S., Jabesa, A., Baskar, G., John, B. I., & Pugazhendhi, A. (2023). Bioethanol production from agricultural residues as lignocellulosic biomass feedstock's waste valorization approach: A comprehensive review. *Science of the Total Environment*, 879, 163158. https://doi.org/10.1016/j.scitotenv.2023.163158

Jayakumar, M., Kuppusamy Vaithilingam, S., Karmegam, N., Gebeyehu, K. B., Boobalan, M. S., & Gurunathan, B. (2022). Chapter 6 – Fermentation technology for ethanol production: Current trends and challenges. In B. Gurunathan & R. Sahadevan (Eds.), *Biofuels and bioenergy* (pp. 105–131). Elsevier. https://doi.org/10.1016/B978-0-323-90040-9.00015-1

Karagoz, P., Khiawjan, S., Marques, M. P. C., Santzouk, S., Bugg, T. D. H., & Lye, G. J. (2024). Pharmaceutical applications of lignin-derived chemicals and lignin-based materials: Linking lignin source and processing with clinical indication. *Biomass Conversion and Biorefinery*, 14(21), 26553–26574. https://doi.org/10.1007/s13399-023-03745-5

Karmanov, A. P., Ermakova, A. V, Raskosha, O. V, Bashlykova, L. A., Rachkova, N. G., & Kocheva, L. S. (2024). Structure and biomedical properties of lignins (a review). *Russian Journal of Bioorganic Chemistry*, 50(7), 2657–2674. https://doi.org/10.1134/S106816202407001X

Kumar, V., Malyan, S. K., Apollon, W., & Verma, P. (2024). Valorization of pulp and paper industry waste streams into bioenergy and value-added products: An integrated biorefinery approach. *Renewable Energy*, 228, 120566. https://doi.org/10.1016/j.renene.2024.120566

Lago, A., Hernando, H., Moreno, J. M., Serrano, D. P., & Fermoso, J. (2021). Valorisation of a lignin-rich residue via catalytic pyrolysis over ZrO_2/ZSM-5 technical catalyst. *Fuel Processing Technology*, 215, 106746. https://doi.org/10.1016/j.fuproc.2021.106746

Lesko, L., Jungova, P., Culenova, M., Thurzo, A., & Danisovic, L. (2025). Polymer-based scaffolds as an implantable material in regenerative dentistry: A review. *Journal of Functional Biomaterials*, 16(3). https://doi.org/10.3390/jfb16030080

Li, J., Guan, D., Xia, S., Fan, Y., Zhao, K., Zhao, Z., & Zheng, A. (2024). Recent advances, challenges, and opportunities in lignin valorization for value-added chemicals, biofuels, and polymeric materials. *Energy Conversion and Management*, 322, 119123. https://doi.org/10.1016/j.enconman.2024.119123

Li, S., Wu, S., Wang, Y., Jiang, J., Zhang, Y., & Wang, L. (2025). Effect of lignin extracted by different methods on the properties and structure of lignin/starch fiber membranes. *Industrial Crops and Products*, 224, 120400. https://doi.org/10.1016/j.indcrop.2024.120400

Lian, Z., & Zhou, X. (2024). Lignocellulosic biomass decomposition and bioconversion. *Fermentation*, *10*(9). https://doi.org/10.3390/fermentation10090480

Liang, Z., Sethupathy, S., Wenqian, D., Jinhao, H., & Zhu, D. (2025). Lignin valorization through microbial production of polyhydroxyalkanoates: Recent trends, challenges and opportunities. *Green Chemistry*, 27(21), 5920–5946. https://doi.org/10.1039/D5GC00370A

Lu, F., Wang, Y., Sun, D., Fu, Y., Zhang, F., & Sun, R. (2025). Research progress on the preparation of lignin micro/nano-particles and their biomedical applications. *Green Chemistry*, 27(19), 5400–5422. https://doi.org/10.1039/D5GC00178A

Lu, X., & Gu, X. (2022). A review on lignin pyrolysis: Pyrolytic behavior, mechanism, and relevant upgrading for improving process efficiency. *Biotechnology for Biofuels and Bioproducts*, 15(1), 106. https://doi.org/10.1186/s13068-022-02203-0

Madigal, J. P. T., Terasaki, M., Takada, M., & Kajita, S. (2025). Synergetic effect of fungal pretreatment and lignin modification on delignification and saccharification: A case study of a natural lignin mutant in mulberry. *Biotechnology for Biofuels and Bioproducts*, 18(1), 13. https://doi.org/10.1186/s13068-025-02611-y

Martínková, L., Grulich, M., Pátek, M., Křístková, B., & Winkler, M. (2023). Bio-based valorization of lignin-derived phenolic compounds: A review. *Biomolecules*, *13*(5). https://doi.org/10.3390/biom13050717

Marturano, V., Marotta, A., Salazar, S. A., Ambrogi, V., & Cerruti, P. (2023). Recent advances in bio-based functional additives for polymers. *Progress in Materials Science*, 139, 101186. https://doi.org/10.1016/j.pmatsci.2023.101186

Mastouri, A., Efhamisisi, D., Lexa, M., Oladi, R., Gholinejad-Pirbazari, A., Torabi, H., Zeidler, A., & Frigione, M. (2025). Biological-degradation, lignin performance and physical-chemical characteristics of historical wood in an ancient tomb. *Fungal Biology*, 129(4), 101588. https://doi.org/10.1016/j.funbio.2025.101588

Melesse, E. Y., Bedru, T. K., & Meshesha, B. T. (2022). Production and characterization of pulp from banana pseudo stem for paper making via soda anthraquinone pulping process. *International Journal of Engineering Research in Africa*, 58, 63–76. https://doi.org/10.4028/www.scientific.net/JERA.58.63

Mujtaba, M., Fernandes Fraceto, L., Fazeli, M., Mukherjee, S., Savassa, S. M., Araujo de Medeiros, G., do Espírito Santo Pereira, A., Mancini, S. D., Lipponen, J., & Vilaplana, F. (2023). Lignocellulosic biomass from agricultural waste to the circular economy: A review with focus on biofuels, biocomposites and bioplastics. *Journal of Cleaner Production*, 402, 136815. https://doi.org/10.1016/j.jclepro.2023.136815

Ning, P., Yang, G., Hu, L., Sun, J., Shi, L., Zhou, Y., Wang, Z., & Yang, J. (2021). Recent advances in the valorization of plant biomass. *Biotechnology for Biofuels*, 14(1), 102. https://doi.org/10.1186/s13068-021-01949-3

Ntsomboh-Ntsefong, G., Rebezov, M., Khairullin, M., & Shariati, M. A. (2025). Cell wall digestibility, diversity, and complexity for developing second-generation bioenergy crops. In R. K. Singhal, A. El Sabagh, & K. K. Dwivedi (Eds.), *Forage crops in the bioenergy revolution: From fields to fuel* (pp. 239–257). Springer Nature Singapore. https://doi.org/10.1007/978-981-96-2536-9_13

Ntunka, M. G., Khumalo, S. M., Makhathini, T. P., Mtsweni, S., Tshibangu, M. M., & Bwapwa, J. K. (2025). Valorization of lignocellulosic biomass to biofuel: A systematic review. *ChemEngineering*, *9*(3). https://doi.org/10.3390/chemengineering9030058

Obrzut, N., Hickmott, R., Shure, L., & Gray, K. A. (2023). The effects of lignin source and extraction on the composition and properties of biorefined depolymerization products. *RSC Sustainability*, *1*(9), 2328–2340. https://doi.org/10.1039/D3SU00262D

Rocha Balbino, T., Sánchez-Muñoz, S., Díaz-Ruíz, E., Moura Rocha, T., Mier-Alba, E., Custódio Inácio, S., Jose Castro-Alonso, M., de Carvalho Santos-Ebinuma, V., Fernando Brandão Pereira, J., César Santos, J., & Silvério da Silva, S. (2023). Lignocellulosic biorefineries as a platform for the production of high-value yeast derived pigments – A review. *Bioresource Technology*, 386, 129549. https://doi.org/10.1016/j.biortech.2023.129549

Saini, J. K., Himanshu, Hemansi, Kaur, A., & Mathur, A. (2022). Strategies to enhance enzymatic hydrolysis of lignocellulosic biomass for biorefinery applications: A review. *Bioresource Technology*, 360, 127517. https://doi.org/10.1016/j.biortech.2022.127517

Saxena, A., Parveen, F., Hussain, A., Khubaib, M., & Ashfaque, M. (2025). Exploring the multifaceted landscape of lignocellulosic biomass-derived nanocellulose and nanolignin: Synthesis and applications. *Polymer Bulletin*. https://doi.org/10.1007/s00289-025-05855-8

Sethupathy, S., Murillo Morales, G., Gao, L., Wang, H., Yang, B., Jiang, J., Sun, J., & Zhu, D. (2022). Lignin valorization: Status, challenges and opportunities. *Bioresource Technology*, 347, 126696. https://doi.org/10.1016/j.biortech.2022.126696

Shakeel, H., Aftab, K., Jannat, F. T., Amin, F., Umbreen, H., & Noreen, R. (2025). Advancing lignocellulosic biomass pretreatment with nanotechnology: A comprehensive bibliometric analysis. *Cellulose*, 32(4), 2167–2193. https://doi.org/10.1007/s10570-025-06403-3

Shorey, R., Salaghi, A., Fatehi, P., & Mekonnen, T. H. (2024). Valorization of lignin for advanced material applications: A review. *RSC Sustainability*, 2(4), 804–831. https://doi.org/10.1039/D3SU00401E

Singh, A. K., Bilal, M., Iqbal, H. M. N., Meyer, A. S., & Raj, A. (2021). Bioremediation of lignin derivatives and phenolics in wastewater with lignin modifying enzymes: Status, opportunities and challenges. *Science of the Total Environment*, 777, 145988. https://doi.org/10.1016/j.scitotenv.2021.145988

Singhania, R. R., Patel, A. K., Raj, T., Chen, C.-W., Ponnusamy, V. K., Tahir, N., Kim, S.-H., & Dong, C.-D. (2022). Lignin valorisation via enzymes: A sustainable approach. *Fuel*, 311, 122608. https://doi.org/10.1016/j.fuel.2021.122608

Sulis, D. B., Lavoine, N., Sederoff, H., Jiang, X., Marques, B. M., Lan, K., Cofre-Vega, C., Barrangou, R., & Wang, J. P. (2025). Advances in lignocellulosic feedstocks for bioenergy and bioproducts. *Nature Communications*, 16(1), 1244. https://doi.org/10.1038/s41467-025-56472-y

Tuler, G. V., Guimarães, T., Marçal, M. T., de Carvalho Bittencourt, R., Demuner, I. F., da Costa, M. M., & Carvalho, A. M. M. L. (2025). Thermochemical conversion of lignocellulosic biomass for sustainable bioproducts. *Biomass Conversion and Biorefinery*. https://doi.org/10.1007/s13399-025-06683-6

Ullah, I., Chen, Z., Xie, Y., Khan, S. S., Singh, S., Yu, C., & Cheng, G. (2022). Recent advances in biological activities of lignin and emerging biomedical applications: A short review. *International Journal of Biological Macromolecules*, 208, 819–832. https://doi.org/10.1016/j.ijbiomac.2022.03.182

Ullah, M., Liu, P., Xie, S., & Sun, S. (2022). Recent advancements and challenges in lignin valorization: Green routes towards sustainable bioproducts. *Molecules*, *27*(18). https://doi.org/10.3390/molecules27186055

Vanisree, G. S., Chandran, A. M., & Aparna, K. (2024). Investigation on thermochemical characteristics and pyrolysis kinetics of lignocellulosic biomass for biofuel production feasibility. *Biomass Conversion and Biorefinery*. https://doi.org/10.1007/s13399-024-05657-4

Vuppaladadiyam, A. K., Vuppaladadiyam, S. S. V., Sahoo, A., Murugavelh, S., Anthony, E., Bhaskar, T., Zheng, Y., Zhao, M., Duan, H., Zhao, Y., Antunes, E., Sarmah, A. K., & Leu, S.-Y. (2023). Bio-oil and biochar from the pyrolytic conversion of biomass: A current and future perspective on the trade-off between economic, environmental, and technical indicators. *Science of the Total Environment*, 857, 159155. https://doi.org/10.1016/j.scitotenv.2022.159155

Vural, D. (2024). Computational study on the impact of linkage sequence on the structure and dynamics of lignin. *European Biophysics Journal*, 53(7), 405–414. https://doi.org/10.1007/s00249-024-01720-0

Wan, Q., Lu, Y., Cheng, L., Shi, J., Xia, L., & Xu, J. (2025). Optimizing energy use in the integrated utilization of woody biomass: Process development and simulation. *Industrial Crops and Products*, 225, 120426. https://doi.org/10.1016/j.indcrop.2024.120426

Wang, H., Chen, J., Pei, Z., Huang, J., Wang, J., Yang, S., & Li, H. (2025). Enhancing lignocellulosic biorefinery sustainability: mechanisms and optimization of microwave-responsive deep eutectic solvents for rapid delignification. *Biofuel Research Journal*, 12(1), 2306–2318. https://doi.org/10.18331/BRJ2025.12.1.3

Zhang, L., Larsson, A., Moldin, A., & Edlund, U. (2022). Comparison of lignin distribution, structure, and morphology in wheat straw and wood. *Industrial Crops and Products*, 187, 115432. https://doi.org/10.1016/j.indcrop.2022.115432

Zheng, L., Lu, G., Pei, W., Yan, W., Li, Y., Zhang, L., Huang, C., & Jiang, Q. (2021). Understanding the relationship between the structural properties of lignin and their biological activities. *International Journal of Biological Macromolecules*, 190, 291–300. https://doi.org/10.1016/j.ijbiomac.2021.08.168

Zhou, L., Meng, X., Li, W., Yu, J., Kemefa, C. O., Dai, S. Y., Ragauskas, A. J., & Yuan, J. S. (2025). Computational modeling-guided design of deep eutectic solvents for tailoring lignin chemistry during lignocellulose pretreatment. *Green Chemistry*, 27(21), 6260–6271. https://doi.org/10.1039/D4GC06120A

8 Role of Lipids in Biofuel Production

Sewunet Abebayehu, Zinnabu Tassew Redda, Desta Getachew Gizaw, Leta Lemma Sharbe, and Selvakumar Periyasamy

8.1 INTRODUCTION

8.1.1 Overview of Lipids: Structure, Classification, and Properties

The term "lipid" refers to a diverse array of substances characterized by distinct physical and chemical properties, distinguished by their solubility in organic non-polar solvents (Asokapandian et al., 2021). Besides carbon, hydrogen, and oxygen atoms, lipids may additionally consist of phosphorus, nitrogen, sulfur, and other elements. The primary lipids found in foods and nutritional processes are fats and oils, with cholesterol and other fatty acids serving as the most emblematic molecules because of their significant nutritional and metabolic roles (Gurr, 2018). In several biological models (cellular, animal, and human), lipids have been extensively studied for their structural, metabolic, and nutritional significance. Lipids have been important to species evolution since they are necessary for tissue growth, development, and maintenance. One obvious example of the importance of fatty acids is the large amount of these molecules found in nerve tissue (Walchuk & Suh, 2020).

Lipids are a broad class of chemical compounds that are insoluble in water but soluble in non-polar organic solvents. They are found naturally in the majority of plants, animals, and microbes. They serve as hormones, insulation, energy storage molecules, and components of cell membranes. The following figure illustrates the broad classifications and structure of lipids (Asokapandian et al., 2021). Lipids, a class of organic compounds with a lot of energy, can be either liquids or non-crystalline solids at room temperature. Generally speaking, pure fats and oils have no color, taste, or odor. They dissolve well in organic solvents like acetone, alcohol, chloroform, and benzene, but are insoluble in water. Lipids don't contain ionic charges like a lot of other macromolecules do. In terms of structure, liquid triglycerols, or oils, are abundant in unsaturated fatty acids, whereas solid triglycerols, also known as fats, contain a large percentage of saturated fatty acids (SAFA) (Kopchak et al., 2021). Figure 8.1 demonstrates an overview of lipid classification.

Lipids are subject to a variety of chemical reactions. Hydrolysis, in which triglycerols combine with water to produce glycerol and fatty acids, is one of the main processes. Esters have this particular reactivity. Saponification, a more specialized type of hydrolysis, is the dissolution of triglycerides (TG) with an alkali present, including potassium (Asokapandian et al., 2021). Glycerol and soap, the sodium or potassium salts of fatty acids, are the products of this process.

Hydrogenation is another significant reaction that occurs when hydrogen combines with unsaturated fatty acids to change C–C double bonds to single bond, hence saturating the fatty acids (Umayangani et al., 2025). In a related addition reaction known as halogenation, halogens like chlorine or bromine react with the double bond in unsaturated fatty acids, frequently causing halogen solution, a popular test for unsaturation, to become decolorized. Finally, lipids can become rancid, which is the term for the emergence of a bad taste or odor. Oxidative rancidity is more prevalent in lipids that contain unsaturated fatty acids; however, hydrolytic rancidity can also happen (Sharma et al., 2022).

DOI: 10.1201/9781003600169-8

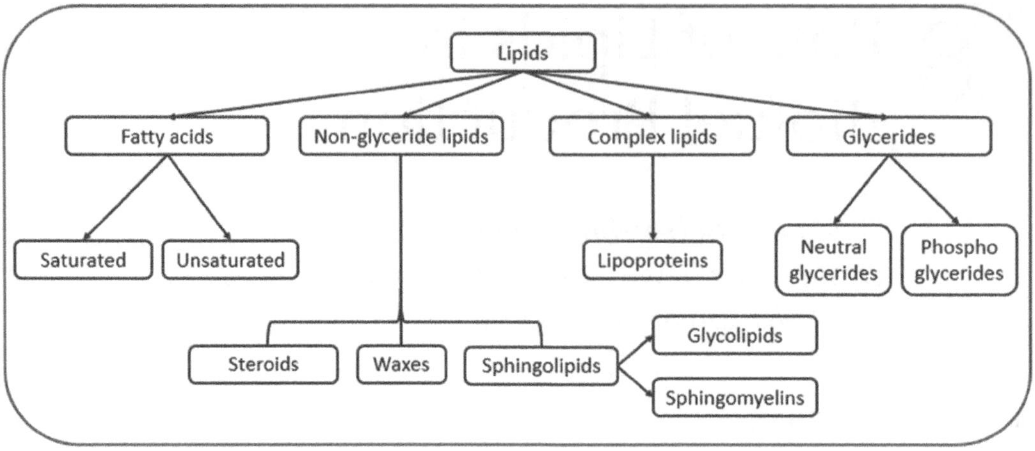

FIGURE 8.1 Overview of lipid classification.

Although lipids are made up of C, H, and O_2, their oxygen (and thus water) content is far lower than that of other macromolecules, such as carbohydrates. Lipids are not polymers, in contrast to proteins and polysaccharides, since they do not include repeated monomeric units (Elzagheid, 2024). Rather, glycerol as well as fatty acids are two main sorts of molecules that make them up. Glycerol is a three-carbon molecule in which hydrogen atoms occupy the remaining bonds and each carbon atom is connected to a hydroxyl group. The carboxylic acid group at one end of fatty acids is joined by a lengthy hydrocarbon chain, which is frequently denoted by the letter "R" (Kyselová et al., 2022). Saturated as well as unsaturated fatty acids are both possible. Since the hydrocarbon chain of a SAFA lacks carbon–carbon double bonds (C=C), hydrogen atoms occupy all of the accessible carbon bonds. There are one or more C=C linkages in unsaturated fatty acids. In contrast to polyunsaturated fatty acids (PUFA), which have several double bonds, monounsaturated fatty acids (MUFA) only have one (Dhull & Punia, 2020).

The most prevalent lipids in food, fats, and oils are triacyl glycerides, which are created when three distinct or comparable fatty acids are joined to tri-alcohol glycerol. At room temperature, typically 20°C, a fat is a combination of triacyl glycerides that is solid or pasty (Sharma et al., 2022). However, the word "oil" describes a room-temperature liquid mixture of TG instead. These compounds often contain, to a lesser degree, phospholipids, sterols, terpenes, fatty alcohols, carotenoids, fat-soluble vitamins, as well as many other minor chemical structures in addition to triacyl glycerides, which make up about 90% of fats and oils (Asokapandian et al., 2021).

A carboxyl group, an acidic functional group, is joined to four or more carbon atoms to generate fatty acids, which are hydrocarbon compounds made up of hydrogen and carbon atoms (Lund & Rustan, 2020). The number of carbon atoms in a fatty acid molecule determines its chemical and physical characteristics, including its melting point and capacity to dissolve in non-polar liquids. The fatty acid's melting point rises with the number of carbon atoms in the chain. Fatty acids are categorized as long-chain if they contain twenty or more carbon atoms, short-chain if they have four to ten carbons, medium-chain if they have 12–14 carbons, and long-chain if they have 16–18 carbons (Kalsoom et al., 2017). The high-water solubility of molecules with less than four carbons means that they are not classified as fatty acids. Conversely, fatty acids with a large number of C atoms are uncommon but found in considerable quantities in the brains of vertebrates, such as humans, as well as mammals. It has been determined that fatty acids with up to 36 carbon atoms exist in the human brain.

The C in fatty acids is linked together by a covalent bond that can be single or double. Within the same molecule, there may be one, six, or more unsaturated bonds. Therefore, a simpler classification of fatty acids separates them into two groups: those without a double bond are called SAFA, and those with one or more double bonds are called unsaturated fatty acids (Jónasdóttir, 2019). In other words, a molecule is categorized as a MUFA if it has one unsaturation, and as a PUFA if it has two to six unsaturation.

8.1.2 The Growing Need for Biofuels: Environmental and Economic Drivers

There has been a surge in the hunt for alternative energy sources due to growing agreement that cheap oil is finite, supply problems resulting from political unrest in key oil-producing regions, and the effects of carbon emissions from fossil fuels (Akca Prill, 2019). The transportation industry is the one where the need for oil substitutes is greatest. Thirty percent of the world's energy is utilized in transportation, with petroleum providing 99% of that energy. Between 2003 and 2030, it is anticipated that transportation will be responsible for 50% of the total growth in world oil consumption.

Furthermore, transportation accounts for 21% of annual global greenhouse gas (GHG) emissions. Although a variety of technologies, such as carbon-free technologies like nuclear and renewable sources like wind and solar photovoltaics, are set to compete with coal and natural gas in the electricity sector, there doesn't seem to be a viable substitute that could significantly lower the cost and improve transportation convenience compared to oil. However, plant-based fuels like ethanol and biodiesel appear to be becoming a viable alternative fuel in the modern day, surpassing innovations like natural gas, electric/hybrid, and fuel cell vehicles (Singh & Sehrawat, 2019). The enthusiasm surrounding biofuels has multiple causes.

The carbon-neutrality of biofuels is one of the factors driving their global introduction; according to the Kyoto Protocol, CO_2 emissions from burning biofuels are balanced by the CO_2 absorbed during plant growth. In other words, burning fossil fuels releases CO_2, one of the global GHGs, but burning biofuels does not emit CO_2, according to the Kyoto Protocol (Sebos, 2022).

Renewable Energy: According to the National Renewable Energy Laboratory, renewable energy is a non-exhaustible resource that is comparable to solar and wind power (Shamshirband et al., 2019). Since biofuel comes from plants and doesn't run out unless it slows down the growth of the plants, it is seen as a renewable energy source. Because agriculture can replenish the stock, biofuels are an endless resource (Agarwal, 2025). Electric vehicles and fuel cell technologies are reliant on the electric grid and hydrogen, respectively, and therefore are essentially reliant on finite resources like coal and natural gas.

Avoiding Air Pollution: Adding ethanol, which includes oxygen, prevents gasoline from burning completely. By incorporating biofuels or bioethanol into gasoline, oxygenated gasoline lowers emissions of carbon monoxide. The International Energy Agency claims that adding bioethanol to gasoline reduces emissions of particulate matter, CO, and C hydride (Roy & Pramanik, 2024). While adding bioethanol to gasoline reduces emissions of nitrogen oxides, the US Environmental Protection Agency says that it also reduces emissions of CO, carbon hydride, and particulate matter (Aslam et al., 2021).

Energy Security Contribution: The nation's energy security may be enhanced by the generation of biofuel from domestically cultivated plants (Sultana et al., 2023). Furthermore, because biofuel is distributed evenly, it can lower the geopolitical risk of energy production, unlike fossil fuels, which frequently carry a significant risk.

Growth of Rural and Agricultural Areas: Biofuel production is anticipated to increase agricultural profit, create jobs in the agricultural sector, and export agricultural output in developing nations. These elements encourage the introduction of biofuel production in numerous nations and areas. In many countries and locations, the introduction of biofuel production is also primarily motivated by the sharp rise in crude oil prices (Naylor & Higgins, 2018). Demand for biofuel rises when its price as a substitute good falls relative to that of crude oil.

8.1.3 Lipid-based Biofuels: A Promising Renewable Energy Source

Fossil fuel burning produces GHGs, which have resulted in major environmental issues that are only expected to get worse over the next few years (Correa et al., 2019). Developing renewable energy sources is of interest to many nations to maintain energy security and lower particulate emissions and climate change due to the consumption of fossil fuel (Sueyoshi et al., 2022). Geothermal, hydroelectric, wind, solar, and ocean energy are examples of common sustainable energy sources. Biofuel is a fuel made using sustainable and renewable methods. Among renewable energy sources, biofuels are particularly helpful because they are compatible with current infrastructure as well as large energy density (Khan et al., 2021).

The technological advancement of biofuel production is significantly influenced by feedstock selection (Lin, 2022), and all significant theoretical advancements in biofuel technology have been intimately related to the characteristics of the substrates with which they are associated. The economic and societal effects of implementing biofuel technology might also vary depending on the feedstock selection (Callegari et al., 2020). Edible oils and sugars, for example, are not the best options for producing low-cost biofuels since the feedstock is expensive and their supply is erratic because of market rivalry. On the other hand, using non-edible lipids, such as leftover animal fat, can help biofuels become more affordable and valuable (Binhweel et al., 2023).

Industrial waste and byproducts are initially less expensive, but pre-processing is necessary for their use in biofuel synthesis, which could cause increased costs. Reduction of the financial and societal expenses associated with acquiring the feedstock and enhancing the product's quality and usefulness are thus important for the advancement of biofuel technology. In order to further enhance the economic viability of producing biofuel, there has been a growing demand for the development of new feedstocks and processing technologies (Perin & Jones, 2019).

Storage lipids, particularly triacylglycerol (TAG), are abundant in plant and microbial lipids, as well as animal fats, and make good feedstocks for the synthesis of biofuel (Wang et al., 2022). The fatty acid moieties' chemical characteristics in these lipids are similar to conventional fuels in terms of branching, saturation level, and carbon chain length of carbon. Because of this, a variety of lipid-based biofuels, including biodiesel and renewable substitutes for traditional gasoline, diesel, and jet fuels, have physicochemical properties in common with traditional fossil fuels, making them highly compatible with current infrastructure (Deora et al., 2022).

One of the several physicochemical properties of oils and fats, a very diverse class of organic substances known as lipids, is the presence of distinct functional groups, which significantly alter the composition of readily available biofuel products. By combining or genetically modifying lipid-producing species, lipid feedstocks can be further optimized for the manufacture of biofuel, improving fuel performance (Wang et al., 2022). Lipid biotechnology advancements were also examined, which could be used to better customize lipid sources to meet the needs of ideal feedstocks for biofuel production, as well as the suitability of animal, plant, and microbial fats for biofuel production (Xue et al., 2021).

8.1.4 Overview of Different Generations of Biofuels

The depletion of fossil fuels as well as the consequences of climate change can be mitigated by using biofuels made from renewable biomass, such as plants and organic waste (Cherwoo et al., 2023). The feedstock and production technique used to produce biofuels allow for their classification into several generations (Singh et al., 2020). Corn, sugarcane, and vegetable oils are employed to make first-generation or traditional biofuels. Wood, municipal refuse, and agricultural waste are examples of nonfood biomass that is utilized for synthesizing second-generation biofuel. The fourth generation of biofuels is made up of oils made from genetically engineered microorganisms and algae, whereas the third generation is more advanced and eco-friendlier and includes algae and unique energy crops. Figure 8.2 shows the broad classifications of biofuel generation.

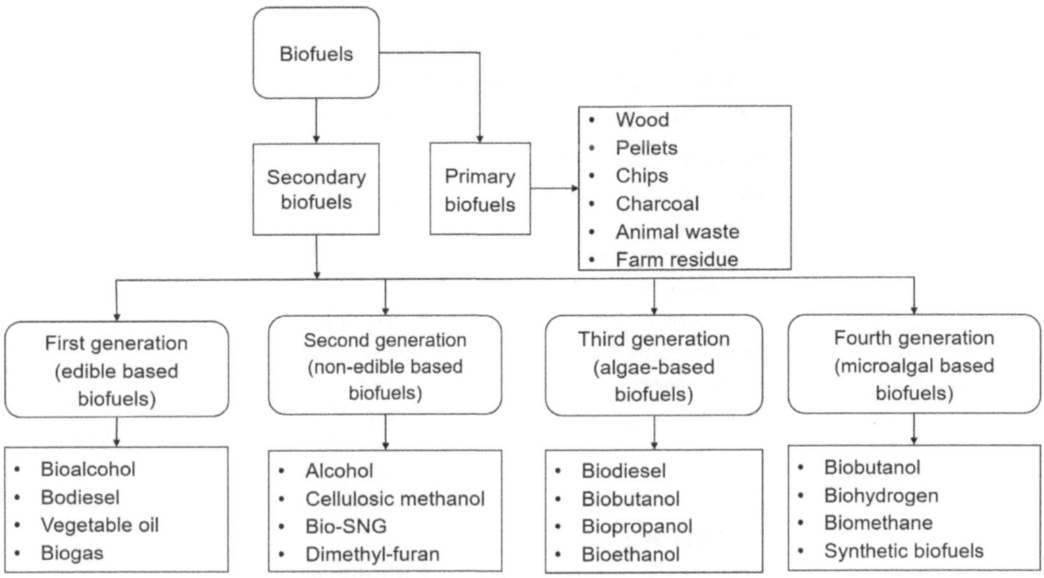

FIGURE 8.2 Types and classification of biofuels. (Reproduced from El-Araby, 2024.)

First-generation Biofuels: Using well-established and economically feasible processes and technologies like fermentation, distillation, and transesterification, first-generation biofuels are made from edible feedstocks. Among these are biodiesel derived from oilseed crops and bioethanol derived from corn and sugarcane (El-Araby, 2024). First-generation biofuels have a marginally lower GHG emission than fossil fuels because of their high energy needs for development, harvesting, and processing. However, they do face certain challenges, such as competition with food production, land usage difficulties, high water use, potential soil degradation from extensive monoculture farming, and a slight decrease in GHG emissions when considering their life cycle (Mahapatra et al., 2021). In an effort to allay worries while preserving the advantages of creating renewable fuels, research on second- and third-generation biofuels has exploded.

In order for achieving demand for fuel to drive their machines, American maize farmers started producing the first generation of biofuels in the late 1990s. Corn, vegetable oils, feedstocks, and sugarcane were the sources of the first generation of biofuel (Hirani et al., 2018). The process used to obtain first-generation biofuels involves first extracting the oils from the plants or feedstocks that have been used as the source. The characteristic values of biodiesel obtained in the first generation are shown in Table 8.1 (Y. Zhang et al., 2022).

This generation was at a disadvantage as the United Nations study made it abundantly evident that these first-generation biofuels could have a negative effect on human life by putting human food supplies in jeopardy and possibly putting many animal species in jeopardy (Bajpai, 2022). Additionally, there will be a crisis of drinkable water because growing maize requires a lot of water for irrigation (1,700 gallons per gallon of ethanol P.A.).

Second-generation Biofuels: The second generation of biofuel entered the market at the start of the 21st century. Advanced biogenesis was also introduced in this field with the second generation. The roots of some sugar-producing or sugar-rich plants were cultivated in a particular way such that oil could be produced straight from the roots for second-generation biofuels (Hirani et al., 2018). In terms of CO_2 emissions throughout the entire process, this generation was nearly carbon-free or carbon negative. Common sources of second-generation biofuels include solid waste, forest leftovers, Jatropha, grasses, lignocellulosic

TABLE 8.1
The Characteristic Value of Biodiesel Obtained in the First Generation

Property	Value
Oxygen concentration (%)	11
Sulfur (ppm)	<1
Heating value (MJ/Kg)	38
Distillation (°C)	340–355
Cloud point (°C)	−5 to +15
Specific gravity	0.88
Cetane number	50–65
Stability	Marginal

TABLE 8.2
The Characteristic Value of Biodiesel Obtained in the Second Generation

Property	Value
Oxygen concentration (%)	0
Sulfur (ppm)	<1
Heating value (MJ/Kg)	44
Distillation (°C)	265–320
Cloud point (°C)	−10 to +20
Specific gravity	0.78
Cetane number	70–90
Stability	Good

feedstocks, seed crops, and leftover vegetable oil (Gautam et al., 2024). Second-generation farmers' primary philosophy is to use crops that are unsuitable for human consumption.

This generation uses a more environmentally friendly method to create biofuels. Burning second-generation biofuels results in either neutral or even negative net carbon emissions. The feedstock consists of lignocellulosic material, which includes the plentiful and reasonably priced non-food biomass found in plants (Rostek, 2016). Due to a number of technological obstacles that must be addressed, the cost-effectiveness of the second generation of biofuels still needs improvement. Researchers have been interested in using waste plant biomass for a number of purposes, such as a raw material for wastewater treatment or as a feedstock to directly burn to create heat as well as power (Mungodla et al., 2019). In contrast, using it as a low-cost biofuel source is highly appealing.

This generation uses two different approaches to turn biomass into liquid biofuels. Thermochemical processing comes first, followed by biochemical processing. Because thermochemical processing could change all of the organic components of biomass, whereas biochemical processing concentrates on converting polysaccharides (Zhang & Zhang, 2019), it is typically favored over biochemical processing. Thermochemical processing, which basically entails heating biomass in an aerobic environment, is the process of changing biomass to different kinds of products through chemical reformation and thermal degradation. The characteristic values of biodiesel obtained in the second generation are indicated in Table 8.2 (Y. Zhang et al., 2022). Despite its many benefits, this generation also has drawbacks, such as a lack of cost-cutting technical solutions, pretreatment of biomass to produce cellulose sugars, unstable economies, and a current unsuitability for mass production (Ganguli et al., 2018). The main issue is that sources are dispersed unevenly throughout the world, making collection challenging.

Third-generation Biofuels: In comparison to earlier generations, third-generation biofuels are thought to have a higher energy density (yield per area of harvest). Compared to land crops like Jatropha, it generates 30 times as much energy. Only 40,000 of the estimated 200,000–800,000 kinds of algae are officially documented (Bagchi et al., 2018). These plants are inexpensive and produce a lot of yield. The crops that are utilized as third-generation fuel sources are algae. Algae have the advantage of being nitrogen-fixing plants that can grow in any kind of water body. This would lessen reliance and ease the demand on freshwater resources and agricultural land. The fact that not all types of algae are suitable for industrial use is another problem. Since the stress conditions necessary to collect lipids cause cell growth and division to stop (Maltsev & Maltseva, 2021), the fatty acid composition of microalgae lipids may not be optimal to use in biodiesel, significantly restricting the amount of biomass that can be produced.

Fourth-Generation Biofuels: Synthesis of this generation of biofuels is still ongoing, as well as no chemicals that can be manufactured in practice have undergone testing. The fourth generation's principles are based on the following theoretical ideas: creation of plants with genetic modifications that will absorb more CO_2 from the atmosphere (Bajpai, 2022). Combination of algae plant with oil seeds for promoting a high-yield cross-breeding species. Only a small number of businesses are now engaged in this. The most well-known is Craig Ventor's Synthetic Genomics, which is presently developing microbes that can directly produce fuel from CO_2. It is anticipated that this generation will reach full maturity and make a substantial contribution to the global power industry by 2050.

8.2 LIPID SOURCES FOR BIOFUEL PRODUCTION

8.2.1 Plant-based Lipid Sources (Oils and Fats)

Biofuels derived from plants are the most abundant renewable energy source. They can convert starch, cellulose, hemicellulose, and oils into long-chain hydrocarbons (for jet fuels or diesel additives) and ethanol and butanol (for gasoline additives) (Kirakosyan et al., 2009; Singh & Singh, 2023). For the next several billion years, plants will continue to take energy from the sun. Burning biofuels releases carbon that is continuously cycled, in contrast to the carbon produced by fossil fuels such as natural gas as well as petroleum.

The most popular commercial method of producing biodiesel involves treating feedstock oil or fat with alcohols and chemical catalysts like potassium or sodium hydroxide. When it comes to business, this strategy can work, but only in certain situations (Maheshwari et al., 2022). First, compared to virgin plant oils, the process frequently calls for more costly feedstocks, such as recycled cooking oil or other spent oils. Second, before chemical reactions may take place, the feedstock oil must either contain a low concentration of free fatty acids. If left behind, they saponify (make soaps) and form emulsions using catalysts, making it difficult to separate the biodiesel from impurities.

Third, because phospholipids create emulsions or foaming that are challenging to eliminate from the final biodiesel, they must also be eliminated. The high energy consumption of the process and the need to cleanse the alkaline wastewater before it can be utilized again are further factors to take into account. According to Nisar et al. (2021), fatty acid methyl esters are the main byproducts of fats and oils. These must be separated from the contaminating glycerol, water, and an acid or base catalyst.

8.2.2 Animal Fats and Byproducts

Animal fats are byproducts of industrialized meat production that are typically utilized to make pet food, animal feed, oleochemicals, and boiler fuel at rendering factories. The usage of Animal fats as feedstocks for the synthesis of renewable as well as biodiesel has grown in recent years (Toldrá-Reig et al., 2020). Additionally, there is significant interest in producing sustainable aviation fuels (SAFs) for use in the aviation industry. The main sources of animal fat byproducts, which are mainly composed of TAG, include tallow, lard, and chicken fat, which are used in the meat manufacturing as well as animal rendering industries. Furthermore, the leather and seafood processing sectors can produce lipids as byproducts. According to Bedoić et al. (2020), the rendering sector is in charge of turning waste animal tissues into useful components, such as lipids, that can be utilized for energy applications like the creation of second-generation biofuels. It has been shown that tallow can be converted into biofuel by hydroprocessing fatty acids and esters to produce diesel and jet fuel equivalents and by using traditional transesterification techniques to produce biodiesel.

Animal fat is predominantly composed of TG, along with phospholipids and sterols. A TG consists of three fatty acids attached to a glycerol molecule, characterized by its oxygen-rich structure. The process of transesterification converts these TG into biodiesel by reacting the fatty acids with alcohol. The key difference between fats and oils lies in their saturation: oils are typically liquid due to higher levels of unsaturated fatty acids, while animal fats are solid because of their SAFA content. Common waste animal fats used in biodiesel production include pork lard, beef tallow, and chicken fat.

According to the risk to human and animal health, animal fats are divided into three groups under the Animal By-Products Regulation. The biggest risk is represented by category 1, which is linked to a particular risk of disease transmission. Since category 2 is unsuitable for human eating but not associated with any particular disease, it presents a medium risk. Fats that were judged suitable for human consumption at the time of slaughter fall under category 3, which is regarded as the least dangerous. Under the EU's Renewable Energy Directive (RED II), all three kinds of animal fats can help meet renewable energy requirements in the transportation sector (Hazards et al., 2020).

In particular, category 1 and 2 animal fats are eligible for double counting and are included with Used Cooking Oil (UCO) in Part B of Annex IX of the RED. This makes it easier for fuel suppliers to reach their RED commitments because two certificates are awarded for each gigajoule of energy provided from these feedstocks. On the other hand, double counting does not apply to category 3 animal fats, which are utilized more frequently in other industries. When combined with food-based biofuels, they can still help achieve the RED objective. Fuel suppliers are therefore financially motivated to prioritize using category 1 and 2 biofuels over category 3 animal fats (Habib et al., 2020). However, category 3 biofuels made from animal fats are becoming much more popular, which has an impact on the environment as well as other rival businesses.

About 3 million tons of animal fats were produced by EFPRA members in 2021; of them, 2.4 million tons were category 3 and edible fats, while 570,000 tons were categories 1 and 2. Additionally, EFPRA offers information on the various applications of animal fat resources from 2006 to 2021 (Malins, 2023). The usage of animal fats for heat and/or power generation has significantly changed from 2006 to 2021, rising from 1% to 46% of all uses for biodiesel synthesis. According to the data, the amount of animal fats claimed to be consumed annually for the manufacturing of biodiesel increased from 30,000 tons in 2006 to 1.4 million tons in 2021. During that year, 540,000 tons of biodiesel (or 96% of all category 1 and 2 fats) and an additional 820,000 tons of category 3 fats (36% of all category 1 and 2 fats) were produced from nearly all category 1 and 2 resources.

The need for these feedstocks in the synthesis of biofuels will not go away. The incentives for biofuels made from animal fats are maintained in the latest RED amendment, which also includes fuels delivered to shipping and aviation in the transport target. Therefore, a larger pool of fuels are subject to the 1.7% cap, which raises overall demand once more. Furthermore, under distinct, industry-specific standards (ReFuelEU6 and Fuel EU Maritime, respectively), Annex IX feedstocks can satisfy new complementary targets for aviation and maritime fuels.

Biodiesel made from animal fats is already being produced by large oil firms. Strata Advisors claims that Neste runs the two largest biodiesel production plants in Europe, with a maximum capacity of 1 million tons annually in the Netherlands and 525,000 tons annually in Finland (Vesanto, 2022), both of which use animal fats as a significant feedstock. Additionally, Neste declared that its operations in Europe and Asia would begin refining SAFs in 2023. Other big oil companies, such as Shell in a new biofuels project in Rotterdam and Total Energies in its Grandpuits refinery in France, also intend to produce SAF from animal fats.

8.2.3 Microbial Lipids (Single-Cell Oils)

A considerable amount of oils of plant and animal origin may be released from microorganisms, whose lipids, after appropriate processing, become acceptable for use in various industrial domains. Three types of lipids are produced during the process of growing microorganisms on various media: simple, complex, and lipid derivatives (Jones et al., 2019). Wax and neutral fats are examples of basic lipids. Glycerol and ether fatty acids make up neutral fats, which are the primary reserve components in a cell. The majority of these are triacyl glycerides, which include mono- and diglycerides. Waxes are long-chain aliphatic alcohols and ethers of fatty acids or monooxyacids that share structural and functional similarities with neutral lipids. Yeasts and filamentous fungi produce the most neutral fats. Bacteria are the primary producers of complex lipids.

Unlike plant fats, microbial lipids are obtained through a technical procedure that involves lipid extraction in a nonpolar solvent and deep destruction of the cellular structure (Hwangbo & Chu, 2020). Two finished products are concurrently produced: a defatted protein compound called bio short and a microbial fat called biofat. Nevertheless, the process of extracting lipids in polar solvents is very combustible and explosive. It is crucial for industrial applications to have the capacity to actively collect lipids. Only a few microorganisms possess this ability; yeasts are one example. There are two distinct phases to the majority of yeasts' lipid production process. Slow lipid accumulation (mostly glycerophosphates and neutral lipids) and rapid protein production under nitrogen-enhanced culture conditions are characteristics of the first stage. It is normal for the second stage to see increased lipid accumulation (mostly neutral) and the cessation of yeast growth.

Typical lipid producers are yeasts from the *Cryptococcus terricolous* species. Under all circumstances, they were able to synthesize large quantities of lipids (up to 60% of dry mass) (Passoth & Müller, 2025). Among other lipid-producing yeasts, *Candida guilliermondii* is intriguing from an industrial perspective since it uses alkanes. It produces phospholipids mostly. *Rhodotorula gracilis* and *Lipomyces lipoferus* are two yeast species that aggressively grow on carbon media (wood hydrolysates, turf, and molasses) and accumulate large amounts of lipids. For this species of yeast, lipogenesis is highly dependent on the conditions of cultivation. Triacyl glycerides are accumulated by these manufacturers in considerable amounts (up to 70%).

Although the composition of fungal fat is similar to that of plant fat, microscopic fungus is not yet widely used to acquire lipids. 51% of dry cell components (DCCs) are fat in *Aspergillus terreus*, for example, when grown on carbon media (Tkachenko et al., 2013). The primary components of the fungal lipid composition are phospholipids and neutral fats. Rhodotorula and Pichia are the most common lipid makers found in nature. Lipids could be produced by them at levels between 30% and 40% of DCCs. The intensity of aeration affects the production of triacyl glycerides, phosphoglycerates, and fatty acids. Lipids with inadequate aeration have eight times more fatty acids, two times more phosphoglycerates (Gupta et al., 2021), and four times fewer triacyl glycerides than those with adequate aeration. The degree of lipid unsaturation and the proportion of each category of unsaturated acids rise as aeration intensifies. The lipids produced by bacteria differ in composition, primarily consisting of complex lipids, while neutral fats make up a negligible portion.

8.2.4 Waste Cooking Oil (WCO) and Other Lipid-Rich Wastes

Trans fat that Indians frequently ingest from waste or used cooking oil (WCO/UCO) contributes to a number of non-communicable diseases, including diabetes, high blood pressure, heart disease, stroke, cancer, and more (Manikandan et al., 2023). WCO's use has been observed to be rising despite numerous medical professionals' warnings about its negative consequences, primarily as a result of India's unstable import cooking oil prices. WCO is imported to meet roughly 60% of the country's oil demand, and because of its questionable pricing, users are encouraged to reuse the oil on a regular basis.

Edible oils used for food frying are utilized to make WCO. When these food wastes are disposed of without being subjected to any treatment procedures, they endanger both human health and the environment. WCO is seen as a promising feedstock for the production of biodiesel since it is abundant and inexpensive in many countries (Singh et al., 2021). Using used cooking oils as raw materials avoids controversial issues like arguments over the rivalry between the fuel and food industries or environmental issues. WCO is seen as a promising feedstock for the production of biodiesel due to its low cost and accessibility in many countries. The utilization of previously used cooking oils as source materials does not pose issues related to the conflict between the food and fuel industries or environmental concerns. The utilization of this raw material for the production of biofuels was first recorded many years ago, and research on the subject dates back to the 1970s. Cooking oil waste can be acquired from residences, lodging facilities, dining establishments, and food enterprises that use frying operations and other comparable food preparation techniques (De Feo et al., 2020). The process of producing biodiesel from waste oils from the commercial, industrial, and residential sectors is straightforward. Making biodiesel from used cooking oil is a well-known and environmentally friendly waste management method. Particularly when it comes to non-edible resources, these fundamental resources are not the only ones that can be highlighted. For example, cow tallow is the second most important raw material used in Brazil to make biofuels, after soybean oil.

8.3 CONVERSION TECHNOLOGIES FOR LIPID-BASED BIOFUELS

8.3.1 Transesterification: The Primary Route to Biodiesel

Transesterification is a crucial chemical process in the manufacturing of biodiesel. Like hydrolysis, it entails substituting another alcohol for the alcohol group in esters, but using alcohol in place of water. The conversion of TG, which are made up of three fatty acid chains (R1, R2, and R3) (Naseef & Tulaimat, 2025), into diacylglycerides (DG) and monoacylglycerides (MG) occurs gradually through a multi-step procedure. The end products of this process are glycerol and fatty acid alkyl esters (biodiesel).

Lipid molecules undergo molecular structural changes during the chemical transesterification process used to produce biodiesel. The physical characteristics alter as a result (Foroutan et al., 2019). Even refined pure plant oil (PPO) can be used in diesel engines that have been converted, but biodiesel, which is produced through a process called transesterification, offers a number of benefits. One benefit is that biodiesel has a lower viscosity than PPO. Diesel engine atomization, pressure, and fuel injection duration are all negatively impacted by increased viscosity. Because biodiesel and fossil diesel are so similar, they can be used in regular diesel engines that require little modification.

Biodiesel is generated by the transesterification of TG using short-chain alcohols along with a catalyst. Because the reaction is reversible, an excess of alcohol must be utilized to shift the equilibrium of the reaction (Salaheldeen et al., 2021). Nonetheless, transesterification can also occur without a catalyst through supercritical process reactions. This method involves three consecutive reversible reactions wherein TG are transformed into diglycerides, diglycerides into monoglycerides, and ultimately, monoglycerides into glycerol. Additionally, for every glyceride that undergoes a reaction, a molecule of ester (biodiesel) is produced. TG and methanol combine in the initial step

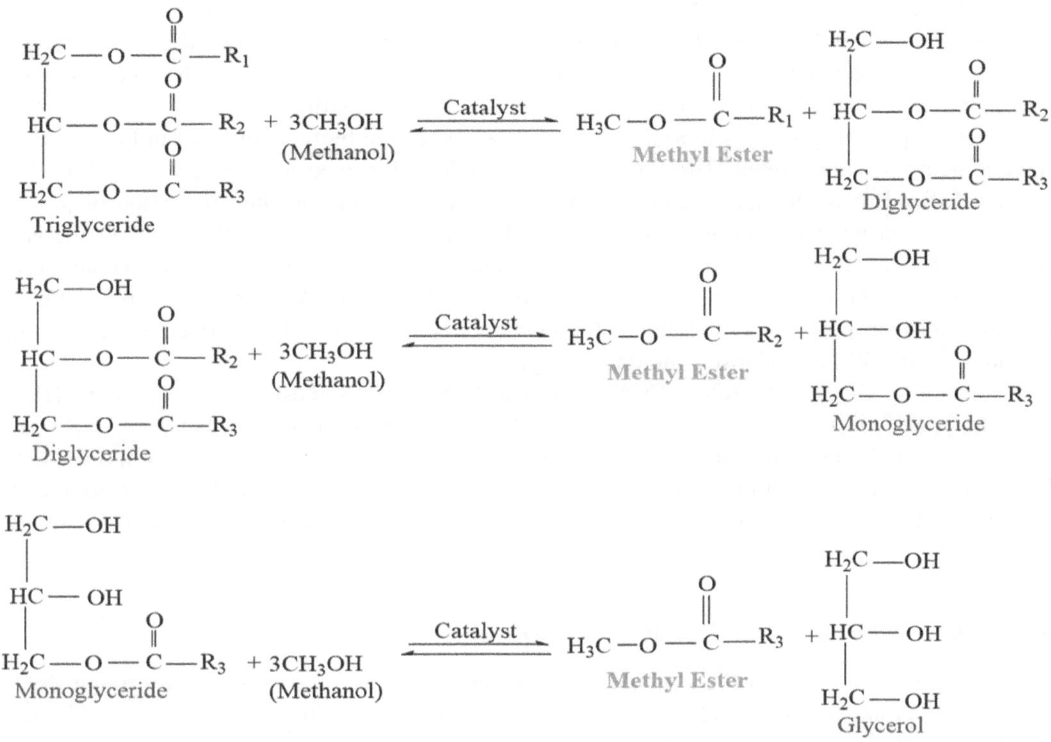

FIGURE 8.3 Transesterification reaction of triglycerides into biodiesel. (Adopted from Naseef & Tulaimat, 2025.)

of transesterification to produce DG and one FAME molecule (Okitsu et al., 2014). The subsequent reaction between DG and methanol yields a second FAME molecule as well as MG. A third FAME molecule is produced when MG finally transforms into glycerol. Figure 8.3 shows the transesterification reaction of TG into biodiesel.

One of the main advantages of this method is its great efficacy and adaptability to a variety of feedstocks with low (FFA) content (less than 0.5 weight percent). Nevertheless, contaminants can affect the procedure. Soap formation brought on by high FFA levels can impede product separation and lower total yield. To avoid unwanted side reactions, reactants and catalysts must have their water content strictly controlled, usually between 0.1 and 0.3 weight percent. When processing high-FFA feedstocks, these constraints call for pretreatment techniques like esterification. The remaining glycerol in biodiesel presents another difficulty. Engine lifetime and performance can be adversely affected by even minute levels of glycerol. It may cause engine knocking or misfiring due to its high viscosity and hygroscopic nature, which can result in damaging deposits in engine parts and lower combustion efficiency (Kashyap et al., 2022). Glycerol and other contaminants are eliminated from biodiesel by a rigorous purification process in order to solve these problems.

8.3.2 Hydrotreatment (Hydrodeoxygenation – HDO) for Renewable Diesel and Jet Fuel

Three primary steps make up the hydrotreating process: separation, hydrocracking/hydroisomerization, and hydrodeoxygenation (HDO) (Ewurum, 2019). In order to extract oxygen as carbon dioxide and water, the TG feedstock and hydrogen enter a first reactor with a solid catalyst. This step is crucial for reaching the bio-jet fuel's freezing point. In order to produce green diesel, long-chain hydrocarbons are produced in the first reactor. These hydrocarbons must be split and isomerized in a second reactor, which also uses hydrogen as a reactant, to reach the boiling range of jet fuel. The second reactor's byproducts include bio-jet fuel, green diesel, naphtha, and light gases. The most common method for purifying these hydrocarbons is distillation.

The HDO process produces bio-jet, which is now one of the biofuels utilized in both commercial and demonstration flights. Additionally, the technology for making renewable diesel has been commercialized. It is anticipated that this pathway will be crucial to the production of jet fuels with low carbon intensity due to its technological maturity (Task, 2021). The steps in the HDO process, hydrogenation, deoxygenation, isomerization, and hydrocracking, are used to convert triacylglycerides (TAG) into hydroprocessed renewable jet (HRJ). All TAG bonds are saturated by the first process, hydrogenation. Following hydrogenation, deoxygenation is comprised of three parallel reactions: HDO, decarbonylation, and decarboxylation.

The resulting alkanes are frequently referred to as hydroprocessed esters and fatty acids (HEFA) synthetic paraffinic kerosene (SPK), or SPK, or HRJ. The main distinction between HDO gasoline products and their petroleum-based counterparts is the absence of aromatic components. The TAG may come from a variety of virgin and used oil sources. The fatty acid concentration of the feedstock may have a substantial impact on the HDO process's output yields, product distribution, hydrogen consumption, and process utilities (Chu et al., 2017).

8.3.3 Other Emerging Conversion Technologies

Globally, more and more biofuels are being produced from diverse bioresources through the use of biological processes and developing technology. Utilizing agricultural crop wastes and biomass waste to create biofuels is anticipated to lessen environmental impact and address several environmental concerns, including waste management (Koul et al., 2022). The synthesis of biofuels from various plants and biomass material originating from microorganisms has been the subject of increased investigation recently due to its carbon neutrality and environmental friendliness. Furthermore, photosynthesis allows these plants and algae to accumulate biomass (Gnanasekaran et al., 2023). Consequently, additional research is being conducted on advanced methods for the production of biofuel as a source of energy.

The novel concept of co-processing, which combines lipids with other refinery products and/or feedstocks to increase gasoline output while lowering carbon intensity as a desired result, has also attracted a lot of attention in recent years. One of the byproducts of vacuum distillation columns in a refinery is vacuum gas oil (VGO), which is commonly utilized in Fluid Catalytic Cracking (FCC) to increase the yield from crude oil to a gasoline or diesel product (Abdulridha et al., 2023). Biomass can be used to create fuels, and in FCC facilities, catalytic cracking of vegetable oils can boost gasoline production. The octane number and aromatic content of the final products from co-processing with VGO are also influenced by the degree of unsaturation of the lipid feedstock, illustrating the influence of feedstock on the production of biofuel (van Dyk et al., 2019).

8.4 PROPERTIES AND QUALITY STANDARDS OF LIPID-BASED BIOFUELS

8.4.1 Biodiesel (Fatty Acid Methyl Esters – FAME) Properties

Fatty Acid Methyl Esters (FAME) are a subset of fatty acid esters. While characteristics differ by kind of vegetable oil, the physical characteristics of fatty acid esters are more like fossil diesel fuels than those of pure vegetable oils (Inam et al., 2019). Biodiesel is a renewable alternative fuel that is commonly described as a mixture of several methyl esters of fatty acids. In terms of physical properties, FAME is similar to ordinary diesel. It is non-toxic and biodegradable as well (Monirul et al., 2015). Additionally, biodiesel differs from fossil fuels in a few ways. FAME must have as few contaminants as possible, such as metals, to be utilized as motor fuel.

The chemical makeup of the feedstock determines the properties of biodiesel. The length of the chain and the level of unsaturation of the fatty acid profile define the characteristics of biodiesel. National (IS 15607-05) and international (ASTM 6751-3 and EN 14214) standards must be met by the calculated and measured biodiesel levels (Meher et al., 2017). A key technique for characterizing biodiesel is FAME analysis, which measures the amount of fatty acid methyl ester in the fuel. This examination, which yields important details regarding the composition and purity of the biodiesel (Torres et al., 2020), is usually carried out by quantifying individual FAME components using gas chromatography (GC) or high-performance liquid chromatography (HPLC).

Viscosity: Because it influences the fuel's performance and flow characteristics, viscosity is an essential criterion for evaluating the quality of biodiesel (Hoang, 2021). In order to ascertain whether biodiesel is suitable for use in engines and transportation applications, dynamic viscosity measurements are frequently carried out to measure the fuel's resistance to flow under particular circumstances. Testing for oxidation stability: Oxidation resistance is a critical component that influences the shelf life and storage stability of biodiesel. Numerous methods, such as the Rancimat test and the PetroOXY test, which gauge the fuel's susceptibility to oxidative degradation over time, are used to assess the oxidation stability of biodiesel.

Cold Flow Properties: Cloud point and pour point are two important variables influencing how well biodiesel performs in cold climates. To ascertain these characteristics and guarantee that biodiesel continues to function at low temperatures (Hazrat et al., 2020), characterization techniques, including differential scanning calorimetry (DSC) and automated cloud and pour point analyzers, are employed. Determining the acid value of biodiesel reveals its degree of acidity, which may affect the integrity of the fuel system and engine parts. Titration techniques are used in acid value determination to measure the quantity of free fatty acids in biodiesel, allowing for the evaluation of its corrosiveness and possible effects on engine performance.

Calorific Values: Also referred to as heating value, calorific value indicates the energy content of biodiesel and is essential for assessing its heat production and combustion efficiency (Kaisan et al., 2020). The calorific value of biodiesel is frequently determined via bomb calorimetry, which offers crucial information for evaluating the fuel's potential for energy production.

Analysis of Sulfur Content: Determining the sulfur content is crucial for guaranteeing adherence to environmental laws and evaluating the effects of burning biodiesel on the environment (Saleh, 2020). Sulfur content in biodiesel samples is measured using methods including ultraviolet fluorescence analysis and X-ray fluorescence (XRF) spectroscopy.

Quantification of Glycerol Content: To guarantee adherence to quality requirements and avoid any problems with fuel stability and engine efficiency, the glycerol content of biodiesel needs to be tracked. Glycerol in biodiesel products is precisely quantified using analytical techniques such as gas chromatography combined with flame ionization detection (GCFID) (Vinoth Kumar et al., 2020). Together, these characterization techniques offer a thorough understanding of the stability, chemical makeup, physical characteristics, and environmental effects of the biodiesel generated, assisting with quality assurance procedures and legal compliance in the biofuel sector.

8.4.2 Renewable Diesel and Jet Fuel Properties

Combustion performance, emission characteristics, and NOx gas production are all significantly impacted by the physical and chemical features of biodiesel fuel (Pires et al., 2018). With the exception of ethyl esters' viscosities, biodiesel's physicochemical characteristics are similar. However, there are significant differences between methyl and ethyl esters in terms of viscosity and pour point.

For instance, compared to ethyl esters, methyl esters generate greater torque per unit of biodiesel utilized. It's also noteworthy that, in terms of biodiesel's physicochemical features, the heat value of combustion may be comparable or the same, but the behavior of combustion, performance, and emissions may change (Maroa & Inambao, 2020). One of the most crucial physicochemical properties of biodiesel, especially for compression ignition engines, is its viscosity.

Cetane Number: The cetane number is a diesel fuel index value that indicates the fuel's quality of ignition. A high cetane number makes it easier for the fuel to ignite (Knothe, 2010b). Besides improving the ignition delay, it also acts as a speed indicator. A diesel engine's auto-igniting speed is indicated when fuel is added. It shows the proportion of hexadecane by volume in an ignitable mixture of hexadecane and 1-methylnaphthalene whose characteristics match those of the diesel fuel that will be tested. The cetane number of biodiesels is not set because they are made from a variety of vegetable oils. Among the plant oil resources utilized to produce biodiesel are palm, karanja, jatropha, peanut, sunflower, rapeseed, and soybean oils. Animal fat has a CN between 56 and 60, while the CN ranges from 68 to 94.8 (Jaiswal et al., 2024). Numerous factors, such as the number of C-atoms that form a long C-chain, the unsaturated bonds, and the ester yield, affect the biodiesel's cetane number.

Density: Density is a number that represents the ratio of an object's mass to its volume. The mass of a fluid divided by its volume is the density of any given material. A substance's density is influenced by its temperature (Siraj et al., 2017). Diesel's density is typically tested at 15°C, while some studies use 40°C. Certain fuel parameters, primarily cetane number, arene structure, viscosity, boiling temperature, and volatile nature, are strongly connected with density. Except for a few, the density decreases and the volume increases whenever the substance's temperature rises. Biodiesel made from various resources, including rapeseed, mahua, palm, and jatropha oils, has a density between 880 and 890 kg/m^3. RD has a density of 775–816 kg/m^3. According to EN590, RD has a lower density than biodiesel and diesel fuel (Riayatsyah, 2018). As a result, the pour speed increases by around 4% to 5% compared to diesel and biodiesel.

Cloud Point (CP): Knowing the fuel's CP is essential because it guides us to its operating conditions in colder climates. CP is the temperature at which crystals form when a fuel's temperature is lowered to the point where clouds emerge (Santos et al., 2023). The oil becomes viscous and clogs the fuel filter and even the injectors because of the crystals. This makes it challenging for the diesel engine to operate.

Even at −5°C, the biofuel can function with ease. The temperature range of biodiesel is between −30°C and −5°C; RD is superior. The CP of the RD can be further lowered to −55°C using the isomerization process, according to a paper by Hartikka et al. (2018). However, it was noted that while isomerization helps the CP get better, it also causes the CN to go worse (Knothe, 2010a). It has been found that short or isomerized HC chains perform better than lengthy HC chains in terms of their flow characteristics. Its chemical structure's saturated bonds is what give the RD its superior operating capability.

On the other hand, saturated fatty esters in biodiesel cause the CP to decrease. In addition to the CP, other characteristics such as the fuel's pour point (PP) and cold flow pour point (CFPP) aid in assessing whether it can function at low temperatures. PP indicates the free pouring of liquid at a lower temperature. In the given time, CFPP determines the oil flow through the filter at the lowest temperature. According to Vozka et al.'s (2018b) research, flow characteristics such as CP and CFPP were comparable to those of regular winter diesel when hydroprocessing was done at 420°C.

Kinematic Viscosity: According to Siraj et al. (2017), viscosity is a measurement of the friction in the nearby layers of flowing materials, such as liquids and gases, that oppose motion. It is the resistance provided by the fluid's two adjacent layers; toothpaste, for instance, has a higher viscosity than oil. The viscosity of biodiesel and RD at 40°C varies from 4.0 to 4.8 mm^2/s and 3 to 4 mm^2/s, respectively, which is comparable to that of petro-diesel (3.6 mm^2/s). According to an engine fuel injector, having the proper fuel viscosity is crucial.

Although biodiesel blends have the advantage of being superior lubricants, their higher viscosity than the normal value results in poor atomization and insufficient flow. When injected, this produces enormous fumes, which result in poor mixing and incomplete burning. Black smoke is another effect of this. According to research, the New European Driving Cycle, the renewable diesel demonstrated lower emissions while operating under the same circumstances (Degraeuwe & Weiss, 2017). On the other hand, less viscosity will cause the fuel pump to deform, increasing the likelihood of internal leaks and inadequate fuel delivery, ultimately leading to engine destruction.

Oxygen Concentration and Stability: A fuel's long-term stability is largely determined by the presence of oxygen in its molecular structure. Biodiesel is FAME, which is aromatic and contains oxygen atoms. However, as an aliphatic saturated straight HC chain devoid of oxygen atoms, renewable diesel is comparable to petro-diesel (Sembiring & Saka, 2019). Because oxygen has a propensity to react, any fuel that contains oxygen atoms is unstable. This makes it extremely difficult for biodiesel to remain stable. The oxygen content of biodiesel ranges from 6% to 12% by weight, while that of renewable diesel is zero. This is an extra benefit of renewable diesel over biodiesel, making it more durable and stable. When biodiesel is stored for an extended period of time, the oxygen stability is hampered by the high concentration of double-bonded fatty acids found in many vegetable oils. Because it can impact the fuel's different qualities, oxygen stability is crucial.

Flash Point: Any organic compound's flash point is the lowest temperature at which it releases enough vapors to catch fire when it comes into contact with a flame. It provides insight into the volatility and flammability of the fuel (Kumar et al., 2022). The fuel's flash point is crucial since it alerts the operator to the need for safety measures when handling it. Every fluid has a distinct vapor pressure that contributes to its temperature and is subject to Boyle's Law (Arunachaleswara et al., 2017). A fuel's vapor concentration expands as a result of rising vapor pressure brought on by rising temperature. This suggests that a key factor in the development of vapor concentration is temperature.

The quantity of gases generated at the flash point is insufficient to maintain the combustion. It won't ignite unless the fuel comes into contact with a flame or an ignition source. There is an inverse link between the two: the more volatile the fuel, the lower the flash point temperature. According to ASTM D9751 criteria, it is safe because the flash point temperature, which varies from 210°C to 290°C, is tabulated for various vegetable oils (Tyagi et al., 2022).

Calorific Value: A substance's calorific value (CV) is the amount of energy it contains or releases when it burns. The CV of BD is reported to be between 33.6 and 43.2 MJ/kg (Tyagi et al., 2022), while that of RD is between 42 and 47 MJ/kg. The oils' presence of O_2 atoms lowers the CV by roughly 10%–12%. On the other hand, RD uses hydroprocessing to remove oxygen, which results in a higher calorific value than biodiesel. With a higher CV, less fuel is used and more energy is released to complete the task. As a result, brake thermal efficiency increases. In terms of CV, RD is comparable to petro-diesel. Because of alternative fuel initiatives, a substantial amount of information about the physical characteristics of jet fuel is now accessible. Few measurements have been conducted in the vapor phase; however, some predicted physical parameters are known for fuels in the vapor phase. Conventional jet fuel, a hydrocarbon distillate in the kerosene boiling range, normally contains liquid fuel data from around −40°C to 120°C, according to the CRC Handbook of Aviation Fuel Properties.

The real kerosine product that is commercially accessible in Ohio usually satisfies the jet fuel requirements and is much simpler to obtain. In fact, the ASTM D3699 specification for kerosine for home heating applications closely mimics the jet fuel specification (Seto & Gaughan, 2018). The physical characteristics of on-specification fuels can differ significantly since the jet fuel specification is fairly loose.

Distillation: The venerable ASTM D86 (Standard Test Method for Distillation of Petroleum Products and Liquid Fuels at Atmospheric Pressure) distillation is commonly used to describe jet fuels using distillation qualities (Santos et al., 2021). Although conversion techniques are available, this straightforward pot-boiling method of characterization does not yield results that are consistent with the true boiling point. Nonetheless, this kind of information is frequently accessible for the majority of jet fuels since ASTM D86 is the reference technique for most jet fuel requirements.

Flash Point: One important feature of jet and diesel fuel standards that makes safe fuel handling possible is the flash point (Vozka et al., 2019). When an ignition source is supplied, it is the lowest temperature at which a substance's vapor will ignite, corrected to 101.5 kPa (Kim et al., 2017). Small compounds with low flash points lower the flash point, which is dependent on the fuel's composition. Fuel standards frequently outline a fuel's minimum flash point. For instance, military jet gasoline has a minimum flash point of 333.15 K (Vozka et al., 2018a). Since fuel's smaller constituents contribute to its low flash point, it's critical to understand how these compounds do so.

Density: The ASTM D4052 (Standard Test Method for Density, Relative Density, and API Gravity of Liquids by Digital Density Meter) is commonly used to measure density for specification reasons at 60°F/15.5°C (Standard, 2018). According to the World Survey and the CRC Handbook, density increases linearly with temperature between −40°C and around 100°C (Lide & Kehiaian, 2020). Moses demonstrates that this linearity is true for both pure hydrocarbons and other fuels. This linearity is demonstrated using information from the three reference fuels, World Survey fuels, and "Jet A" from the CRC Handbook. Table 8.3 illustrates jet fuel specifications.

8.4.3 Quality Standards and Specifications

For biofuels to be used successfully and without long-term technical problems, they must be of a high grade. Quality is influenced by several factors that reflect the physical and chemical characteristics

TABLE 8.3
Jet Fuel Specifications

Property	AN-F-32, 1944 "JP-1"	ASTM D1655, 1959 "Jet A"	ASTM D1655, 2018 "Jet A"	JP-8, 2018	World Survey Avg
Density at 15°C (g/cm³)	0.85 max	0.775–0.83	0.775–0.84	0.775–0.84	0.80
Flash point (°C)	43 min	44–66	38 min	38 min	49.5
Viscosity (cSt)	10 at −40°C, max	15 at −34°C, max	8 at −20°C, max	8 at −20°C, max	4.3
Freeze point (°C)	−60°C max	−40 max (Jet A), −50 (Jet A-1)	−40 max (Jet A), −47 (Jet A-1)	−47 °C, max	−52.1
Aromatics (vol %)	20, max	20, max	25, max	25, max	17.5
Heat of combustion (MJ/kg)	-	42.8, min	42.8, min	42.8, min	43.05
Sulfur content (mass %)	0.2, max	0.3	0.3	0.3	0.049
Acid number (mg KOH/g)	-	0.1	0.1	0.015	0.006
Smoke point (mm)	-	20 min	18 min	18 min	-

Source: Rumizen (2021).

of biodiesel. Many factors can impact the quality of biodiesel, such as the quality of the feedstock, the fatty acid composition of the parent vegetable oil or animal fat, the postproduction parameters, handling, and storage, as well as the production process and other materials used in it. Given that diesel fuel is designed to power most contemporary diesel engines, the physicochemical properties of biodiesel should be similar to those of diesel oil (He, 2016). The primary criterion for evaluating the quality of biodiesel is whether or not its physical and chemical characteristics meet a suitable standard (Barabás & Todoruţ, 2011; Mofijur et al., 2017). Because of the development of compression ignition engines, increasingly stringent pollution regulations, reassessment of the suitability of feedstocks used in biodiesel production, etc., quality criteria for biodiesel are constantly being updated.

Numerous elements that may be expressed in the chemical and physical characteristics of biodiesel can affect its quality. In order for biodiesel to be sold, it must meet the quality requirements set by organizations such as the American Society for Testing and Materials (ASTM) and the European Committee for Standardization (ISO) (Di Serio et al., 2017). These rules, which are subject to change and require regular reviews, outline both the test procedures used and the quality standards. Tables 8.4 and 8.5 show different biodiesel specifications in several standards of testing methods (Barabás et al., 2010).

Features of the current diesel fuel standards include the majority of the most common diesel engine types in the region, the emissions' regulations governing those engines, the climate and development stage of the country or region where biodiesel is produced and/or used, and finally, the justification for using biodiesel. Although there is a substantial fleet of cars with diesel engines in Europe, the United States and Brazil specifically use diesel engines in trucks (Ogunkunle & Ahmed, 2019). The most commonly used feedstocks are rapeseed and sunflower oil in Europe, palm, jatropha, and coconut oil in Asia, palm oil and soybean oil in Australia, waste vegetable oil and animal fat in New Zealand, and soybean oil and waste vegetable oil in the United States and Canada. As a result, it is not surprising that regional requirements vary somewhat; it is and will never be able to establish a worldwide biodiesel quality standard.

The biodiesel standards in Brazil and the United States apply to both types of biodiesels: FAME and FAEE, while the current European biodiesel standard only covers FAME. Furthermore, the

TABLE 8.4
European (EN 14214) Specifications of Biodiesel Fuels

		Limits		
Property Specifications	**Test Methods**	**Min**	**Max**	**Unit**
Ester content	EN 14103	96.5	-	% (m/m)
Density at 15°C	EN ISO 3675, EN ISO 12185	860	900	kg/m^3
Viscosity at 40°C	EN ISO 3104, ISO 3105	3.5	5.0	mm^2/s
Flash point	EN ISO 3679	120	-	°C
Sulfur content	EN ISO 20846, EN ISO 20884	-	10.0	mg/kg
Cetane number	EN ISO 5165	51	-	-
Water content	EN ISO 12937	-	500	mg/kg
Oxidative stability, 110°C	EN 14112 6.0	6.0	-	h
Acid value	EN 14104	-	0.5	mg KOH/g
Iodine value	EN 14111	-	120	g L/100 g
Methanol content	EN 14110	-	0.2	% (m/m)
Monoglyceride content	EN 14105	-	0.8	% (m/m)
Triglyceride content	EN 14105	-	0.25	% (m/m)
Alkali metals (Na+K)	EN14108; EN 14109	-	0.5	mg/kg
Phosphorus content	EN 14107	-	10.0	mg/kg

TABLE 8.5
ASTM Specifications of Biodiesel Fuels

Property Specifications	Test Methods	Limits Min	Max	Unit
Ester content	ASTM D6751	96.5	-	% (m/m)
Density at 15°C	ASTM D1298	860	890	kg/m^3
Viscosity at 40°C	ASTM D445	3.5	5.0	mm^2/s
Flash point	ASTM D93	120	-	°C
Sulfur content	D5443/P83	-	10.0	mg/kg
Cetane number	ASTM D613	51	-	-
Water content	D2709/P40	-	500	mg/kg
Oxidative stability,110°C	ASTM D6751	3	-	hours
Acid value	ASTM D664	-	0.5	mgKOH/g
Monoglyceride content	ASTM D6584	-	1.0	% (m/m)
Triglyceride content	ASTM D6584	-	0.25	% (m/m)
Phosphorus content	STM D4951	-	10.0	mg/kg
Kinematic viscosity at 40°C	D 445	1.9	6.0	mm^2/s

biodiesel standards in Australia, Brazil, India, Japan, South Africa, and the United States describe a product that represents a blending component in conventional hydrocarbon-based diesel fuel, in contrast to the European standard, which describes a product that can be used as a stand-alone fuel for diesel engines or as a blending component in conventional diesel fuel (Etim, 2022; Ozkan, 2019). There are feedstock-neutral biodiesel requirements as well as ones that are based on locally accessible feedstock. The main causes of the variation in these technical requirements are the feedstock's place of origin and the features of regional markets.

8.5 SOCIOECONOMIC AND ENVIRONMENTAL CONSIDERATIONS

8.5.1 Life Cycle Assessment of Lipid-Based Biofuel Production

The potential for advanced biofuels depends on technologies that are both economically and environmentally sound. The demand for biofuels is rapidly increasing globally due to its captivating and environmentally friendly benefits, including (i) being renewable, (ii) improving carbon fixation and indirectly lowering carbon dioxide, (iii) fostering local economic growth, (iv) reducing air pollution from burning biomass in fields and biomass rotting in fields, (v) bringing energy security and reducing reliance on imported oil, and (vi) creating jobs for farmers and high-tech jobs for engineers, specialists, process bioengineers, and scientists (Azad & Rasul, 2019; Mounica et al., 2024).

However, there are a number of important constraints that must be properly managed, including economics, limited resources, disputes over food security, health and safety, water demand, land use difficulties, and several environmental repercussions that could affect those in the vicinity. These issues can usually be addressed via a life cycle assessment (LCA) (Prasad et al., 2020). LCA can help evaluate the potential environmental impacts of producing biofuel throughout the course of its entire life cycle. The first stage in using the LCA approach for advanced biofuels is to comprehend the needs of farmers, the types of feedstocks and their potential, and changes in land usage.

A LCA is frequently used to ascertain if a biomass biofuel system leads to an improved energy balance (input–output energy ratio) or a net reduction in GHG emissions. "Compilation and evaluation of the inputs, outputs, and the potential environmental impacts of a product system throughout its life cycle" is what ISO 14040 defines as an LCA (Silva, 2021). All input and output data from every stage of the product's life cycle, including the cultivation of biomass, the storage and transportation of feedstock, the manufacture and transportation of biofuel, and the end usage, must be included in an LCA. Additionally, all outputs including byproducts and gases that are either caught or leak are taken into consideration. Although there are numerous documented LCA techniques, the majority concentrate on two main environmental criteria: GHG emissions and energy balance, as well as a small number of important input categories (Zhang et al., 2017); few LCAs take into account extra factors like soil damage or water use.

Depending on the type of feedstock used and how it is produced and processed, LCAs (Pikula et al., 2020) of the environmental effects of biofuel production and consumption have revealed a wide range of results, from a net decrease in GHG emissions to a net increase, as well as the possibility of unanticipated negative environmental effects (Osman et al., 2021). LCA analyses are difficult not just because they need a lot of data, but also because they try to mix different quantities in ways that need a lot of interpretation and explanation (Lesage et al., 2018). A product's energy consumption, for instance, could be the subject of an LCA analysis that combines energy inputs such as local heat generated by burning natural gas, electricity generated by a nuclear power plant, and the power from a truck that runs on diesel fuel to deliver the product to the market. Solar heat is one example of an energy source that is thought to be free and environmentally benign.

8.5.2 Land Use Change and Its Impact

When evaluating the possible emission impacts of increasing the production of biofuels, it is important to know whether the increased production will be achieved through increased cultivated area or improved land productivity (Shrestha et al., 2019; Taheripour et al., 2017); in the latter case, the type of land is also important. Moreover, agricultural production methods influence GHG balances. These two elements will also influence other environmental effects on water, soils, and biodiversity. Around 80% of the rise in worldwide agricultural commodity production over the previous 50 years has come from yield gains, with the remaining portion coming from increasing cultivated area and cultivation frequency. Over the last few years, the demand for biofuels has grown at a rate that is significantly faster than historical rates of growth in crop yields and demand for agricultural commodities.

In light of both first- and second-generation technology, this implies that land use change and the resulting environmental effects may become a more significant concern. This demand may be met mainly in the short term by expanding the land area planted to biofuel crops (Bello et al., 2021; Van Der Hilst et al., 2018), but in the medium and long term, new technologies (such as cellulosic conversion), agronomic methods, and the creation of enhanced biofuel crop varieties may start to take center stage. The sustainable production of biofuel feedstocks will require significant yield gains and technology advancements to reduce the conversion of land not now used for crop production, such as grassland or forest land (Zuo et al., 2018), and the rapid land use shift in areas already under cultivation.

Expanding crop production in forests, wetlands, or other land uses could be harmful to the ecosystem because these regions provide important environmental services like carbon sequestration, water filtration, and biodiversity preservation. Between 250 and 800 million hectares of land, the majority of which is in tropical Latin America or Africa, are thought to be potentially available for increased crop production after removing land required to meet the growing demand for food crops and livestock, protected areas, and forest land.

While some of this land might be utilized directly to produce biofuel feedstock, more biofuel production on already-existing cropland could also lead to more non-biofuel crops being produced elsewhere. In the central United States, for instance, greater production of maize for ethanol has replaced soybeans on some existing cropland. This might lead to increased production of soybeans and the conversion of grassland or forest land elsewhere. In order to fully comprehend the possible environmental effects, it is necessary to take into account both the direct and indirect land use changes brought about by increased biofuel production.

8.5.3 Greenhouse Gas Emissions Reduction Potential

Prior to the development of renewable energy sources, fossil fuels dominated the energy landscape for a number of years. At first, coal and oil were utilized more frequently than natural gas. Research on switching from fossil fuels to renewable energy sources has increased in response to growing environmental concerns (Ebhota & Jen, 2020). The use of renewable energy sources, particularly geothermal, biomass, and hydro fuel, started to increase around 1965, despite the fact that fossil fuel-dominated energy production has historically predominated. Since the 1990s, wind and solar energy have grown significantly as the environmental impact of fossil fuels has come to light (Balali et al., 2017). Renewable energy sources, including solar, wind, hydroelectric, and geothermal power, are essential for addressing climate change and offering a more environmentally friendly substitute for fossil fuels.

More than 90% of the required reductions in energy-related emissions can be accounted for by efficiency and renewable energy, according to the analysis. Companies and governments have crucial roles in facilitating the switch to renewable energy sources in order to build a low-carbon, sustainable future. The particular technology and lifetime considerations can affect how renewable energy sources affect the environment. Certain biomass conversion processes, for instance, do emit GHGs, but usually not to the same extent as ones that require fossil fuels. Comparing renewables to traditional fossil fuel energy sources, however, reveals substantial environmental benefits overall (Maradin, 2021). Renewable energy comes from infinitely abundant, naturally replenished sources like biomass, geothermal heat, wind, water, and sunshine.

Comparing renewables to traditional fossil fuel energy sources, however, reveals substantial environmental benefits overall. Throughout their life cycle, renewables often have a significantly smaller carbon footprint, improve air quality, and provide long-term, sustainable energy solutions. Because they have the potential to drastically cut GHG emissions, biofuels like ethanol and biodiesel are becoming more and more popular. According to Datta et al. (2024), ethanol can lower emissions by up to 65%, while biodiesel derived from cooking oil can reduce emissions by up to 87% when compared to petroleum diesel.

8.5.4 Economic Aspects and Market Analysis

Concerns of bioenergy are growing in the recent decade at the worldwide level as a consequence of expanding global demand for energy, predicted fossil fuel shortages, and the impact of fossil energy consumption on our environment (Duarah et al., 2022). However, energy prices are also thought to be one of the main causes of the growing interest in energy-saving alternatives. These prices have an impact on the food and agriculture industries in a number of ways, including opening up new markets for goods that can be used as biomass feedstock to produce biofuel in place of petroleum-based fuels (diesel, petrol).

While there are many benefits to bioenergy, there are also some drawbacks, such as the conversion of land from food to fuel commodities, which raises food costs and decreases food supply in some areas. The FAO concurred that the present growth in bioenergy production in developing nations had both possible drawbacks and advantages; while it might jeopardize food security and harm the environment, it also presented substantial chances for sustainable development and the eradication of poverty in rural areas. Governments of rich and developing economies are fast to respond to the energy challenge by establishing and putting in place bioenergy policies and programs. Plant-based fuel would significantly lower carbon dioxide emissions and, in certain cases, lessen dependency on imported oil.

When evaluating the economic viability and feasibility of biodiesel manufacturing processes, techno-economic analysis, or TEE, is essential. It entails assessing the total costs, income, and profitability of producing biodiesel while taking into account a number of variables, including feedstock costs, capital expenditures, operating costs, and market pricing. TEA has been used in recent research to evaluate and improve the processes involved in the production of biodiesel, offering important information for process design and decision-making. S. Zhang et al.'s (2022) study of several feedstocks and process configurations for biodiesel synthesis is one instance of TEA in biodiesel production.

Market prices, feedstock availability, conversion efficiency, capital expenditures, and operating expenses were all taken into account in the analysis. The study emphasized how crucial feedstock selection and process optimization are for producing biodiesel at a reasonable cost. In a different study, Tasić (2020) employed TEA to create biodiesel from used cooking oil. Energy use, environmental effects, and capital and operating cost estimation were all included in the investigation. The study identified key factors influencing the process's overall economics and proved that producing biodiesel from used cooking oil is economically feasible. The investigation took into account a number of possibilities, such as alternative conversion methods and cultivation techniques (Bhatia et al., 2021).

The study evaluated the economic feasibility of producing biodiesel from microalgae by taking into account variables including lipid content, biomass productivity, capital investment, and operating costs. This new study emphasizes how important TEA is for evaluating the economics of biodiesel production. Process optimization and decision-making are aided by TEA's analytical information on the sustainability, profitability, and cost-effectiveness of biodiesel production processes, which is provided by considering a wide range of variables.

8.5.5 Social and Ethical Considerations

The generation of biofuel has significant and varied social ramifications. The decrease in reliance on foreign oil is one important advantage. Both sides of the Atlantic have acknowledged the potential economic and security benefits of this change. The government of the United States has recognized that biofuels can improve economic stability and national security by reducing the need to import petroleum fuels (Jeswani et al., 2020). Biofuels can also lower GHG emissions and provide supply security for the EU, according to the European Commission (Pryshliak & Tokarchuk, 2020).

Another essential component of the social advantages connected to the manufacturing of biofuel is the development of jobs. Numerous job possibilities are generated by the building and operation of biorefineries in a variety of industries, such as feedstock production, design and construction, research and development, biorefinery operation, sales, and distribution (Malik et al., 2024). In addition to promoting economic growth, this local employment helps communities build their skills. It is difficult to estimate how many jobs biofuel initiatives would create, as estimates differ from study to study (Bacovsky et al., 2016). But it's obvious that the production of biofuel might lead to a sizable increase in direct, indirect, and induced employment. There are a number of social costs associated with the shift to biofuel production, especially for the aviation and maritime industries, which need to be properly evaluated. These expenses impact related industries and the environment in addition to being direct, such as those related to labor and production supplies.

A major worry is the competition for resources. There may be a conflict in the distribution of resources because biofuels demand materials that may be utilized for animal feed or compost. Although biofuels might lower GHG emissions, their production method may result in additional environmental problems, making their environmental impact still another crucial consideration (Kostidi & Lyridis, 2024). One possible outcome of land acquisition for biofuel crops is social displacement as communities may be uprooted.

Similarly, when agricultural byproducts are redirected from food production to biofuel production, it may put strain on food supply networks, raising concerns about food security. Jobs in traditional energy sectors may be lost to the establishment of jobs in biofuel production (Harvey-Sánchez, 2022). Furthermore, it is impossible to ignore the health effects on nearby communities because industrial operations may pollute the air or water, which could have serious health repercussions.

The production of biofuel from nonfood wastes, such as pellets, straw, pruning from vineyards and olive groves, and organic waste is expected to have minimal societal costs due to the nature of the feedstock and its small scale. However, in addition to its environmental benefits, the manufacturing of biofuel offers several socioeconomic benefits. Because they reduce GHG emissions, create jobs, and lessen dependency on foreign oil, biofuels present a sustainable path forward for society and the planet. Economic growth, environmental conservation, and social well-being must all be balanced for implementation to be successful (Santos et al., 2024; Solarte-Toro et al., 2023).

8.6 CHALLENGES AND FUTURE PERSPECTIVES IN LIPID-BASED BIOFUEL PRODUCTION

8.6.1 Feedstock Sustainability and Scalability

The sustainability and accessibility of appropriate feedstocks present another major obstacle to the manufacturing of bioethanol. Addressing environmental issues more sustainably can be achieved by encouraging the production of bioethanol from non-food sources like algae or LCB (Devi et al., 2022). One urgent problem is the shortage of feedstock brought on by the world's expanding population and demand. Furthermore, the production of first-generation bioethanol frequently interferes with the food supply chain and competes with the food sector for crops that contain starch, such as corn, sugarcane, beets, sorghum, barley, and fruits. More focus is being placed on the manufacture of second- and third-generation bioethanol in order to counteract this. However, if implemented carefully, second-generation bioethanol has tremendous promise given the feedstock's availability.

Using vegetable oil to produce biodiesel (first generation) has several obstacles, such as competition between fuel and food production, changes in land use, and more expensive fuel. However, the overuse of water and land is a risk associated with second-generation biodiesel. However, lipids from LCB produced by microorganisms like bacteria, yeast, and fungi can be utilized efficiently to produce sustainable biodiesel. The third (algal) and fourth generations of biodiesel have been used to try to address these issues with the first two, although they have some drawbacks.

Despite mounting scientific proof of the feasibility of alternative biofuel production pathways, legislative and regulatory restrictions exacerbate these challenges and prevent further advancement in the global process of de-escalating the use of conventional fuels. The production of biodiesel is expanding quickly worldwide despite these obstacles because of the fuel's security and significantly lowers environmental impact (Naylor & Higgins, 2018). Additionally, governments are under growing pressure to embrace and put into place more environmentally friendly production methods. Therefore, it is anticipated that the demand for biofuel consumption will increase dramatically over the next several years as a result of rising petroleum-based fuel prices, growing concerns about pollutants, and government policies in some countries that encourage a move toward the use of renewable energy.

The conflict between using land to grow biomass and maintaining food security is a major obstacle to the production of biofuel and its sustainability. There is a chance that additional land usage and deforestation will occur to satisfy the expanding demand for agricultural raw materials, particularly in regions with significant biodiversity (Kline et al., 2017). Indeed, the indirect shift in land usage must be taken into account. When food supplies are insufficient for everyone, the development of biofuels from raw materials like sugarcane, corn, wheat, and soybeans competes with food production. In fact, over time, the usage of food crops has sparked discussions about food safety and fueled the "food versus fuel" debate. The demand for food commodities will undoubtedly rise as a result, which will directly raise their cost. Because of this circumstance, scientists are searching for non-food substitutes to use as feedstocks in order to produce affordable biofuels that are comparable to traditional fuel (Malode et al., 2022).

8.6.2 Improving Conversion Efficiency and Reducing Costs

In recent years, biodiesel, a clean and renewable substitute for traditional diesel fuel, has advanced significantly (Maheshwari et al., 2022). These advancements have sought to expand the range of feedstock options, enhance biodiesel's overall sustainability as a viable energy source, and strengthen the manufacturing processes' efficiency. The use of sophisticated catalysts in the manufacture of biodiesel is one noteworthy recent advancement. The process known as transesterification, which turns vegetable or animal fats into biodiesel, depends heavily on catalysts (Babadi et al., 2022). To increase reaction rates, lower energy consumption, and improve the quality of biodiesel, researchers have been investigating a variety of catalysts, including solid acid catalysts, enzyme catalysts, and heterogeneous catalysts.

According to Garcia-Silvera et al. (2023), these catalysts have benefits such as slower reaction times, higher conversion rates, and simpler catalyst-product separation. The generation of biodiesel using unconventional feedstocks is another noteworthy development. Although rapeseed and soybean oils are common feedstocks for biodiesel, scientists have been looking at other sources like algae, used cooking oil, and non-food crops like camelina and jatropha. The capacity of algae to flourish in a variety of conditions and their high oil content have drawn attention to them. By using non-traditional feedstocks, biodiesel's overall sustainability is improved, and competition with food production is lessened (Garg et al., 2023).

Additionally, attempts have been undertaken to lessen the environmental impact of biodiesel manufacturing in order to increase its sustainability. This includes lowering GHG emissions, putting waste management plans into place, and streamlining production procedures to use less energy and water. Researchers have also been investigating the idea of "second-generation" biodiesel, which entails making biodiesel from waste materials, including lignocellulosic biomass and agricultural wastes. This strategy optimizes resource use while minimizing waste (Makepa, 2023).

The conversion of biomass into biofuels is greatly influenced by its physical properties, including its bulk density (BD), microstructure, particle size, moisture content, and elastic properties. The final volume that is acceptable will therefore rely on how the processing system is used, and a high degree of size reduction is necessary for a high conversion through the biochemical process. However, biomass conversion through the thermochemical process is less impacted by biomass particle size because of the quick rate at which liquid media are heated (Das et al., 2021). This makes particle size less sensitive. However, before being pumped into a continuous system, the biomass sludge needs to be greatly reduced in size. According to a different study, particles smaller than 0.5 mm are best suited for the biomass-converting process known as pyrolysis because they can heat up faster and generate less char. The optimal size for burning biomass depends on the type of feedstock; in general, some particle size reduction is necessary for any conversion process.

The other physical attribute that most significantly affects the conversion of biomass waste into biofuel is moisture content (Kenney et al., 2013). It may affect the supply of feedstock and the biorefining procedures directly or indirectly. Few studies, for instance, have demonstrated that a higher moisture content during the pretreatment stage of the biological process may result in poorer bio-oil quality and thermochemical conversion, as well as reduced combustion thermal efficiency. In addition to moisture content and particle size, other physical attributes that affect biomass conversion to biofuels include microstructure, elastic properties, and BD (Shojaeiarani et al., 2019). The primary source of fuel is biomass, which can be either plant seeds or crops like sugarcane but have limited product yield and compete with food crops (first generation); lignocellulosic biomass, kitchen waste, and crop residue, but are unsuccessful due to high processing costs caused by lignin and costly lipid extraction technologies (second generation); microalgae show increased oil yield and rapid growth, but are still costly for large-scale production (third generation); genetically modified algae with sophisticated extraction techniques have recently been used for large-scale production of biofuel but also require a significant initial investment and a lot of research (fourth generation).

The cost of producing bioethanol is still higher than that of producing ethanol from fossil fuels or starch crops, underscoring the necessity of cost competitiveness to maintain economic viability (Mohapatra et al., 2019). In order to get over these obstacles, using microalgae as a high-lipid, non-edible feedstock has become a viable strategy for developing third-generation bioethanol production. Algal biomass can be converted into a variety of biofuels, including biogas, bioethanol, biodiesel, and bio-oils, using techniques such as anaerobic digestion, fermentation, transesterification, liquefaction, and pyrolysis. The utilization of algal biomass for production is economically challenging due to the high costs of pond culture, downstream processing, and energy inputs.

8.6.3 Enhancing Biofuel Properties and Performance

First-generation biodiesel production from vegetable oil faces a number of challenges, including competition between fuel and food production, shifting land usage, and high fuel costs (Babadi et al., 2022). However, second-generation biodiesel presents a risk of overwatering and land exploitation. However, LCB microorganisms, including bacteria, yeast, and fungi, can create lipids that can be used to produce sustainable biodiesel. Third- (algal) and fourth-generation biodiesel have been used to try to address these issues with both generations of biodiesel, but they're not without limitations.

The substantial cost of extracting oil from microalgae and turning it into biodiesel has been one of the main challenges in the manufacture of algal biodiesel (Kim et al., 2022). In addition to this, the use of biodiesel in the combustion process affects the engine's operational efficiency and indirectly affects the noise and vibration levels of the engine. These elements play a critical role in determining how comfortable car occupants are overall. According to one definition, noise is the aural manifestation of irregular vibrations and is a serious health risk that follows air pollution.

A variety of everyday human tasks may be negatively impacted by this phenomenon, and it may even contribute to the emergence of physical illnesses such as neurological and cardiovascular conditions. This biofuel also has several disadvantages, such as corrosion and oxidative deterioration (Ogunkunle & Ahmed, 2021). Acids, alcohols, aldehydes, and polymers are examples of oxidation byproducts that may develop during the breakdown of biodiesel. These wastes can clog the tiny pores of filters and nozzles and could affect the corrosion resistance of materials used in the fuel delivery system. The transesterification process uses a variety of homogeneous and heterogeneous catalysts to create biodiesel.

There are several challenges with using homogeneous catalysts in the biodiesel synthesis process, including the formation of sludge and the intricate product separation. The development of heterogeneous catalysts with high catalytic efficiency in the production of biodiesel is therefore still heavily favored. Heterogeneous catalysts also have a number of drawbacks, including complex reaction processes, difficult catalyst manufacturing, extended reaction times, active site leaching, and lower biodiesel yield. Because of the aforementioned issues, scientists are concentrating on creating solid heterogeneous catalysts with improved environmental friendliness, cost-effectiveness, and efficiency. The constraints have been attempted to be addressed by a number of recent technologies, such as nanotechnology.

Utilizing bioethanol as a complete biofuel presents a number of challenges. The biggest problems are with the engine, specifically with corrosion and wear (Matbouei, 2018). Bioethanol's high oxygen concentration and hygroscopic nature cause corrosion in the electrical fuel pump and fuel injector, which results in engine wear and startup issues, particularly in cold climates. Second, higher ethanol mixes can damage components and dissolve contaminants, which can lead to blocked filters and injectors, which can cause problems with fuel system compatibility, especially in older cars. It also requires more frequent refueling due to its inferior fuel economy, due to its lower energy density as compared to gasoline. Furthermore, although bioethanol lowers emissions of hydrocarbons and carbon monoxide, it may raise emissions of nitrogen oxides (NOx), which would contribute to air pollution.

8.6.4 Integration with Existing Fuel Infrastructure

While the petroleum industry's current infrastructure, which includes storage, blending, transportation, and dispensing, can be utilized by renewable liquid fuels, there are still certain problems that must be resolved. The majority of biofuels can be blended with gasoline and diesel quite simply because of their similar qualities. In addition to creating challenges for its storage and transportation, cold weather can cause engine problems due to increased viscosity and gel formation when utilizing some biodiesels.

Coexisting oil-product infrastructure with biofuels may result in hydrous ethanol-induced water contamination, which could necessitate the use of new materials to prolong the equipment and pipeline's useful lives. If ethanol is delivered in the same lines as oil products, moisture from condensation in the pipelines may raise the water content of the ethanol. Further distillation may be necessary after delivery if it surpasses the bioethanol's technical criteria. Additionally, contaminants found in multi-product pipeline systems can be dissolved and carried by ethanol and biodiesel, which could be detrimental to ICEs. Therefore, if better cleanup techniques between items being sent over multi-product pipelines prove unsuccessful, a dedicated pipeline might be preferable.

Elevated levels of ethanol can cause rapid stress corrosion cracking (SCC) in storage tanks and steel pipes, particularly at bends and weld joints. Using selective post-weld heat treatments, covering internal important zones (such as pipeline weld points), and adding tank liners can all help to prevent this. All of them raise system costs, though. Seals and valves in pipelines, terminals, and some engines may require repair because ethanol can break down various elastomers and polymers. Ethanol-compatible polymers might be used in the valves, gaskets, and seals of new pipelines, which would reduce SCC.

Biofuel may be easily incorporated into practically any fossil fuel, including gasoline, diesel, and kerosene, and it doesn't require significant upgrades to the current infrastructure. In addition, biofuel can cut carbon dioxide emissions by 78% when compared to fossil fuels, making it a carbon-neutral fuel. Moreover, it has been demonstrated that biodiesel is adaptable; after 30 days, it ranges from 80.4% to 91.2% biodegradability, whereas fossil diesel only has 24.5%. Biofuel is a more promising option to replace depleting fossil fuels and lessen the effects of global warming because of all these factors.

Despite its seeming promise, the implementation of biofuel in numerous industries was not proceeding as planned. Biofuels provide a novel and unknown risk to the environment and the current infrastructure, notwithstanding any potential benefits of their use (Oloyede et al., 2024). Furthermore, even in underdeveloped nations, the widespread use of biofuel made directly from edible organisms may raise more questions about food security. The study found that, despite its simplicity, biofuel made from edible food has a greater influence on society as a whole and raises issues related to food availability and security. As feedstock costs continue to rise, this situation may worsen and ultimately reduce the economics of producing biofuel.

8.7 CONCLUSIONS

Lipids are essential biological components and prospective renewable energy sources, as the chapter Introduction to Lipids and Biofuels concludes. It clarifies the variety of lipid structures and their roles in energy storage and cellular functions, especially those of TAGs. This chapter goes on to examine the potential of several lipid sources, such as waste-derived fats, microbial lipids, and plant oils, in the environmentally friendly manufacturing of biofuels like biodiesel. It draws attention to the benefits of using waste and non-food biomass, which can lessen competition for food supplies and have a positive environmental impact. This chapter also discusses the difficulties and breakthroughs in lipid extraction and conversion technology, highlighting the necessity of efficient and economical ways to improve the viability of biofuels. Through the integration of biochemical insights and energy innovation, this chapter provides a thorough framework for comprehending how lipid science and sustainable fuel development connect.

COMPETING INTERESTS

The authors declare that they have no competing interest.

DECLARATION OF GENERATIVE AI AND AI-ASSISTED TECHNOLOGIES IN THE WRITING PROCESS

During the preparation of this work, the book chapter author(s) have not taken any assistance from AI, and the book chapter author(s) reviewed and edited the content as needed and take(s) full responsibility for the content of the publication.

BIBLIOGRAPHY

Abdulridha, S. G., Radhi, A. J., Abbas, M. N., Shehab, A. K., & Aluaibi, W. M. (2023). Maximizing high value light products by catalytic cracking of VGO produced by Iraqi refinery over formulated FCC catalyst. *Engineering Chemistry*, 4, 7–16.

Agarwal, R. (2025). *Biofuels and biodiesels: Renewable energy explained*. Educohack Press.

Akca Prill, M. (2019). *Energy dependency, the potential supply of renewable energies and the political responses in Turkey in the decades since the oil-crisis*. Dissertation, Darmstadt, Technische Universität Darmstadt.

Arunachaleswara, P., Kanna, R., Nair, S. G., Francis, C., Bhaktan, N., & Kuruvila, A. (2017). Effect of free water and rust on flash point of diesel. *International Journal of Engineering Science*, 7, 21–24.

Aslam, A., Ibrahim, M., Mahmood, A., Mubashir, M., Sipra, H. F. K., Shahid, I., Ramzan, S., Latif, M. T., Tahir, M. Y., & Show, P. L. (2021). Mitigation of particulate matters and integrated approach for carbon monoxide remediation in an urban environment. *Journal of Environmental Chemical Engineering*, 9(4), 105546.

Asokapandian, S., Sreelakshmi, S., & Rajamanickam, G. (2021). Lipids and oils: An overview. In *Food biopolymers: Structural, functional and nutraceutical properties* (pp. 389–411). Springer Nature.

Azad, K. A., & Rasul, M. (2019). *Advanced biofuels: Applications, technologies and environmental sustainability*. Woodhead Publishing.

Babadi, A. A., Rahmati, S., Fakhlaei, R., Barati, B., Wang, S., Doherty, W., & Ostrikov, K. K. (2022). Emerging technologies for biodiesel production: Processes, challenges, and opportunities. *Biomass and Bioenergy*, 163, 106521.

Bacovsky, D., Ludwiczek, N., Pointner, C., & Verma, V. K. (2016). *IEA bioenergy countries' report: Bioenergy policies and status of implementation*. www.researchgate.net/institution/BIOENERGY_2020_GmbH/members

Bagchi, S. K., Patnaik, R., & Mallick, N. (2018). Algal biodiesel production: An overview. In *Bioenergy and biofuels* (pp. 145–172). Taylor and Francis.

Bajpai, P. (2022). *Fourth generation biofuels*. Springer.

Balali, M. H., Nouri, N., Omrani, E., Nasiri, A., & Otieno, W. (2017). An overview of the environmental, economic, and material developments of the solar and wind sources coupled with the energy storage systems. *International Journal of Energy Research*, 41(14), 1948–1962.

Barabás, I., & Todoruţ, I.-A. (2011). *Biodiesel quality, standards and properties* (Vol. 2). InTech, Pleasanton, CA.

Barabás, I., Todoruţ, A., & Băldean, D. (2010). Performance and emission characteristics of an CI engine fueled with diesel–biodiesel–bioethanol blends. *Fuel*, 89(12), 3827–3832.

Bedoić, R., Špehar, A., Puljko, J., Čuček, L., Ćosić, B., Pukšec, T., & Duić, N. (2020). Opportunities and challenges: Experimental and kinetic analysis of anaerobic co-digestion of food waste and rendering industry streams for biogas production. *Renewable and Sustainable Energy Reviews*, 130, 109951.

Bello, S., Salim, I., Feijoo, G., & Moreira, M. T. (2021). Inventory review and environmental evaluation of first-and second-generation sugars through life cycle assessment. *Environmental Science and Pollution Research*, 28, 27345–27361.

Bhatia, S. K., Mehariya, S., Bhatia, R. K., Kumar, M., Pugazhendhi, A., Awasthi, M. K., Atabani, A., Kumar, G., Kim, W., & Seo, S.-O. (2021). Wastewater based microalgal biorefinery for bioenergy production: Progress and challenges. *Science of the Total Environment*, 751, 141599.

Binhweel, F., Hossain, M. S., & Ahmad, M. I. (2023). Recent trends, potentials, and challenges of biodiesel production from discarded animal fats: A comprehensive review. *BioEnergy Research*, 16(2), 778–800.

Callegari, A., Bolognesi, S., Cecconet, D., & Capodaglio, A. G. (2020). Production technologies, current role, and future prospects of biofuels feedstocks: A state-of-the-art review. *Critical Reviews in Environmental Science and Technology*, 50(4), 384–436.

Cherwoo, L., Gupta, I., Flora, G., Verma, R., Kapil, M., Arya, S. K., Ravindran, B., Khoo, K. S., Bhatia, S. K., & Chang, S. W. (2023). Biofuels an alternative to traditional fossil fuels: A comprehensive review. *Sustainable Energy Technologies and Assessments*, 60, 103503.

Chu, P. L., Vanderghem, C., MacLean, H. L., & Saville, B. A. (2017). Process modeling of hydrodeoxygenation to produce renewable jet fuel and other hydrocarbon fuels. *Fuel*, 196, 298–305.

Correa, D. F., Beyer, H. L., Fargione, J. E., Hill, J. D., Possingham, H. P., Thomas-Hall, S. R., & Schenk, P. M. (2019). Towards the implementation of sustainable biofuel production systems. *Renewable and Sustainable Energy Reviews*, 107, 250–263.

Das, P., Chandramohan, V., Mathimani, T., & Pugazhendhi, A. (2021). A comprehensive review on the factors affecting thermochemical conversion efficiency of algal biomass to energy. *Science of the Total Environment*, 766, 144213.

Datta, D., Biswas, S., Barman, D., Mandal, E., & Das, B. (2024). Evolution, challenges and benefits of biofuel production and its potential role in meeting global energy demands. In *Emerging sustainable technologies for biofuel production* (pp. 595–632). Springer.

De Feo, G., Di Domenico, A., Ferrara, C., Abate, S., & Sesti Osseo, L. (2020). Evolution of waste cooking oil collection in an area with long-standing waste management problems. *Sustainability*, 12(20), 8578.

Degraeuwe, B., & Weiss, M. (2017). Does the New European Driving Cycle (NEDC) really fail to capture the NOX emissions of diesel cars in Europe? *Environmental Pollution*, 222, 234–241.

Deora, P. S., Verma, Y., Muhal, R. A., Goswami, C., & Singh, T. (2022). Biofuels: An alternative to conventional fuel and energy source. *Materials Today:* Proceedings, 48, 1178–1184.

Devi, A., Bajar, S., Kour, H., Kothari, R., Pant, D., & Singh, A. (2022). Lignocellulosic biomass valorization for bioethanol production: A circular bioeconomy approach. *BioEnergy Research*, 15(4), 1820–1841.

Dhull, S. B., & Punia, S. (2020). Essential fatty acids: Introduction. In *Essential fatty acids* (pp. 1–18). CRC Press.

Di Serio, M., Turco, R., Vitiello, R., Tesser, R., Riazi, M., & Chiaramonti, D. (2017). Biofuels characterization and standard methods for quality control. In *Biofuels production and processing technology* (pp. 183–198). CRC Press.

Duarah, P., Haldar, D., Patel, A. K., Dong, C.-D., Singhania, R. R., & Purkait, M. K. (2022). A review on global perspectives of sustainable development in bioenergy generation. *Bioresource Technology*, 348, 126791.

Ebhota, W. S., & Jen, T.-C. (2020). Fossil fuels environmental challenges and the role of solar photovoltaic technology advances in fast tracking hybrid renewable energy system. *International Journal of Precision Engineering and Manufacturing-Green Technology*, 7, 97–117.

El-Araby, R. (2024). Biofuel production: Exploring renewable energy solutions for a greener future. *Biotechnology for Biofuels and Bioproducts*, 17(1), 129.

Elzagheid, M. (2024). *Biomacromolecules: Carbohydrates, Lipids, Proteins and Nucleic Acids*. Walter de Gruyter GmbH & Co KG.

Etim, A. O. (2022). *Experimental and computational exploration of advanced biodiesel fuels and hybridization process evaluation of feedstocks and their chemical combinations* (Doctoral thesis). Durban University of Technology. https://doi.org/10.51415/10321/4409.

Ewurum, C. E. (2019). *Techno-economic analysis of micro-algae bio-jet fuel production processes* (the degree of doctor of philosophy). University of Nottingham.

Foroutan, R., Esmaeili, H., Mousavi, S. M., Hashemi, S. A., & Yeganeh, G. (2019). The physical properties of biodiesel-diesel fuel produced via transesterification process from different oil sources. *Physical Chemistry Research*, 7(2), 415–424.

Ganguli, S., Somani, A., Motkuri, R. K., & Bloyd, C. N. (2018). *India alternative fuel infrastructure: The potential for second-generation biofuel technology*. Pacific Northwest National Laboratory (PNNL).

Garcia-Silvera, E., Melendez-Mogollon, I., Pérez-Arias, A., & de la Rosa, Y. A. (2023). *Enzymatic biodiesel production: Challenges and opportunities*. Congress on Research, Development, and Innovation in Renewable Energies: Selected Papers from CIDiER 2022.

Garg, J., Barnwal, P. P., Suman, J., & Rakshit, A. (2023). Bioenergy crops: Challenges and opportunities. In *Handbook of energy management in agriculture* (pp. 435–450). Springer Nature.

Gautam, A., Pant, M., Pant, G., & Kumar, G. (2024). Second-generation biofuels: Concepts, applications, and challenges. In *Microbial applications for environmental sustainability* (pp. 277–304). Research Gate.

Gnanasekaran, L., Priya, A., Thanigaivel, S., Hoang, T. K., & Soto-Moscoso, M. (2023). The conversion of biomass to fuels via cutting-edge technologies: Explorations from natural utilization systems. *Fuel*, 331, 125668.

Gupta, R., Gupta, N., Gupta, R., & Gupta, N. (2021). Lipid biosynthesis and degradation. In *Fundamentals of bacterial physiology and metabolism* (pp. 491–523). Research Gate.

Gurr, M. I. (2018). Lipids and nutrition. In *Lipid technologies and applications* (pp. 79–112). Routledge.

Habib, M. S., Tayyab, M., Zahoor, S., & Sarkar, B. (2020). Management of animal fat-based biodiesel supply chain under the paradigm of sustainability. *Energy Conversion and Management*, 225, 113345.

Hartikka, T., Kuronen, M., & Kiiski, U. (2018). *Technical performance of HVO (hydrotreated vegetable oil) in diesel engines*. SAE Technical Paper 2012-01-1585, 2012. *Diesel Engines*.

Harvey-Sánchez, A. (2022). *The history and activism surrounding the University of Toronto's (U of T) decision to divest from fossil fuels*. Climate Justice Organizing HUB Core Team, University of Toronto.

Hazards, E. P. o. B., Koutsoumanis, K., Allende, A., Bolton, D., Bover-Cid, S., Chemaly, M., Davies, R., De Cesare, A., Herman, L., & Hilbert, F. (2020). Evaluation of an alternative method for production of biodiesel from processed fats derived from Category 1, 2 and 3 animal by-products (submitted by College Proteins). *EFSA Journal*, *18*(4), e06089.

Hazrat, M., Rasul, M. G., Mofijur, M., Khan, M., Djavanroodi, F., Azad, A., Bhuiya, M. M., & Silitonga, A. (2020). A mini review on the cold flow properties of biodiesel and its blends. *Frontiers in Energy Research*, 8, 598651.

He, B.-Q. (2016). Advances in emission characteristics of diesel engines using different biodiesel fuels. *Renewable and Sustainable Energy Reviews*, 60, 570–586.

Hirani, A. H., Javed, N., Asif, M., Basu, S. K., & Kumar, A. (2018). A review on first-and second-generation biofuel productions. In *Biofuels: Greenhouse gas mitigation and global warming: Next generation biofuels and role of biotechnology* (pp. 141–154). Springer.

Hoang, A. T. (2021). Prediction of the density and viscosity of biodiesel and the influence of biodiesel properties on a diesel engine fuel supply system. *Journal of Marine Engineering & Technology*, 20(5), 299–311.

Hwangbo, M., & Chu, K.-H. (2020). Recent advances in production and extraction of bacterial lipids for biofuel production. *Science of the Total Environment*, 734, 139420.

Inam, S., Khan, S., & Nadeem, F. (2019). Impacts of derivatization on physiochemical fuel quality parameters of fatty acid methyl esters (FAME) – A comprehensive review. *International Journal of Chemical and Biochemical Science*, 15, 42–49.

Jaiswal, K. K., Chowdhury, C. R., Dutta, S., Banerjee, I., Jaiswal, K. S., Nisansala, H., Sangmesh, B., & Sirimuthu, N. (2024). Synthesis of renewable diesel as a substitute for fossil fuels. In *Renewable diesel* (pp. 1–31). Elsevier.

Jeswani, H. K., Whiting, A., & Azapagic, A. (2020). Environmental and economic sustainability of biomass heat in the UK. *Energy Technology*, 8(11), 1901044.

Jónasdóttir, S. H. (2019). Fatty acid profiles and production in marine phytoplankton. *Marine Drugs*, 17(3), 151.

Jones, A. D., Boundy-Mills, K. L., Barla, G. F., Kumar, S., Ubanwa, B., & Balan, V. (2019). Microbial lipid alternatives to plant lipids. In *Microbial lipid production: Methods and protocols* (pp. 1–32). Springer.

Kaisan, M., Anafi, F., Nuszkowski, J., Kulla, D., & Umaru, S. (2020). Calorific value, flash point and cetane number of biodiesel from cotton, jatropha and neem binary and multi-blends with diesel. *Biofuels*, *11*(3), 321–327. https://doi.org/10.1080/17597269.2017.1358944.

Kalsoom, M., El Zerey-Belaskri, A., Nadeem, F., & Shehzad, M. R. (2017). Fatty acid chain length optimization for biodiesel production using different chemical and biochemical approaches – A comprehensive. *International Journal of Chemical and Biochemical Sciences*, 11, 75–94.

Kashyap, D., Das, S., & Kalita, P. (2022). Influence of oxygenated fuel and additives in biofuel run compression ignition engine. In *Potential and challenges of low carbon fuels for sustainable transport* (pp. 183–243). Springer Singapore.

Kenney, K. L., Smith, W. A., Gresham, G. L., & Westover, T. L. (2013). Understanding biomass feedstock variability. *Biofuels*, 4(1), 111–127.

Khan, M. A. H., Bonifacio, S., Clowes, J., Foulds, A., Holland, R., Matthews, J. C., Percival, C. J., & Shallcross, D. E. (2021). Investigation of biofuel as a potential renewable energy source. *Atmosphere*, 12(10), 1289.

Kim, D., Martz, J., Abdul-Nour, A., Yu, X., Jansons, M., & Violi, A. (2017). A six-component surrogate for emulating the physical and chemical characteristics of conventional and alternative jet fuels and their blends. *Combustion and Flame*, 179, 86–94.

Kim, J. Y., Jung, J.-M., Jung, S., Park, Y.-K., Tsang, Y. F., Lin, K.-Y. A., Choi, Y.-E., & Kwon, E. E. (2022). Biodiesel from microalgae: Recent progress and key challenges. *Progress in Energy and Combustion Science*, 93, 101020.

Kirakosyan, A., Kaufman, P. B., Cseke, L. J., Podila, G. K., Kirakosyan, A., & Kaufman, P. B. (2009). Plants as sources of energy. In *Recent advances in plant biotechnology* (pp. 163–210). Springer US.

Kline, K. L., Msangi, S., Dale, V. H., Woods, J., Souza, G. M., Osseweijer, P., Clancy, J. S., Hilbert, J. A., Johnson, F. X., & McDonnell, P. C. (2017). Reconciling food security and bioenergy: Priorities for action. *Gcb Bioenergy*, 9(3), 557–576.

Knothe, G. (2010a). Biodiesel and renewable diesel: A comparison. *Progress in Energy and Combustion Science*, 36(3), 364–373.

Knothe, G. (2010b). Biodiesel derived from a model oil enriched in palmitoleic acid, macadamia nut oil. *Energy & Fuels*, 24(3), 2098–2103.

Kopchak, N., Lisovska, T., Pokotylo, O., & Vichko, O. (2021). *Biochemistry (structural biochemistry)*. Ministry of Science and Education of Ukraine Ternopil Ivan Puluj National Technical University, Ternopil..

Kostidi, E., & Lyridis, D. (2024). Social impact assessment of biofuel production for maritime and aviation sectors: A case study of a pilot biorefinery project. *Frontiers in Energy Research*, 12, 1454862.

Koul, B., Yakoob, M., & Shah, M. P. (2022). Agricultural waste management strategies for environmental sustainability. *Environmental Research*, 206, 112285.

Kumar, S., Singla, A., & Vidwans, R. R. (2022). Fire investigation and ignitable liquid residue analysis. In *Technologies to advance automation in forensic science and criminal investigation* (pp. 91–118). IGI Global.

Kyselová, L., Vítová, M., & Řezanka, T. (2022). Very long chain fatty acids. *Progress in Lipid Research*, 87, 101180.

Lesage, P., Mutel, C., Schenker, U., & Margni, M. (2018). Uncertainty analysis in LCA using precalculated aggregated datasets. *The International Journal of Life Cycle Assessment*, 23, 2248–2265.

Lide, D. R., & Kehiaian, H. V. (2020). *CRC handbook of thermophysical and thermochemical data*. CRC Press.

Lin, C.-Y. (2022). The influences of promising feedstock variability on advanced biofuel production: A review. *Journal of Marine Science and Technology*, 29(6), 714–730.

Lukić, I., Kesić, Ž., Zdujić, M., & Skala, D. (2016). Vegetable oil as a feedstock for biodiesel synthesis. In Vegetable oil*: Properties, uses and benefits* (pp. 83–127). Nova Pbl. Brittany Holt.

Lund, J., & Rustan, A. C. (2020). Fatty acids: Structures and properties. In *Encyclopedia of life sciences* (Issue October, pp. 283–292).

Mahapatra, S., Kumar, D., Singh, B., & Sachan, P. K. (2021). Biofuels and their sources of production: A review on cleaner sustainable alternative against conventional fuel, in the framework of the food and energy nexus. *Energy Nexus*, 4, 100036.

Maheshwari, P., Haider, M. B., Yusuf, M., Klemeš, J. J., Bokhari, A., Beg, M., Al-Othman, A., Kumar, R., & Jaiswal, A. K. (2022). A review on latest trends in cleaner biodiesel production: Role of feedstock, production methods, and catalysts. *Journal of Cleaner Production*, 355, 131588.

Makepa, D. C. (2023). *Techno-economic analysis and life-cycle assessment of bio-oil production from microwave-assisted pyrolysis of pine sawdust.* Chinhoyi University of Technology.

Malik, M. A. I., Zeeshan, S., Khubaib, M., Ikram, A., Hussain, F., Yassin, H., & Qazi, A. (2024). A review of major trends, opportunities, and technical challenges in biodiesel production from waste sources. *Energy Conversion and Management:* X, *23*, 100675.

Malins, C. (2023). *The fat of the land*. Transport and Environment: Ixelles, Belgium.

Malode, S. J., Gaddi, S. A. M., Kamble, P. J., Nalwad, A. A., Muddapur, U. M., & Shetti, N. P. (2022). Recent evolutionary trends in the production of biofuels. *Materials Science for Energy Technologies*, 5, 262–277.

Maltsev, Y., & Maltseva, K. (2021). Fatty acids of microalgae: Diversity and applications. *Reviews in Environmental Science and Bio/Technology*, 20, 515–547.

Manikandan, G., Kanna, P. R., Taler, D., & Sobota, T. (2023). Review of waste cooking oil (WCO) as a feedstock for biofuel—Indian perspective. *Energies*, 16(4), 1739.

Maradin, D. (2021). Advantages and disadvantages of renewable energy sources utilization. *International Journal of Energy Economics and Policy*, 11(3), 176–183.

Maroa, S., & Inambao, F. (2020). *Biodiesel, combustion, performance and emissions characteristics*. Springer Nature.

Matbouei, M. (2018). *The effect of biofuel on the corrosion and wear of automotive engine components.* University of Leicester.

Meher, L. C., Chandra, K., Sharma, N., Mittal, M., & Nasim, M. (2017). Assessment of fuel qualities of methyl esters from Camelina sativa seed oil for biofuel applications. *Journal of Biofuels*, *8*, 1.

Mofijur, M., Rasul, M., Hassan, N., Masjuki, H., Kalam, M., & Mahmudul, H. (2017). Assessment of physical, chemical, and tribological properties of different biodiesel fuels. In *Clean energy for sustainable development* (pp. 441–463). Elsevier.

Mohapatra, S., Ray, R. C., & Ramachandran, S. (2019). Bioethanol from biorenewable feedstocks: Technology, economics, and challenges. In *Bioethanol production from food crops* (pp. 3–27). Elsevier.

Monirul, I., Masjuki, H., Kalam, M., Zulkifli, N., Rashedul, H., Rashed, M., Imdadul, H., & Mosarof, M. (2015). A comprehensive review on biodiesel cold flow properties and oxidation stability along with their improvement processes. *RSC Advances*, 5(105), 86631–86655.

Mounica, P., Karuppiah, N., Tamilselvi, S., & Jarin, T. (2024). Feasibility and possibility of biofuel-based electric vehicles in the electric vehicle market. In *Renewable energy for plug-in electric vehicles* (pp. 111–127). Elsevier.

Mungodla, S. G., Linganiso, L. Z., Mlambo, S., & Motaung, T. (2019). Economic and technical feasibility studies: Technologies for second generation biofuels. *Journal of Engineering, Design and Technology*, 17(4), 670–704.

Naseef, H. H., & Tulaimat, R. H. (2025). Transesterification and esterification for biodiesel production: A comprehensive review of catalysts and palm oil feedstocks. *Energy Conversion and Management:* X, *26*, 100931.

Naylor, R. L., & Higgins, M. M. (2018). The rise in global biodiesel production: Implications for food security. *Global Food Security*, 16, 75–84.

Nisar, S., Hanif, M. A., Rashid, U., Hanif, A., Akhtar, M. N., & Ngamcharussrivichai, C. (2021). Trends in widely used catalysts for fatty acid methyl esters (Fame) production: A review. *Catalysts*, *11*(9), 1085.

Ogunkunle, O., & Ahmed, N. A. (2019). A review of global current scenario of biodiesel adoption and combustion in vehicular diesel engines. *Energy Reports*, 5, 1560–1579.

Ogunkunle, O., & Ahmed, N. A. (2021). Overview of biodiesel combustion in mitigating the adverse impacts of engine emissions on the sustainable human–environment scenario. *Sustainability*, 13(10), 5465.

Okitsu, K., Maeda, Y., & Bandow, H. (2014). Ultrasound assisted production of fatty acid methyl esters from transesterification of triglycerides with methanol in the presence of KOH catalyst: Optimization, mechanism and kinetics. *Ultrasonics Sonochemistry*, 21(2), 467–471.

Oloyede, C. T., Itabiyi, O. E., Popoola, O. A., Jekayinfa, S. O., Olaniyan, M. A., Adebayo, A. O., Ogunkunle, O., Zamri, M. F. M. A., & Fattah, I. M. R. (2024). Navigating prospects and challenges for green fuels for achieving economical, environmental and ecological resilience: A scientific review. *Biofuels*, 15(7), 929–941.

Osman, A. I., Mehta, N., Elgarahy, A. M., Al-Hinai, A., Al-Muhtaseb, A. a. H., & Rooney, D. W. (2021). Conversion of biomass to biofuels and life cycle assessment: A review. *Environmental Chemistry Letters*, 19, 4075–4118.

Ozkan, S. (2019). *Development of renewable based on fuels for aviation, using heterogeneous catalysis.* Dissertação de Mestrado, Instituto Superiror de Engenharia de Lisboa do ….

Passoth, V., & Müller, B. (2025). Lipids of yeasts and filamentous fungi and their importance to biotechnology. In A. A. Sibirny (Ed.), *Biotechnology of yeasts and filamentous fungi* (pp. 163–227). Springer.

Perin, G., & Jones, P. R. (2019). Economic feasibility and long-term sustainability criteria on the path to enable a transition from fossil fuels to biofuels. *Current Opinion in Biotechnology*, 57, 175–182.

Pikula, K., Zakharenko, A., Stratidakis, A., Razgonova, M., Nosyrev, A., Mezhuev, Y., Tsatsakis, A., & Golokhvast, K. (2020). The advances and limitations in biodiesel production: Feedstocks, oil extraction methods, production, and environmental life cycle assessment. *Green Chemistry Letters and Reviews*, 13(4), 275–294.

Pires, A. P., Han, Y., Kramlich, J., & Garcia-Perez, M. (2018). *Chemical composition and fuel properties of alternative jet fuels* (No. BioRes_13_2_2632). Federal Aviation Administration. Center of Excellence for Alternative Jet Fuels and Environment.

Prasad, S., Potdar, V., Cherian, S., Abraham, P., Basu, A., & Team, I. C. (2020). Transmission electron microscopy imaging of SARS-CoV-2. *Indian Journal of Medical Research*, 151(2–3), 241–243.

Pryshliak, N., & Tokarchuk, D. (2020). Socio-economic and environmental benefits of biofuel production development from agricultural waste in Ukraine. In *Environmental & socio-economic studies*. Department of Administrative Management and Alternative Energy Sources, Vinnytsia National Agrarian University.

Riayatsyah, T. M. I. (2018). *Techno-economics analysis of biodiesel production from non-edible reutealis trisperma oil.* University of Malaya, Malaysia.

Rostek, E. (2016). Biofuels of first and second generation. *Journal of KONES*, *23*, 413–420.

Roy, A., & Pramanik, S. (2024). A review of the hydrogen fuel path to emission reduction in the surface transport industry. *International Journal of Hydrogen Energy*, 49, 792–821.

Rumizen, M. A. (2021). Qualification of alternative jet fuels. *Frontiers in Energy Research*, 9, 760713.

Salaheldeen, M., Mariod, A. A., Aroua, M. K., Rahman, S. A., Soudagar, M. E. M., & Fattah, I. R. (2021). Current state and perspectives on transesterification of triglycerides for biodiesel production. *Catalysts*, *11*(9), 1121.

Saleh, T. A. (2020). Characterization, determination and elimination technologies for sulfur from petroleum: Toward cleaner fuel and a safe environment. *Trends in Environmental Analytical Chemistry*, 25, e00080.

Santos, A. d. M., Becker, A. M., Ayala, N. F., & Sant'Anna, Â. M. O. (2024). Industry 4.0 as an enabler of sustainability for small and medium enterprises. *Academia Revista Latinoamericana de Administración*, 37(2), 204–226.

Santos, R. N., Lima, E. R., & Paredes, M. L. (2021). ASTM D86 distillation curve: Experimental analysis and premises for literature modeling. *Fuel*, 284, 118958.

Santos, S. M., Wolf-Maciel, M. R., & Fregolente, L. V. (2023). Cold flow properties: Applying exploratory analyses and assessing predictive methods for biodiesel and diesel-biodiesel blends. *Sustainable Energy Technologies and Assessments*, 57, 103220.

Sebos, I. (2022). Fossil fraction of CO_2 emissions of biofuels. *Carbon Management*, 13(1), 154–163.

Sembiring, K. C., & Saka, S. (2019). Renewable hydrocarbon fuels from plant oils for diesel and gasoline. *Journal of the Japan Petroleum Institute*, 62(4), 157–172.

Seto, S., & Gaughan, R. (2018). Chapter 6. Aviation fuels. In J. R. Salvatore & A. W. Verstuyft (Eds.), *Significance of tests for petroleum products*. ASTM Digital liberary.

Shamshirband, S., Rabczuk, T., & Chau, K.-W. (2019). A survey of deep learning techniques: Application in wind and solar energy resources. *IEEE Access*, 7, 164650–164666.

Sharma, S. K., Barthwal, R., Saini, D., & Rawat, N. (2022). Chemistry of food fats, oils, and other lipids. In *Advances in food chemistry: Food components, processing and preservation* (pp. 209–254). Springer.

Shojaeiarani, J., Bajwa, D. S., & Bajwa, S. G. (2019). Properties of densified solid biofuels in relation to chemical composition, moisture content, and bulk density of the biomass. *BioResources*, 14(2), 4996–5015.

Shrestha, D., Staab, B., & Duffield, J. (2019). Biofuel impact on food prices index and land use change. *Biomass and Bioenergy*, 124, 43–53.

Silva, D. A. L. (2021). Life cycle assessment (LCA)—Definition of goals and scope. In *Life cycle engineering and management of products: Theory and practice* (pp. 45–69). Springer International Publishing.

Singh, D., Sharma, D., Soni, S., Inda, C. S., Sharma, S., Sharma, P. K., & Jhalani, A. (2021). A comprehensive review of biodiesel production from waste cooking oil and its use as fuel in compression ignition engines: 3rd generation cleaner feedstock. *Journal of Cleaner Production*, 307, 127299.

Singh, D., Sharma, D., Soni, S. L., Sharma, S., Sharma, P. K., & Jhalani, A. (2020). A review on feedstocks, production processes, and yield for different generations of biodiesel. *Fuel*, 262, 116553.

Singh, R., & Sehrawat, P. (2019). Transformative innovations and economic parameters for use of biofuels with sustainable marketing for growth. *Available at SSRN 3408793*.

Singh, S., & Singh, S. (2023). Plant-based biofuels: An overview. In M. P. Shah (Ed.), *Green approach to alternative fuel for a sustainable future* (pp. 433–442). Elsevier.

Siraj, S., Kale, R., & Deshmukh, S. (2017). Effects of thermal, physical, and chemical properties of biodiesel and diesel blends. *American Journal of Mechanical and Industrial Engineering*, 2(1), 24–31.

Solarte-Toro, J. C., Ortiz-Sanchez, M., & Cardona Alzate, C. A. (2023). Environmental life cycle assessment (E-LCA) and social impact assessment (SIA) of small-scale biorefineries implemented in rural zones: The avocado (*Persea americana* var. Americana) case in Colombia. *Environmental Science and Pollution Research*, 30(4), 8790–8808.

Standard, A. (2018). D4052. Standard test method for density, relative density, and API gravity of liquids by digital density meter. In *2011 ASTM annual book of standards*. American National Standards Institute.

Sueyoshi, T., Mo, F., & Wang, D. D. (2022). Sustainable development of countries all over the world and the impact of renewable energy. *Renewable Energy*, 184, 320–331.

Sultana, S., Nur, A. A., Khan, S., Ambati, R. R., & Gokare Aswathanarayana, R. (2023). Biofuels as economic security for the poor. In *Biofuels in circular economy* (pp. 201–218). Springer.

Taheripour, F., Zhao, X., & Tyner, W. E. (2017). The impact of considering land intensification and updated data on biofuels land use change and emissions estimates. *Biotechnology for Biofuels*, 10, 1–16.

Tasić, R. U. (2020). *Ispitivanje sindroma sagorevanja na poslu kod medicinskih sestara vaspitača u predškolskim ustanovama u Beogradu*. University of Belgrade, Serbia.

Task, I. B. (2021). Progress in commercialization of biojet/sustainable aviation fuels (SAF): technologies, potential and challenges. In *IEA bioenergy task* (Vol. 39).

Tkachenko, A., Tigunova, O., & Shulga, S. (2013). Microbial lipids as a source of biofuel. *Cytology and Genetics*, 47, 343–348.

Toldrá-Reig, F., Mora, L., & Toldrá, F. (2020). Trends in biodiesel production from animal fat waste. *Applied Sciences*, 10(10), 3644.

Torres, A., Fuentes, B., Rodríguez, K. E., Brito, A., & Díaz, L. (2020). Analysis of the content of fatty acid methyl esters in biodiesel by Fourier-transform infrared spectroscopy: Method and comparison with gas chromatography. *Journal of the American Oil Chemists' Society*, 97(6), 651–661.

Tyagi, U., Aslam, M., & Sarma, A. K. (2022). Characterization of green diesel: Existing standards and beyond. In *Green diesel: An alternative to biodiesel and petrodiesel* (pp. 249–263). Springer.

Umayangani, S., De Silva, I., Munasinghe, D., Deshika, C., Chandrapala, J., & Silva, M. (2025). Elimination of trans fatty acids from foods using novel technologies. *Food Reviews International*, 1–29.

Van Der Hilst, F., Verstegen, J. A., Woltjer, G., Smeets, E. M., & Faaij, A. P. (2018). Mapping land use changes resulting from biofuel production and the effect of mitigation measures. *Gcb Bioenergy*, 10(11), 804–824.

van Dyk, S., Su, J., Mcmillan, J. D., & Saddler, J. (2019). Potential synergies of drop-in biofuel production with further co-processing at oil refineries. *Biofuels, Bioproducts and Biorefining*, 13(3), 760–775.

Vesanto, V. (2022). *Evaluating biodiversity impacts of renewable raw materials for biofuel production: Case Neste* (Master's thesis). Lappeenranta–Lahti University of Technology LUT, LUT School of Energy Systems.

Vinoth Kumar, R., Ganesh Moorthy, I., Goswami, L., Pugazhenthi, G., Pakshirajan, K., Silva, A. M., & Morales-Torres, S. (2020). Analytical methods in biodiesel production. In *Biomass valorization to bioenergy* (pp. 197–219). Springer.

Vozka, P., Mo, H., Šimáček, P., & Kilaz, G. (2018a). Middle distillates hydrogen content via GC× GC-FID. *Talanta*, 186, 140–146.

Vozka, P., Modereger, B. A., Park, A. C., Zhang, W. T. J., Trice, R. W., Kenttämaa, H. I., & Kilaz, G. (2019). Jet fuel density via GC× GC-FID. *Fuel*, 235, 1052–1060.

Vozka, P., Simacek, P., & Kilaz, G. (2018b). Impact of HEFA feedstocks on fuel composition and properties in blends with jet A. *Energy & Fuels*, 32(11), 11595–11606.

Walchuk, C., & Suh, M. (2020). Nutrition and the aging retina: A comprehensive review of the relationship between nutrients and their role in age-related macular degeneration and retina disease prevention. *Advances in Food and Nutrition Research*, 93, 293–332.

Wang, J., Singer, S. D., Souto, B. A., Asomaning, J., Ullah, A., Bressler, D. C., & Chen, G. (2022). Current progress in lipid-based biofuels: Feedstocks and production technologies. *Bioresource Technology*, 351, 127020.

Xue, J., Balamurugan, S., Li, T., Cai, J.-X., Chen, T.-T., Wang, X., Yang, W.-D., & Li, H.-Y. (2021). Biotechnological approaches to enhance biofuel producing potential of microalgae. *Fuel*, 302, 121169.

Zhang, J., & Zhang, X. (2019). The thermochemical conversion of biomass into biofuels. In *Biomass, biopolymer-based materials, and bioenergy* (pp. 327–368). Elsevier.

Zhang, S., Li, K., Zhu, P., Dai, M., & Liu, G. (2022). An efficient hydrogen production process using solar thermo-electrochemical water-splitting cycle and its techno-economic analyses and multi-objective optimization. *Energy Conversion and Management*, 266, 115859.

Zhang, X., Bauer, C., Mutel, C. L., & Volkart, K. (2017). Life cycle assessment of power-to-gas: Approaches, system variations and their environmental implications. *Applied Energy*, 190, 326–338.

Zhang, Y., Zhong, Y., Lu, S., Zhang, Z., & Tan, D. (2022). A comprehensive review of the properties, performance, combustion, and emissions of the diesel engine fueled with different generations of biodiesel. *Processes*, 10(6), 1178.

Zuo, L., Zhang, Z., Carlson, K. M., MacDonald, G. K., Brauman, K. A., Liu, Y., Zhang, W., Zhang, H., Wu, W., & Zhao, X. (2018). Progress towards sustainable intensification in China challenged by land-use change. *Nature Sustainability*, 1(6), 304–313.

9 Role of Glycerol in Biofuel Production

Mohd Azhar, Moonisah Aftab, Mohd Arsh Khan, Naseem Ahmad, Qazi Inamur Rahman, and Nafees Ahmad

9.1 INTRODUCTION

Amid growing environmental concerns and declining fossil fuel reserves, the demand for sustainable and cleaner energy sources is more required of hour than ever before. Biofuels derived from renewable biological sources, such as plant oils, animal fats, and waste cooking oils, have emerged now as promising alternatives. Among all, biodiesel stands out due to its compatibility with existing diesel engines and potential to significantly reduce emission of greenhouse gases. Reported studies have shown biodiesel feedstocks and production pathways could reduce emissions up to 90% compared to petroleum diesel (Farouk et al. 2024; Leung et al., 2010). Biodiesel is typically produced through transesterification, a chemical reaction in which triglycerides from oils or fats react with an alcohol (methanol) in the presence of a catalyst, sodium or potassium hydroxide. This reaction yields two main products: biodiesel (a mixture of fatty acid esters) and glycerol (Figure 9.1). On average, for every 10 kg of biodiesel produced, about 1 kg of glycerol is generated as a byproduct (Yang et al., 2012; Kumar et al., 2019). The global scale-up of biodiesel production has led to a significant increase in raw glycerol output; estimated million tons of raw glycerol are produced annually, exceeding the capacity of traditional markets to absorb it easily; as a result, glycerol prices dropped sharply in some regions from around $0.70 per pound in the early 2000s (Costa et al., 2022).

Raw glycerol is not simply a diluted form of pure glycerol; rather, it is a complex mixture of methanol, catalysts, salts, water, soaps, and unreacted triglycerides. These impurities can make up more than 50% of its composition, depending on the production process (Kumar et al., 2019). As a result, raw glycerol is unsuitable for direct use in industries that require high-purity glycerol, such as pharmaceuticals, cosmetics, and food. Although purification techniques—such as distillation, neutralization, and membrane filtration—are available, they are often costly and technically demanding, particularly for small-scale producers (Fairooz et al., 2010; Muniru et al., 2019).

Instead of viewing raw glycerol as waste, there is a growing interest in converting it into a valuable co-product—a process known as glycerol valorization (Figure 9.2). This concept aligns with the broader vision of integrated biorefineries, where every component of biomass is utilized to produce energy, fuels, or chemicals, thereby reducing waste and improving the overall economics of biofuel production (da Costa et al., 2022). Economically, integrating glycerol valorization into biodiesel plants can significantly enhance profitability. Studies suggest that the revenue generated from upgraded glycerol products precisely offsets production costs and reduces the environmental impact associated with glycerol disposal (Elsayed et al., 2024). The growing volume of glycerol byproducts presents both a challenge and an opportunity, with global production expected to exceed 60 billion liters by 2025. This chapter explores the various pathways through which raw glycerol can be refined or converted, examining their technical feasibility, economic viability, and potential role in shaping a more circular and sustainable biofuel industry.

DOI: 10.1201/9781003600169-9

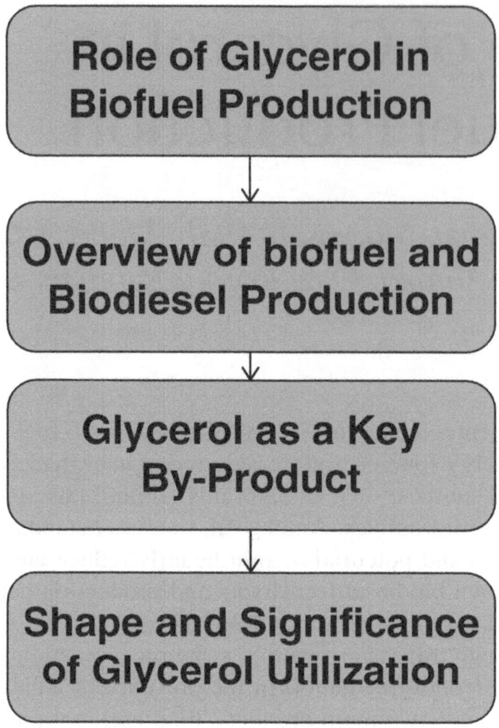

FIGURE 9.1 An overview of glycerol valorization.

9.1.1 An Overview of Biofuels and Biodiesel Production

The global energy system has undergone a significant transformation in recent decades, driven by concerns about climate change, the depletion of fossil fuel reserves, and energy security. Among various alternatives, biofuels have attracted considerable attention due to their renewability, lower carbon footprint, and compatibility with existing energy infrastructure. Biomass—organic materials such as agricultural residues, plant oils, and animal fats—can be converted into energy carriers like ethanol, biodiesel, and advanced liquid fuels (Christ, 2007; Demirbas, 2009).

Biofuels are typically classified into first-, second-, and third-generation categories based on the feedstock used and the maturity of the technology. First-generation biofuels are produced from food crops such as sugarcane, corn, and soybean oil. Although these technologies are well-established, their use raises concerns about food versus fuel competition (Hill et al., 2006). Second-generation biofuels are derived from lignocellulosic biomass (straw, wood chips), which avoids the food conflict but involves more complex conversion processes. Third-generation biofuels focus on microalgae and other high-yield biomass sources that require minimal land use (Geng et al., 2025).

Among liquid biofuels, biodiesel stands out for its ease of production, renewable nature, and chemical compatibility with diesel engines. Biodiesel typically refers to fatty acid methyl esters (FAMEs), which are mainly synthesized through the transesterification of triglycerides found in oils and fats with short-chain alcohols (usually methanol) using acid, base, or enzyme catalysts. The standard transesterification reaction yields two products: biodiesel and glycerol.

Common biodiesel feedstocks include virgin vegetable oils (soybean, rapeseed, palm), used cooking oils, animal fats, and non-edible sources such as Jatropha and algae oil (Atabani et al., 2012). While edible oils still dominate global biodiesel production, there is an increasing interest in non-food feedstocks to enhance sustainability and avoid competition with food markets (Gui et al., 2008). The biodiesel industry has grown significantly over the past two decades. According to the

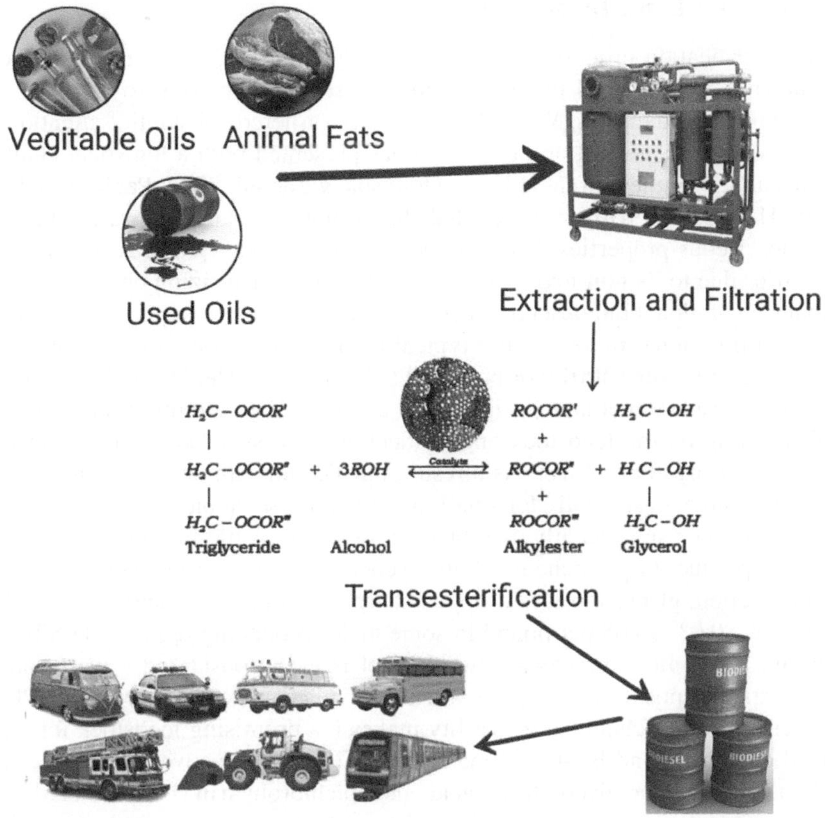

FIGURE 9.2 An overview of transesterification process and product utilization.

International Energy Agency (IEA), global biodiesel production surpassed 47 billion liters in 2020, with leading producers including the European Union, the United States, Brazil, and Argentina (IEA, 2021). This growth is largely driven by increased environmental awareness, blending mandates, and concerns over climatic change.

From a technical perspective, biodiesel offers several advantages: it is biodegradable, non-toxic, and has a higher flash point and better lubricity than petroleum diesel. It can be used directly or in blends such as B5 (5% biodiesel, 95% diesel) or B20 (20% biodiesel), often without requiring engine modifications (Knothe et al., 2005). Moreover, life cycle analyses have consistently shown that biodiesel results in significantly lower net carbon emissions compared to fossil diesel, especially when produced from waste oils or residues (Sheehan et al., 1998).

However, biodiesel production also faces challenges. Raw materials account for 60%–80% of the total production cost, making the process economically vulnerable to fluctuations in feedstock prices (Zhang et al., 2003). Additionally, the generation of raw glycerol as a byproduct presents both a waste management issue and a value-adding opportunity—depending on how it is handled (Yang et al., 2012). In recent years, attention has shifted toward integrated biorefineries, where biodiesel production is combined with the conversion of byproducts like glycerol into value-added chemicals and materials. These systems aim to maximize resource efficiency and profitability, aligning with the principles of a circular bioeconomy (Cherubini, 2010).

In conclusion, biodiesel plays a vital role in the transition to renewable energy. While its production pathways are mature and scalable, long-term sustainability depends on strategic feedstock selection, effective co-product utilization, and integration into broader bio-based value chains.

9.1.2 Glycerol as a Large Byproduct

The core transesterification process for biodiesel production yields two main products: FAMEs, commonly known as commercial biodiesel, and glycerol (also referred to as glycerin), which is generated in significant quantities. While biodiesel has garnered substantial attention as a renewable transport fuel, the accumulation of glycerol has presented both a logistical challenge and an economic opportunity for the biofuel industry (Johnson & Taconi, 2007; Pagliaro & Rossi, 2008).

Glycerol ($C_3H_5OH_3$) is a trihydroxy alcohol that is highly soluble in water and valued for its hygroscopic and viscous properties. It is widely used in the food, pharmaceutical, cosmetic, and polymer industries due to its non-toxic, biodegradable, and moisturizing characteristics. However, the rapid expansion of global biodiesel production has created an oversupply of raw glycerol, which has overwhelmed traditional markets. In a typical biodiesel reaction, approximately 10 kg of raw glycerol is produced for every 100 kg of biodiesel (Thompson & He, 2006). This crude glycerol is impure, containing residual methanol, salts, soaps, catalysts (e.g., sodium or potassium hydroxide), and water. Depending on the feedstock and production process, crude glycerol purity can range from 40% to 88% (Yang et al., 2012). As a result, purification can be technically demanding and economically unfeasible—especially for small- and medium-scale biodiesel producers. The global glycerol market has been significantly disrupted. Prior to the biodiesel boom, glycerol was mainly produced as a co-product in petrochemical and oleochemical industries. However, with the surge in biodiesel production, glycerol prices dropped sharply—from over \$0.60 per pound in the early 2000s to as low as \$0.03–\$0.05 per pound in some major producing regions (da Silva et al., 2009; Ciriminna et al., 2014). In some cases, raw glycerol is even considered a waste product, incurring disposal costs and posing environmental risks if mismanaged (Meher et al., 2006). Despite these challenges, glycerol's chemical versatility makes it a promising feedstock for the production of value-added chemicals and bio-based materials. Valuable derivatives include 1,3-propanediol, glycerol carbonate, propylene glycol, lactic acid, and epichlorohydrin (Pagliaro et al., 2007). These transformations not only reduce waste but also enhance the sustainability and economic viability of biodiesel production systems. The biological conversion of glycerol has also gained attention due to its low energy requirements and environmental compatibility. Several microbial strains—including *Klebsiella, Clostridium, and Enterobacter*—can metabolize glycerol to produce commercially valuable biochemicals such as succinic acid, ethanol, and hydrogen (Yazdani & Gonzalez, 2007; Clomburg & Gonzalez, 2013). These pathways are particularly attractive when utilizing crude glycerol as the feedstock. Ultimately, glycerol valorization is central to the development of integrated biorefineries, where every byproduct of biomass conversion is treated as a resource rather than waste. This approach aligns with the principles of the circular economy, promoting energy efficiency, resource optimization, and economic resilience (Kamm & Kamm, 2004).

9.1.3 The Scope and Importance of Glycerol Use

The increasing global production of biodiesel has led to a corresponding rise in glycerol output, positioning this compound not merely as a byproduct but as a valuable raw material for industrial applications. Crude glycerol, once considered waste, now offers significant potential for use in chemical, biochemical, and bioenergy processes. The effective integration of glycerol into sustainable biorefinery systems is recognized as a key strategy for enhancing both the environmental and economic performance of large-scale biodiesel production (Pagliaro et al., 2007; Clomburg & Gonzalez, 2013).

Glycerol is considered a platform chemical, meaning it serves as a precursor to a wide range of industrially valuable compounds (Table 9.1). Its three hydroxyl groups confer high reactivity, enabling various chemical transformations such as oxidation, hydrogenolysis, esterification, and dehydration. Through these processes, glycerol can be converted into valuable derivatives including 1,3-propanediol, glycerol carbonate, acrolein, lactic acid, dihydroxyacetone, and even syngas (Zhou et al., 2008). These compounds are key building blocks in the production of polymers, solvents,

TABLE 9.1
Key Pathways for Glycerol Valorization

Pathway	Conversion Method	Main Product	Applications	Advantages
Biochemical conversion	Using bacteria or yeasts	Hydrogen 1,3-propanediol, organic acids	Biofuels, polymers, food additives	Mild conditions, high selectivity
Thermochemical conversion	Steam reforming, pyrolysis, gasification	Hydrogen syngas, aromatics	Fuel additive solvents	Enhance fuel quality, low emissions
Chemical conversion	Esterification, etherification	Solketal, glycerol carbonate,GTBE	Biofuels, nutraceuticals	CO_2 capture, bioremediation
Direct fuel use	Blending with diesel	Crude glycerol–diesel blends	Formic acid, glycolic acid	Simple implementation
Electrochemical conversion	Electro-oxidation	Electro-oxidation	Fuel generation clean energy	Renewable energy utilization

Sources: Yazdani and Gonzalez (2007), Zhou et al. (2008), Pagliaro and Rossi (2008), Meher et al., (2006), Clomburg and Gonzalez (2013).

resins, and pharmaceuticals, significantly expanding the industrial relevance of glycerol beyond its traditional uses. For instance, glycerol carbonate, a non-toxic and biodegradable compound, is used in cosmetics, as an electrolyte in lithium-ion batteries, and as a green solvent. It can be synthesized directly from glycerol and urea, or via transesterification with dimethyl carbonate, offering high atom efficiency and a low environmental footprint (Yang et al., 2012).

Glycerol also has both direct and indirect potential as a benzene alternative. Thermochemical pathways, such as steam reforming or aqueous-phase reforming of glycerol, produce hydrogen-rich gas streams that can be used in fuel cells or for chemical synthesis (Davda et al., 2005). Alternatively, glycerol can be blended with diesel to improve fuel lubricity and reduce NOx emissions, or converted into oxygenated additives such as solketal, which enhances the combustion efficiency of gasoline engines (da Silva et al., 2010).

In addition, glycerol fermentation using engineered microbes can yield bioethanol, butanol, succinic acid, and even polyhydroxyalkanoates (PHAs)—biodegradable plastics (Yazdani & Gonzalez, 2007; Clomburg & Gonzalez, 2011). These microbial routes are especially attractive because they allow the use of crude glycerol without the need for extensive purification, thereby reducing production costs.

Utilizing raw glycerol in value-added applications directly addresses the environmental impacts associated with waste management in biodiesel production. Improper disposal of glycerol-rich waste can result in high chemical oxygen demand (COD) and lead to potential soil and water pollution (Meher et al., 2006). Therefore, converting glycerol into useful products supports zero-waste strategies and enhances the overall sustainability profile of biofuels. From an economic perspective, establishing secondary income streams through glycerol valorization is crucial to the long-term viability of the biodiesel industry. Glycerol prices, which once plummeted due to oversupply, can stabilize if new markets and applications are developed (Johnson & Taconi, 2007). The integration of glycerol processing units into existing biodiesel plants is viewed as a strategic component of modern biorefinery development, promoting more efficient resource utilization and improving economic resilience (Kamm & Kamm, 2004) (Table 9.1).

9.1.3.1 Research and Innovation Landscape

Current research is focused on catalyst development, process intensification, and metabolic engineering to make glycerol conversion pathways more efficient and economically viable. Green chemistry approaches—such as solvent-free systems, recyclable catalysts, and enzyme-based catalysis—are increasingly employed to meet both regulatory and environmental standards (Pagliaro & Rossi, 2008).

Ultimately, the potential applications of glycerol are vast and continually expanding. Once considered a low-value byproduct, glycerol has now emerged as a key component in the circular bioeconomy, offering solutions that bridge renewable energy, green chemistry, and industrial biotechnology. Advancing the science and infrastructure for glycerol valorization is not only a technical challenge but also a strategic necessity for sustainable development.

9.2 GLYCEROL: PROPERTIES AND SOURCES

Glycerol, also known as glycerin or propane-1,2,3-triol, is a colorless, odorless, viscous liquid with a sweet taste and a wide range of applications in medicine, food, cosmetics, and biofuels. Glycerol is a colorless, odorless, and viscous liquid at room temperature, with a sweet taste. It has a melting point of approximately 18.2°C and a relatively high boiling point of about 290°C under standard atmospheric pressure. These thermal properties make it well-suited for high-temperature reactions without degradation, a beneficial feature in many catalytic processes (Katryniok et al., 2009). Chemically classified as a trihydroxy sugar alcohol, glycerol is a highly versatile molecule due to its three hydroxyl groups, which make it suitable for various biological and chemical transformations (da Silva et al., 2009). Chemically, glycerol is a three-carbon triol, consisting of three hydroxyl (–OH) groups attached to a linear propane backbone. Glycerol has the molecular formula $C_3H_5OH_3$ and a molar mass of 92.09 g/mol. Its structure results in high polarity and a strong capacity for hydrogen bonding, making it miscible with water and other polar solvents. The hydroxyl groups not only enable extensive hydrogen bonding but also serve as reactive sites for various chemical modifications, including esterification, etherification, oxidation, and dehydration (Table 9.2); this abundance of functional groups makes glycerol an exceptionally versatile platform molecule for synthetic and industrial chemistry (Pagliaro & Rossi, 2008). The dynamic chemical structural properties with three hydroxyl groups (Figure 9.3) contribute to glycerol's high hydrophilicity, hygroscopicity, and viscosity (Anitha et al., 2016).

TABLE 9.2
Physicochemical Properties of Glycerol

Property	Value
Molecular formula	$C_3H_8O_3$
Molar mass	92.09 g/mol
Melting point	18.2°C
Boiling point	290°C
Density (20°C)	1.26 g/cm³
Viscosity (25°C)	~1,200 mPa·s
Solubility in water	Fully miscible
Appearance	Clear, viscous liquid
Refractive index (20°C)	1.473

Source: Compiled from standard chemical reference sources (PubChem; National Center for Biotechnology Infor)

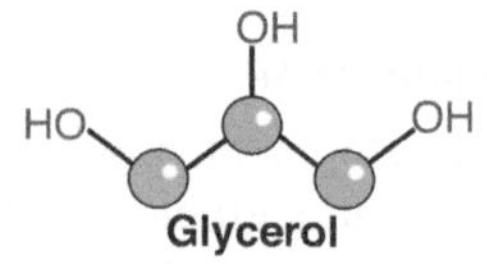

FIGURE 9.3 Simplified structural formula of glycerol.

The high viscosity of glycerol arises from strong intermolecular hydrogen bonding, which also accounts for its hygroscopic nature and high boiling point. While these properties can pose challenges in certain processing stages—such as pumping or atomization in combustion systems—they also make glycerol valuable as a solvent, plasticizer, and intermediate in thermal or catalytic conversion processes (Clomburg & Gonzalez, 2011). Additionally, its thermal stability and low volatility make glycerol a desirable medium for high-temperature reactions. Its refractive index and optical clarity also make it suitable for use in pharmaceutical and cosmetic formulations (da Silva et al., 2009). The high viscosity of glycerol arises from strong intermolecular hydrogen bonding, which also accounts for its hygroscopic nature and high boiling point. While these properties can pose challenges in certain processing stages—such as pumping or atomization in combustion systems—they also make glycerol valuable as a solvent, plasticizer, and intermediate in thermal or catalytic conversion processes (Clomburg & Gonzalez, 2011). Additionally, its thermal stability and low volatility make glycerol a desirable medium for high-temperature reactions. Its refractive index and optical clarity also make it suitable for use in pharmaceutical and cosmetic formulations (da Silva et al., 2009).

Due to its high polarity, glycerol readily dissolves in water, methanol, and ethanol, forming strong hydrogen bond networks. However, it is largely insoluble in hydrocarbons and other non-polar solvents. Its polarity and multiple hydroxyl groups also enable interactions with enzymes and other biomolecules, making glycerol highly compatible with biotechnological processes such as fermentation, enzyme stabilization, and cryopreservation (Yazdani & Gonzalez, 2007).

From a chemical standpoint, the structural richness of glycerol offers numerous pathways for its transformation into value-added chemicals, such as glycerol carbonate, acrolein, and 1,2-propanediol. Each of these conversions exploits the reactivity of the hydroxyl groups and the flexibility of the propane backbone under acidic, basic, or catalytic conditions (Zhou et al., 2008). This ability to generate diverse derivatives makes glycerol an ideal bio-based platform molecule for applications in green chemistry and sustainable development.

9.2.1 Sources of Glycerol, Especially from Biodiesel Processes

Glycerol is obtained from both natural and industrial sources, but the biodiesel industry has been the most prominent contributor to the global glycerol supply over the past two decades. Understanding the different sources of glycerol is essential to assess its availability, purity, and suitability for downstream processing and valorization.

9.2.1.1 Natural Sources

In nature, glycerol serves as the structural backbone of triglycerides and phospholipids, playing a central role in lipid metabolism across plants, animals, and microorganisms (Table 9.3). Triglyceride molecules, composed of three fatty acids esterified to a glycerol backbone, are hydrolyzed during digestion or energy mobilization, releasing free glycerol into biological systems (Yang et al., 2012). Additionally, glycerol functions as a compatible solute in osmoregulation in microorganisms such as *Saccharomyces cerevisiae* (Albertyn et al., 1994).

9.2.1.2 Petrochemical Synthesis

Before the expansion of biodiesel production, synthetic glycerol was primarily produced via the hydrolysis of epichlorohydrin, a process based on petroleum derivatives. However, due to concerns over costs, toxicity, and environmental impact, this method has been largely replaced by bio-based production on an industrial scale (Pagliaro et al., 2007).

9.2.1.3 Biodiesel Industry as the Primary Source

Today, the most prominent source of glycerol is the transesterification of triglycerides in biodiesel production. The production of biodiesel involves treating vegetable oils or animal fats with an

TABLE 9.3
Glycerol Yield Per Feedstock from Natural Biodiesel Production Sources

Feed Stock	Glycerol Yield (kg/ton Biodiesel)
Soybean oil	~100
Palm oil	~95
Animal fat	~105
Waste cooking oil	~90–100

Source: Thompson & He, 2006, Quispe et al., 2013 & Canakci & Van Gerpen, 1999

alcohol, such as methanol or ethanol, in the presence of a base catalyst like sodium hydroxide (NaOH) or potassium hydroxide (KOH). This reaction yields fatty acid methyl esters (FAME), which make up biodiesel, and generates crude glycerol as a by-product (Thompson & He, 2006).

The response is usually:

$$\text{Triglyceride} + \text{methanol} \rightarrow \text{FAME}\ (\text{biodiesel}) + \text{glycerol}$$

For every 100 kg of biodiesel produced, about 10 kg of raw glycerol (Yazdani & Gonzalez, 2007) is generated. This raw glycerol contains impurities such as methanol, salts, soap, and water, which require purification for many downstream applications. The exponential growth of the biodiesel industry, driven by policies supporting renewable energy and carbon reduction, has consequently led to a significant global surplus of glycerol (Johnson & Taconi, 2007).

9.2.2 Global Production and Surplus Trends

The emergence of the biodiesel sector has significantly reshaped the global glycerol market. Glycerol, once considered a specialty chemical, has become a widely available byproduct. This shift has brought both opportunities and challenges, particularly in managing the resulting surplus of glycerol.

According to the International Energy Agency (IEA), over 40 billion liters of biodiesel were produced globally in 2022, generating approximately 4 million tonnes of raw glycerol as a byproduct (IEA, 2023). The major biodiesel-producing regions—the United States, the European Union, Brazil, and Argentina—are also the largest contributors to global glycerol production.

The sharp increase in glycerol supply has disrupted the market, leading to significant price drops. The price of glycerol fell by approximately 80%, particularly for low-grade or unrefined forms (Quispe et al., 2013). This price collapse has raised economic and environmental concerns, especially regarding the feasibility of large-scale glycerol management and the development of value-added conversion pathways.

In response to the surplus, both academic and industrial efforts have focused on upgrading glycerol into fuels, solvents, polymers, and biochemical intermediates. Downstream applications of glycerol (Macedo et al., 2019) rely on advanced treatment technologies such as vacuum distillation, ion exchange, and membrane separation to enhance economic viability. Moreover, ongoing research into glycerol fermentation, hydrogenolysis, and catalytic conversion is exploring ways to transform glycerol into high-value compounds such as 1,3-propanediol, acrolein, and glycerol carbonate (Clomburg & Gonzalez, 2011).

9.3 GLYCEROL IN BIODIESEL PRODUCTION

Glycerol is an integral part of the biodiesel production process—not as a fuel, but as a significant byproduct formed during the transesterification of triglycerides. Understanding its formation, quantity, and quality is essential for the design of biodiesel plants, as well as for assessing their economic viability and operational stability. This section explores the formation mechanism, functional role, recovery methods, and industrial implications of glycerol in biodiesel production.

9.3.1 Pathway of Glycerol Generation in the Trans Esterification

Transesterification is the primary reaction in biodiesel production, where triglycerides react with an alcohol—typically methanol—in the presence of a catalyst such as sodium hydroxide (NaOH) or potassium hydroxide (KOH). This reaction converts the triglycerides into FAMEs, which constitute biodiesel, while producing glycerol as a byproduct. The reaction occurs in three reversible steps:

I. Triglycerides are first converted into diglycerides.
II. Diglycerides are then converted into monoglycerides.
III. Monoglycerides are lastly converted into glycerol.

Each step releases one molecule of FAME. The resulting reaction mixture separates into two layers due to differences in density: the upper layer contains FAME (biodiesel), while the denser lower layer contains glycerol. This separation can be achieved by gravity settling or centrifugation (Freedman et al., 1984).

Transesterification can be catalyzed by acids, bases, or enzymes:

i. **Alkaline Catalysts:** These are the most commonly used in industrial processes. Examples include sodium hydroxide (NaOH) and potassium hydroxide (KOH). They are highly effective but are sensitive to the presence of free fatty acids and water, which can lead to soap formation (Freedman et al., 1984).
ii. **Acid Catalysts:** Sulfuric acid (H_2SO_4) is often used, especially when the feedstock contains high levels of free fatty acids. Although acid catalysis is more tolerant of impurities, the reaction is slower and more corrosive (Canakci & Van Gerpen, 1999).
iii. **Enzymatic Catalysts:** Lipase enzymes offer a more environmentally friendly and selective alternative. They operate under mild conditions and are suitable for feedstocks with high free fatty acid content. However, high costs currently limit their commercial viability (Fukuda et al., 2001)

Many parameters influence the efficiency of the transesterification process:

i. **Molar Ratio of Alcohol to Oil:** A molar ratio of 6:1 is commonly used in base-catalyzed reactions. While increasing the ratio can enhance conversion efficiency, it may also complicate the separation of products (Encinar et al., 2005).
ii. **Temperature:** The reaction typically occurs at temperatures between 50°C and 60°C in methanol-based systems, close to the boiling point of methanol, to accelerate the conversion rate (Zheng et al., 2006).
iii. **Catalyst Concentration:** A catalyst concentration of approximately 0.5%–1.0% w/w is typical. Excessive catalyst amounts can lead to soap formation and emulsion issues, which hinder product separation.
iv. **Reaction Time:** Under batch conditions, optimal transesterification is usually achieved within 1–2 hours.

Commercial transesterification is typically carried out in batch reactors for small-scale production. For large-scale or continuous operations, continuous stirred-tank reactors (CSTRs) or plug flow reactors (PFRs) are commonly used. After the reaction, the mixture is allowed to settle. Due to differences in density, glycerol forms the heavier bottom phase, while biodiesel forms the lighter upper phase and can be separated accordingly. When using high-purity triglyceride feedstock and optimized reaction conditions, biodiesel yields can reach up to 98% conversion, with glycerol constituting approximately 10% of the total biodiesel mass as a byproduct. However, crude glycerol typically contains impurities such as methanol, soaps, catalyst residues, and water, requiring further purification before it can be used in value-added applications.

9.3.2 Crude Glycerol: Formation and Chemical Profile

In the industrial production of biodiesel through transesterification, crude glycerol is an inevitable and significant byproduct. For every 10 kg of biodiesel produced, approximately 1 kg of glycerol is generated (Thompson & He, 2006). This byproduct is not pure glycerol but rather a dark, sticky, and heterogeneous mixture commonly referred to as crude glycerol.

The formation of crude glycerol occurs during the final phase of biodiesel synthesis. Upon completion of the transesterification reaction, the reaction mixture undergoes phase separation due to the difference in density between the non-polar methyl esters (biodiesel) and the polar glycerol-based compounds. Glycerol, which has a higher density (~1.26 g/cm³), settles at the bottom of the reactor or separation tank, while biodiesel forms the upper layer (Freedman et al., 1984; Dasari et al., 2005). The quantity and quality of crude glycerol depend on several factors:

i. The type and purity of the feedstock.
ii. Catalyst concentration.
iii. Alcohol-to-oil molar ratio.
iv. The efficiency of separation and washing steps.

Unlike pharmaceutical- or food-grade glycerol (≥99.5% purity), crude glycerol typically contains only 30%–80% glycerol, along with a complex mixture of contaminants (Table 9.4). The exact composition varies depending on the feedstock and production method used (Johnson & Taconi, 2007).

The methanol content in crude glycerol poses flammability and safety risks, and it must be recovered through distillation prior to safe disposal or valorization. Alkaline catalysts (e.g., NaOH) can react with free fatty acids to form soaps, which complicate downstream processing. Additionally,

TABLE 9.4
Typical Composition Profile of Raw Glycerol

Component	Approximate Range (%)
Glycerol	30–80
Methanol	5–20
Water	5–20
Soap (sodium/potassium)	5–15
Catalyst residues (NaOH/KOH)	1–5
Free fatty acids	1–5
Methyl esters	1–10
Traces of feedstock impurities	<1

Source: Thompson and He (2006); Johnson and Taconi (2007)

water, often introduced during the biodiesel washing stage, dilutes the glycerol and negatively affects both its calorific value and its suitability for further chemical reactions.

The weird composition of raw glycerol presents many handling and valorization challenges (Samul et al., 2014):

- Low market value due to impurities and over task.
- Difficulty in cleansing, especially in small-scale biodiesel plants.
- Environmental considerations, if discharged without treatment.
- Corrosive and unstable behavior when stored for a long time.

Despite these challenges, raw glycerol has emerged as a platform substrate for biotechnological and chemical valorization provided that proper cleansing or pretension.

As biodiesel production expands globally, the accumulation of crude glycerol has become a significant byproduct of the transesterification process. While refined glycerol holds value in various industrial applications, the crude form generated during biodiesel production presents substantial management difficulties. These arise primarily from its impure composition, low commercial value, and the growing imbalance between supply and demand (Yang et al., 2012).

Crude glycerol is a highly variable waste stream, typically containing a mixture of glycerol, water, methanol, salts, soaps, and other organic impurities. These characteristics lead to numerous environmental, technical, and economic challenges related to its disposal, treatment, and valorization (Thompson & He, 2006).

9.3.3 Environmental and Health Risks

One of the major concerns in the determination and handling of raw glycerol is its potential environmental impact. Its high COD and biological oxygen demand (BOD) can lead to oxygen depletion in aquatic environments if it is improperly discharged. Additionally, residual methanol in raw glycerol is toxic, flammable, and poses health risks to both humans and animals if not properly removed or managed (Johnson & Tachoni, 2007).

Improper disposal practices, such as land spreading or direct discharge, can result in:

i. soil pollution due to salt and catalyst residues,
ii. water pollution from runoff, and
iii. air pollution from incomplete combustion or methanol volatilization.

Due to the overproduction of biodiesel, the market has become saturated with low-grade raw glycerol. As a result, its market value has declined, often shifting from a potential revenue stream to a disposal liability (Quispe et al., 2013). In some biodiesel operations, glycerol management can account for up to 15% of production costs, particularly for small-scale producers lacking purification infrastructure (Meher et al., 2006). Furthermore, regulatory bodies such as the Environmental Protection Agency (EPA) and the European Union impose strict documentation and treatment protocols for glycerol handling, especially when it contains hazardous substances such as methanol or heavy metals (da Silva et al., 2009).

Although several studies have demonstrated the potential use of raw glycerol as a substrate for microbial fermentation, combustion, or chemical conversion, its impure nature presents significant limitations (Yazdani & Gonzalez, 2007).

Some major technical challenges include:

i. The inhibitory effects of methanol and salts on microbial cultures used in bioconversion processes.

ii. Corrosive properties due to residual catalysts, which can damage storage tanks and processing equipment.
iii. Phase separation issues, which can complicate process consistency and material flow.

Additionally, purifying raw glycerol for pharmaceutical or food-grade applications requires costly distillation and advanced cleaning stages, which are often economically unfeasible for small-scale producers (Vasudevan & Briggs, 2008).

9.4 GLYCEROL-DERIVED FUELS AND ADDITIVES—APPLICATION

As global biodiesel production continues to rise, the resulting surplus of glycerol has sparked significant interest in converting it into value-added products, particularly renewable fuels and fuel additives. Rather than treating glycerol as merely a waste byproduct, recent research has highlighted its potential as a key intermediate in the development of sustainable energy solutions. One of the most promising applications involves its use in clean combustion fuels, either directly or as a performance-enhancing additive in conventional fuels. Glycerol can be converted into hydrogen—a clean and efficient energy carrier—through thermochemical processes such as steam reforming, aqueous-phase reforming, and autothermal reforming. Among these, steam reforming is widely studied due to its relatively high hydrogen yield. Catalysts such as nickel or platinum supported on alumina are commonly used to enhance efficiency and minimize coke deposition. Another thermochemical route involves the gasification of glycerol to produce syngas—a mixture of hydrogen and carbon monoxide—which can subsequently be converted into synthetic fuels via Fischer–Tropsch synthesis (Pagliaro et al., 2007). Beyond these conversion pathways, glycerol also serves as a precursor for the synthesis of fuel additives such as ethers, esters, and acetals. These compounds not only improve key fuel properties like octane and cetane numbers but also contribute to lower emissions, making them environmentally advantageous alternatives.

9.4.1 Fuel Additives Taken from Glycerol

One of the most extensively studied glycerol-derived compounds is solketal, formed through the ketalization of glycerol with acetone (Figure 9.4). This reaction typically occurs in the presence

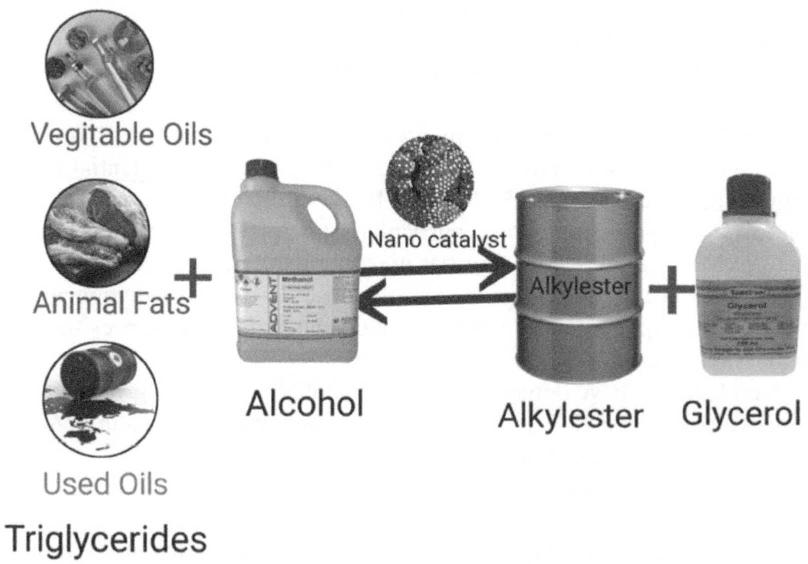

FIGURE 9.4 Glycerol-derived fuels and additives.

of solid acid catalysts such as p-toluenesulfonic acid or ion-exchange resins (Checa et al., 2020). Solketal serves as a valuable oxygenate in gasoline and diesel blends, improving octane rating, enhancing combustion efficiency, and reducing carbon monoxide and hydrocarbon emissions (da Silva et al., 2009). Its low water solubility and high boiling point make it suitable for diverse climatic conditions.

Another important class of additives includes glycerol ethers, particularly glycerol tert-butyl ether (GTBE), synthesized via etherification with isobutylene or GTBE, attributed improves octane rating and exhibits excellent solubility in gasoline blends. Additionally, it enhances combustion, resulting in lower emissions of particulates and sulfur oxides (Estevez et al., 2019).

Glycerol esters, such as monoglycerides and diglycerides, are produced through esterification with fatty acids. These compounds contribute to lubricant performance and enhance lubricity in diesel engines (Liu et al., 2006). Their thermal stability and compatibility with petroleum diesel enable effective performance under a wide range of operating conditions.

9.4.2 Effect on Engine Performance and Discharge

Comprehensive engine tests have demonstrated the functional benefits of incorporating glycerol derivatives into fuel. Improved fuel atomization leads to a more homogeneous air–fuel mixture, enhancing combustion dynamics and increasing thermal efficiency while significantly reducing CO and HC emissions (Talebian-Kiakalaieh et al., 2012). In some cases, a slight increase in NOx emissions has been observed, likely due to higher combustion temperatures (Pinzi et al., 2013).

In spark-ignition engines, research on GTBE shows improvements in octane rating (RON), reducing knocking and enabling higher compression ratios, which in turn enhances fuel economy and engine longevity (Alptekin & Canakci, 2008). Diesel engines benefit from the oxygen content in glycerol derivatives, promoting more complete combustion and reducing soot formation (Gui et al., 2008). The lubricating properties of glycerol esters also reduce engine wear, lowering maintenance frequency and extending service life.

The use of these additives is particularly appealing in regions with strict emissions standards, offering a renewable, biodegradable alternative to enhance conventional fuels. Their compatibility with various engine types and fuel systems supports their practical adoption across diverse transportation applications.

9.4.3 Comparative Benefits and Prospects

Glycerol-derived fuel additives offer a compelling combination of environmental sustainability, economic viability, and functional efficiency. Unlike fossil-based compounds, these bio-based additives are renewable, non-toxic, and biodegradable. The surplus of raw glycerol from biodiesel production presents not only a waste management challenge but also an opportunity to advance a circular, low-waste fuel economy (Kratzeisen & Müller, 2010).

Additives such as solketal and GTBE enhance fuel ignition, reduce emissions, and improve engine performance under various conditions. They can be integrated into existing fuel infrastructure with minimal modifications, enabling a smooth transition to greener alternatives. Additionally, their synthesis often relies on mild conditions and cost-effective catalysts, making them suitable for industrial-scale production (Vasudevan & Briggs, 2008).

Despite these advantages, several challenges remain—including phase separation in fuel blends, purification costs of raw glycerol, and regulatory hurdles for new fuel components. Addressing these issues requires continued research, improved catalyst systems, optimized production processes, and comprehensive life cycle assessments.

Looking ahead, integrating glycerol valorization into biorefineries could significantly boost the sustainability of the biodiesel sector. With collaboration among researchers, industry, and policymakers, glycerol-derived additives are well-positioned to support the clean energy transition.

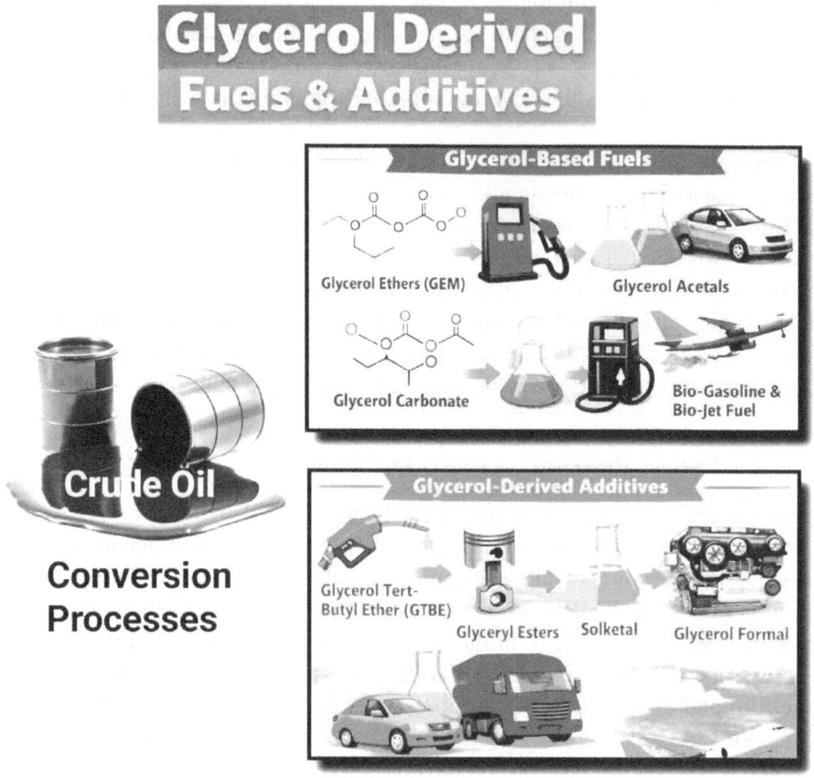

FIGURE 9.5 Economic and environmental perspectives of glycerol.

9.5 ECONOMIC AND ENVIRONMENTAL PERSPECTIVES OF GLYCEROL VALORIZATION

Glycerol, a major byproduct of biodiesel production, poses both a challenge and an opportunity. Its oversupply demands sustainable and profitable uses to enhance the biofuel supply chain's economic and environmental performance. Techno-economic analyses show that converting glycerol into valuable products like hydrogen, solketal, or epichlorohydrin can improve plant profitability, especially with low-cost catalysts and modular systems suited for small producers (Figure 9.5).

Environmental assessments reveal that glycerol-based processes generally reduce greenhouse gas emissions and resource use compared to fossil alternatives. For example, biohydrogen from glycerol emits less carbon than methane reforming, while solketal additives cut fuel life cycle emissions by about 20%. However, environmental impacts depend on factors like catalyst type and glycerol purity, as shown in Figure 9.5.

9.6 FUTURE PROSPECTS

The ongoing challenges in biodiesel production have highlighted the urgent need for glycerol valorization to ensure environmental sustainability and economic viability. The future of glycerol utilization depends on technological innovation, integration with renewable energy systems, and supportive research and policies. This chapter explores promising advancements, cross-sector integration, and policy frameworks that can accelerate the transition to a bio-based circular economy.

9.6.1 New Technologies and Innovation

Recent advances in catalysis, biopolymers, and materials science have broadened glycerol valorization pathways. Notably, heterogeneous catalysts—such as nickel-supported or bimetallic cobalt-copper systems—have improved selectivity and thermal stability for glycerol reforming into hydrogen (Almeida et al., 2012). The electrochemical conversion of glycerol into valuable chemicals like formic and glycolic acids is rapidly evolving, operating under mild, green conditions. Synthetic biology and metabolic engineering have enabled microbes like *Escherichia coli* and *Klebsiella pneumoniae* to convert glycerol into high-value products such as 1,3-propanediol, ethanol, and succinic acid (Zhou et al., 2019). Additionally, genetically modified microalgae show promise for producing biofuels and bioplastics from glycerol derivatives.

9.6.2 Integration with Renewable Energy Systems

Glycerol valorization is increasingly linked with renewable energy technologies. Solar-assisted steam reforming of glycerol can lower the carbon footprint and boost energy efficiency in hydrogen production (Gujar et al., 2025). Direct glycerol fuel cells (DGFC) offer clean, portable power by converting glycerol to electricity at low temperatures (Table 9.5). These fuel cells are especially promising for remote areas and off-grid applications, providing high energy density from a renewable, low-cost substrate and supporting decentralized energy models.

9.6.3 Research and Policy Directions

A comprehensive roadmap for glycerol valorization should combine focused research priorities with supportive policy measures. Research collaboration is essential to address technical challenges like raw material variability, catalyst performance, and scale-up issues. Developing standardized methods for glycerol purification and conversion will improve process stability and encourage industrial adoption. Policy frameworks should promote glycerol valorization through tax incentives, subsidies for bio-based products, and mandates for renewable fuel content. Incorporating glycerol valorization into national bioeconomy strategies can boost investment and innovation. International cooperation and knowledge-sharing platforms are crucial to accelerate technology transfer, particularly in developing countries expanding their biodiesel sectors.

TABLE 9.5
Summary of the Most Promising Technologies Integrating Glycerol Utilization with Renewable Energy Platforms

Technology	Description	Advantages
Solar-assisted glycerol reforming	Utilizes solar energy to drive the reforming of glycerol into hydrogen gas	Low greenhouse gas emissions; harnesses renewable solar input
Electrochemical conversion	Converts glycerol into value-added organic acids under mild electrochemical conditions	High product selectivity; energy-efficient and scalable
Direct glycerol fuel cells (DGFCs)	Directly oxidizes glycerol to generate electricity in a fuel cell	Compact and portable energy generation; low emissions; renewable fuel use
Bio-photoreactors with microalgae	Employs microalgae to convert glycerol into biomass and biofuels	Integrates carbon capture; produces biofuels and value-added byproducts

Source: Electrochemical oxidation of glycerol also produces organic acids and co-generated hydrogen using renewable electricity (da Silva et al., 2025; From waste to value). Direct glycerol fuel cells (DGFCs) combine glycerol oxidation with electricity generation (Huang et al., 2022; Direct Liquid Fuel Cells). Additionally, bio-photoreactors employing microalgae can integrate glycerol utilization with biomass and biofuel production (Chisti, 2007).

Additionally, regulatory alignment and clear environmental assessment protocols are needed to ensure glycerol-derived products meet safety and performance standards. Mechanisms such as carbon pricing and green procurement policies can help create market demand for glycerol-based fuels and chemicals.

9.7 CONCLUSION AND CHALLENGES

The growing demand for renewable energy has driven the biodiesel industry and increased interest in glycerol—a once-discarded byproduct now valued for its role in the bioeconomy. Glycerol's unique properties and abundance have made it essential in linking biodiesel production with byproduct valorization. While impurities pose challenges, advanced purification and processing methods enable their conversion into valuable chemicals, fuels, and additives that improve engine performance and reduce emissions.

Economically and environmentally, glycerol valorization supports biodiesel viability by lowering costs and emissions, fitting well within circular bioeconomy goals. Future progress relies on innovations in catalysis, synthetic biology, and supportive policies to expand applications and markets. Ultimately, glycerol is poised to evolve from waste to a key driver of a sustainable and resilient bioeconomy.

9.7.1 Key Challenges

I. **Raw Glycerol Impurities:** Raw glycerol from biodiesel processes often contains methanol, salts, soap and water, which directly prevents the application. The purification processes are energy intensive and increase the total costs.

II. **The Market Saturation and Value Instability:** Glycerol oversight suppresses the market value from biodiesel production. Without scalable alcoholism routes, manufacturers face financial losses from storage or disposal.

III. **Technical Limitations and Scalability:** Many well-organized methods are still in the laboratory or pilot stage with challenges in scaling while maintaining efficiency, safety and profitability.

IV. **Catalyst Costs and Durability:** Skilled glycerol conversion often requires expensive or short-term catalysts (e.g., platinum, nickel-based), which can increase the process costs and the environmental impact.

V. **Regulatory and Certification Barriers:** New glycerol-rich fuel and chemicals must undergo extensive testing and approval in order to meet safety and environmental regulations, which can delay the entry into the market.

VI. **Separation and Mixture of Stability Problems:** Some glycerol-based additives show poor errors with traditional fuel, causing technical challenges in fuel combination and long-term storage.

VII. **Lack of Delivery Infrastructure:** Unlike fossil fuels, glycerol-based fuel components may lack the infrastructure established for mass distribution and usage in current engines or industrial processes.

COMPETING INTERESTS

The authors declare that they have no competing interest.

DECLARATION OF GENERATIVE AI AND AI-ASSISTED TECHNOLOGIES IN THE WRITING PROCESS

During the preparation of this work, the book chapter author(s) have not taken any assistance from AI, and the book chapter author(s) reviewed and edited the content as needed and take(s) full responsibility for the content of the publication.

REFERENCES

Albertyn, J., Hohmann, S., Thevelein, J. M., & Prior, B. A. (1994). GPD1, which encodes glycerol-3-phosphate dehydrogenase, is essential for growth under osmotic stress in *Saccharomyces cerevisiae*, and its expression is regulated by the high-osmolarity glycerol response pathway. *Molecular and Cellular Biology, 14*(6), 4135–4144. https://doi.org/10.1128/mcb.14.6.4135-4144.1994

Almeida, J. R. M., Fávaro, L. C. L. & Quirino, B. F. (2012). Biodiesel biorefinery: opportunities and challenges for microbial production of fuels and chemicals from glycerol waste. *Biotechnol Biofuels, 5*, 48. https://doi.org/10.1186/1754-6834-5-48

Alptekin, E. (2013). Characteristics and composition of glycerol-based oxygenated fuel additives: A review. *Energy Education Science and Technology Part A: Energy Science and Research*, 31(3), 1347–1358.

Anitha, M., Kamarudin, S. K., & Kofli, N. T. (2016). The potential of glycerol as a value-added commodity. *Chemical Engineering Journal, 295*, 119–130. https://doi.org/10.1016/j.cej.2016.03.012

Atabani, A. E., Silitonga, A. S., Badruddin, I. A., Mahlia, T. M. I., Masjuki, H. H., & Mekhilef, S. (2012). A comprehensive review on biodiesel as an alternative energy resource and its characteristics. *Renewable and Sustainable Energy Reviews, 16*(4), 2070–2093. https://doi.org/10.1016/j.rser.2012.01.003

Canakci, M., & Van Gerpen, J. (1999). Biodiesel production via acid catalysis. *Transactions of the ASAE*, 42(5), 1203–1210. https://doi.org/10.13031/2013.13210

Checa, M., Nogales-Delgado, S., Montes, V., & Encinar, J. M. (2020). Recent advances in glycerol catalytic valorization: A review. *Catalysts, 10*(11), 1279. https://doi.org/10.3390/catal10111279

Cherubini, F. (2010) The biorefinery concept: Using biomass instead of oil for producing energy and chemicals. *Energy Conversion and Management, 51*(7), 1412–1421. https://doi.org/10.1016/j.enconman.2010.01.015

Chisti, Y. (2007). Biodiesel from microalgae. *Biotechnology Advances, 25*(3), 294–306. https://doi.org/10.1016/j.biotechadv.2007.02.001

Ciriminna, R., Della Pina, C., Rossi, M., & Pagliaro, M. (2014). Understanding the glycerol market. *European Journal of Lipid Science and Technology, 116*(10), 1432–1439. https://doi.org/10.1002/ejlt.201400229

Clomburg, J. M., & Gonzalez, R. (2011). Biofuel production in Escherichia coli: The role of metabolic engineering and synthetic biology. *Applied Microbiology and Biotechnology, 89*(2), 419–434. https://doi.org/10.1007/s00253-010-2446-1

Clomburg, J. M., & Gonzalez, R. (2013). Anaerobic fermentation of glycerol: A platform for renewable fuels and chemicals. *Trends in Biotechnology, 31*(1), 20–28. https://doi.org/10.1016/j.tibtech.2012.10.006

Costa, A. A. F. d., de Oliveira, A. d. N., Esposito, R., Len, C., Luque, R., Noronha, R. C. R., Rocha Filho, G. N. d., & Nascimento, L. A. S. d. (2022). Glycerol and catalysis by waste/low-cost materials—A review. *Catalysts, 12*(5), 570. https://doi.org/10.3390/catal12050570

da Silva, G. P., Mack, M., & Contiero, J. (2009).Glycerol: A promising and abundant carbon source for industrial microbiology. *Biotechnology Advances, 27*(1), 30–39. https://doi.org/10.1016/j.biotechadv.2008.07.006

da Silva, L. M. G., Costa, L. G. A., Santos, J. E. L., Costa, E. C. T. de A., de Morais Araújo, A. M., Gondim, A. D., Cavalcanti, L. N., Quiroz, M. A., dos Santos, E. V., & Martínez-Huitle, C. A. (2025). Electro-refinery in organics to produce energy carriers: Co-generation of green hydrogen and carboxylic acids by glycerol electrooxidation using dimensionally stable anode. *Catalysts, 15*(4), 333. https://doi.org/10.3390/catal15040333

Dasari, M. A., Kiatsimkul, P.-P., Sutterlin, W. R., & Suppes, G. J. (2005). Low-pressure hydrogenolysis of glycerol to propylene glycol. Applied Catalysis A: *General, 281*(1–2), 225–231. https://doi.org/10.1016/j.apcata.2004.11.033

Davda, R. R., Shabaker, J. W., Huber, G. W., Cortright, R. D., & Dumesic, J. A. (2005). A review of catalytic issues and process conditions for renewable hydrogen and alkanes by aqueous-phase reforming of oxygenated hydrocarbons over supported metal catalysts. *Applied Catalysis B: Environmental, 56*(1–2), 171–186. https://doi.org/10.1016/j.apcatb.2004.04.027

Demirbas, A. (2009). Biofuels securing the planet's future energy needs. *Energy Conversion and Management, 50*(9), 2239–2249.https://doi.org/10.1016/j.enconman.2009.05.010

Elsayed, M., Eraky, M., Osman, A. I., Wang, J., Farghali, M., Rashwan, A. K., & Abomohra, A. (2024). Sustainable valorization of waste glycerol into bioethanol and biodiesel through biocircular approaches: a review. *Environmental Chemistry Letters*, *22*(2), 609–634. https://doi.org/10.1007/s10311-023-01671-6

Encinar, J. M., González, J. F., & Rodríguez-Reinares, A. (2005). Biodiesel from used frying oil. Variables affecting the yields and characteristics of the biodiesel. *Industrial & Engineering Chemistry Research, 44*(15), 5491–5499. https://doi.org/10.1021/ie040214f

Estevez, R., Aguado-Deblas, L., Luna, D., & Bautista, F. M. (2019). An overview of the production of oxygenated fuel additives by glycerol etherification, either with isobutene or tert-butyl alcohol, over heterogeneous catalysts. *Energies, 12*(12), 2364. https://doi.org/10.3390/en12122364

Farouk, S. M., Tayeb, A. M., Abdel-Hamid, S. M. S., & Osman, R. M.(2024). Recent advances in transesterification for sustainable biodiesel production, challenges, and prospects: A comprehensive review. *Environmental Science and Pollution Research, 31*, 12722–12747. https://doi.org/10.1007/s11356-024-32027-4

Freedman, B., Pryde, E. H., & Mounts, T. L. (1984). Variables affecting the yields of fatty esters from transesterified vegetable oils. *Journal of the American Oil Chemists' Society, 61*(10), 1638–1643. https://doi.org/10.1007/BF02541649

Fukuda, H., Kondo, A., & Noda, H. (2001). Biodiesel fuel production by transesterification of oils. *Journal of Bioscience and Bioengineering, 92*(5), 405–416. https://doi.org/10.1016/S1389-1723(01)80288-7

Geng, Y., Shaukat, A., Azhar, W., et al. (2025). Microalgal biorefineries: A systematic review of technological trade-offs and innovation pathways. *Biotechnology for Biofuels and Bioproducts, 18*, 93. https://doi.org/10.1186/s13068-025-02694-7

Gui, M. M., Lee, K. T., & Bhatia, S. (2008). Feasibility of edible oil vs. non edible oil vs. waste edible oil as biodiesel feedstock. *Energy, 33*(11), 1646–1653. https://doi.org/10.1016/j.energy.2008.06.002

Gujar, J. P., Verma, A., & Modhera, B. (2025). Optimizing glycerol conversion to hydrogen: A critical review of catalytic reforming processes and catalyst design strategies. *International Journal of Hydrogen Energy, 109*, 823–850. https://doi.org/10.1016/j.ijhydene.2025.02.077

Hill, J., Nelson, E., Tilman, D., Polasky, S., & Tiffany, D. (2006). Environmental, economic, and energetic costs and benefits of biodiesel and ethanol biofuels. *Proceedings of the National Academy of Sciences of the United States of America, 103*(30), 11206–11210. https://doi.org/10.1073/pnas.0604600103

Huang, L., Yu, X., Huang, L., Zhang, X., Gu, L., Cao, Y., Li, W., Hu, J., & Cao, X. (2022). Co generation of electricity and formate in glycerol fuel cells with a bifunctional PdPtAg alloy nanowire electrocatalyst. *Green Chemistry, 24*, 9721–9733. https://doi.org/10.1039/D2GC03405K

International Energy Agency (IEA). (2021). *Renewable energy market update 2021: Transport biofuels*. IEA. https://www.iea.org/reports/renewable-energy-market-update-2021/transport-biofuels

Johnson, D. T., & Taconi, K. A. (2007).The glycerin glut: Options for the value-added conversion of crude glycerol resulting from biodiesel production. *Environmental Progress, 26*(4), 338–348. https://doi.org/10.1002/ep.10225

Kamm, B., & Kamm, M. (2004). Principles of biorefineries. *Applied Microbiology and Biotechnology, 64*, 137–145. https://doi.org/10.1007/s00253-003-1537-7

Katryniok, B., Paul, S., Capron, M., & Dumeignil, F. (2009) Towards the sustainable production of acrolein by glycerol dehydration, towards the sustainable production of acrolein by glycerol dehydration. *ChemSusChem*, 2(8), 719–730. https://doi.org/10.1002/cssc.200900134

Knothe, G. (2005). Dependence of biodiesel fuel properties on the structure of fatty acid alkyl esters. *Fuel Processing Technology, 86*(10), 1059–1070. https://doi.org/10.1016/j.fuproc.2004.11.002

Kratzeisen, M., & Müller, J. (2010). Influence of phosphorus content of coconut oil on deposit and performance of plant oil pressure stoves. *Renewable Energy, 35*(11), 2585–2589. https://doi.org/10.1016/j.renene.2010.04.001

Kumar, L. R., Yellapu, S. K., Tyagi, R. D., & Zhang, X. (2019). A review on variation in crude glycerol composition, bio-valorization of crude and purified glycerol as carbon source for lipid production. *Bioresource Technology, 293*, 122155. https://doi.org/10.1016/j.biortech.2019.122155

Liu, Y., Lotero, E., & Goodwin, J.G., Jr. (2006). A comparison of the esterification of acetic acid with methanol using heterogeneous versus homogeneous acid catalysis. *Journal of Catalysis, 242*(2), 278–286. https://doi.org/10.1016/j.jcat.2006.05.026

Leung, D. Y. C., Wu, X., & Leung, M. K. H. (2010). A review on biodiesel production using catalyzed transesterification. *Applied Energy, 87*(4), 1083–1095. https://doi.org/10.1016/j.apenergy.2009.10.006

Macedo, M.S., Soria, M.A., Madeira, L.M., (2019). Glycerol steam reforming for hydrogen production: Traditional versus membrane reactor. *International Journal of Hydrogen Energy, 44*(45), 24719–24732. https://doi.org/10.1016/j.ijhydene.2019.07.046

Meher, L. C., Sagar, D. V., & Naik, S. N. (2006). Technical aspects of biodiesel production by transesterification—A review. *Renewable and Sustainable Energy Reviews, 10*(3), 248–268. https://doi.org/10.1016/j.rser.2004.09.002

Mendieta López, J., Pérez Flores, F. J., Castillo Rosales, E., Ortiz Muñoz, E., Hernández-Anzaldo, S., Vázquez Lima, H., & Reyes Ortega, Y. (2022). A theoretical and experimental study of liquid-liquid equilibrium to refine raw glycerol obtained as a by-product of biodiesel production. *Chemical Engineering Journal Advances, 10*, 100257. https://doi.org/10.1016/j.ceja.2022.100257

National Center for Biotechnology Information (2026). PubChem Compound Summary for CID 753, Glycerol. Retrieved January 7, 2026 from https://pubchem.ncbi.nlm.nih.gov/compound/Glycerol

Pagliaro, M., & Rossi, M. (2008). *The future of glycerol: New usages for a versatile raw material*. Royal Society of Chemistry (RSC Publishing). ISBN: 978-0-85404-124-4. https://doi.org/10.1039/9781847558305

Pagliaro, M., Ciriminna, R., Kimura, H., Rossi, M., & Della Pina, C. (2007). From glycerol to value added products. *Angewandte Chemie International Edition, 46*(24), 4434–4440. https://doi.org/10.1002/anie.200604694

Pinzi, S., Leiva Candia, D. E., López García, I., Redel Macías, M. D., & Dorado, M. P. (2013). Latest trends in feedstocks for biodiesel production. *Biofuels, Bioproducts & Biorefining*, 8(2), 126–143. https://doi.org/10.1002/bbb.1435

Quispe, C. A. G., Coronado, C. J. R., & Carvalho Jr., J. A. (2013).Glycerol: Production, consumption, prices, characterization and new trends in combustion. *Renewable and Sustainable Energy Reviews, 27*, 475–493. https://doi.org/10.1016/j.rser.2013.06.017

Samul, D., Leja, K. & Grajek, W.(2014). Impurities of crude glycerol and their effect on metabolite production. *Ann Microbiol 64*, 891–898. https://doi.org/10.1007/s13213-013-0767-x

Sheehan, J., Camobreco, V., Duffield, J., Graboski, M., & Shapouri, H. (1998). *Life cycle inventory of biodiesel and petroleum diesel for use in an urban bus (NREL/SR 580 24089)*. National Renewable Energy Laboratory. https://doi.org/10.2172/658310

Talebian-Kiakalaieh, A., Amin, N. A. S., & Mazaheri, H.(2012). A review on novel processes of biodiesel production from waste cooking oil. *Applied Energy, 104*, 683–710. https://doi.org/10.1016/j.apenergy.2012.11.061

Thompson, J. C., & He, B. B. (2006).Characterization of crude glycerol from biodiesel production from multiple feedstocks. *Applied Engineering in Agriculture, 22*(2), 261–265. https://doi.org/10.13031/2013.20272

Vasudevan, P. T., & Briggs, M. (2008). Biodiesel production current state of the art and challenges. *Journal of Industrial Microbiology & Biotechnology, 35*(5), 421–430. https://doi.org/10.1007/s10295-008-0312-2

Yang, F., Hanna, M. A., & Sun, R. (2012).Value added uses for crude glycerol – A byproduct of biodiesel production. *Biotechnology for Biofuels, 5*, 13. https://doi.org/10.1186/1754-6834-5-13

Yazdani, S. S. & Gonzalez, R. (2007). Anaerobic fermentation of glycerol: A path to economic viability for the biofuels industry. *Current Opinion in Biotechnology, 18*(3), 213–219. https://doi.org/10.1016/j.copbio.2007.05.002

Zhang, Y., Dube, M. A., McLean, D. D., & Kates, M. (2003). Biodiesel production from waste cooking oil: 2. Economic assessment and sensitivity analysis. *Bioresource Technology, 90*(3), 229–240. https://doi.org/10.1016/S0960-8524(03)00150-0

Zhou, C. H., Beltramini, J. N., Fan, Y. X., & Lu, G. Q. (2008). Chemoselective catalytic conversion of glycerol as a biorenewable source to valuable commodity chemicals. *Chemical Society Reviews, 37*(3), 527–549. https://doi.org/10.1039/B707343G

Zhou, J.-J., Shen, J.-T., Wang, X.-L., Sun, Y.-Q., & Xiu, Z.-L. (2019). Bioconversion of raw glycerol from waste cooking oil based biodiesel production to 1,3-propanediol and lactate by a microbial consortium. *Frontiers in Bioengineering and Biotechnology, 7*, Article 14. https://doi.org/10.3389/fbioe.2019.00014

10 Techniques of Biofuel Production

An Overview

Ashok Pundir, Karan Singh, Suraj Prakash, Neeraj Kumari, and Radha

10.1 INTRODUCTION

With industrialization, urbanization and increasing population growth, there is an increase in global demand for energy. At present, fossil fuels account for about 90% of the world's energy supply, and they have been linked to adverse environmental effects like ecosystem degradation and emission of greenhouse gases. These have raised serious concerns about their use for various applications like agriculture, transportation, household, industry, etc. (Wang et al., 2016). It is estimated that by 2040, renewable sources will contribute about 50% of the global energy, highlighting the urgent need to move toward more sustainable energy systems (Hussein, 2015; Guin, Kundu, and Singh, 2022). However, challenges like climate change, price, energy insecurity and increasing consumption demand to explore or investigate alternative sources of fuels or energy that are cost-effective, environment friendly and renewable (Srivastava and Prasad, 2000). In view of this, fuels derived from natural resources known as "biofuels" like plant biomass, as well as agricultural and organic residues, have attracted lots of attention in the last few decades. They have emerged as an promising alternative to fossil fuels (Serrano, Rus, and García-Martínez, 2009; Mehejabin et al., 2024). They can directly replace petrol, diesel and jet fuel (Reddy et al., 2008; Bandyopadhyay, 2015). Biofuels are non-toxic, carbon neutral and biodegradable, and they support the rural economies by utilizing natural sources (Reddy et al., 2008; Serrano, Rus, and García-Martínez, 2009; Aransiola et al., 2012; Bandyopadhyay, 2015). Their production involves the use of variety of feedstocks, including animal fats, vegetable oils and plant residues, and yields gaseous, liquid or solid fuels that lead to reduction in emission of sulfur oxides, carbon dioxide, nitrogen oxides and particulate matter (Pugh et al., 2011; Puri, Abraham, and Barrow, 2012). Additionally, the efficiency of biofuel production depends on range of parameters like catalyst, feedstock source, temperature and processing time (Puri, Abraham, and Barrow, 2012). Biomass, especially agricultural waste, is made up of easily accessible, renewable, eco-friendly, carbon-neutral (Interdepartmental Program Directorate Energy Transition, The Hague (Netherlands), & SenterNovem, Utrecht (Netherlands), 2006) raw resources that don't raise CO_2 levels. It also releases less nitrogen and sulfur chemicals into the atmosphere (National Renewable Energy Laboratory (NREL), 2009). Depending on where it comes from and how it is processed, biomass is typically divided into two categories: conventional and advanced. Conventional biomass, also known as traditional biomass, is energy produced from non-sustainable sources and is frequently characterized by low efficiency and the emission of large quantities of hazardous pollutants. Modern or advanced biomass, on the other hand, produces high-quality biofuels by using effective conversion methods and appropriate management techniques. They come from a variety of sources, including vegetable oils, industrial waste and organic waste, as well as bioethanol, biogas, bio-oil and biodiesel (Ravindranath et al., 2011). Modern biomass is converted via a variety of physical, chemical and biological processes to create biofuels, which are effective energy carriers. Among them, liquid biofuels present a viable substitute for traditional fuels like petrol, diesel and oil (Sivagurunathan et al., 2016; Singhvi, Zinjarde, and Kim, 2022). Among the different types of feedstock sources,

 DOI: 10.1201/9781003600169-10

lignocellulosic biomass and algal sources have been explored due to their renewability, abundance and minimal competition with food crops (Government of India (GOI), 2016; Shu et al., 2020). For instance, biomass derived from agricultural and forestry waste, known as lignocellulosic biomass, is especially used for the production of second-generation biofuel. However, the complex structure of cellulose and lignin poses challenges for the conversion, highlighting the need to use of advanced technologies such as nanotechnology-assisted and consolidated bioprocessing methods to improve yields and limit the processing cost (Singh et al., 2015; Singhvi and Kim, 2022). Biohydrogen is another type of biofuel derived from lignocellulosic biomass. Its ability to produce no greenhouse gas emissions during combustion is one of its main advantages, which makes it a perfect clean energy source (Sivagurunathan et al., 2016). But the issues like cost, conversion efficiency and storage are limiting its adoption at large scale (Singhvi, Zinjarde, and Kim, 2022). Therefore, there is a need to explore further in this field for efficient utilization of natural sources for fuel production. The current chapter provides an overview of biofuel production from natural resources.

10.2 METHODS OF BIOFUEL PRODUCTION

The production process for biofuel involves different techniques or methods. The selection of method depends on the feedstock or biomass used, required product and economic consideration. Thus, the techniques are broadly classified into three categories, viz., mechanical, biochemical and thermochemical, and they utilize chemical, physical and biological methods for the conversion of biomass or raw material to biofuel (Figure 10.1). These conversion technologies generate three products: heat, power and biofuel, which can be used for electricity generation (Bhatia et al., 2020; El-Araby, 2024).

10.2.1 Mechanical Method

Mechanical treatment like grinding and chipping of wood material, followed by densification of material by pelletizing the biomass, comes under the mechanical process for the conversion of biomass into biofuel; briquetting technology can be used to mechanically convert or densify biomass into solid biofuel (Lei et al., 2013).

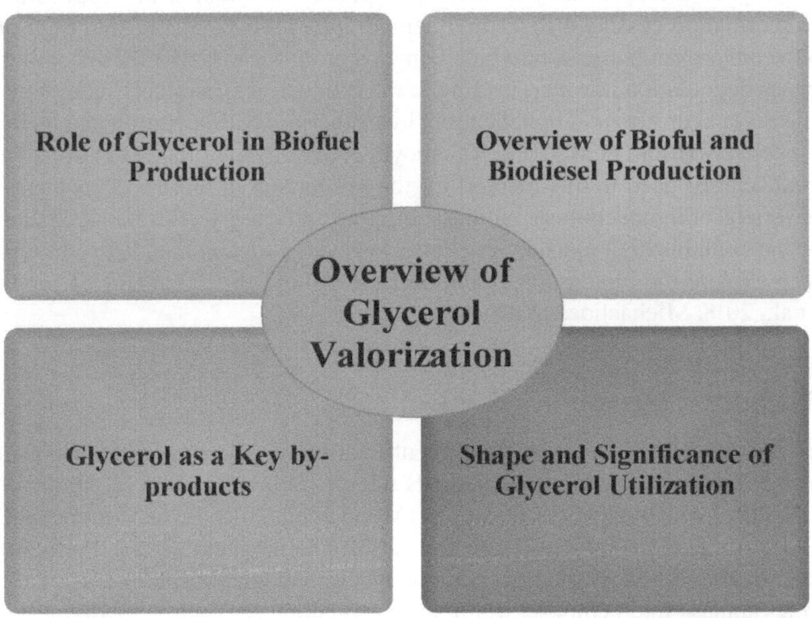

FIGURE 10.1 Production of biofuel via different methods using biomass or feedstock and generation of different types of biofuels. (Adapted from El-Araby, 2024.)

10.2.2 Thermochemical Method

This technique involves the use of extreme temperature resulting in chemical reformation, breaking of bond and the reforming of organic particles into gas, biochar and highly oxygenated biofuel (Lee et al., 2019). Processes like gasification, liquefaction, combustion, pyrolysis and hydrothermal processing are involved in thermochemical techniques of biofuel production.

Combustion is one of the conventional and basic ways of biomass conversion. In this, organic materials or raw material are burnt in air and converting chemical energy into mechanical power, heat or electricity. The whole process is carried out using furnaces, stoves, steam turbines, boilers and turbo generators (Escalante et al., 2022). The particle size of feedstock, combustion environment and temperature have major impact on biomass combustion. The fundamental limitation of this process is that it involves the emission of nitric oxide and carbon dioxide, along with discharge of ashes and particulates (Correa et al., 2019; Escalante et al., 2022). On the other hand, the breakdown process that involves conversion of biomass into charcoal, syngas and bio-oil is known as pyrolysis. It breaks the long chain molecules to short chain molecules, resulting in generation of concentrated fuel oils (Liu et al., 2022). Generally, biomass or waste is used as a feedstock for producing syngas and other liquid fuel by varying the processing conditions (Ahmed et al., 2022; Ray et al., 2022). Pyrolysis of waste oils has attracted attention as a method for producing high-grade biofuel (Mohanty et al., 2024). One advantage of pyrolysis is that it converts solid materials into gaseous or vapor form that are easy to handle, transport and store. On the other hand, the limitation of this method is that it requires high heat input to carry out the chemical reaction involving formation of syngas (Dasan et al., 2019; Ramos et al., 2022). In gasification, biomass is heated (500°C–1400°C) under oxygen-limited conditions, producing syngas, mixture of carbon monoxide, nitrogen, methane, hydrogen and other hydrocarbons that can be used to generate power, heat or biofuel (Bhargavi et al., 2018; Lee et al., 2019). It is one of the most efficient methods for the production of hydrogen gas from biomass (Ahmad et al., 2016). Furthermore, liquefication process converts biomass to liquid fuels using heat (250°C–350°C) and a catalyst (Ambaye et al., 2021). It produces complex mixture of volatile organic acids and hydrocarbons (Bhargavi et al., 2018). Catalytic liquefication, assisted by high hydrogen partial pressure or catalyst, effectively generates products with higher energy density in the liquid phase (Muritala et al., 2020; Mat Aron et al., 2020). Hydrothermal processing utilizes high temperature (180°C–375°C) and high-pressure water (2–22 MPa) to convert wet biomass into biochemicals or biofuels. Under these conditions, the feedstock undergoes degradation in water, leading to the production of biofuels such as biocrude, hydro char and syngas via hydrothermal liquefication, hydrothermal carbonization and hydrothermal gasification, respectively (Libra et al., 2011; Yakaboylu et al., 2015; Gollakota et al., 2018). Algae and waste materials can be used in this method for the production of biofuel. This approach offers an effective conversion of lignocellulosic biomass into fuel and energy, addressing challenges of utilizing biomass that contains high moisture content. However, to understand hydrothermal conversion, it is crucial to explore the biomass degradation kinetics and the chemistry of product formation (Bhargavi et al., 2018; Michaelides, 2018).

10.2.3 Biochemical Method

Biological or biochemical process involves the utilization of microorganisms (like fungi, bacteria and enzymes) and chemical or physical methods to convert biomass into economically viable gaseous products with a wide range of applications (Mood et al., 2013; Perez-Cantu et al., 2013; Huang et al., 2016; Leonard et al., 2020; Hoffman et al., 2021; Olatunji et al., 2021). Pretreatment methods (physical, chemical and biological) are required in the biological process for breaking down the lignocellulosic biomass into cellulose, lignin and hemicellulose. Factors involving feedstock type, chemical composition, biomass structure enzyme loading and creation of inhibitory compounds generally affect the pretreatment process. Fermentation and anaerobic digestion are the primary methods used in biological conversion of biomass. In fermentation, bioethanol from starch and sugar-rich biomass (like sugarcane, maize) is produced. This process involves the conversion of saccharides into sugars or glucose to ethanol and other byproducts (Srivastava et al., 2019; González et al., 2020; Leonard et al., 2020; Muritala et al., 2020). The efficiency of fermentation process depends on various factors including type of microorganism, concentration of sugar and temperature (Kucharska et al., 2018). Anaerobic digestion involves the use of bacteria to break down organic matter to produce methane-rich biogas, without oxygen. Agricultural residues, sewage sludge and food waste are generally used as feedstock in anaerobic digestion (Thakare et al., 2024). This is an energy-efficient and eco-friendly technology method used for electricity generation and fuel production (Deng et al., 2020; Abdullah et al., 2021). Transesterification, another biochemical method, is an alcoholysis process that involves the reaction between 3 and 4 mol of alcohol and 1 mol of triglyceride molecule, producing esters and glycerol. It involves the use of several acids and bases as catalysts like H_2SO_4, K_2CO_3, KOH, NaOH, sodium methoxide, sodium ethoxide and sulfonic acid (Ganesan et al., 2020). Transesterification is basically an exchange of alcohol from an ester by another alcohol like hydrolysis with the exception of using alcohol in place of water (Abbaszaadeh et al., 2012). Microalgae and other microbes are used as feedstock which produce lipids that are then converted into biodiesel through transesterification (Voloshin et al., 2016; Saha et al., 2019).

10.2.4 Emerging Technologies

With continuous rise in global energy demand, biofuels have been considered as a potential alternative to conventional fossil fuels, having the ability to reduce greenhouse gas emissions and improve energy security (Rodionova et al., 2017; Osman et al., 2021; Ranjitha et al., 2021; Sundaram et al., 2025). In production of biofuels, nanotechnology plays an important role in increasing efficiency of enzymes, improving substances movement involved in chemical reactions and creating new types of catalysts for biofuel-related process (Arya et al., 2021; Bhuyan et al., 2021; Phillips, 2022). Using genetic engineering, these enzymes can be made to use different types of feedstocks, which improve the yield and make the process more sustainable (Adegboye et al., 2021; Mohanty et al., 2024).

10.3 TYPES OF BIOFUELS

Depending on the kind of raw material used, production method and their impact on environment, biofuels are grouped into two categories including conventional biofuels, also called first generation, and advanced biofuels, which include both second- and third-generation types (Bisht et al., 2021; Sadaqat et al., 2025).

10.3.1 Conventional Biofuels

Conventional biofuels, also known as first-generation biofuels, are produced from food-based materials such as animal fats and agricultural crops using standard production methods in which bioethanol and biodiesel are mostly produced (Tuli and Kasture, 2022; Almeida et al., 2023; Malik et al., 2024) (Table 10.1). In the case of the United States, corn is the main crop used for making bioethanol, while in Brazil, sugarcane is the major source (Wandscher Busanello et al., 2024). Biodiesel, on the other hand, is produced through a chemical process called transesterification, where oils and fats such as vegetable oil, animal fat and recycled greases are converted into fatty acid methyl esters (FAME) (Almeida et al., 2023; Arista et al., 2024). Common feedstocks used in biodiesel production include soybean oil, palm oil and rapeseed oil (Lee et al., 2021). These biofuels serve as renewable alternatives to conventional fossil fuels that are compatible with existing fuel infrastructure, thereby reducing dependency on finite petroleum-based resources (Wandscher Busanello et al., 2024). Concerns have been raised about the long-term use of biofuels because of their competition with food crops, possible changes in land use and associated environmental impacts

TABLE 10.1
Production of Conventional and Advanced Biofuels from Various Feedstocks, Summarizing Key Feedstocks, Fuel Types and Yield Ranges

Source	Biofuel	Yield Range	Production (2022–2024)	References
Corn (grain)	Ethanol	350–420 L/ton corn (typical)	United States: ~15 billion gallons (2022)	Mohammadi and Saddler (2023)
Sugarcane	Ethanol	70–85 L/ton cane	Brazil: ~7 billion gallons (2022)	Mohammadi and Saddler (2023)
Wheat, barley, sorghum	Ethanol	350–400 L/ton grain	EU: growing use in ethanol plants	Sarwal et al. (2021); Mohammadi and Saddler (2023)
Soybean oil	Biodiesel	950–1,000 L/ton oil	Brazil: ~8.9 billion liters (2024; 69% from soybean oil)	Mohammadi and Saddler (2023); Hayashi (2024)
Palm oil	Biodiesel	1000–1,100 L/ton oil	Indonesia significant producer	Lorne (2023); Mohammadi and Saddler (2023)
Animal fats	Biodiesel, HVO	800–900 L/ton	Brazil: ~16% biodiesel feedstock share (2023)	Hayashi (2024)
Lignocellulosic biomass	Cellulosic ethanol	250–350 L/ton (depends on process and feedstock)	Commercial scale limited; United States, Brazil active pilots	Mohammadi and Saddler (2023; Hayashi (2024)
Wood chips (pyrolysis)	Bio-oil	44.6% yield by weight (from dry matter, monthly avg. in solar-thermal pyrolysis, 2022 study)	R&D stage; industrial demo	Ghenai et al. (2022)
Cover crops (various)	Ethanol, Bio-oil	0.10–0.30 g ethanol/g dry matter; 0.45–0.55 g bio-oil/g dry matter	Yield depends on species	Yang et al., 2023)

(Maia et al., 2020). Although these challenges remain, conventional biofuels have continued to play an important role in the renewable energy sector and have encouraged ongoing research to improve their environmental and economic sustainability. With increasing global energy demand, the future of conventional biofuels continues to be explored with a focus on achieving energy security without compromising food availability and ecological balance (Rodionova et al., 2017; Callegari et al., 2020). The production of conventional biofuels follows systematic and standardized steps in the case of both bioethanol and biodiesel (Figure 10.2).

Bioethanol production starts with feedstock preparation where crops including corn, sugarcane and other sugar-containing materials are collected and processed to get fermentable sugars (Luque et al., 2008). After extraction of sugars, fermentation process is carried out using yeast or selected bacterial strains, which help to convert sugars into ethanol (Lackner, 2021). This is then separated from the fermentation broth through the process of distillation, followed by a dehydration step to remove water and obtain anhydrous ethanol suitable for blending with conventional fuels such as gasoline (Luque et al., 2008). In contrast, biodiesel production is based on the transesterification reaction where vegetable oils, animal fats or waste cooking oils are first refined and

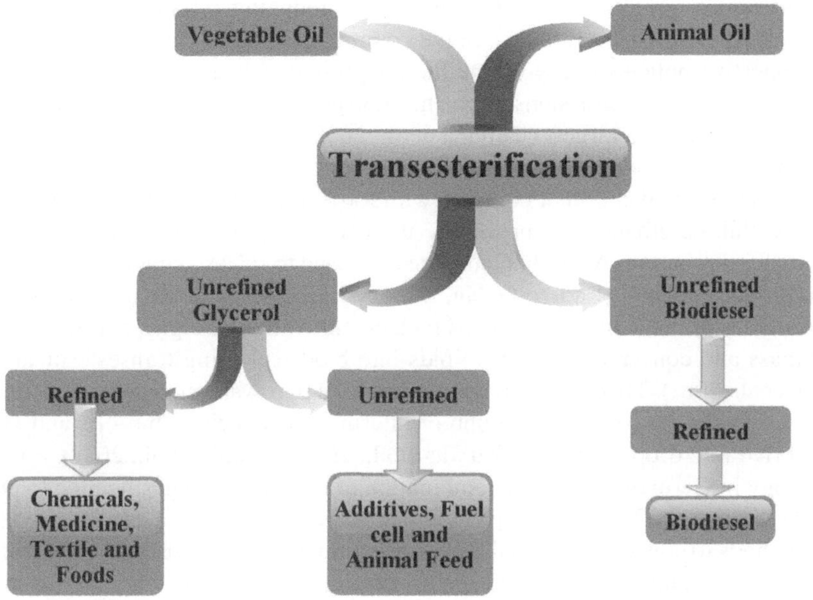

FIGURE 10.2 Flowchart showing stepwise process of conventional biodiesel production following preliminary investigation, pretreatment, post-treatment and characterization. (Kosuru et al., 2024.)

pretreated to remove moisture and impurities (Singh et al., 2022). The refined oil or fat is mixed with methanol in the presence of a chemical catalyst such as sodium hydroxide, which leads to the formation of biodiesel along with glycerol as a byproduct. In the final step, biodiesel is separated and purified to ensure that it meets the required fuel quality standards (Arista et al., 2024). Although both bioethanol and biodiesel production methods are well established, several recent studies have highlighted the need to optimize the individual steps to improve fuel yield, enhance process sustainability and reduce negative environmental impacts (Jung et al., 2021; Dos Santos et al., 2021).

10.3.2 Advanced Biofuels

Advanced biofuels, also known as second- and third-generation biofuels, are produced from non-edible plant materials including agricultural waste, lignocellulosic biomass, algae and other alternative sources (Cherwoo et al., 2023). These biofuels were developed to address the problems linked with first-generation biofuels, mainly the issue of food and fuel competition (Mizik, Igbeghe, and Deák, 2025). Some of the commonly reported advanced biofuels include cellulosic ethanol, algae-based biodiesel, biomethanol, dimethyl ether (bio-DME), Fischer–Tropsch diesel and biohydrogen (Filimonau et al., 2020). Cellulosic ethanol is obtained from sources like crop residues, wood and energy plants, which are not used as food (García-Sancho, Cecilia, and Luque, 2021). Algae-based biofuels, particularly from microalgae, are rich in lipid content and can be cultivated on non-arable lands, even in wastewater, making them a promising sustainable option (Preradovic and Papuga, 2021). Thermochemical routes such as pyrolysis and gasification can convert lignocellulosic biomass into syngas and bio-oils, which are upgraded into advanced fuels like synthetic diesel, methanol and DME (Sikarwar et al., 2016; Usmani, 2020). Another example is Hydrogenated Vegetable Oil (HVO), also called renewable diesel or cleaner-burning diesel fuel, which is produced by hydrogenating vegetable oils, animal fats, or used cooking oils (Alobaid et al., 2022). These fuels can be used directly in engines without requiring changes to the existing fuel infrastructure, which helps in lowering environmental pressure and reducing reliance on fossil-based energy sources (Khalid, Hussain, and Ali, 2021). Some countries, especially in the European Union, have introduced supportive policies that promote the adoption of advanced biofuels. These efforts are mainly focused on reducing emissions from the transport sector and reaching long-term climate goals like achieving net zero targets (Yuliana et al., 2020). Some industrial organizations, such as ExxonMobil, have started working on the production of biofuels using sources such as algae, corn stover and switchgrass, which do not compete with food crops or freshwater use (Suo et al., 2021). In the case of cellulosic ethanol, the procedure usually starts with pretreatment of lignocellulosic biomass to break its structure. After that, enzymes are used to release simple sugars, which are then fermented to produce ethanol (Ladero Galán, 2024). Similarly, algae are used in the production of biodiesel through different steps. These steps include cultivation of algae, extraction of lipids from the algal biomass and conversion of those lipids into biodiesel using transesterification or related methods (Lee et al., 2021). Thermochemical methods like pyrolysis and gasification involve heating biomass in low-oxygen environments to obtain intermediates such as bio-oils and syngas, which are then converted into drop-in fuels (Sikarwar et al., 2016; Alobaid et al., 2022). For example, the Bioliq technology in Germany utilizes a combination of pyrolysis and gasification to produce DME and methanol (Karki et al., 2021). Similarly, companies like Enerkem in Canada have adopted waste gasification methods to convert municipal solid waste into ethanol and methanol (Karki et al., 2021). Advanced molecular techniques like genetic engineering and metabolic engineering are commonly applied to enhance biofuel production, improve the utilization of raw materials and modify the fuel characteristics based on the requirement of different applications (Yuliana et al., 2020; Banu, Kumar, and Chattopadhyay, 2021). Along with this, studies are also being carried out on the use of new catalysts, nanomaterials and specially developed microbial strains, which help in increasing the overall efficiency of the process and improving the performance of the produced fuel (Yuliana et al., 2020). Although production of advanced biofuels remains cost-intensive and technologically challenging compared to first-generation biofuels, their significant environmental benefits and potential for long-term sustainability make them an essential constituent in achieving low-carbon energy future (Mizik and Gyarmati, 2022).

10.4 CONCLUSION

The shift from fossil fuels to biofuels has now become necessary due to increasing environmental concerns. Biofuels obtained from lignocellulosic biomass, algae, vegetable waste and organic residues are known to reduce greenhouse gas emissions and can support the energy demands of growing populations. These bioresources also help in strengthening the rural economy by creating new livelihood opportunities. Research and developments in thermochemical and biochemical methods have improved the production efficiency, reduced energy inputs and allowed the use of various types of raw materials. Life cycle assessments confirm that when sustainably sourced and efficiently processed, biofuels outperform petroleum fuels in terms of carbon intensity, air-pollutant profiles and resource circularity. Yet, economic competitiveness, land use trade-offs and scalability remain critical obstacles; resolving them will require integrated biorefinery strategies, waste-to-fuel valorization, genetic and catalytic innovations and supportive policy frameworks that internalize environmental costs. Recent efforts in biofuel research have focused on pilot and commercial-level production. Such initiatives, along with proper sustainability certification, can help in gaining public trust and encourage private investments. With suitable government policies and adoption of new technologies, the biofuel sector has the potential to play an important role in meeting clean energy targets. For countries like India, where agriculture and biomass resources are in abundance, biofuel production can offer an effective solution for long-term energy security and carbon reduction. The coming decade will decide whether biofuels fulfil their promise of delivering clean, reliable and inclusive energy without undermining food or ecological integrity. Stakeholder collaboration today will determine that low-carbon future tomorrow.

REFERENCES

Abbaszaadeh, Ahmad, Barat Ghobadian, Mohammad Reza Omidkhah, and Gholamhassan Najafi. "Current biodiesel production technologies: A comparative review." *Energy Conversion and Management* 63 (2012): 138–148.

Abdullah, Mohd Azmuddin, and Hanaa Ali Hussein. "Integrated algal and oil palm biorefinery as a model system for bioenergy co-generation with bioproducts and biopharmaceuticals." *Bioresources and Bioprocessing* 8, no. 1 (2021): 40.

Adegboye, Mobolaji Felicia, Omena Bernard Ojuederie, Paola M. Talia, and Olubukola Oluranti Babalola. "Bioprospecting of microbial strains for biofuel production: Metabolic engineering, applications, and challenges." *Biotechnology for Biofuels* 14, no. 1 (2021): 5.

Ahmad, Anis Atikah, Norfadhila Abdullah Zawawi, Farizul Hafiz Kasim, Abrar Inayat, and Azduwin Khasri. "Assessing the gasification performance of biomass: A review on biomass gasification process conditions, optimization and economic evaluation." *Renewable and Sustainable Energy Reviews* 53 (2016): 1333–1347.

Ahmed, Shams Forruque, M. Mofijur, Samiha Nuzhat, Nazifa Rafa, Afla Musharrat, Su Shiung Lam, and Alberto Boretti. "Sustainable hydrogen production: Technological advancements and economic analysis." *International Journal of Hydrogen Energy* 47, no. 88 (2022): 37227–37255.

Almeida, Eugênia Leandro, José Eduardo Olivo, and Cid Marcos Gonçalves Andrade. "Production of biofuels from glycerol from the biodiesel production process - A brief review." *Fermentation* 9, no. 10 (2023): 869. doi:10.3390/fermentation9100869.

Alobaid, Falah, Naser Almohammed, Massoud Massoudi Farid, Jan May, Philip Rößger, Andreas Richter, and Bernd Epple. "Progress in CFD simulations of fluidized beds for chemical and energy process engineering." *Progress in Energy and Combustion Science* 91 (2022): 1–213. doi:10.1016/j.pecs.2021.100930.

Ambaye, Teklit Gebregiorgis, Mentore Vaccari, Adrián Bonilla-Petriciolet, Shiv Prasad, Eric D. van Hullebusch, and Sami Rtimi. "Emerging technologies for biofuel production: A critical review on recent progress, challenges and perspectives." *Journal of Environmental Management* 290 (2021): 112627.

Aransiola, Elizabeth, Betiku Eriola, Ikhuomoregbe Dio, and Tunde Victor Ojumu. "Production of biodiesel from crude neem oil feedstock and its emissions from internal combustion engines." *African Journal of Biotechnology* 11, no. 22 (2012): 6178–6186. doi:10.5897/ajb11.2301.

Arista, Amanda Elliora, Gilang Maulana Dwi Putra, Nuno Vansa Adiaksa, Debora Engelien Christa Jaya, and Heri Septya Kusuma. "Bibliometric analysis and visualization for revealing publication trends research mapping of solid acid catalyst for biodiesel and biofuel production based on VOSviewer." *Science Talks* 9 (March 2024): 100295. doi:10.1016/j.sctalk.2024.100295.

Arya, Indrajeet, Asha Poona, Pritam Kumar Dikshit, Soumya Pandit, Jatin Kumar, Himanshu Narayan Singh, and Niraj Kumar Jha, Hassan Ahmed Rudayni, Anis Ahmad Chaudhary, and Sanjay Kumar. "Current trends and future prospects of nanotechnology in biofuel production." *Catalysts* 11 (2021): 1308.

Bandyopadhyay, Kaushik Ranjan. "Biofuel promotion in India for transport: Exploring the grey areas." TERI Policy Brief. Mithapur, India: The Energy and Resources Institute, TERI, 2015. https://www.teriin.org/policy-brief/biofuel-promotion-india-transport-exploring-grey-areas.

Banu, J. Rajesh, Gopalakrishnan Kumar, and Indranil Chattopadhyay. "Management of microbial enzymes for biofuels and biogas production by using metagenomic and genome editing approaches." *3 Biotech* 11, no. 10 (2021): 1–15. doi:10.1007/s13205-021-02962-x.

Bhargavi, Gunturu, Sudhakar Uppalapati, and Renganathan Sahadevan. "Review on the extraction methods of crude oil from all generation biofuels in last few decades." In *International Conference on Recent Advances in Materials, Mechanical and Civil Engineering, 1–2 June 2017, Hyderabad, India*, vol. 330, 012024. IOP Publishing, 2018.

Bhatia, Shashi Kant, Sujit Sadashiv Jagtap, Ashwini Ashok Bedekar, Ravi Kant Bhatia, Anil Kumar Patel, Deepak Pant, J. Rajesh Banu, Christopher V. Rao, Yun-Gon Kim, and Yung-Hun Yang. "Recent developments in pretreatment technologies on lignocellulosic biomass: Effect of key parameters, technological improvements, and challenges." *Bioresource Technology* 300 (2020): 122724.

Bhuyan, Nilutpal, Anurag Dutta, Rajkamal Mohan, Neelam Bora, and Rupam Kataki. "Advances in nanotechnology for biofuel production." In *Nanomaterials*, Eds. R. Praveen Kumar and B. Bharathiraja, 533–562. Amsterdam: Academic Press, 2021.

Bisht, Sarita, Amit Kumar, Narendra Kumar, Hukum Singh, and Parmanand Kumar. "Biofuel production by using biomass and its application." In *Renewable Energy and Green Technology*, Eds. Narender Kumar, Hukum Singh, and Amit Kumar, 85–103. Boca Raton: CRC Press, 2021.

Callegari, Arianna, Silvia Bolognesi, Daniele Cecconet, and Andrea G. Capodaglio. "Production technologies, current role, and future prospects of biofuels feedstocks: A state-of-the-art review." *Critical Reviews in Environmental Science and Technology* 50, no. 4 (2020): 384–436. doi:10.1080/10643389.2019.1629801.

Cherwoo, Lubhan, Ishika Gupta, G. Flora, Ritu Verma, Muskaan Kapil, Shailendra Kumar Arya, Balasubramani Ravindran, et al. "Biofuels an alternative to traditional fossil fuels: A comprehensive review." *Sustainable Energy Technologies and Assessments* 60 (December 2023): 103503. doi:10.1016/j.seta.2023.103503.

Correa, Diego F., Hawthorne L. Beyer, Joseph E. Fargione, Jason D. Hill, Hugh P. Possingham, Skye R. Thomas-Hall, and Peer M. Schenk. "Towards the implementation of sustainable biofuel production systems." *Renewable and Sustainable Energy Reviews* 107 (2019): 250–263.

Dasan, Yaleeni Kanna, Man Kee Lam, Suzana Yusup, Jun Wei Lim, and Keat Teong Lee. "Life cycle evaluation of microalgae biofuels production: Effect of cultivation system on energy, carbon emission and cost balance analysis." *Science of the Total Environment* 688 (2019): 112–128.

Deng, Chen, Richen Lin, Xihui Kang, Benteng Wu, Richard O'Shea, and Jerry D. Murphy. "Improving gaseous biofuel yield from seaweed through a cascading circular bioenergy system integrating anaerobic digestion and pyrolysis." *Renewable and Sustainable Energy Reviews* 128 (2020): 109895.

Dos Santos, Alexandre Soares, Lílian Pantoja, Kele Aparecida Costa Vespermann, Mayara Barcelos, and Gustavo Molina. "Strategies for obtaining biofuels through the fermentation of C5-raw materials: Part s2." In *Bioenergy Research*, Eds. Neha Srivastava and Manish Shrivastava, 1st ed., 85–103. Hoboken, NJ: Wiley, 2021. doi:10.1002/9781119772125.ch5.

El-Araby, R. "Biofuel production: Exploring renewable energy solutions for a greener future." *Biotechnology for Biofuels and Bioproducts* 17, no. 1 (2024): 129.

Escalante, Jamin, Wei-Hsin Chen, Meisam Tabatabaei, Anh Tuan Hoang, Eilhann E. Kwon, Kun-Yi Andrew Lin, and Ayyadurai Saravanakumar. "Pyrolysis of lignocellulosic, algal, plastic, and other biomass wastes for biofuel production and circular bioeconomy: A review of thermogravimetric analysis (TGA) approach." *Renewable and Sustainable Energy Reviews* 169 (2022): 112914.

Filimonau, Viachaslau, Jorge Matute, Magdalena Kubal-Czerwińska, Kinga Krzesiwo, and Mirosław Mika. "The determinants of consumer engagement in restaurant food waste mitigation in Poland: An exploratory study." *Journal of Cleaner Production* 247 (February 2020): 1–62. doi:10.1016/j.jclepro.2019.119105.

Ganesan, Ramya, S. Manigandan, Melvin S. Samuel, Rajasree Shanmuganathan, Kathirvel Brindhadevi, Nguyen Thuy Lan Chi, Pham Anh Duc, and Arivalagan Pugazhendhi. "A review on prospective production of biofuel from microalgae." *Biotechnology Reports* 27 (2020): e00509.

García-Sancho, Cristina, Juan A. Cecilia, and Rafael Luque. "Conversion of lignocellulosic biomass to biofuels." In *Heterogeneous Catalysts*, Eds. Wey Yang Teoh, Atsushi Urakawa, Yun Hau Ng, and Patrick Sit, 593–616.Germany: Wiley, February 2021. doi:10.1002/9783527813599.ch33.

Ghenai, Chaouki, Ruqiya Abdullah Farah, Ola Al Saidi, Ayesha Al Suwaidi, Oussama Rejeb, and Abrar Inayat. "Performance analysis and biofuels conversion yield correlations for solar-thermal wood chips pyrolysis reactor using response surface methodology." *Case Studies in Thermal Engineering* 36 (2022): 1–15. doi:10.1016/j.csite.2022.102225.

Gollakota, Anjani Ravi Kiran, Nanda Kishore, and Sai Gu. "A review on hydrothermal liquefaction of biomass." *Renewable and Sustainable Energy Reviews* 81 (2018): 1378–1392.

González, Carlos A. Díaz, and Leonardo Pacheco Sandoval. "Sustainability aspects of biomass gasification systems for small power generation." *Renewable and Sustainable Energy Reviews* 134 (2020): 110180.

Government of India (GOI). *Report of the Committee on Development of Bio-Fuel*. New Delhi, India: Planning Commission, Government of India, 2016. https://fas.usda.gov/.

Guin, Mridula, Tanaya Kundu, and Riya Singh. "Applications of nanotechnology in biofuel production." In *Clean Energy Production Technologies*, Ed. Neha Srivastava, 297–332. Singapore: Springer Nature Singapore, 2022. doi:10.1007/978-981-16-8094-6_17.

Hayashi, Thiemi. "Biofuels Annual." Biofuels BR2024-0022. Brazil: USDA, Foreign Agricultural Service, 2024.

Hoffman, Shannon, M., Maria Alvarez, Gilad Alfassi, Dmitry M. Rein, Sergio Garcia-Echauri, Yachin Cohen, and Jose L. Avalos. "Cellulosic biofuel production using emulsified simultaneous saccharification and fermentation (eSSF) with conventional and thermotolerant yeasts." *Biotechnology for Biofuels* 14, no. 1 (2021): 157.

Huang, Yu-Fong, Pei-Te Chiueh, Wen-Hui Kuan, and Shang-Lien Lo. "Microwave pyrolysis of lignocellulosic biomass: Heating performance and reaction kinetics." *Energy* 100 (2016): 137–144.

Hussein, Ahmed Kadhim. "Applications of nanotechnology in renewable energies — A comprehensive overview and understanding." *Renewable and Sustainable Energy Reviews* 42 (February 2015): 460–476. doi:10.1016/j.rser.2014.10.027.

Interdepartmental Program Directorate Energy Transition, The Hague (Netherlands), & SenterNovem, Utrecht (Netherlands). "Criteria for sustainable biomass production." Final Report of the Project Group Sustainable Production of Biomass. SenterNovem, 2006. https://www.globalbioenergy.org/bioenergyinfo/bioenergy-and-food-security/detail/en/c/1488/.

Jung, Sungyup, Nagaraj P. Shetti, Kakarla Raghava Reddy, Mallikarjuna N. Nadagouda, Young-Kwon Park, Tejraj M. Aminabhavi, and Eilhann E. Kwon. "Synthesis of different biofuels from livestock waste materials and their potential as sustainable feedstocks – A review." *Energy Conversion and Management* 236 (May 2021): 114038. doi:10.1016/j.enconman.2021.114038.

Karki, Renisha, Wachiranon Chuenchart, K. C. Surendra, Shilva Shrestha, Lutgarde Raskin, Shihwu Sung, Andrew Hashimoto, and Samir Kumar Khanal. "Anaerobic co-digestion: Current status and perspectives." *Bioresource Technology* 330 (2021): 1–12. doi:10.1016/j.biortech.2021.125001.

Khalid, Khaizran, Bilal Hussain, and Shafaqat Ali. "Evaluating eco-efficiency in consumption and production through sustainable utilization of resources: A panel analysis of APAC by population." *Renewable Energy* 170 (June 2021): 1096–1106. doi:10.1016/j.renene.2021.01.018.

Kosuru, Sai Mani Yogesh, Yashraj Delhiwala, Prasad Babu Koorla, and Mallaiah Mekala. "A review on the biodiesel production: Selection of catalyst, pre-treatment, post treatment methods." *Green Technologies and Sustainability* 2 (2024): 1–13. doi:10.1016/j.grets.2023.100061.

Kucharska, Karolina, Iwona Hołowacz, Donata Konopacka-Łyskawa, Piotr Rybarczyk, and Marian Kamiński. "Key issues in modeling and optimization of lignocellulosic biomass fermentative conversion to gaseous biofuels." *Renewable Energy* 129 (2018): 384–408.

Lackner, Maximilian. "Third-generation biofuels: Bacteria and algae for better yield and sustainability." In *Handbook of Climate Change Mitigation and Adaptation*, Ed. Walter Leal Filho, 1–41. New York: Springer, 2021. doi:10.1007/978-1-4614-6431-0_90–2.

Ladero Galán, Miguel. "Upstream bioprocesses to biomass-based platform chemicals and derivatives." *Fermentation* 10, no. 59 (2024): 1–5. doi:10.3390/fermentation10010059.

Lee, Sze Ying, Ianatul Khoiroh, Dai-Viet N. Vo, P. Senthil Kumar, and Pau Loke Show. "Techniques of lipid extraction from microalgae for biofuel production: A review." *Environmental Chemistry Letters* 19, no. 1 (2021): 231–251. doi:10.1007/s10311-020-01088-5.

Lee, Sze Ying, Revathy Sankaran, Kit Wayne Chew, Chung Hong Tan, Rambabu Krishnamoorthy, Dinh-Toi Chu, and Pau-Loke Show. "Waste to bioenergy: A review on the recent conversion technologies." *BMC Energy* 1, no. 1 (2019): 1–22.

Lei, Tingzhou, Zhiwei Wang, Zaifeng Li, Jiufeng Xu, Xiaofeng He, and Jinling Zhu. "A biomass briquetting fuel machine and its large-scale operation system." *Journal of Renewable and Sustainable Energy* 5, no. 1 (2013): 013107.

Leonard, Matthew D., Efstathios E. Michaelides, and Dimitrios N. Michaelides. "Energy storage needs for the substitution of fossil fuel power plants with renewables." *Renewable Energy* 145 (2020): 951–962.

Libra, Judy A., Kyoung S. Ro, Claudia Kammann, Axel Funke, Nicole D. Berge, York Neubauer, Maria-Magdalena Titirici et al. "Hydrothermal carbonization of biomass residuals: A comparative review of the chemistry, processes and applications of wet and dry pyrolysis." *Biofuels* 2, no. 1 (2011): 71–106.

Liu, Yuchen, Yan Yuan, Ganesan Ramya, Shiv Mohan Singh, Nguyen Thuy Lan Chi, Arivalagan Pugazhendhi, Changlei Xia, and Thangavel Mathimani. "A review on the promising fuel of the future–biobutanol; the hindrances and future perspectives." *Fuel* 327 (2022): 125166.

Lorne, Daphne. *Biofuels in the Road Transport Sector*. Economic Outlook: Biofuels Dashboard 2022. IFP Energies Nouvelles, 2023.

Luque, Rafael, Lorenzo Herrero-Davila, Juan M. Campelo, James H. Clark, Jose M. Hidalgo, Diego Luna, Jose M. Marinas, and Antonio A. Romero. "Biofuels: A technological perspective." *Energy & Environmental Science* 1, no. 5 (2008): 542–564. doi:10.1039/b807094f.

Maia, Jorge Lucas Da, Jéssica Soares Cardoso, Duna Joanol Da Silveira Mastrantonio, Caroline Krause Bierhals, Juliana Botelho Moreira, Jorge Alberto Vieira Costa, and Michele Greque De Morais. "Microalgae starch: A promising raw material for the bioethanol production." *International Journal of Biological Macromolecules* 165 (December 2020): 2739–2749. doi:10.1016/j.ijbiomac.2020.10.159.

Malik, Kamla, Sergio C. Capareda, Baldev Raj Kamboj, Shweta Malik, Karmal Singh, Sandeep Arya, and Dalip Kumar Bishnoi. "Biofuels production: A review on sustainable alternatives to traditional fuels and energy sources." *Fuels* 5, no. 2 (2024): 157–175. doi:10.3390/fuels5020010.

Mat Aron, Nurul Syahirah, Kuan Shiong Khoo, Kit Wayne Chew, Pau Loke Show, Wei-Hsin Chen, and The Hong Phong Nguyen. "Sustainability of the four generations of biofuels – A review." *International Journal of Energy Research* 44, no. 12 (2020): 9266–9282.

Mehejabin, Fatema, Afla Musharrat, Shams Forruque Ahmed, Zobaidul Kabir, T. M. Yunus Khan, and C. Ahamed Saleel. "Sustainable biofuel production utilizing nanotechnology: Challenges and potential solutions." *GCB Bioenergy* 16, no. 10 (2024): 1–14. doi:10.1111/gcbb.70001.

Michaelides, Efstathios E. *Energy, the Environment, and Sustainability*. Boca Raton: CRC Press, 2018.

Mizik, Tamás, and Gábor Gyarmati. "Three pillars of advanced biofuels' sustainability." *Fuels* 3, no. 4 (2022): 607–626. doi:10.3390/fuels3040037.

Mizik, Tamás, Christian Barika Igbeghe, and Zsuzsanna Deák. "Production efficiency of advanced liquid biofuels: Prospects and challenges." *Energies* 18, no. 4 (2025): 1–17. doi:10.3390/en18041008.

Mohammadi, Hana, and Jack Saddler. *Compare-and-Contrast Transport Biofuels Policies (2021–2023 Update)*. IEA Bioenergy Task 39- Implementation Agendas: 2021–2023 Triennium Update. IEA Bioenergy, 2023.

Mohanty, Akansha, Siddhika Ajmera, Sampath Chinnam, Vineet Kumar, Ranjeet Kumar Mishra, and Bishnu Acharya. "Pyrolysis of waste oils for biofuel production: An economic and life cycle assessment." *Fuel Communications* 18 (2024): 100108.

Mohanty, Akansha, Siddhika Ajmera, Sampath Chinnam, Vineet Kumar, Ranjeet Kumar Mishra, and Bishnu Acharya. "Pyrolysis of waste oils for biofuel production: An economic and life cycle assessment." *Fuel Communications* 18 (2024): 100108.

Mood, Sohrab Haghighi, Amir Hossein Golfeshan, Meisam Tabatabaei, Gholamreza Salehi Jouzani, Gholam Hassan Najafi, Mehdi Gholami, and Mehdi Ardjmand. "Lignocellulosic biomass to bioethanol, a comprehensive review with a focus on pretreatment." *Renewable and Sustainable Energy Reviews* 27 (2013): 77–93.

Muritala, Ibrahim Kolawole, Dorottya Guban, Martin Roeb, and Christian Sattler. "High temperature production of hydrogen: Assessment of non-renewable resources technologies and emerging trends." *International Journal of Hydrogen Energy* 45, no. 49 (2020): 26022–26035.

National Renewable Energy Laboratory (NREL). "Biomass research." 2009. https://www.nrel.gov/.

Olatunji, Kehinde Oladoke, Noor A. Ahmed, and Oyetola Ogunkunle. "Optimization of biogas yield from lignocellulosic materials with different pretreatment methods: A review." *Biotechnology for Biofuels* 14, no. 1 (2021): 159.

Osman, Ahmed I., Neha Mehta, Ahmed M. Elgarahy, Amer Al-Hinai, Ala'A. H. Al-Muhtaseb, and David W. Rooney. "Conversion of biomass to biofuels and life cycle assessment: A review." *Environmental Chemistry Letters* 19, no. 6 (2021): 4075–4118.

Perez-Cantu, Lilia, Andreas Schreiber, Fokko Schütt, Bodo Saake, Christian Kirsch, and Irina Smirnova. "Comparison of pretreatment methods for rye straw in the second generation biorefinery: Effect on cellulose, hemicellulose and lignin recovery." *Bioresource Technology* 142 (2013): 428–435.

Phillips, Enosh. "Nanotechnological interventions in biofuel production." In *Handbook of Biofuels*, Ed. Sanjay Sahay, 593–604. London: Academic Press, 2022.

Preradović, Milica, and Saša Papuga. "Third generation biofuels: Cultivation methods and technologies for processing of microalgal biofuels." *Zastita Materijala* 62, no. 4 (2021): 249–261. doi:10.5937/zasmat2104249p.

Pugh, Shawn, Rebekah McKenna, Richard Moolick, and David R. Nielsen. "Advances and opportunities at the interface between microbial bioenergy and nanotechnology." *The Canadian Journal of Chemical Engineering* 89, no. 1 (2011): 2–12. doi:10.1002/cjce.20434.

Puri, Munish, Reinu E. Abraham, and Colin J. Barrow. "Biofuel production: prospects, challenges and feedstock in Australia." *Renewable and Sustainable Energy Reviews* 16, no. 8 (2012): 6022–6031. doi:10.1016/j.rser.2012.06.025.

Ramos, Ana, Eliseu Monteiro, and Abel Rouboa. "Biomass pre-treatment techniques for the production of biofuels using thermal conversion methods – A review." *Energy Conversion and Management* 270 (2022): 116271.

Ranjitha, J., Shreya Subedi, M. Anand, R. Shobana, S. Vijayalakshmi, and Bhaskar Das. "Microbial-based biofuel production: A green sustainable approach." In *Biofuel from Microbes and Plants*, Ed. Nitish Kumar, 285–298. Boca Raton: CRC Press, 2021.

Ravindranath, Nijavalli H., C. Sita Lakshmi, Ritumbra Manuvie, and P. Balachandra. "Biofuel production and implications for land use, food production and environment in India." *Energy Policy* 39, no. 10 (2011): 5737–5745. doi:10.1016/j.enpol.2010.07.044.

Ray, Ayusmita, Manoranjan Nayak, and Amit Ghosh. "A review on co-culturing of microalgae: A greener strategy towards sustainable biofuels production." *Science of the Total Environment* 802 (2022): 149765.

Reddy, Belum V. S., S. Ramesh, A. Ashok Kumar, S. P. Wani, R. Ortiz, H. Ceballos, and T. K. Sreedevi. "Bio-fuel crops research for energy security and rural development in developing countries." *BioEnergy Research* 1, nos. 3–4 (2008): 248–258. doi:10.1007/s12155-008-9022-x.

Rodionova, Margarita V., Roshan Sharma Poudyal, Indira Tiwari, Roman A. Voloshin, Sergei K. Zharmukhamedov, Hong Gil Nam, Bolatkhan K. Zayadan, Barry D. Bruce, Harvey J. M. Hou, and Suleyman I. Allakhverdiev. "Biofuel production: Challenges and opportunities." *International Journal of Hydrogen Energy* 42, no. 12 (2017): 8450–8461.

Sadaqat, Beenish, Mudasir Ahmad Dar, Rongrong Xie, and Jianzhong Sun. "Drawbacks of first-generation biofuels: Challenges and paradigm shifts in technology for second- and third-generation biofuels." In *Biofuels and Sustainability*, Eds. Zhu Daochen, Mudasir Ahmad Dar, and Mohd Shahnawaz, 33–47. Amsterdam: Elsevier, 2025. doi:10.1016/b978-0-443-21433-2.00003-7.

Saha, Rituparna, Debalina Bhattacharya, and Mainak Mukhopadhyay. "Biofuel production from algal biomass." In *Sustainable Biofuel and Biomass* Ed. Kuila Arindam, 119–144. New York: Apple Academic Press, 2019.

Sarwal, Rakesh, Sunil Kumar, Amit Mehta, Amit Varadan, Subodh Kumar Singh, S. S. V. Ramakumar, and Reji Mathai. "Roadmap for ethanol blending in India 2020–25." Report of the Expert Committee. Delhi: NITI Aayog | Ministry of Petroleum and Natural Gas, 2021.

Serrano, Elena, Guillermo Rus, and Javier García-Martínez. "Nanotechnology for sustainable energy." *Renewable and Sustainable Energy Reviews* 13, no. 9 (2009): 2373–2384. doi:10.1016/j.rser.2009.06.003.

Shu, Riyang, Rongxuan Li, Biqin Lin, Chao Wang, Zhengdong Cheng, and Ying Chen. "A review on the catalytic hydrodeoxygenation of lignin-derived phenolic compounds and the conversion of raw lignin to hydrocarbon liquid fuels." *Biomass and Bioenergy* 132 (2020): 1–14. doi:10.1016/j.biombioe.2019.105432.

Sikarwar, Vineet Singh, Ming Zhao, Peter Clough, Joseph Yao, Xia Zhong, Mohammad Zaki Memon, Nilay Shah, Edward J. Anthony, and Paul S. Fennell. "An overview of advances in biomass gasification." *Energy & Environmental Science* 9, no. 10 (2016): 2939–2977. doi:10.1039/c6ee00935b.

Singh, Anoop, Surajbhan Sevda, Ibrahim Abu Reesh, Karolien Vanbroekhoven, Dheeraj Rathore, and Deepak Pant. "Biohydrogen production from lignocellulosic biomass: Technology and sustainability." *Energies* 8, no. 11 (2015): 13062–13080. doi:10.3390/en81112357.

Singh, R. S., Taranjeet Singh, Muhammad Hassan, and Christian Larroche. "Biofuels from inulin-rich feedstocks: A comprehensive review." *Bioresource Technology* 346 (February 2022): 1–12. doi:10.1016/j.biortech.2021.126606.

Singhvi, Mamata, and Beom Soo Kim. "Green hydrogen production through consolidated bioprocessing of lignocellulosic biomass using nanobiotechnology approach." *Bioresource Technology* 365 (December 2022): 128108. doi:10.1016/j.biortech.2022.128108.

Singhvi, Mamata, Smita Zinjarde, and Beom-Soo Kim. "Sustainable strategies for the conversion of lignocellulosic materials into biohydrogen: Challenges and solutions toward carbon neutrality." *Energies* 15, no. 23 (2022): 1–13. doi:10.3390/en15238987.

Sivagurunathan, Periyasamy, Gopalakrishnan Kumar, Péter Bakonyi, Sang-Hyoun Kim, Takurou Kobayashi, Kai Qin Xu, Gábor Lakner, Gábor Tóth, Nándor Nemestóthy, and Katalin Bélafi-Bakó. "A critical review on issues and overcoming strategies for the enhancement of dark fermentative hydrogen production in continuous systems." *International Journal of Hydrogen Energy* 41, no. 6 (2016): 3820–3836. doi:10.1016/j.ijhydene.2015.12.081.

Srivastava, Anjana, and Ram Prasad. "Triglycerides-based diesel fuels." *Renewable and Sustainable Energy Reviews* 4, no. 2 (2000): 111–133. doi:10.1016/s1364-0321(99)00013-1.

Srivastava, Neha, Rahul Kumar Kharwar, and Pradeep Kumar Mishra. "Cost economy analysis of biomass-based biofuel production." In *New and Future Developments in Microbial Biotechnology and Bioengineering*. Eds. Neha Srivastava, Manish Srivastava, Pradeep Kumar Mishra, Pramod W. Ramteke, and Ram Lakhan Singh, 1–10. Amsterdam: Elsevier, 2019.

Sundaram, Thanigaivel, R. Kamalesh, A. Saravanan, P. R. Yaashikaa, A. S. Vickram, R. Ancy Teena, and S. Thiruvengadam. "Harnessing algal biomass for renewable energy and biofuel production: Current strategies and future insights." *International Journal of Biological Macromolecules* 319 (2025): 145487.

Suo, C., Y. P. Li, H. Mei, J. Lv, J. Sun, and S. Nie. "Towards sustainability for China's energy system through developing an energy-climate-water nexus model." *Renewable and Sustainable Energy Reviews* 135 (2021): 1–17. doi:10.1016/j.rser.2020.110394.

Thakare, Mayur, Soumya Pandit, Chetan Pandit, Subhasree Ray, Abdullah M. Alkahtani, Khulood Fahad Alabbosh, Nishant Ranjan, S. J. Geetha, and Sanket J. Joshi. "A concise review on biorefinery remedies for biomethanol production and processing." *Biocatalysis and Agricultural Biotechnology* 57 (2024): 103061.

Tuli, Deepak, and Sangita Kasture. "An overview of some futurist advanced biofuels and their conversion technologies." In *Advanced Biofuel Technologies*, Eds. Deepak Tuli, Sangita Kasture, and Arindam Kuila, 1–20. Elsevier, 2022. doi:10.1016/b978-0-323-88427-3.00001-5.

Usmani, Rahil Akhtar. "Advanced biofuels: Review of effects on environment and socioeconomic development." *European Journal of Sustainable Development Research* 4, no. 4 (2020): 1–8. doi:10.29333/ejosdr/8245.

Voloshin, Roman A., Margarita V. Rodionova, Sergey K. Zharmukhamedov, T. Nejat Veziroglu, and Suleyman I. Allakhverdiev. "Biofuel production from plant and algal biomass." *International Journal of Hydrogen Energy* 41, no. 39 (2016): 17257–17273.

Wandscher Busanello, Francieli, Maria Carolina Sérgi Gomes, Sirlei Marques Paschoal, Tatiana Rodrigues Da Silva Baumgärtner, Gersiane Barp, and Leila Denise Fiorentin-Ferrari. "The influence of membrane separation technique in the biodiesel and bioethanol production process: A review." *Biofuels* 15, no. 7 (2024): 903–928. doi:10.1080/17597269.2023.2294593.

Wang, Tao, Dong Zhang, Lingling Dai, Yinguang Chen, and Xiaohu Dai. "Effects of metal nanoparticles on methane production from waste-activated sludge and microorganism community shift in anaerobic granular sludge." *Scientific Reports* 6, no. 1 (2016): 25857–25862. doi:10.1038/srep25857.

Yakaboylu, Onursal, John Harinck, K. G. Smit, and Wiebren De Jong. "Supercritical water gasification of biomass: A literature and technology overview." *Energies* 8, no. 2 (2015): 859–894.

Yang, Liangcheng, Lucas D. Lamont, Shan Liu, Chunchun Guo, and Shelby Stoner. "A review on potential biofuel yields from cover crops." *Fermentation* 9 (October 2023): 1–15. doi:10.3390/fermentation9100912.

Yuliana, Maria, Shella Permatasari Santoso, Felycia Edi Soetaredjo, Suryadi Ismadji, Aning Ayucitra, Artik Elisa Angkawijaya, Yi-Hsu Ju, and Phuong Lan Tran-Nguyen. "A one-pot synthesis of biodiesel from leather tanning waste using supercritical ethanol: Process optimization." *Biomass and Bioenergy* 142 (2020): 1–9. doi:10.1016/j.biombioe.2020.105761.

11 Plant Cell Wall Engineering for Higher Biofuel Production

Ashok Pundir, Karan Singh, Radha, and Suraj Prakash

11.1 INTRODUCTION

With the rising consumption of plant-based resources for food and fuel, it is estimated that demand will increase by nearly 50% by 2050 (Alexandratos and Bruinsm, 2012). To reduce the dependency on fossil fuels, which contribute significantly to greenhouse gas emissions, there is a gradual shift towards sources of renewable energy including wind, hydroelectric power, solar, and particularly, biomass (Charamba et al., 2025). Among these, biomass stands out due to its potential to produce liquid biofuels including bioethanol, biodiesel, butanol, and bio-oil. However, the use of food crops like corn, sugarcane, and palm oil in first-generation biofuel production has led to food versus fuel conflicts and increased food prices globally. Therefore, second-generation biofuels derived from lignocellulosic biomass, particularly from non-edible parts of plants rich in cellulose, hemicellulose, and lignin, have received significant attention (Searchinger et al., 2015). These structural components of plant cell wall have potential as a sustainable source for biofuel production and can help in significantly reducing CO_2 emissions. For improving biofuel yield and conversion efficiency, plant cell walls must be engineered through approaches like selective breeding, genetic modification, and process optimisation. The integration of efforts between energy and agricultural sectors is essential to ensure a sustainable bioenergy supply (Alexandratos and Bruinsm, 2012; Searchinger et al., 2015; Welker et al., 2015). Although the transition to bio-based fuels offers several environmental benefits, it also requires technological advancement and a fundamental rethinking of how biological resources are utilised for energy production. One of the major challenges faced in this direction is the natural resistance of plant biomass towards degradation, which is primarily due to the complex structural arrangement of the plant cell wall (Fiorentino et al., 2019). Lignocellulosic biomass, which includes the fibrous and non-edible parts of plants such as agricultural residues, woody biomass, and energy crops, is considered to be the most abundant and economically feasible renewable resource for biofuel generation (Rai et al., 2022). In contrast to first-generation biofuels, second-generation biofuels rely on this non-food plant material, offering advantages from both ecological and economic perspectives (Khan et al., 2021). The plant cell wall, developed through evolution to resist microbial attack and physical damage, is composed of cellulose microfibrils surrounded by a matrix of hemicellulose and further strengthened by lignin, a hydrophobic phenolic polymer (Rai et al., 2022). This architectural complexity, referred to as "biomass recalcitrance", hinders the enzymatic breakdown of biomass into fermentable sugars, thereby limiting the efficiency and increasing the cost of biofuel production. The present study aims to review the current advancements in plant cell wall engineering strategies for improving biomass utilisation in biofuel production.

11.2 PLANT CELL WALL

Land plants produce an estimated 150–170 billion tonnes of cell walls per year (Pauly and Keegstra, 2008). To put this in context, the annual production land plants manufacture cell walls in just 2 days, accounting for around 1.2 billion tonnes of the global chemical industry (Loqué et al., 2015). Plants use photosynthesis to absorb and store carbon dioxide from the atmosphere, making the wall the

DOI: 10.1201/9781003600169-11

primary carbon-sequestration mechanism on Earth. The wall is more than just a carbon storage mechanism; it's a complex material composite comprising many polymer networks that play a crucial role in plant life cycles. Plant cells are enclosed by primary wall composed of polysaccharides, including glucans, substituted polysaccharides, hemicelluloses, and pectins, which are placed in water (60%–70% of the mass) (Albersheim et al., 2010). After cell elongation, some cells develop a secondary wall. Secondary walls contain polysaccharides such as hemicelluloses and cellulose microfibrils but have lower water content of 5% due to the presence of hydrophobic polyphenol lignin (Albersheim et al., 2010). Plant wall composition and components vary by species, tissue type, and developmental stage (Pauly, 2010). There are differences in proportional abundance of cellulose, lignin, and non-cellulosic polysaccharides. Hardwood has more glucuronosylated xylans and mannose-rich polymers than arabinosylated xylans, while grass straw contains *p*-coumaric acid and ferulic acid in its lignin. Plant cell walls, or lignocellulosics, are abundant and considered a renewable resource. Lignocellulosic feedstocks are a valuable resource for producing biofuels and chemicals, as they are not food source for humans. Biorefineries use chemical pretreatment at high temperatures to break down lignocellulosic material, then convert polysaccharides to monosaccharides using enzymes. Sugars can be fermented using bacteria or reformed through chemical catalysis to produce desired fuel and chemicals. To make biorefining commercially viable, several challenges must be overcome. Optimising quantitative agronomic features, such as improving lignocellulosic biomass yield per ha, can be achieved by selecting appropriate feedstocks for given environmental weather/growth circumstances. Qualitative aspects of lignocellulosic biomass, such as making carbohydrates more accessible to enzymes or optimising fermentation composition, can also be a challenge (Pauly and Keegstra, 2010; Somerville, 2010; Meng and Ragauskas, 2014; Sorek et al., 2014; Loqué et al., 2015). The cell wall is a dynamic interface that undergoes modifications as the cell grows and differentiates. It is critical for many of a plant's vital processes, including giving cells structure, supporting the tissues and organs of the plant, and creating interface between neighbouring cells. Additionally, cell wall has a role in cellular response for external stimuli such as intercellular communication and pathogens (Bellincampi et al., 2014; Tameshige et al., 2015). A combination of proteins, polysaccharides, and polymers come together to form stiff network that makes up the very complex structure known as the plant cell wall, which is finely and dynamically organised (Alberts et al., 2002). Growing or dividing cells are surrounded by the main wall, which must extend to permit cell to grow and divide even while it provides mechanical strength. The majority of carbohydrates in biomass are found in secondary, thicker cell walls that enclose specialised cells like fibre cells or vessel elements (Cosgrove and Jarvis, 2012). Since every differentiated cell has its own unique wall composition, dividing into two types of cell wall groups is suggestive. This creates a variety where secondary and primary walls were regarded as two extremes. Cellulose and hemicelluloses are found in primary and secondary walls; however, while pectins, enzymes, and structural proteins are found in primary walls, lignin and a trace amount of pectin and protein are often found in secondary walls (Somerville et al., 2004; Cosgrove and Jarvis, 2012).

11.2.1 Polysaccharide Composition

When selecting wall materials for biofuel production, it's important to note that higher plant walls vary significantly in both quality and quantity. During cell elongation, a main wall forms. According to Brett and Waldron (1996), primary walls are made up of cellulose and hydrated 65% of water matrix comprising hemicellulose, pectins, and structural proteins. Type II and I main cell walls are often distinguished by their polysaccharide composition. In addition to cellulose, type I walls were found in dicots and non-commelinid monocots; they also typically contain large levels of pectic polysaccharides and xyloglucan, the primary hemicellulose (Carpita and Gibeaut, 1993). The main hemicellulose found in type II walls, walls of poales like grasses, is arabinoxylan. Furthermore, type II walls have a much larger cellulose content and very small levels of proteins and pectins (Carpita, 1996). After cell elongation stops, secondary walls are produced; they can be deposited in multiple layers and are often thicker than primary walls. Cellulose, arabinoxylan, and/or glucomannans are the main hemicelluloses found in secondary walls (Brett and Waldron, 1996). More significantly, lignin mostly replaces water in secondary walls, rendering them practically impervious to solutes and enzymes. Almost all of the roughly 35 distinct cell types can be identified at the single-plant level by their differing wall structures, as shown by the analysis of chemical composition (Richmond and Somerville, 2001), microscopy (Carpita and McCann, 2000), and labelling of wall polymers with particular antibodies (Willats et al., 2000). Therefore, even though the components primarily come from secondary walls, plant cell wall feedstocks could be utilised to produce biofuel, which can vary greatly in composition. Some broad characteristics emerge even if the numbers cannot be compared because they were determined using separate methodologies. Cellulose, which makes up 40.6% to 51.2% of wall material, is the most prevalent polysaccharide in these walls. Hemicelluloses make up the second biggest fraction, accounting for 28.5%–37.2% of walls. The amount of lignin in cell wall is smaller (13.6%–28.1%), and the difference between softwood and switchgrass walls, for instance, is more than double. When examined more closely, the composition of the hemicelluloses is found to differ greatly (Pauly and Keegstra, 2008). Although arabinoxylan makes up the majority of the grasses, there is a wide range in the degree of arabinosylation. Sorghum xylans have extraordinarily high degree of substitution, while wheat straw has the lowest degree (Verbruggen et al., 1995). Hardwood, which is produced by angiosperm trees, is primarily composed of xylan with very little arabinosylation. Here, 4-*O*-methyl-glucuronic acid residues primarily replace the xylans (Ebringerova and Heinze, 2000). On the other hand, mannans like *O*-acetylated galactoglucomannans are the primary hemicellulose found in softwood (derived from conifers), while xylans are also present in significant proportions (Capek et al., 2002). Most significantly, the variations in the tissues from which feedstocks are obtained, such as corn stover (Chundawat et al., 2007), also contribute to the variations and ranges of wall components and their fine structures.

11.3 PLANT CELL WALLS AS RENEWABLE RESOURCE

The most plentiful renewable resource on Earth is the cell walls of plants. Even though they are quite abundant, humans only use 2% of this resource at the moment. Therefore, it is desirable to conduct research on the viability of employing plant cell walls to produce affordable biofuels. The primary obstacle to use wall materials is their resistance to effective breakdown into fermentable sugars (Tan et al., 2016). Customising wall composition, architecture to increase sugar yields on wall digestion for biofuel fermentation should be feasible by the inclusion of chemicals that affect wall structure or modification of the wall polysaccharide biosynthesis machinery. Research on the biosynthetic machinery and how it is regulated is still in early stage and poses significant technical and scientific challenge. Given the various functions walls play in a plant's existence, any modification to the wall structure to enable more economical biofuel generation may, of course, have a negative impact on plant growth and development. Nonetheless, there is a hope that this difficulty can be overcome due to the variety and number of wall structures found in plant kingdom (Pauly and Keegstra, 2008).

11.3.1 Plant Cell Wall Modification

The potential for functional "failure" of cell wall having harmful effect on plant growth, resulting in a concurrent decrease in wall biomass and ultimately endangering plant's survival, must be considered in any plan to enhance plant wall materials as feedstocks for biofuels. In order to compete for sunlight with other plants, land plants have expanded their capacity to absorb sunlight by increasing width and height of their plant canopy through evolutionary introduction of polyphenol incorporation into wall. Several instances have shown that changing walls can result in mortality (Goubet et al., 2003) and structural compromises like dwarfism (Desprez et al., 2007). Since the shape of the cell is determined by its cell walls, changing them may result in morphological alterations that could be detrimental to water transport, such as uneven xylem (Turner et al., 2007). Cell attachment is guaranteed by walls, especially the pectinaceous intermediate lamellae. It has been demonstrated that some mutants with modified pectic polysaccharides exhibit decreased cell adhesion (Bouton et al., 2002; Iwai et al., 2002; Krupkova et al., 2007), likely reducing plant resistance to specific mechanical forces. However, wall components derived from these plants may be easier for wall-degrading enzymes to reach, which would speed up, simplify, and lower the cost of processing these plants. Plant pathogens like bacteria and fungus are kept away from nutrient-rich cytosolic substance of the cells by wall, which serves another purpose. Walls have signalling chemicals that enable plant cell to detect pathogen attack and respond with many lines of defence in addition to basic mechanical line of defence (Vorwerk et al., 2004). Changes in the composition and design of the wall may therefore also increase vulnerability to infections, or illness symptoms may result from the endogenous release of oligosaccharides derived from the wall (Table 11.1).

For instance, apoplastic arabinan-oligosaccharides release caused a highly stressed plant morphology when fungal arabinanase was expressed in potato tubers (Skjot et al., 2002). When enzyme was directed towards Golgi apparatus rather than the apoplast, this morphology was resolved. When the arabinanase in the Golgi was targeted and kept there, the amount of pectic arabinans in plants and tubers decreased significantly while their appearance remained same. Genetic engineering of glycan synthases was used in a few instances to inhibit polysaccharide biosynthesis; however, these changes did not reduce pathogen resistance, and they even increased resistance (Hernandez-Blanco et al., 2007). During cell elongation and differentiation, the dynamic plant cell wall experiences subtle metabolic changes. The cell must strike a balance between releasing the wall and preserving the wall's cohesion and turgor pressure during the elongation process. It is believed that the metabolism of hemicelluloses permits cellulose microfibrils to slip and, consequently, regulate cell elongation by interlacing them with wall loosening enzymes including expansins, xyloglucantransglycosylases, and endoglucanases (Cosgrove, 2001a,b). New cell wall material is deposited when cells lengthen,

TABLE 11.1
Recent Studies on Plant Cell Wall Engineering for Enhanced Biofuel Production

Study	Organism/System	Key Findings	Key References
Alteration of lignin to improve cell wall digestibility	Tall fescue, barley	Reduced/altered lignin led to enhanced enzymatic access and increased saccharification	Loque et al. (2015)
Genetic/biochemical modification of wall polysaccharides	Various cereals/ crops	Targeted wall changes improved processability for biofuel and health benefits	Burton et al. (2011)
Genetic modification of cell wall polymers	Grasses, crops, trees	Changes in cellulose/hemicellulose/lignin enhanced biomass and saccharification	Wang et al. (2016)
Glycome profiling for bioprospecting	Himalayan plants	Potential for novel polysaccharides to enhance bioenergy conversion	Rather et al. (2022)
RNAi/downregulation of lignin pathways	Switchgrass	Increased ethanol yield (up to 38%) and improved saccharification efficiency	Kalluri et al. (2014)
Deep eutectic solvent pretreatment with engineered lignin	*Arabidopsis*	Reduced lignin polymerisation and increased fermentable sugar yields	Fan et al. (2020)
Overexpression of bacterial hydratase-lyase	*Arabidopsis*	Improved saccharification and reduced lignin polymerisation	Eudes et al. (2012)
Introduction of trans-genes from algae, fungi, viruses	Transgenic plants	Polymer "Trojan Horses": Increased cell wall porosity, significantly higher cellulose hydrolysis	Abramson et al. (2011)

most likely strengthening the wall. Therefore, altering the quantity or composition of wall polymers may cause wall to become so rigid that cell cannot extend efficiently, which may result in mechanical failure and cell bursting during elongation process. Plant cell's heretofore unidentified mechanism for assessing wall integrity and adapting to change has come to light (Pilling and Hofte, 2003; Humphrey et al., 2007; Pauly and Keegstra, 2008).

11.3.2 Plant Cell Walls as Energy Sources

Plant biology researchers are working to develop a cost-effective method for producing alternative fuels in response to the world's expanding energy needs and worries about the consumption of fossil fuels. The most prevalent biofuel is bioethanol, which makes up over 90% of proportion of the total amount of biofuel used in 2013 (IEA-ETSAP). The ethanol-fuel economy is currently based on first-generation bioethanol (1G), which is made primarily in Brazil and the United States from sucrose or starch (Table 11.2). In 2014, ethanol is reduced by 39.6 million metric tonnes of CO_2-equivalent greenhouse gas emissions from transportation (RFA, 2025).

First-generation bioenergy's sustainability raises severe questions; for instance, the FAO established the Bioenergy and Food Security project (BEFS) for determining the role of bioenergy innovations without compromising food security. Although they do not rule out the possibility of causing excessive competition for land and water needed for production of food and fibre, the majority of authorities concur that certain first-generation biofuels have contributed to recent increases in global food and animal feed costs. The creation of biofuels derived from non-food crop feedstocks and agricultural and forest leftovers can solve many of the issues with first-generation biofuels. According to Pauly and Keegstra (2008), plant biomass is now most plentiful and most wasted accessible organic raw material. Crop wastes such as switchgrass, sorghum, and sugarcane are also regarded as a significant biomass resource for the generation of biofuel. According to Carroll and Somerville (2009), maize and miscanthus offer promising biotechnological potential as feedstocks for the synthesis of second-generation bioethanol. None of them, meanwhile, meet every requirement for the perfect crop for biofuel. The well-known conventional method of producing bioethanol involves fermenting six-carbon sugars or enzymatically converting starchy biomass into sugars, followed by ethanol to fuel distillation (Canilha et al., 2012). Second-generation process uses available lignocellulosic materials to improve diversity and quantity of appropriate feedstock, whereas conventional processes only use the biomass components of sugar and starch. Several conversion routes have been tried in an effort to provide the most promising choice for the future. There are

TABLE 11.2
Global Production of Ethanol by Nation in Million Gallons

Region	2020	2021	2022	2023	2024
Argentina	210	270	310	310	310
Brazil	8,100	7,320	7,400	8,470	8,780
Canada	429	434	447	454	464
China	940	900	960	1,070	1,200
European Union	1,310	1,380	1,420	1,390	1,440
India	530	950	1,220	1,510	1,630
Rest of World	630	690	722	806	807
Thailand	390	350	380	340	360
United States	13,941	15,016	15,361	15,580	16,219
Total	26,480	27,310	28,220	29,930	31,210

Source: RFA (2025).

two primary conversion paths: Biochemical – the method relies on the lignocellulosic material being hydrolysed by a variety of enzymes, converting it into sugars. Sugars undergo fermentation in subsequent stage for alcohol production, which is subsequently distilled to produce ethanol. Thermochemical – feedstock is first gasified into a synthesis gas at a high temperature. This petrol can then be converted into other liquid or gaseous fuels, sometimes known as "synthetic fuels" (bio-SNG). (IEA, 2008; RFA, 2025).

11.4 ADVANCING BIOFUEL TECHNOLOGIES AND HUMAN HEALTH THROUGH DIETARY FIBRE INNOVATION

Figure 11.1 shows the cell wall engineering for industrial and human health uses. Non-cellulosic polysaccharides and cellulosic microfibrils found in plant cell walls contain significant amount of biomass sources, including wheat straw and lumber. Improved knowledge of wall design, constituent polysaccharide fine structures, and wall polysaccharide interactions can help the biofuels industry create more effective biomass deconstruction and conversion technology. This raises the question of whether "feedstock agnostic" technologies like pyrolysis and hydrothermal liquefaction could replace the current enzymatic and pretreatments conversion technologies for production of ethanol, which were best marginally viable and require adjustment for individual feedstocks. This would allow for the simultaneous conversion of carbon from lignin and polysaccharides to hydrocarbons and the use of multiple feedstocks in integrated biofuel production facilities. It is still true that the most plentiful renewable carbon sources on Earth are lignin and cell wall polysaccharides, and it is equally certain that we will keep working to produce biofuel from these sources and understanding of basic science of cell wall biology that will allow us to design new plant feedstocks to meet the requirements of the production system. As more data on health advantages and modes of action of dietary fibre becomes available, it is also reasonable to assume that cell wall polysaccharides will remain significant in human diets. Through the use of metagenomics technology, we will be able to better understand how dietary wall components impact compositions of microbial populations and the metabolic activities that take place in large intestine, and how factors ultimately impact intestinal health. Growing evidence about the protective effects of short-chain fatty acids against inflammatory conditions like asthma and emphysema will further encourage the inclusion

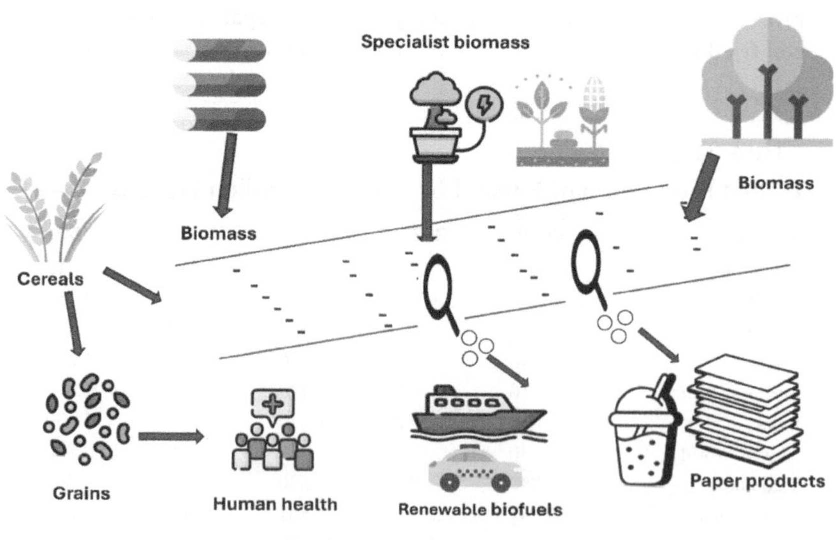

FIGURE 11.1 Schematic overview illustrating the dual role of engineered plant cell walls enhancing biomass deconstruction for biofuel production and improving dietary fibre intake for better human health. (Adapted from *Current Opinion in Biotechnology*.)

of more dietary fibre in human diets. Additionally, it may be feasible to modify advantageous large intestine microbes to more effectively compete for incoming carbon from various polysaccharides of cell wall (Burton and Fincher, 2014).

11.5 CELL WALL COMPOSITION MODIFICATION FOR THE PRODUCTION OF BIOFUEL, ANIMALS, AND PAPER

Our growing understanding of the genes underlying the formation of cell wall polysaccharides offers potential for altering wall constituents to aid in wall dismantling and, consequently, enhance mass/energy balances during the generation of biofuel, if this is possible without materially sacrificing the strength and morphology of the plants. Cellulases and other enzymes may be able to penetrate and hydrolyse different wall polysaccharides in brittle mutants, which are frequently identified by decreased cellulose levels (Figure 11.2). Brittle mutants will also require less energy to mill vegetative biomass. Moreover, cellulose crystallinity has been used as a measure of cellulose's vulnerability to enzymatic depolymerisation (Hall et al., 2010).

It has proven challenging to manipulate the levels of total cellulose, though. When *CesA* gene was overexpressed in transgenic aspens (*Populus tremuloides*), transgene and native *CesA* genes were silenced, and the amount of cellulose was significantly decreased. Severe phenotypic abnormalities and extensive pleiotropic effects were noted (Joshi et al., 2011). Partially disrupting cellulose synthesis results in more subtle phenotypes. A retroelement in first intron of *HvCesA4* gene is present in the barley brittle stem mutant fs2. The retroelement lowers but does not eliminate *HvCesA4* transcript levels by interfering with transcription. Although mutant line produces viable seed, cellulose crystallinity and stem strength are decreased. Similar to this, a brittle rice mutant with an *OsCesA4* gene lesion displays several phenotypic traits, such as reduced cellulose levels and straw that may be more nutritious for ruminant animals (Burton et al., 2011; Su et al., 2012).

Normal cellulose levels are present in the *Arabidopsis anisotropy1* mutant, which has a dwarf phenotype and decreased cell wall crystallinity due to a single amino acid mutation in the major cell wall *CesA* gene. Similarly, decreased cellulose crystallinity and higher rates of enzymatic degradation of mutant lines are linked to single amino acid alterations in the COOH-terminal transmembrane domain of cellulose synthases in primary cell wall (Harris et al., 2012; Fujita et al., 2013). Apart from methods that utilise *CesA* genes, growing understanding of the transcription

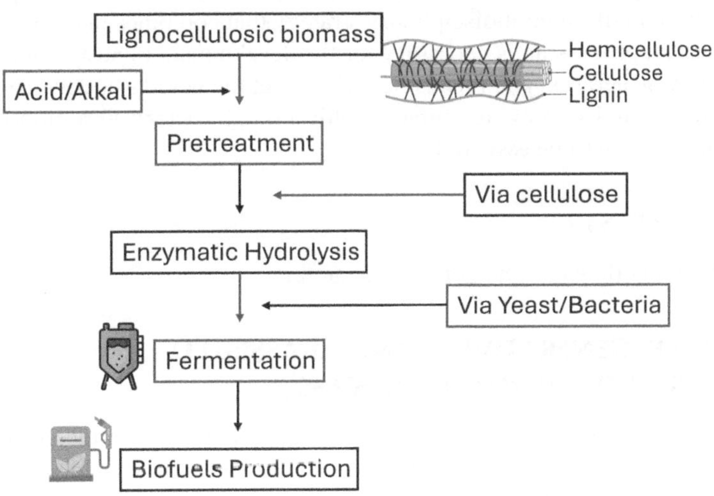

FIGURE 11.2 Diagrammatic representation of biofuel production using lignocellulosic biomass via pretreatment, enzymatic hydrolysis, and fermentation.

factors that control cellulose synthase expression opens up new avenues for adjusting crop plant cellulose levels. There may be more room to manipulate the amounts of the wall's non-cellulosic polysaccharides, even if the research now available indicates plants can tolerate very minor alterations of cellulose content. Transgenic lines were produced by transforming barley with the *HvCslF6* gene driven by the 35S promoter, increasing the amount of (1,3;1,4)-*b*-glucan in the leaves by three to four times. Conversion efficiencies may be improved by raising the quantities of (1,3;1,4)-*b*-glucan in cereal crop wastes and specialised biofuel grasses, particularly if doing so comes at the expense of heteroxylans. In experiments comparing the biomass of poplar and switchgrass (*Panicum virgatum*), it was discovered that while lignin removal was more advantageous for poplar, xylan removal was crucial in lowering recalcitrance in the grass species. Following hot water pretreatment, the saccharification yield of engineered Arabidopsis lines with 18% and 23% decrease in lignin and xylosyl residue. Additionally beneficial is the lack of MeGlcA residues, as this uronic acid does not ferment well and is transformed in the pulp and paper sectors to residues of hexenuronosyl, which reduce the final paper product's brightness. Furthermore, strongly substituted xylans make animal feed comparatively indigestible, which presents chances to raise the value of animal fodder and feed (Shatalov and Pereira, 2009; Petersen et al., 2012; Burton and Fincher, 2014; DeMartini et al., 2013).

11.6 CONCLUSION

An excellent chance to meet the growing demand for renewable biofuels worldwide is to engineer plant cell walls. Plant biomass, which includes cellulose, hemicellulose, and lignin, is the most plentiful source of renewable carbon and has enormous potential for the manufacture of biofuel. Nonetheless, the intrinsic complexity of lignocellulosic biomass and its resistance to enzymatic degradation continue to be significant challenges. Targeted changes in cell wall design are now possible thanks to recent developments in molecular genetics, plant biotechnology, and breeding. These include reducing lignin content to enhance polysaccharide accessibility, modifying cellulose crystallinity, and increasing hexose-rich polysaccharides, all of which contribute to improved biomass digestibility and higher biofuel yields. To prevent negative effects on plant development, mechanical strength, and disease resistance, such changes must be properly balanced. The economic viability of lignocellulosic biofuels is being further improved by novel pretreatment methods, enzyme optimisation, and the fusion of thermochemical and biochemical pathways. In order to maintain environmental sustainability and avoid rivalry with food crops, it is also essential to prioritise non-food biomass sources. Essentially, a multidisciplinary strategy that combines plant biology, genetic engineering, and industrial processing is necessary for the development of sustainable bioenergy in the future. To create next-generation plant feedstocks that meet the world's energy needs while preserving ecological balance and improving human health through dietary fibre improvements, further research and cooperation will be essential.

COMPETING INTERESTS

The authors declare that they have no competing interest.

DECLARATION OF GENERATIVE AI AND AI-ASSISTED TECHNOLOGIES IN THE WRITING PROCESS

During the preparation of this work, the book chapter author(s) have not taken any assistance from AI, and the book chapter author(s) reviewed and edited the content as needed and take(s) full responsibility for the content of the publication.

REFERENCES

Abramson, M., Shoseyov, O., Shani, Z. (2011) A novel approach to increase cell wall saccharification for efficient biofuel production. *BMC Proc.* https://doi.org/10.1186/1753-6561-5-s7-p104

Albersheim, P., Darvill, A.G., Roberts, K., Sederoff, R.R., Staehelin, L.A. (2010) *Plant Cell Walls*. New York: Garland Science.

Alberts, B.J.A., Lewis, J., Raff, M., Roberts, K., Walter, P. (2002) *Molecular Biology of the Cell*, fourth ed. New York: Garland Science.

Alexandratos, N., Bruinsm, J. (2012) *World Agriculture Towards 2030/2050: The 2012 Revision*; ESA working paper No: 12-03. Rome, Italy: Food and Agriculture Organization of the United Nations (FAO).

Bellincampi, D., Cervone, F., Lionetti, V. (2014) Plant cell wall dynamics and wall-related susceptibility in plant-pathogen interactions. *Front. Plant Sci.* 5, 228. https://doi.org/10.3389/fpls.2014.00228.

Bouton, S., Leboeuf, E., Mouille, G., Leydecker, M.T., Talbotec, J., Granier, F., Lahaye, M., Hofte, H., Truong, H.N. (2002) QUASIMODO1encodes a putative membrane-bound glycosyltransferase required for normal pectin synthesis and cell adhesion in Arabidopsis. *Plant Cell* 14(10), 2577–2590.

Brett, C., Waldron, K. (1996) *Physiology and Biochemistry of Plant Cell Walls. Topics in Plant Functional Biology*. London: Chapman and Hall.

Burton, R.A., Collins, H.M., Kibble, N.A.J., Smith, J.A., Shirley, N.J., Jobling, S.A., Henderson, M., Singh, R.R., Pettolino, F., Wilson, S.M., Bird, A.R., Topping, D.L., Bacic, A., Fincher, G.B. (2011). Over-expression of specific HvCslF cellulose synthase-like genes in transgenic barley increases the levels of cell wall (1,3;1,4)-b-d-glucans and alters their fine structure. *Plant Biotechnol. J.* 9, 117–135.

Burton, R.A., Fincher, G.B. (2014) Plant cell wall engineering: Applications in biofuel production and improved human health. *Curr. Opin. Biotechnol.* 26(1), 79–84. https://doi.org/10.1016/j.copbio.2013.10.007.

Canilha, L., Kumar Chandel, A., dos Santos Milessi, T.S., Fernandes Antunes, F.A., da Costa Freitas, W.L., das Gracas Almeida Felipe, M., da Silva, S.S. (2012) Bioconversion of sugarcane biomass into ethanol: An overview about composition, pretreatment methods, detoxification of hydrolysates, enzymatic saccharification, and ethanol fermentation. *J. Biomed. Biotechnol.* 989572. https://doi.org/10.1155/2012/989572.

Capek, P., Alfoldi, J., Liskova, D. (2002) An acetylated galactoglucomannan from *Picea abies* L. *Karst. Carbohydr. Res.* 337, 1033–1037.

Carpita, N., McCann, M. (2000) The cell wall. In *Biochemistry and Molecular Biology of Plants* (Buchanan, B.B., Gruissem, W., Jones, R.L., eds). Rockville, MD: American Society of Plant Physiologists, pp. 52–108.

Carpita, N.C. (1996) Structure and biogenesis of the cell walls of grasses. *Annu. Rev. Plant Physiol. Plant Mol. Biol.* 47, 445–476.

Carpita, N.C., Gibeaut, D.M. (1993) Structural models of primary cell walls in flowering plants: Consistency of molecular structure with the physical properties of the walls during growth. *Plant J.* 3, 1–30.

Carroll, A., Somerville, C. (2009) Cellulosic biofuels. *Annu. Rev. Plant Biol.* 60, 165–182. https://doi.org/10.1146/annurev.arplant.043008.092125.

Charamba, A.N., Kumba, H., Makepa, D.C. (2025) Assessing the opportunities and obstacles of Africa's shift from fossil fuels to renewable sources in the southern region. *Clean Energy* 9, 74–93.

Chundawat, S.P.S., Venkatesh, B., Dale, B.E. (2007) Effect of particle size based separation of milled corn stover on AFEX pretreatment and enzymatic digestibility. *Biotechnol. Bioeng.* 96, 219–231.

Cosgrove, D.J. (2001a) Wall structure and wall loosening. A look backwards and forwards. *Plant Physiol.* 125, 131–134.

Cosgrove, D.J. (2001b) *Enhancement of accessibility of cellulose by expansins*. US Patent No. 6326470.

Cosgrove, D.J., Jarvis, M.C. (2012) Comparative structure and biomechanics of plant primary and secondary cell walls. *Front. Plant Sci.* 3, 204. https://doi.org/10.3389/fpls.2012.00204.

DeMartini, J.D., Pattathil, S., Miller, J.S., Li, H., Hahn, M.G., Wyman, C.E. (2013) Investigating plant cell wall components that affect biomass recalcitrance in poplar and switchgrass. *Energy Environ. Sci.* 6, 898–909.

Desprez, T., Juraniec, M., Crowell, E.F., Jouy, H., Pochylova, Z., Parcy, F., Höfte, H., Gonneau, M., Vernhettes, S. (2007) Organization of cellulose synthase complexes involved in primary cell wall synthesis in Arabidopsis thaliana. *Proc. Natl. Acad. Sci. USA* 104(39), 15572–15577. https://doi.org/10.1073/pnas.0706569104.

Ebringcrova, A., Heinze, T. (2000) Xylan and xylan derivatives – biopolymers with valuable properties. 1 – Naturally occurring xylans: Structures, isolation procedures and properties. *Macromol. Rapid Commun.* 21, 542–556.

Eudes, A., George, A., Mukerjee, P., Kim, J.S., Pollet, B., Benke, P.I., Yang, F., Mitra, P., Sun, L., Cetinkol, O.P., Chabout, S., Mouille, G., Soubigou-Taconnat, L., Balzergue, S., Singh, S., Holmes, B.M., Mukhopadhyay, A., Keasling, J.D., Simmons, B.A., Lapierre, C., Ralph, J., Loque, D. (2012) Biosynthesis and incorporation of side-chain-truncated lignin monomers to reduce lignin polymerization and enhance saccharification. *Plant Biotechnol. J.* 10, 609–620.

Fan, D., Li, C., Fan, C., Hu, J., Li, J., Yao, S., Lu, W., Yan, Y., Luo, K. (2020) MicroRNA6443-mediated regulation of FERULATE 5-HYDROXYLASE gene alters lignin composition and enhances saccharification in Populus tomentosa. *New Phytol.* 226, 410–425.

Fiorentino, G., Zucaro, A., Ulgiati S (2019) Towards an energy efficient chemistry. Switching from fossil to bio-based products in a life cycle perspective. *Energy* 170, 720–729.

Fujita, M., Himmelspach, R., Ward, J., Whittington, A., Hasenbein, N., Liu, C., Truong, T.T., Galway, M.E., Mansfield, S.D., Hocart, C.H., Wasteneys, G.O. (2013) The anisotropy1 D604N mutation in the Arabidopsis cellulose synthase1 catalytic domain reduces cell wall crystallinity and the velocity of cellulose synthase complexes. *Plant Physiol.* 162, 74–85.

Goubet, F., Misrahi, A., Park, S.K., Zhang, Z.N., Twell, D., Dupree, P. (2003) AtCSLA7, a cellulose synthase-like putative glycosyltransferase, is important for pollen tube growth and embryogenesis in Arabidopsis. *Plant Physiol.* 131, 547–557.

Hall, M., Bansal, P., Lee, J.H., Realff, M.J., Bommarius, A.S. (2010) Cellulose crystallinity—A key predictor of the enzymatic hydrolysis rate. *FEBS J.* 277, 1571–1582.

Harris, D.M., Corbin, K., Wang, T., Gutierrez, R., Bertolo, A.L., Petti, C., Smilgies D-M, Estevez, J.M., Bonetta, D., Urbanowicz, B.R., Ehrhardt, D.W., Somerville, C.R., Rose, J.K., Hong, M., Debolt, S. (2012) Cellulose microfibril crystallinity is reduced by mutating Cterminal transmembrane region residues CESA1A903V and CESA3T942I of cellulose synthase. *Proc. Natl. Acad. Sci. USA* 109, 4098–4103.

Hernandez-Blanco, C., Feng, D.X., Hu, J., Sánchez-Vallet, A., Deslandes, L., Llorente, F., Berrocal-Lobo, M., Keller, H., Barlet, X., Sánchez-Rodríguez, C., Anderson, L.K., Somerville, S., Marco, Y., Molina, A. (2007) Impairment of cellulose synthases required for Arabidopsis secondary cell wall formation enhances disease resistance. *Plant Cell* 19, 890–903.

Humphrey, T.V., Bonetta, D.T., Goring, D.R. (2007) Sentinels at the wall: Cell wall receptors and sensors. *New Phytol.* 176, 7–21.

IEA (2008) *From 1st- to Second-generation Biofuel Technologies: An Overview of Current Industry and RD&D Activities*. Paris: OECD/IEA. https://www.iea.org/reports/from-1st-to-2nd-generation-biofuel-technologies.

Iwai, H., Masaoka, N., Ishii, T., Satoh, S. (2002) A pectin glucuronyltransferase gene is essential for intercellular attachment in the plant meristem. *Proc. Natl Acad. Sci. USA* 99, 16319–16324.

Joshi, C.P., Thammannagowda, S., Fujino, T., Gou J-Q, Avci, U., Haigler, C.H., McDonnell, L.M., Mansfield, S.D., Mengesha, B., Carpita, N.C. (2011). Perturbation of wood cellulose synthesis causes pleiotropic effects in transgenic aspen. *Mol. Plant* 4, 331–345.

Kalluri, U.C., Yin, H., Yang, X., Davison, B.H. (2014) Systems and synthetic biology approaches to alter plant cell walls and reduce biomass recalcitrance. *Plant Biotechnol. J.* 12, 1207–1216.

Khan, N., Sudhakar, K., Mamat, R. (2021) Role of biofuels in energy transition, green economy and carbon neutrality. *Sustainability* 13, 12374.

Krupkova, E., Immerzeel, P., Pauly, M., Schmulling, T. (2007) The tumorous shoot development2 gene of Arabidopsis encoding a putative methyltransferase is required for cell adhesion and co-ordinated plant development. *Plant J.* 50, 735–750.

Loque, D., Scheller, H.V., Pauly, M. (2015) Engineering of plant cell walls for enhanced biofuel production. *Curr. Opin. Plant Biol.* 25, 151–161. https://doi.org/10.1016/j.pbi.2015.05.018.

Meng, X., Ragauskas, A.J. (2014) Recent advances in understanding the role of cellulose accessibility in enzymatic hydrolysis of lignocellulosic substrates. *Curr. Opin. Biotechnol.* 27, 150–158.

Pauly, M., Keegstra, K. (2008a) Cell-wall carbohydrates and their modification as a resource for biofuels. *Plant J.* 54(4), 559–568. https://doi.org/10.1111/j.1365-313X.2008.03463.x.

Pauly, M., Keegstra, K. (2008b) Harnessing plant biomass for biofuels and biomaterials: Cell-wall carbohydrates and their modification as a resource for biofuels. *Plant J.* 54(4), 559–568. doi:10.1111/j.1365-313X.2008.03463.x

Pauly, M., Keegstra, K. (2010) Plant cell wall polymers as precursors for biofuels. *Curr. Opin. Plant Biol.* 13, 304–311.

Petersen, P., Lau, J., Ebert, B., Yang, F., Verhertbruggen, Y., Kim, J., Varanasi, P., Suttangkakul, A., Auer, M., Loqué, D., Scheller, H.V. (2012) Engineering of plants with improved properties as biofuels feedstocks by vessel-specific complementation of xylan biosynthesis mutants. *Biotechnol. Biofuels* 5, 1–19.

Pilling, E., Hofte, H. (2003) Feedback from the wall. *Curr. Opin. Plant Biol.* 6, 611–616.

Rai, A.K., Al Makishah, N.H., Wen, Z., Gupta, G., Pandit, S., Prasad, R. (2022) Recent developments in lignocellulosic biofuels, a renewable source of bioenergy. *Fermentation* 8, 161.

Rather, M.A., Dolley, A., Hazarika, N., Ritse, V., Sarma, K., Jamir, L., Satapathy, S.S., Ray, S.K., Deka, R.C., Biswal, A.K., Doley, R., Mandal, M., Namsa, N.D. (2022) Glycome profiling and bioprospecting potential of the Himalayan Buddhist handmade paper of Tawang region of Arunachal Pradesh. *Front. Plant Sci.* 13, 1–13. https://doi.org/10.3389/fpls.2022.831589

RFA (2025) Annual ethanol production. *Renewable Fuels Association*. https://ethanolrfa.org/markets-and-statistics/annual-ethanol-production

Richmond, T.A., Somerville, C.R. (2001) Integrative approaches to determining Csl function. *Plant Mol. Biol.* 47, 131–143.

Searchinger, T., Edwards, R., Mulligan, D., Heimlich, R., Plevin, R. (2015) Do biofuel policies seek to cut emissions by cutting food? *Science* 347, 1420–1422.

Shatalov, A.A., Pereira, H. (2009) Impact of hexenuronic acids on xylanase-aided bio-bleaching of chemical pulps. *Bioresour. Technol.* 100, 3069–3075.

Skjot, M., Pauly, M., Bush, M.S., Borkhardt, B., McCann, M.C., Ulvskov, P. (2002) Direct interference with rhamnogalacturonan I biosynthesis in *Golgi vesicles*. *Plant Physiol.* 129, 95–102.

Somerville, C., Bauer, S., Brininstool, G., Facette, M., Hamann, T., Milne, J., Osborne, E., Paredez, A., Persson, S., Raab, T., Vorwerk, S., Youngs, H. (2004) Toward a systems approach to understanding plant cell walls. *Science* 306(5705), 2206–2211. https://doi.org/10.1126/science.1102765.

Somerville, C., Youngs, H., Taylor, C., Davis, S.C., Long, S.P. (2010) Feedstocks for lignocellulosic biofuels. *Science* 329, 790–792.

Sorek, N., Yeats, T.H., Szemenyei, H., Youngs, H., Somerville, C.R. (2014) The implications of lignocellulosic biomass chemical composition for the production of advanced biofuels. *BioScience* 64, 192–201.

Su, Y., Zhao, G., Wei, Z., Yan, C., Liu, S. (2012) Mutation of cellulose synthase gene improves the nutritive value of rice straw. *Asian-Aust Assoc. Anim. Prod. Soc.* 25, 800–805.

Tameshige, T., Hirakawa, Y., Torii, K.U., Uchida, N. (2015) Cell walls as a stage for intercellular communication regulating shoot meristem development. *Front. Plant Sci.* 6, 324. https://doi.org/10.3389/fpls.2015.00324.

Tan, H.-T., Corbin, K.R., Fincher, G.B. (2016) Emerging technologies for the production of renewable liquid transport fuels from biomass sources enriched in plant cell walls. *Front. Plant Sci.* 7, 1–18.

Turner, S., Gallois, P., Brown, D. (2007) Tracheary element 1differentiation. *Annu. Rev. Plant Biol.* 58, 407–433.

Verbruggen, M.A., Beldman, G., Voragen, A.G.J. (1995) The selective extraction of glucuronoarabinoxylans from sorghum endosperm cell-walls using barium and potassium hydroxide solutions. *J. Cereal Sci.* 21, 271–282.

Vorwerk, S., Somerville, S., Somerville, C. (2004) The role of plant cell wall polysaccharide composition in disease resistance. *Trends Plant Sci.* 9, 203–209.

Wang, Y., Fan, C., Hu, H., Li, Y., Sun, D., Wang, Y., Peng L (2016) Genetic modification of plant cell walls to enhance biomass yield and biofuel production in bioenergy crops. *Biotechnol. Adv.* 34, 997–1017.

Welker, C.M., Balasubramanian, V.K., Petti, C., Rai, K.M., DeBolt, S., Mendu, V. (2015, Jul 27) Engineering plant biomass lignin content and composition for biofuels and bioproducts. *Energies* 8(8), 7654–7676.

Willats, W.G.T., Steele-King, C.G., McCartney, L., Orfila, C., Marcus, S.E., Knox, J.P. (2000) Making and using antibody probes to study plant cell walls. *Plant Physiol. Biochem.* 38, 27–36.

12 Biotechnological Advances in Biofuel Production

Strategies and Innovations from Lower and Higher Plant Species

Ambreen Ali, Kashif Ali Khan, Zul Kamal, Bashir Ullah, and Yi Chen

12.1 INTRODUCTION

Biofuel production represents a pivotal intersection of biotechnology, plant sciences, and renewable energy research. Technological advancements have dramatically transformed our capacity to generate sustainable energy from plant biomass, addressing critical challenges in global energy sustainability and climate change mitigation. Each and every solution is crucial to shift toward future with a sustainable energy resources and healthy goods production. Hence, the energy source has to be more efficient, and processes for the production of sustainable energy resource have to be improved with completely new technologies. Biotechnological applications with its clean and chicanery use of bio-based technologies can play a vital role in making biofuels more sustainable (Accord, 2008).

Fossil fuels are hydrocarbons, primarily coal, fuel oil, or natural gas, formed from the remains of dead plants and animals. Because of their high heating potential, availability, and ignition properties, fossil fuels have been used as a source for transportation and other energy purposes (Hassan et al., 2013). Through sustainable biofuel production and continuous speedy improvements in both bio feedstock and processes for biofuel production, it is possible to reduce the dependency on non-renewable fossil fuels and to provide national energy security (Wolt, 2009). Biofuel is any solid, liquid, or gaseous fuels derived from organic biomass, which is any living matter such as field crops, wood products, water plants, and municipal solid waste that is converted into energy (Mussatto et al., 2010).

As an alternative for fossil fuel in the transportation part, biofuel can become critical for solving environmental troubles because of these reasons: it minimizes greenhouse gas (GHG) emission and does not need much engine modifications; it also improves national security, provides employment, and, finally, supports rural development (Hassan et al., 2013). To move successfully toward biofuel future, the use of modern biotechnology is extremely important, and the cost of conversion from bio feedstock into biofuel has already been reduced using modern biotechnology (Wolt, 2009, Lynd et al., 2008).

12.2 BIOTECHNOLOGICAL APPROACHES

12.2.1 Microalgal Biotechnology

Algae are well-known photosynthetic organisms of marine and freshwaters habitats. They are considered among the fastest growing creatures of nature. Due to their ability to fix atmospheric carbon dioxide through their various inherent pathways and to produce high-quality biomass enriched

 DOI: 10.1201/9781003600169-12

in various products of industrial uses, they are the main focus of different industries. In recent years, algal industries have been focusing toward these organisms for fuel, food, and nutraceuticals production (Dolan et al., 2019). Algal biofuel including bioethanol, biogas, biohydrogen, biodiesel, and bio-oil has been emerged as an eco-friendly, clean, and cost-effective solution to other fuels. The interest of various industries toward food and nutraceuticals production from algae cannot be ignored. They are being explored for the production of pigments, cosmetics, proteins, bioactive substances, and pharmaceuticals (Ariede et al., 2017). Using the tools and applying the modern approaches of bioengineering, inherent capabilities of algae for production of high-quality biomass can be improved significantly (Wang et al., 2016).

Presently, most of the industries are preferring clustered regularly interspaced short palindromic repeats (CRISPR) technology for algal genome engineering because it can be used as versatile tool for multiple gene editing at a time, allowing a single gene or even the complete metabolic pathway to be manipulated in shorter duration than other classical genome engineering approaches (Dolan et al., 2019). CRISPR are unique arrays of short repeats interspaced with distinct spacer DNA sequences within different Eubacteria and Archaea genomes. CRISPR technology has empowered the proficient algal genome editing with its comparatively higher efficiency, accuracy, and simplicity compared to other nucleases (Jeon et al., 2017). Microalgae have emerged as important platforms for the production of biofuels and other biomolecules, and genetic engineering of microalgae is, thus, one of the fastest growing biotechnology fields. In addition to overexpression of genes of interest, genome editing is essential for the suppression of genes interfering with the production of target molecules. However, progress in this field has been hampered by multiple layers of difficulties inherent to microalgae (Balamurugan et al., 2021).

Microalgae can accumulate as high as 30%–70% of lipids in their cell dry weight under certain conditions. Lipids in most microalgae are constituted mostly by saturated and monounsaturated C14–C20 fatty acids, which has been regarded as a promising alternative source for the production of lipid-based biodiesel. In addition, some microalgae species have the ability to synthesize polyunsaturated fatty acids (PUFAs) with more than 20 carbon atoms, especially eicosapentaenoic acid (EPA) and docosahexaenoic acid (DHA) (Manning, 2022). In recent years, these PUFAs have attracted extensive attention from people due to Oleaginous eukaryotic microorganisms (i.e., fungi, yeasts, and microalgae), and some species of autotrophic and heterotrophic bacteria are at the forefront of the biotechnological research thanks to their ability to accumulate oil (triacylglycerol, TAGs), the so-called single cell oil (SCO), and thus the potential to be used as feedstock in the biodiesel manufacture. Although the first works on oleaginous microorganisms date back to the 1960s, they have positive effects on human health (Manning, 2022).

Metabolic engineering strategies for increasing lipid accumulation in *Yarrowia lipolytica* have expanded considerably. The early efforts focused on:

i. Maximizing the flux toward lipid biosynthesis, such as deleting genes related to β-oxidation involved in the lipid degradation pathway, tuning the activity of transcription factors involved in lipid metabolism, overexpressing fatty acid (FA) and triacylglyceride (TAG) synthesis genes, and overexpressing functional heterologous lipid synthetic related genes in *Y. lipolytica* (Park, 2020)
ii. Minimizing the flux toward competing metabolic pathways such as the inhibition of glycogen and citric acid biosynthesis pathways (Sagnak et al., 2018).

In the past two decades, fossil fuel consumption and global warming have strongly accelerated the development and research of microalgae biodiesel. Microalgae biodiesel has been recognized as a favorable renewable and environmentally friendly biofuel with its non-toxic, carbon-neutral, and environmental-friendly renewable advantages (Brindhadevi et al., 2021). Microalgae could convert light energy into renewable energy through photosynthesis to complete the growth cycle. It has many advantages like rapid growth rate, wide distribution, desired fatty acid profile, and suitability for various climates. Researches have proved that microalgae could absorb pollutants (e.g., nitrogen and phosphorus) from wastewater, which reduces fresh water and converts pollutants to bioresource. These advantages make microalgae biodiesel more competitive and occupy the main position of sustainable energy development (Maltsev et al., 2021).

Microalgae, as a kind of micro-cell factory that fixes carbon dioxide efficiently, has the following unique advantages: (i) solar energy can be directly utilized to save a lot of energy compared with physical and chemical methods; (ii) high photosynthetic efficiency. The efficiency of microalgae using solar energy to fix carbon dioxide is 10–50 times higher than that of other terrestrial plants; (iii) growth rate is fast. Microalgae can multiply at a rapid rate (can multiply in a few hours), much higher than higher plants; (iv) CO_2 can be recycled. CO_2 can be turned into bioenergy through photosynthesis by microalgae (Pal et al., 2023; Song et al., 2024).

At present, research on high-efficiency CO_2 fixation technology of microalgae mainly focuses on the screening, cultivation, and development of reactor and system. The significant improvement of CO_2 fixation and conversion efficiency of industrial oil-producing microalgae is one of the bottlenecks to realize the economic feasibility of the energy industry of microalgae and has always been the focus of attention in the academic and industrial circles. In single-cell photosynthetic organisms such as Cyanobacteria and *Chlamydomonas reinhardtii*, the previous academia mainly modified the carbon-fixing enzyme Rubisco itself in photosynthesis or regulated the carbon anhydrase to increase the CO_2 concentration around Rubisco. However, in industrial oleaginous microalgae, molecular breeding for carbon sequestration has not been successful (Chen et al., 2019).

In order to significantly increase photosynthetic CO_2 fixation capability, a breakthrough in the control of the Calvin cycle is essential. A sequence of enzymatic processes is carried out by 11 enzymes in the Calvin cycle, which is a fundamental part of plant and algal metabolism. Carboxylation (carbon fixation), reduction, and regeneration are the three key phases of this cycle, which provide substrates for carbohydrate biosynthesis by utilizing ATP and NADPH. The Calvin cycle is initiated by ribulose 1,5-bisphosphate carboxylase/oxygenase (Rubisco) carboxylating the CO_2 acceptor molecule ribulose 1,5-bisphosphate (RuBP). Carbon assimilation is primarily determined by its catalytic properties. This enzyme, which plays a critical role in the Calvin cycle, has been the subject of several engineering efforts in recent years to increase photosynthesis (Mustafa & Hayder, 2021).

In order to improve the above basic method, many modifications have been adopted by researchers. The most common modification is the addition of 1 M NaCl instead of water to prevent binding of acidic lipids to denatured lipids. It was also reported that the addition of 0.2 M phosphoric acid and HCl to the salt solution improves lipid recovery with a shorter separation time in comparison with the earlier extraction methods. Similarly, the addition of 0.5% (v/v) acetic acid to the water phase increased the recovery of acidic phospholipids. A more recent report suggests that Hajra's method of lipid extraction was demonstrated to be the most efficient procedure for the extraction of plant sphingolipids (Hajra, 1974).

Successful commercialization of the technology relies on optimization of microalgal growth, effective lipid extraction, and conversion of the same to biodiesel. However, efficient lipid extraction and highest recovery remain the vital downstream processing difficulties in the algal biofuel industry. At present, solvent extraction methods are most commonly used for lipid extraction as they provide the highest lipid recovery. The use of mechanical methods, though environment friendly and cheap, is not a wise option because of poor recovery and the possible degradation of lipids. Solvent-free methods appear promising at the laboratory scale at present, and more research has to be carried out for minimal use of solvents for large-scale commercialization. A more promising way for effective and efficient lipid extraction could be to use combinative methods such as enzymatic and mechanical/solvent extraction methods. Solvent-free methods such as enzymatic degradation when combined with other methods will reduce solvent usage/energy consumption and also increase recovery efficiency.

12.2.2 Cyanobacterial Biofuel Production

The depletion and use of fossil resources, together with global environmental concerns related to CO_2 emissions, requires a transition toward carbon-neutral, sustainable bioenergy. Photosynthetic organisms have great potential as cell factories for directly converting CO_2 into a broad range of value-added biochemicals, especially biofuels. Cyanobacteria are particularly attractive as platforms for bio-production due to relatively fast growth rate, genetic tractability, and low nutrient requirements. Synthetic biology and metabolic engineering pave the way to develop cyanobacteria as green chassis for production of various biofuels including hydrogen, ethanol, butanol, and isoprene. However, engineered cyanobacteria show relatively low productivity compared with model heterotrophic microorganisms. Significant advances of biofuel production using cyanobacterial chassis are expected by implementing recently developed system biology tools, as well as optimized cultivation systems with newly developed photobioreactors (Abernathy et al., 2017; Xie et al., 2022).

Photosynthetic organisms are favorable for the production of biofuels, mainly because of their low cost of cultivation, but biofuel yields obtained under normal conditions are not satisfactory (Wang et al., 2013). In addition to the production of biodiesel, microalgae and cyanobacteria serve as attractive feedstock for the production of bioethanol, although the scientific and technological knowledge on this context is still scarce (Santander et al., 2014; Sawaengsak & Gheewala, 2017).

The technologies for the first (sugar or starch feedstock) and second generation (lignocellulosic feedstock) of bioethanol basically involve two stages: (i) the conversion of sunlight into chemical energy (such as carbohydrates and lipids) and (ii) the conversion of chemical energy into biofuel. These two stages are related to each other and result in increased production costs. As an improvement to this process, the use of a single-stage system that is capable of capturing sunlight directly and converting it into biofuel (bioethanol) would avoid one step, thereby reducing the cost of production and increasing the sustainability of the bioethanol production process (de Farias Silva & Bertucco, 2016).

A wide range of fermentative and photosynthetic organisms have been explored as cell factories for hydrogen production. Within these biological systems, hydrogen production acts as a valve to maintain redox poise by relieving the cell of excess reducing equivalents. Biological hydrogen production processes include dark fermentation, photofermentation, and biophotolysis (Srirangan et al., 2012). Dark fermentation, typically performed by chemoheterotrophic species of Clostridium and Enterobacter, involves the light-independent anaerobic fermentation of carbohydrates or other organic substrates. Through this process, protons accept electrons generated from substrate catabolism to form hydrogen. Hydrogen can also be produced fermentatively in a light-dependent process, termed photofermentation, by non-oxygenic photosynthetic bacteria, such as purple non-sulfur and green sulfur bacteria. Rather than carbohydrates, photofermentation involves the use of alternative reduced compounds, such as organic acids and hydrogen sulfide, as electron donors. Additionally, solar energy is required in order to photoheterotrophically convert the substrate compounds to CO_2 and hydrogen as the only fermentation products. Biophotolysis, a third mode of biohydrogen production, is unique to photoautotrophic organisms, with green algae and cyanobacteria exhibiting the greatest propensity for hydrogen production. Light is required in this process for the functioning of oxygenic photosynthesis, in which water, rather than carbohydrates or organic acids, serves as the sole electron donor. Through the splitting of water, protons are generated and subsequently reduced to molecular hydrogen (Srirangan et al., 2011).

The future success of an industrial biological process for hydrogen production is strictly dependent upon the ability to compete economically with current petrochemical methods. However, the paramount problem of feedstock cost, including transportation, handling, and pretreatment of raw materials, has blighted the large-scale production of many biofuels (Pfromm et al., 2010), with feedstock being the single greatest cost contributor, representing up to 95% of total production costs.

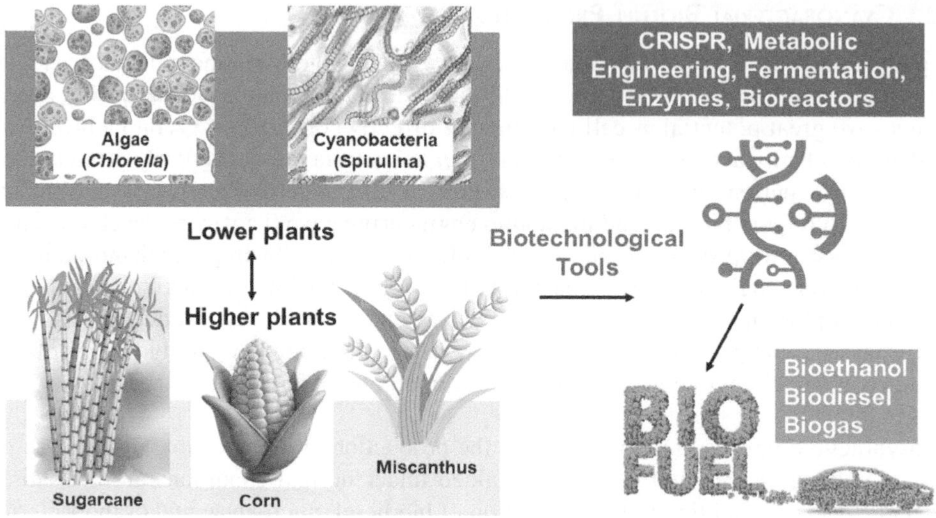

FIGURE 12.1 Biofuel production from lower (algae, cyanobacteria) and higher (sugarcane, corn) plants via biotechnological tools (genetic engineering, enzymes, fermentation), leading to bioethanol, biodiesel, and biogas for transportation, electricity, and industrial applications.

This salient limitation gravely jeopardizes the feasibility of large-scale hydrogen bioprocesses based upon dark fermentation and photofermentation, both of which entail consumption of costly carbonaceous feedstocks. As a result, biophotolysis, relying upon free solar energy and atmospheric CO_2 as feedstocks, is ostensibly identified as a feasible avenue due to a significant reduction in feedstock costs (Srirangan et al., 2011) (Figure 12.1).

12.3 TECHNOLOGICAL INNOVATIONS AND RESEARCH FRONTIERS

12.3.1 Synthetic Genomics

12.3.1.1 Custom Organism Design Methodologies

It refers to sophisticated techniques employed to genetically engineer or alter microorganisms to boost their biofuel-producing abilities. These methods aim to optimize metabolic pathways, thereby enhancing the yield and efficiency of biofuel synthesis, and often utilize genetic engineering, metabolic engineering, and synthetic biology strategies (Jiang et al., 2019). Various strains of algae, such as green algae (*C. reinhardtii*) and cyanobacteria *Anabaena cylindrica* and *Anabaena variabilis*, are renowned for producing hydrogen upon exposure to sunlight.

The organisms mentioned earlier are having the ability to extract and channel electrons and protons from water to produce hydrogen. This process is catalyzed by chloroplast hydrogenases, namely HydA1 and A2 (Hydrogenase) (Jagadevan et al., 2018). For commercial biofuel production, an efficient cellulolytic organism must be capable of fully hydrolyzing cellulosic biomass, fermenting all sugars in the hydrolysate simultaneously, and tolerating toxic lignin compounds without compromising productivity (Vinuselvi et al., 2011). To achieve energy-positive production of fuels for these applications, an ideal organism for biofuel production would possess excellent substrate utilization and processing capabilities, particularly rapid and rotatable sugar transport pathways, and demonstrate good tolerance to inhibitors and products, maintaining high metabolic flux in producing a single fermentation product. Whether this will ultimately be a native isolated strain of this organism that is characterized or a developed recombinant model organism remains to be seen (Alper & Stephanopoulos, 2009).

12.3.1.2 Metabolic Flux Analysis Techniques

The metabolism of engineered organisms is fine-tuned to maintain homeostasis while enabling adaptation to changing environmental conditions. These unique metabolic traits position them as essential microbial producers in various industries, including pharmaceuticals, biofuel production, and bioremediation. Specifically, they are extensively utilized for the production of recombinant eukaryotic proteins, polyphenols, and isoprenoids (Gopalakrishnan & Maranas, 2015). Metabolic flux analysis demonstrated that a reduction in ethanol yield at RQ 6 was associated with the cyclic, oxidative functioning of TCA cycle reactions. The model suggested that respiratory dissimilation of glucose takes place only when the oxygen uptake rate is sufficiently high to completely replace glycerol formation.

The balances of cytosolic and mitochondrial NADH were crucial in determining flux distributions. The smallest discrepancies between estimated and actual product yields were observed when the unknown co-factor requirements for a limited number of biosynthetic reactions were adjusted to minimize the excess NADH produced during biosynthesis (Franzén, 2003). Modern fermentation engineers must adopt advanced metabolic flux analysis tools that incorporate kinetics, hydrodynamics, and 13C-proteomics to uncover the dynamic physiologies of microbial hosts in large bioreactor settings. Using these metabolic insights, engineers can implement rational pathway modifications, synthetic biology circuits, and bioreactor control algorithms to enhance large-scale biofuel production efficiency (Hollinshead et al., 2014).

12.3.1.3 Artificial Intelligence Integration in Strain Development

The idea of artificial intelligence (AI) was first brought forward in the 1950s (Choi et al., 2020). The core of artificial learning is machine learning (ML) (Janiesch et al., 2021). Integrating AI into strain development for biofuel production utilizes ML techniques to streamline processes and boost efficiency. ML relies on three fundamental components: data, algorithms (models), and computing power (arithmetic) (Ma et al., 2023). Particularly in tasks involving high-dimensional data, such as classification, regression, and clustering, ML demonstrates significant utility. By analyzing previous computations and identifying patterns within large datasets, ML helps to produce consistent and dependable decisions (Naveed et al., 2024). For instance, in the realm of optimizing ethanol production, researchers have utilized ML algorithms to enhance *Saccharomyces cerevisiae* strains for ethanol production from lignocellulosic biomass. This approach has led to notable improvements in fermentation efficiency (Goshisht, 2019).

A similar example can be found in predictive analytics for fermentation. A study showcased the application of AI-driven predictive analytics to optimize fermentation conditions for biofuel production, leading to enhanced yields from engineered strains (Damian et al., 2024). Concisely, ML is an integral part of artificial learning since 1950s. Both of these aid in developing process of biofuel production by making predictions using various predictive analysis. Both of these show a significant amount of increase in biofuel production.

12.3.2 Sustainability Framework

12.3.2.1 Water and Land Use Optimization

Sustainability will be a strong factor in the regulatory environment and investments in biofuels. With numerous countries, including the United States, mandating significant increases in production, there is a strong need for society to assess and comprehend the sustainability of biofuels. The production of biomass feedstock plays a significant role in the environmental, social, and economic impacts of biofuels (Gopalakrishnan et al., 2009). The water demands for energy generated from biomass are approximately 70–400 times greater than those for other energy carriers like fossil fuels, wind, and solar. Over 90% of this water is consumed during the production of the feedstock, with only a minor portion used in biomass processing.

Biodiesel and ethanol from conventional food crops generally require more water compared to biofuels derived from lignocellulosic crops, although water usage varies widely across all alternative feedstocks Utilizing feedstocks that require less water, enhancing water productivity, and implementing improved water management strategies designed to support various ecosystem services can mitigate some of the negative impacts of biofuel systems on water resources (de Fraiture & Berndes, 2009).

The use of land for biofuel production has reaped significant attention in recent years. Biofuel production significantly impacts global land use. Natural forests and pastures are the most frequently reduced land types to accommodate biofuels, while cropland, managed forests, and natural grasslands experience minimal net change (Gurgel et al., 2008). Looking at the big picture, the amount of land and water used for growing biofuel crops isn't enormous, somewhere between 11 and 12 million ha. Brazil, which is the top producer of bioethanol, uses about 2.5 million ha for this purpose almost 5% of its cultivated land, and they get around 6,200 L of ethanol per hectare, mostly from sugarcane. The United States, which comes in second, dedicates nearly 4 million ha to biofuel crops. This accounts for 4% of its total farmland, and they produce roughly 3,300 L/ha, mainly from maize (De Fraiture et al., 2007).

12.3.2.2 Carbon Footprint Reduction Strategies

Many nations have introduced policies to boost energy security and aid rural development while lowering carbon emissions impact within a span of 12 years, from 2000 to 2012. During this period, ethanol production surged from 17 billion to more than 83 billion liters. Biodiesel output skyrocketed from under 1 billion to reach 22.5 billion liters. Currently, biofuels constitute around 4% of the world's transport fuel consumption. It is anticipated to rise to 27% by 2050 (Jeswani, 2016). Usually, plants (though not exclusively) are used as a renewable energy source called biomass, the mass of living organisms. It can be utilized as a solid, liquid, or gaseous fuel for various applications including biofuels for vehicles, production of energy, and heating and cooking of household (Lobus et al., 2023). Biofuels help lessen fossil fuel use by serving as renewable alternatives, which in turn reduces carbon dioxide emissions. They achieve this by utilizing organic materials like crops and waste, which absorb CO_2 as they grow, offsetting the emissions produced when fuels are burned (Hanaki & Portugal-Pereira, 2018).

12.3.2.3 Circular Bioeconomy Model Development

Sustainability issues are becoming a more integral part of both policymakers' agendas and corporate strategies. The concept of sustainability comes from the French verb *soutenir*, which means "to hold up or support" (Brundtland, 1985). Ellen MacArthur Foundation has provided a highly acclaimed definition of the Circular Economy, describing it as "an industrial economy that is restorative or regenerative by intention and design." (Geissdoerfer et al., 2017). As defined by the European Commission, the bioeconomy encompasses sectors of the economy that use renewable biological resources to generate food, materials, and energy (Ronzon et al., 2016). A circular bioeconomy is viewed as a strategy to stimulate economic growth in developed countries by merging the appealing "what" of a circular economy with the practical, viable, and attractive "how" of a bioeconomy (Giampietro, 2019). Producing biofuels from photosynthetic microalgae involves a complex web of interconnected upstream and downstream processes. Despite notable advancements in this field, significant challenges remain in the downstream aspects, particularly in dewatering and harvesting biomass. These processes are currently both energy-intensive and costly (Mohan et al., 2016). The circular bioeconomy model has gained attention for its focus on sustainable production, the use of renewable resources, and the transformation of these resources into value-added products (Moreira et al., 2023). There is an increasing demand for bio-based products, which points to the growth of a new economic sector that is moving towards sustainability. In 2016, biofuels generated $3.69 billion in revenue and is expected to grow at an annual rate of 16.43% by 2025, offering alternatives to conventional petrochemicals (Chandel et al., 2020).

12.4 TECHNOLOGICAL CHALLENGES AND FUTURE PERSPECTIVES

12.4.1 Research Challenges

12.4.1.1 Technological Scaling Limitations

Microalgae cultivation and biofuel production systems are inherently complex, governed by a web of interdependent variables that challenge their scalability and commercial feasibility (Goswami et al., 2022). Success in both technical efficiency and market viability depends on meticulously harmonizing interconnected components, such as bioreactor design, environmental controls, and process optimization, while balancing biological, technological, and operational factors at every stage of production (Kasani et al., 2022). In contrast to traditional waste treatment and waste-to-energy procedures, microalgal production and biofuel synthesis technologies are far more complex and diverse. For example, the activated sludge technique, which is commonly employed to treat wastewater, is based on design principles that are extrapolated from a small number of crucial characteristics (Su et al., 2022). These include the bioreactor's oxygen levels, hydraulic load, specific organic load rate of the activated sludge, and the desired concentration of bacterial microflora (Sun et al., 2022). Such processes' straightforward use as well as effectiveness provide a significant benefit, which contributes to understanding why many large-scale institutions have embraced and employed them (Burch et al., 2022).

12.4.1.2 Economic Feasibility Constraints

The price that we pay at the gas station is the consumer price of fuel. However, this price only includes direct costs such as profit, tax, and all well-to-pump chain expenses. Indirect costs, or those resulting from the effects of fuel production and consumption on the environment and human health, make up a second portion of the overall cost (Coady et al., 2015). We all end up paying the indirect costs, whether or not we're the ones using the resource. The direct costs have an immediate impact on production, but it's the indirect costs that really matter for the long-term health of our planet and all our activities. When we compare biofuels to fossil fuels in terms of cost, we usually only look at the upfront expenses. Unfortunately, we often overlook the bigger picture of indirect costs (Coady et al., 2015). Right now, economic factors are driving how we produce fuel, but sustainability plays a huge role in how we manage our planet's resources over time. Both are crucial in order to sustain our current way of life. A high Energy Return on Investment (EROI) and efficient resource use are two requirements that must be met for fuels to be long-term sustainable (Carneiro et al., 2017). Biofuel precursors are locked inside microalgae biomass, so we have to harvest and process the cells to extract the chemicals. To make biofuels more cost-effective and sustainable over time, one approach is to boost the rate at which algae biomass grows (Khan et al., 2018). The ongoing interest in biofuels stems from the growing need to offset the significant negative impact that the transportation sector has on both the environment and human health. One of the biggest challenges for biofuels is meeting both economic and long-term sustainability standards. Developing production methods that use resources efficiently, such as microalgae, is crucial for overcoming this challenge (Perin & Jones, 2019).

12.4.1.3 Conversion Efficiency Improvements

Studying energy efficiency in different ways of using biomass will help us make smarter choices when building transportation systems that rely on biomass in the future (Huang & Zhang, 2011). To overcome limitations and boost productivity in large-scale outdoor algae cultivation, one approach is to genetically modify these organisms to better adapt to the light conditions found in photobioreactors. In natural environments, algae cells are optimized to capture light efficiently for survival, leaving no evolutionary advantage for passing radiation energy to layers beneath them (Simionato et al., 2013). Even though algae are incredibly diverse, it's unlikely that any naturally occurring species will have all the perfect traits for large-scale cultivation. Therefore, genetically improving them will probably be essential to maximize productivity (Radakovits et al., 2012).

12.4.1.4 Plant Variety Adaptability

Selecting the right plants for their environments is crucial for developing efficient and eco-friendly farming systems focused on biofuel production. The article highlights that numerous plant species, when adapted to local soil conditions and weather patterns, can effectively serve as biofuel sources. This approach ensures that biofuel production is both sustainable and regionally optimized. By focusing on these species, we not only encourage biodiversity and greener practices but also open doors to growing energy crops on underused or degraded land. This way we avoid conflicts with food-producing farmland, ensuring that energy needs don't compete with food security (a wide range of plants offer cellulosic biofuel potential). Building on this, scientists have identified variant straw plants with more digestible cell walls for biofuel production. These plants maintain normal size and strength, promising cheaper and faster biofuel production from agricultural residues. This discovery eases the pressure on using food crops for biofuels (Marriott et al., 2014). Further, modifying plants through genomics offers new strategies. Breeding high-yield crops with efficient conversion rates minimizes land competition and supports sustainable biomass production for biofuels and biomaterials (Furtado et al., 2014). Therefore, adapting dedicated energy crops to abiotic stresses from climate change is vital. Understanding plant responses to these stresses will aid in selecting crops that efficiently adapt, ensuring sustainable biomass production for bioenergy under future climate conditions (Freitas et al., 2021).

12.4.2 Emerging Research Trajectories

12.4.2.1 Multi-omics Investigative Approaches

Multi-omics approaches have proved to be an important key in altering biofuel production thanks to its ability to enhance feedstocks and microbial systems. By uncovering genes like biomass yield and stress tolerance, transcriptomics may facilitate the development of high-performance cultivars suited for marginals for crops like sugarcane and switchgrass (Li et al., 2023). Integrative omics has identified lipid biosynthesis pathways in microalgae, thereby enabling genetic engineering to increase the production of biodiesel while leveraging bioinformatics tools to model metabolic networks (Misra et al., 2013). Microbial systems benefit from 13C-MFA and proteomics, which map carbon fluxes in engineered *Escherichia coli* strains to resolve ATP shortages and improve biofuel titers during scale-up (Hollinshead et al., 2014). Cutting-edge synthetic biology, guided by multi-omics data, now employs CRISPR and ML to design algae and bacteria with tailored metabolic pathways, accelerating the transition from lab-scale discoveries to industrial bioreactors (Kim et al., 2021). Collectively, these advances demonstrate how multi-omics links genetic potential with useful biofuel applications, safeguarding cost-effectiveness and sustainability.

12.4.2.2 Advanced Plant–Microbe Interaction Studies

Microbiomes are important in order to increase the sustainability of biofuel production because they play an important role in improving seed germination, increasing the ability to resist stress, and improving the absorption of nutrients and protection against infections. By taking advantage of these microbial connections, biomass output can rise even in nutrient-poor soils (Zhalnina et al., 2021). Advanced plant–microbe interaction studies are transforming biofuel production by leveraging the natural symbiosis between plants and microbes to enhance crop growth, resilience, and yield. For bioenergy crops like switchgrass and miscanthus, microbiome management is being explored to improve nutrient uptake and stress tolerance, particularly on marginal lands where conventional crops struggle (Bandopadhyay et al., 2024). Microbial metabolic pathways, such as those in cellulolytic bacteria, are being engineered to optimize the breakdown of lignocellulosic biomass into fermentable sugars, enabling more efficient biofuel production (Kumar & Kumar, 2017). Beneficial bacteria living freely in the soil boost bioenergy crops as they produce hormones like gibberellic acid, and these hormones directly enhance the growth and metabolism of plants (Wu et al., 2009). These advancements reveal the potential of integrating plant–microbe interactions into sustainable biofuel production systems that address both environmental challenges and energy demands (Paul et al., 2021).

12.4.2.3 Integrated Biotechnological Solutions

Integrated biotechnological solutions refers to transforming substrates into biofuels and other valuable side products through the use of microbial processes, hence maximizing economic value while minimizing waste and making biofuel production more interesting and competitive. Some of the major processes involve bacteria, yeast, and microalgae (Da Silva et al., 2014). Developments in synthetic biology have enabled engineering new pathways in microorganisms such as *E. coli* and Cyanobacteria, which are useful in turning lignocellulosic sugars into drop-in hydrocarbons (Dellomonaco et al., 2010). The Synthetic Biology Foundry, Department of Energy for the United States has further accelerated these efforts by developing certain enzymes for the deconstruction of biomass and microbes that are capable of synthesizing fuels directly, thus relating lab innovations with industrial scalability (Jegannathan et al., 2009). Microalgae illustrate the benefits of integration; various strains are being optimized genetically for lipid overproduction, whereas residual biomass is valorized into fertilizers or biogas, thus maximizing resource efficiency (Sundaram et al., 2023). Some hybrid systems like hydrothermal liquefaction help in combining biochemical and thermochemical processes converting lignin and microbial residues into biocrude, contributing to the importance of integrating multi-pathway in reducing costs and carbon footprints (Biddy et al., 2020). The above stated strategies demonstrate the role of biotechnology in creating synergistic, scalable biofuel ecosystem.

12.4.2.4 Next-Generation Biofuel Production Technologies

Next-generation biofuel production methods are shifting towards non-food feedstocks and integrated systems in order to improve sustainability and scalability. Various processes such as advanced thermochemical gasification and hydrothermal liquefaction convert lignocellulosic waste such as wheat straw, and forestry residues into biofuels while minimizing land use struggles (O'Neill et al., 2024). Currently, pyrolysis and fermentation workflows are refined by AI-driven optimization, boosting yields of bioethanol and diesel from algae and engineered microbes (Lourinho & Brito, 2015). Various regional models demonstrate how marginal lands can sustainably supply biomass without compromising food security, one of which is in the US Midwest. These in turn supported robust supply chains linking decentralized biorefineries (Poore, 2024). Emerging cascading systems further help in closing the loop by linking biofuel production with carbon capture and waste valorisation, positioning next-gen biofuels as a linchpin for achieving net-negative emissions (Kraljić & Desmarais, 2021).

12.5 CONCLUSION AND FUTURE REMARKS

Biotechnological innovations have opened promising avenues for sustainable biofuel production from both lower and higher plant species. Techniques such as microalgal and cyanobacterial engineering, synthetic genomics, and AI-assisted metabolic optimization are reshaping the bioenergy landscape. These advances not only improve yield and efficiency but also align with environmental and economic sustainability goals. Looking ahead, addressing key challenges, such as technological scalability, economic viability, and conversion efficiency, remains essential. Future research should focus on integrated multi-omics approaches, enhanced plant–microbe synergies, and the development of robust, climate-resilient bioenergy crops. By fostering interdisciplinary collaboration and adopting a circular bioeconomy model, the transition toward next-generation biofuels can be accelerated, offering a clean, renewable solution to global energy demands.

COMPETING INTERESTS

The authors declare that they have no competing interest.

DECLARATION OF GENERATIVE AI AND AI-ASSISTED TECHNOLOGIES IN THE WRITING PROCESS

During the preparation of this work, the book chapter author(s) have not taken any assistance from AI, and the book chapter author(s) reviewed and edited the content as needed and take(s) full responsibility for the content of the publication.

REFERENCES

Abernathy, M. H., Yu, J., Ma, F., Liberton, M., Ungerer, J., Hollinshead, W. D., Gopalakrishnan, S., He, L., Maranas, C. D., Pakrasi, H. B., Allen, D. K., & Tang, Y. J. (2017). Deciphering cyanobacterial phenotypes for fast photoautotrophic growth via isotopically nonstationary metabolic flux analysis. *Biotechnology for biofuels*, *10*(1), 273.

Accord, A. (2008). United Nations conference on trade and development. *Agenda,* 1(5).

Alper, H., & Stephanopoulos, G. (2009). Engineering for biofuels: Exploiting innate microbial capacity or importing biosynthetic potential? *Nature Reviews Microbiology*, 7(10), 715–723.

Ariede, M. B., Candido, T. M., Jacome, A. L. M., Velasco, M. V. R., de Carvalho, J. C. M., & Baby, A. R. (2017). Cosmetic attributes of algae – A review. *Algal Research*, 25, 483–487.

Balamurugan, S., Sathishkumar, R., & Li, H.-Y. (2021). Biotechnological perspectives to augment the synthesis of valuable biomolecules from microalgae by employing wastewater. *Journal of Water Process Engineering*, 39, 101713.

Bandopadhyay, S., Li, X., Bowsher, A. W., Last, R. L., & Shade, A. (2024). Disentangling plant-and environment-mediated drivers of active rhizosphere bacterial community dynamics during short-term drought. *Nature Communications*, 15(1), 6347.

Biddy, M. J., Davis, R., Dutta, A., Singh, A., Tao, L., Tan, E. C., Jones, S. B., Collett, J. R., Snowden-Swan, L. J., & Webb, E. (2020). *Integrated strategies to enable lower-cost biofuels* (No. DOE/EE-2079). Golden, CO: National Renewable Energy Lab. (NREL); Richland, WA: Pacific Northwest National Lab. (PNNL); Oak Ridge, TN: Oak Ridge National Lab. (ORNL); Argonne, IL: Argonne National Lab. (ANL); Idaho Falls, ID: Idaho National Lab. (INL).

Brindhadevi, K., Mathimani, T., Rene, E. R., Shanmugam, S., Chi, N. T. L., & Pugazhendhi, A. (2021). Impact of cultivation conditions on the biomass and lipid in microalgae with an emphasis on biodiesel. *Fuel*, 284, 119058.

Brundtland, G. H. (1985). World commission on environment and development. *Environmental Policy and Law*, 14(1), 26–30.

Burch, T. R., Firnstahl, A. D., Spencer, S. K., Larson, R. A., & Borchardt, M. A. (2022). *Fate and seasonality of antimicrobial resistance genes during full-scale anaerobic digestion of cattle manure across seven livestock production facilities* (0047-2425).

Carneiro, M. L. N., Pradelle, F., Braga, S. L., Gomes, M. S. P., Martins, A. R. F., Turkovics, F., & Pradelle, R. N. (2017). Potential of biofuels from algae: Comparison with fossil fuels, ethanol and biodiesel in Europe and Brazil through life cycle assessment (LCA). *Renewable and Sustainable Energy Reviews*, 73, 632–653.

Chandel, A. K., Garlapati, V. K., Jeevan Kumar, S., Hans, M., Singh, A. K., & Kumar, S. (2020). The role of renewable chemicals and biofuels in building a bioeconomy. *Biofuels, Bioproducts and Biorefining*, 14(4), 830–844.

Chen, F., Xiao, Z., Yue, L., Wang, J., Feng, Y., Zhu, X., Wang, Z., & Xing, B. (2019). Algae response to engineered nanoparticles: Current understanding, mechanisms and implications. *Environmental Science: Nano*, 6(4), 1026–1042.

Choi, R. Y., Coyner, A. S., Kalpathy-Cramer, J., Chiang, M. F., & Campbell, J. P. (2020). Introduction to machine learning, neural networks, and deep learning. *Translational Vision Science & Technology*, 9(2), 14–14.

Coady, M. D., Parry, I. W., Sears, L., & Shang, B. (2015). *How large are global energy subsidies?* International Monetary Fund.

Da Silva, T. L., Gouveia, L., & Reis, A. (2014). Integrated microbial processes for biofuels and high value-added products: The way to improve the cost effectiveness of biofuel production. *Applied Microbiology and Biotechnology*, 98, 1043–1053.

Damian, C. S., Devarajan, Y., Thandavamoorthy, R., & Jayabal, R. (2024). Harnessing artificial intelligence for enhanced bioethanol productions: A cutting-edge approach towards sustainable energy solution. *International Journal of Chemical Reactor Engineering*, 22(7), 719–727.

de Farias Silva, C. E., & Bertucco, A. (2016). Bioethanol from microalgae and cyanobacteria: A review and technological outlook. *Process Biochemistry*, *51*(11), 1833–1842.

de Fraiture, C., & Berndes, G. (2009). *Biofuels and water*. Cornell University Library's Initiatives in Publishing (CIP).

De Fraiture, C., Giordano, M., & Yongsong, L. (2007). *Biofuels: Implications for agricultural water use*. Colombo, Sri Lanka: International Water Management Institute.

Dellomonaco, C., Fava, F., & Gonzalez, R. (2010). The path to next generation biofuels: Successes and challenges in the era of synthetic biology. *Microbial Cell Factories*, 9, 1–15.

Dolan, A. E., Hou, Z., Xiao, Y., Gramelspacher, M. J., Heo, J., Howden, S. E., Freddolino, P. L., Ke, A., & Zhang, Y. (2019). Introducing a spectrum of long-range genomic deletions in human embryonic stem cells using type I CRISPR-Cas. *Molecular Cell*, 74(5), 936–950. e935.

Franzén, C. J. (2003). Metabolic flux analysis of RQ-controlled microaerobic ethanol production by Saccharomyces cerevisiae. *Yeast*, 20(2), 117–132.

Freitas, E. N. d., Salgado, J. C. S., Alnoch, R. C., Contato, A. G., Habermann, E., Michelin, M., Martínez, C. A., & Polizeli, M. D. L. T. (2021). Challenges of biomass utilization for bioenergy in a climate change scenario. *Biology*, *10*(12), 1277.

Furtado, A., Lupoi, J. S., Hoang, N. V., Healey, A., Singh, S., Simmons, B. A., & Henry, R. J. (2014). Modifying plants for biofuel and biomaterial production. *Plant Biotechnology Journal*, 12(9), 1246–1258. https://doi.org/10.1111/pbi.12300.

Geissdoerfer, M., Savaget, P., Bocken, N. M., & Hultink, E. J. (2017). The circular economy–A new sustainability paradigm? *Journal of Cleaner Production*, 143, 757–768.

Giampietro, M. (2019). On the circular bioeconomy and decoupling: Implications for sustainable growth. *Ecological Economics*, 162, 143–156.

Gopalakrishnan, G., Negri, M. C., Wang, M., Wu, M., Snyder, S. W., & Lafreniere, L. (2009). Biofuels, land, and water: A systems approach to sustainability. *Environmental Science & Technology*, 43(15), 6094–6100.

Gopalakrishnan, S., & Maranas, C. D. (2015). Achieving metabolic flux analysis for *S. cerevisiae* at a genome-scale: Challenges, requirements, and considerations. *Metabolites*, 5(3), 521–535.

Goshisht, M. K. (2024). Machine learning and deep learning in synthetic biology: Key architectures, applications, and challenges. *ACS Omega*, *9*(9), 9921–9945.

Goswami, R. K., Agrawal, K., Upadhyaya, H. M., Gupta, V. K., & Verma, P. (2022). Microalgae conversion to alternative energy, operating environment and economic footprint: An influential approach towards energy conversion, and management. *Energy Conversion and Management*, 269, 116118.

Gurgel, A. C., Reilly, J. M., & Paltsev, S. (2008). *Potential land use implications of a global biofuels industry*. MIT Joint Program on the Science and Policy of Global Change.

Hajra, A. K. (1974). On extraction of acyl and alkyl dihydroxyacetone phosphate from incubation mixtures. *Lipids*, 9(8), 502–505.

Hanaki, K., & Portugal-Pereira, J. (2018). The effect of biofuel production on greenhouse gas emission reductions. In *Biofuels and sustainability: Holistic perspectives for policy-making* (pp. 53–71). Tokyo: Springer Japan.

Hassan, M. H., & Kalam, M. A. (2013). An overview of biofuel as a renewable energy source: Development and challenges. *Procedia Engineering*, *56*, 39–53.

Hollinshead, W., He, L., & Tang, Y. J. (2014). Biofuel production: An odyssey from metabolic engineering to fermentation scale-up. *Frontiers in Microbiology*, 5, 344.

Huang, W.-D., & Zhang, Y. P. (2011). Energy efficiency analysis: Biomass-to-wheel efficiency related with biofuels production, fuel distribution, and powertrain systems. *PLoS One*, 6(7), e22113.

Jagadevan, S., Banerjee, A., Banerjee, C., Guria, C., Tiwari, R., Baweja, M., & Shukla, P. (2018). Recent developments in synthetic biology and metabolic engineering in microalgae towards biofuel production. *Biotechnology for Biofuels*, 11, 1–21.

Janiesch, C., Zschech, P., & Heinrich, K. (2021). Machine learning and deep learning. *Electronic Markets*, 31(3), 685–695.

Jegannathan, K. R., Chan, E.-S., & Ravindra, P. (2009). Harnessing biofuels: A global renaissance in energy production? *Renewable and Sustainable Energy Reviews*, 13(8), 2163–2168.

Jeon, S., Lim, J.-M., Lee, H.-G., Shin, S.-E., Kang, N. K., Park, Y.-I., Oh, H.-M., Jeong, W.-J., Jeong, B.-r., & Chang, Y. K. (2017). Current status and perspectives of genome editing technology for microalgae. *Biotechnology for Biofuels*, 10, 1–18.

Jeswani, H. (2016). Carbon footprint of biofuels. In *Life cycle approaches to sustainable regional development* (pp. 245–252). Routledge.

Jiang, Y., Wu, R., Zhou, J., He, A., Xu, J., Xin, F., Zhang, W., Ma, J., Jiang, M., & Dong, W. (2019). Recent advances of biofuels and biochemicals production from sustainable resources using co-cultivation systems. *Biotechnology for Biofuels*, 12(1), 155. https://doi.org/10.1186/s13068-019-1495-7

Kasani, A. A., Esmaeili, A., & Golzary, A. (2022). Software tools for microalgae biorefineries: Cultivation, separation, conversion process integration, modeling, and optimization. *Algal Research*, 61, 102597.

Khan, M. I., Shin, J. H., & Kim, J. D. (2018). The promising future of microalgae: Current status, challenges, and optimization of a sustainable and renewable industry for biofuels, feed, and other products. *Microbial Cell Factories*, 17, 1–21.

Kim, Y.-M., Petzold, C. J., Kerkhoven, E. J., & Baker, S. E. (2021). *Multi-omics technologies for optimizing synthetic biomanufacturing* (Vol. 9, p. 818010). Frontiers Media SA.

Kraljić, D., & Desmarais, M. (2021). Cascading bioenergy systems for sustainable biofuel production by integrating biological and thermochemical technologies. In *Waste biorefinery* (pp. 475–489). Elsevier.

Kumar, R., & Kumar, P. (2017). Future microbial applications for bioenergy production: A perspective. *Frontiers in Microbiology*, 8, 450.

Li, Y., Liu, W., Zhang, X., Wang, S., Yadegari, R., & Wang, J. (2023). Advances in crop biomass production based on multi-omics approach. *Frontiers in Plant Science*, 14, 1155442.

Lobus, N. V., Knyazeva, M. A., Popova, A. F., & Kulikovskiy, M. S. (2023). Carbon footprint reduction and climate change mitigation: A review of the approaches, technologies, and implementation challenges. C, 9(4), 120. https://www.mdpi.com/2311-5629/9/4/120

Lourinho, G., & Brito, P. (2015). Advanced biodiesel production technologies: Novel developments. *Reviews in Environmental Science and Bio/Technology*, 14, 287–316.

Lynd, L. R., Laser, M. S., Bransby, D., Dale, B. E., Davison, B., Hamilton, R., Himmel, M., Keller, M., McMillan, J.D., Sheehan, J., & Wyman, C. E. (2008). How biotech can transform biofuels: For cellulosic ethanol to become a reality, biotechnological solutions should focus on optimizing the conversion of biomass to sugars. *Nature Biotechnology*, *26*(2), 169–173.

Ma, X., Guo, G., Wu, X., Wu, Q., Liu, F., Zhang, H., Shi, N., & Guan, Y. (2023). Advances in integration, wearable applications, and artificial intelligence of biomedical microfluidics systems. *Micromachines*, 14(5), 972.

Maltsev, Y., Maltseva, K., Kulikovskiy, M., & Maltseva, S. (2021). Influence of light conditions on microalgae growth and content of lipids, carotenoids, and fatty acid composition. *Biology*, *10*(10), 1060.

Manning, S. R. (2022). Microalgal lipids: Biochemistry and biotechnology. *Current Opinion in Biotechnology*, 74, 1–7.

Marriott, P. E., Sibout, R., Lapierre, C., Fangel, J. U., Willats, W. G., Hofte, H., Gómez, L. D., & McQueen-Mason, S. J. (2014). Range of cell-wall alterations enhance saccharification in *Brachypodium distachyon* mutants. *Proceedings of the National Academy of Sciences*, 111(40), 14601–14606.

Misra, N., Panda, P. K., & Parida, B. K. (2013). Agrigenomics for microalgal biofuel production: An overview of various bioinformatics resources and recent studies to link OMICS to bioenergy and bioeconomy. *Omics: A Journal of Integrative Biology*, 17(11), 537–549.

Mohan, S. V., Nikhil, G. N., Chiranjeevi, P., Reddy, C. N., Rohit, M., Kumar, A. N., & Sarkar, O. (2016). Waste biorefinery models towards sustainable circular bioeconomy: Critical review and future perspectives. *Bioresource Technology*, 215, 2–12.

Moreira, J. B., Santos, T. D., Duarte, J. H., Bezerra, P. Q. M., de Morais, M. G., & Costa, J. A. V. (2023). Role of microalgae in circular bioeconomy: From waste treatment to biofuel production. *Clean Technologies and Environmental Policy*, 25(2), 427–437.

Mussatto, S. I., Dragone, G., Guimarães, P. M., Silva, J. P. A., Carneiro, L. M., Roberto, I. C., Vicente, A., Domingues, L., & Teixeira, J. A. (2010). Technological trends, global market, and challenges of bio-ethanol production. *Biotechnology Advances*, *28*(6), 817–830.

Mustafa, H. M., & Hayder, G. (2021). Recent studies on applications of aquatic weed plants in phytoremediation of wastewater: A review article. *Ain Shams Engineering Journal*, 12(1), 355–365.

Naveed, M. H., Khan, M. N. A., Mukarram, M., Naqvi, S. R., Abdullah, A., Haq, Z. U., Ullah, H., & Al Mohamadi, H. (2024). Cellulosic biomass fermentation for biofuel production: Review of artificial intelligence approaches. *Renewable and Sustainable Energy Reviews*, 189, 113906.

O'Neill, E. G., Geissler, C. H., & Maravelias, C. T. (2024). Large-scale spatially explicit analysis of carbon capture at cellulosic biorefineries. *Nature Energy*, 9(7), 828–838.

Pal, A., Srivastava, I., Sahu, S., & Singh, K. R. (2023). Algae's promise to sequester carbon sheds new insight on changing climate. *International Journal of Current Science Research and Review*, 6(5), 2904–2913.

Park, Y. K. (2020). *Metabolic engineering of the yeast Yarrowia lipolytica for the production of even-and odd-chain fatty acids.* Université Paris-Saclay.

Paul, D., Arora, A., & Verma, M. L. (2021). *Advances in microbial biofuel production* (Vol. 12, p. 746216). Frontiers Media SA.

Perin, G., & Jones, P. R. (2019). Economic feasibility and long-term sustainability criteria on the path to enable a transition from fossil fuels to biofuels. *Current Opinion in Biotechnology*, 57, 175–182.

Pfromm, P. H., Amanor-Boadu, V., Nelson, R., Vadlani, P., & Madl, R. (2010). Bio-butanol vs. bio-ethanol: A technical and economic assessment for corn and switchgrass fermented by yeast or Clostridium acetobutylicum. *Biomass and Bioenergy*, *34*(4), 515–524.

Poore, C. (2024). Researchers provide a supply chain model to chart a pathway to next-gen biofuels. *Tecdh Xplore*. https://techxplore.com/news/2024-05-chain-pathway-gen-biofuels.html

Radakovits, R., Jinkerson, R. E., Fuerstenberg, S. I., Tae, H., Settlage, R. E., Boore, J. L., & Posewitz, M. C. (2012). Draft genome sequence and genetic transformation of the oleaginous alga Nannochloropsis gaditana. *Nature Communications*, 3(1), 686.

Ronzon, T., Lusser, M., Landa, L., Mbarek, R., Giuntoli, J., Cristobal Garcia, J., Parisi, C., Ferrari, E., Marelli, L., Torres De Matos, C., Gomez Barbero, M., & Rodriguez Cerezo, E. (2017). *Bioeconomy report 2016* (EUR 28468 EN). Luxembourg: Publications Office of the European Union.

Sagnak, R., Cochot, S., Molina-Jouve, C., Nicaud, J.-M., & Guillouet, S. E. (2018). Modulation of the glycerol phosphate availability led to concomitant reduction in the citric acid excretion and increase in lipid content and yield in *Yarrowia lipolytica*. *Journal of Biotechnology*, 265, 40–45.

Santander, C., Robles, P. A., Cisternas, L. A., & Rivas, M. (2014). Technical–economic feasibility study of the installation of biodiesel from microalgae crops in the Atacama Desert of Chile. *Fuel Processing Technology*, *125*, 267–276.

Sawaengsak, W., & Gheewala, S. H. (2017). Analysis of social and socio-economic impacts of sugarcane production: A case study in Nakhon Ratchasima province of Thailand. *Journal of Cleaner Production*, *142*, 1169–1175.

Simionato, D., Basso, S., Giacometti, G. M., & Morosinotto, T. (2013). Optimization of light use efficiency for biofuel production in algae. *Biophysical Chemistry*, 182, 71–78.

Song, Y., Xie, L., Zhang, X., Hu, Z., Li, S., Zhang, P., & Yang, X. (2024). Enhancement of biomass, lipid accumulation, and carbon sequestration potential in microalgae via cultivation with aggregation-induced emission light-conversion films. *Chemical Engineering Journal*, 483, 149148.

Srirangan, K., Akawi, L., Moo-Young, M., & Chou, C. P. (2012). Towards sustainable production of clean energy carriers from biomass resources. *Applied Energy*, *100*, 172–186.

Su, B., Lin, Y., Wang, J., Quan, X., Chang, Z., & Rui, C. (2022). Sewage treatment system for improving energy efficiency based on particle swarm optimization algorithm. *Energy Reports*, 8, 8701–8708.

Sun, H., Huang, K., Zhang, X., Ren, H., & Ye, L. (2022). Stable isotope probing reveals specific assimilating bacteria of refractory organic compounds in activated sludge. *Water Research*, 212, 118105.

Sundaram, T., Rajendran, S., Gnanasekaran, L., Rachmadona, N., Jiang, J.-J., Khoo, K. S., & Show, P. L. (2023). Bioengineering strategies of microalgae biomass for biofuel production: Recent advancement and insight. *Bioengineered*, 14(1), 2252228.

Vinuselvi, P., Park, J. M., Lee, J. M., Oh, K., Ghim, C.-M., & Lee, S. K. (2011). Engineering microorganisms for biofuel production. *Biofuels*, 2(2), 153–166.

Wang, H., Gao, L., Chen, L., Guo, F., & Liu, T. (2013). Integration process of biodiesel production from filamentous oleaginous microalgae Tribonema minus. *Bioresource Technology*, *142*, 39–44.

Wang, Q., Lu, Y., Xin, Y., Wei, L., Huang, S., & Xu, J. (2016). Genome editing of model oleaginous microalgae *Nannochloropsis* spp. by CRISPR/Cas9. *The Plant Journal*, 88(6), 1071–1081.

Wide range of plants offer cellulosic biofuel potential, e. d., August 10. *ScienceDaily*. https://www.sciencedaily.com/releases/2010/08/100812151636.htm.

Wolt, J. D. (2009). Advancing environmental risk assessment for transgenic biofeedstock crops. *Biotechnology for Biofuels*, 2(1), 27.

Wu, C. H., Bernard, S. M., Andersen, G. L., & Chen, W. (2009). Developing microbe–plant interactions for applications in plant-growth promotion and disease control, production of useful compounds, remediation and carbon sequestration. *Microbial Biotechnology*, 2(4), 428–440.

Xie, D., Ji, X., Zhou, Y., Dai, J., He, Y., Sun, H., Guo, Z., Yang, Y., Zheng, X., & Chen, B. (2022). Chlorella vulgaris cultivation in pilot-scale to treat real swine wastewater and mitigate carbon dioxide for sustainable biodiesel production by direct enzymatic transesterification. *Bioresource Technology*, *349*, 126886.

Zhalnina, K., Hawkes, C., Shade, A., Firestone, M. K., & Pett-Ridge, J. (2021). *Managing plant microbiomes for sustainable biofuel production* (Vol. 5, pp. 3–13). Am Phytopath Society.

13 Augmented Role of Nanotechnology for Biofuel Production from Lower and Higher Plant Species

Sandra Fabiola Velasco Ramírez,
Maria Adriana Delgado Armas, Leslie Andrea Colorado Torres,
José de Jesús Luna Díaz, and Ana Cristina Ramírez Anguiano

13.1 INTRODUCTION

The global energy transition requires sustainable alternatives to fossil fuels, with biofuels being essential in the decarbonisation of the energy sector. Derived from bio-based materials, they offer significant environmental and socioeconomic benefits, such as reduced greenhouse gas emissions, increased energy security and alignment with the circular bioeconomy (Akhtar, 2023; Malhotra, 2024; Rial, 2024). These span several technology generations: from starch- and sugar-rich food crops (first generation), through lignocellulosic biomass (second generation), to algae and cyanobacteria (third generation), each with different sustainability and efficiency implications (Khan et al., 2021; Mignogna et al., 2024).

However, their adoption faces technical and social barriers. Major challenges include land use conflicts, limited conversion efficiency, availability of sustainable raw materials and economic viability of emerging technologies (Gupta et al., 2022). Lignocellulosic biomass, although abundant, suffers from structural challenges that hinder its efficient decomposition due to the recalcitrance of its cell walls and the high cost of pretreatment and hydrolysis (Balan, 2014; Zhao L. et al., 2020). In addition, there is a need to improve algal cultures and strains without increasing pressure on resources such as water or nutrients (Karp & Shield, 2008; Kour et al., 2019).

In this scenario, nanotechnology emerges as a disruptive tool to overcome key limitations in biofuel production. Its application at the molecular scale can significantly improve conversion processes through nanocatalysts, functionalised materials and nanosensors, which increase the efficiency and selectivity of biochemical and thermochemical pathways. Nanotechnology applied to plant systems can also increase photosynthesis, nutrient uptake and stress tolerance, thereby improving biomass yield and its energy density.

In addition to these technical functions, nanotechnology converges with biotechnology and synthetic biology, facilitating the design of optimised bioenergy systems in both lower species (algae, cyanobacteria) and higher plants (sorghum, miscanthus). This interdisciplinary integration creates new opportunities for scaling up the production of advanced biofuels. Nevertheless, their implementation requires rigorous life cycle analysis (LCA), toxicity studies and the establishment of robust regulatory frameworks to ensure environmental sustainability and public safety.

This chapter analyses the transformative role of nanotechnology in bioenergy, focusing on innovations that improve biomass availability, optimise energy conversion and ensure system sustainability. Specific strategies for different types of plants are examined, highlighting how these technologies can strengthen biofuels as a strategic component in the global energy transition (Figure 13.1).

DOI: 10.1201/9781003600169-13

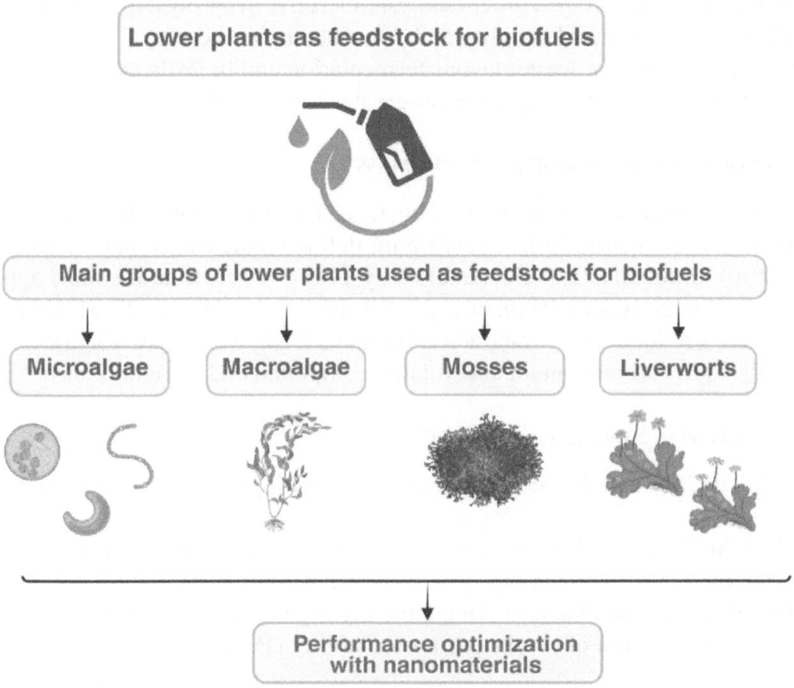

FIGURE 13.1 Illustrated diagram of the use of lower plants as raw material for biofuel production. The main taxonomic groups (microalgae, macroalgae, mosses, and liverworts) are identified, along with their bioenergy applications through pathways such as lipid accumulation and carbohydrate fermentation. Additionally, performance enhancement is illustrated through optimized cultivation strategies and the use of nanomaterials, within the framework of sustainable biorefinery systems.

13.2 BASIS OF NANOTECHNOLOGY APPLIED TO BIOENERGY

Nanotechnology has become a very important tool to optimise biofuel production (Antunes et al., 2017; Palaniappan, 2017). Nanomaterials such as metal nanoparticles, nanotubes, nanofibres or metal oxides have been shown to improve the pretreatment, hydrolysis and conversion of lignocellulosic biomass, algae and wastewater (Khoo et al., 2020). Magnetic nanoparticles stand out in enzyme immobilisation and biogas production (Rahman et al., 2016), as well as nanosensors and nanobiotechnology to optimise microorganisms (Muthumari & Subbaiyan, 2024). However, they face challenges including cost, safety and sustainability (Kamla et al., 2021).

13.2.1 Classification of Nanomaterials

By Dimensionality (Awogbemi & Kallon, 2024):

- **0D:** nanoparticles, quantum dots, nanoclusters. Highly active as catalysts.
- **1D:** nanotubes, nanowires. Favour enzyme immobilisation and transesterification.
- **2D:** nanolayers. Promote charge transfer.
- **3D:** combined structures. High efficiency and stability.

By Chemical Composition (Awogbemi & Kallon, 2024):

- **Nanometals (Au, Ag, Ni, Sn):** improve catalysis and fermentation.
- **Magnetic Nanoparticles:** enzyme immobilisation/reutilisation, efficient hydrolysis.
- **CNTs:** stable, conductive, catalyse key reactions.

- **Graphene and its Derivatives:** green catalysts, effective in pyrolysis and transesterification.
- **Metal Oxides:** high thermal stability and efficiency in biodiesel.
- **Nanofibres:** derived from waste, favour fermentation and biohydrogen production.
- **Nanocomposites:** combine properties for greater functionality.

13.2.2 Physicochemical Properties of Biomass

Bioenergy efficiency depends on properties such as porosity, lignocellulosic composition and calorific value. Cellulose, hemicellulose and lignin define recalcitrance and conversion efficiency (Alvira et al., 2010). Immediate and elemental analysis allows selecting the most appropriate technology and minimising emissions (Naik et al., 2010). Techniques such as SEM, XRD, FTIR, TGA/DSC and HPLC are key for accurate characterisation. Calorific value (GCV) guides biomass selection, which is related to carbon content and volatile compounds (Capareda, 2011).

13.2.3 Nanomaterial Synthesis Methods

- **Conventional Chemistry:** precise control of morphology, yet involves toxic reagents and high energy.
- **Green Synthesis:** uses plant extracts and microorganisms as reducing and stabilising agents. Compatible with bioenergy systems (Osman et al., 2024).
- **Microbial Biosynthesis:** bacteria, fungi and algae generate biocompatible nanomaterials. Requires standardisation of conditions for scalability (Parveen et al., 2016).

Current research seeks to balance efficiency, reproducibility and low environmental impact to integrate nanomaterials into sustainable bioenergy technologies (Huston et al., 2021) (Figure 13.2).

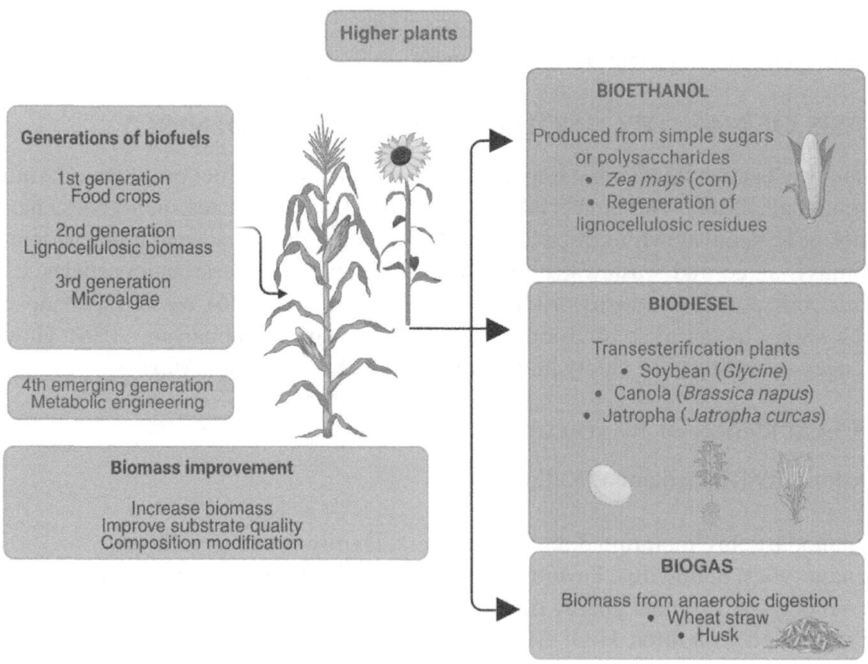

FIGURE 13.2 Illustrated diagram of the role of higher plants in the production of bioethanol, biodiesel, and biogas. The main generations of biofuels and the plant species used are shown: food crops such as *Zea mays* (corn) and *Saccharum officinarum* (sugarcane) for bioethanol; oilseeds such as *Glycine max* (soybean), *Brassica napus* (canola), and *Jatropha curcas* for biodiesel; and lignocellulosic residues such as wheat straw and corn stover for biogas. The diagram also highlights emerging strategies such as metabolic engineering and the enhancement of plant biomass through biotechnology.

13.3 LOWER PLANTS AS A RAW MATERIAL FOR BIOFUELS

Lower plants (microalgae, macroalgae, mosses and liverworts) offer a promising and sustainable alternative for third-generation biofuels. Their fast growth, ability to fix CO_2 and cultivation in wastewater places them as non-competitive feedstocks with food crops. In particular, microalgae stand out for their high photosynthetic efficiency and lipid content, facilitating the production of biodiesel, bioethanol and biohydrogen. The integration of green nanotechnology has enabled the synthesis of functional nanoparticles to catalyse key bioenergy processes (Alsaiari et al., 2023; Osman et al., 2024).

13.3.1 Main Groups of Lower Plants

13.3.1.1 Microalgae

Species such as *Chlorella vulgaris* and *Scenedesmus accumulate* up to 50% lipids and show high biomass productivity, allowing the production of biodiesel, bioethanol, biogas and biohydrogen (Li et al., 2008). Their cultivation in waters unsuitable for agriculture, their ability to fix CO_2 and their use in water treatment make them multifunctional bioenergy platforms. The application of biorefinery allows the extraction of valuable compounds along with energy, thus enhancing profitability (Gouveia & Oliveira, 2009).

13.3.1.2 Macroalgae

Green, brown and red algae exhibit a high productivity, carbohydrate richness and lignin-free chemical compositions, making them easy to process. *Ulva lactuca*, *Padina boryana* and *Chlorella* offer potential in biodiesel and bioethanol (Abomohra et al., 2018). Biorefinery improves economic viability through integrated biomass utilization (Lopez-Contreras et al., 2017).

13.3.1.3 Bryophytes

Mosses and liverworts, even with low productivity, synthesise valuable metabolites (terpenoids, long-chain fatty acids) relevant for advanced biofuels (Lu et al., 2019). Metabolic engineering and in vitro cultivation are emerging as strategies to overcome biomass limitations (Martínez-Abaigar & Núñez-Olivera, 2021).

13.3.2 Crop Optimisation and Nanotechnology Application

Improvements include strain selection, stress induction, genetic engineering and environmental control in bioreactors. The use of nanomaterials (ZnO, TiO_2, CNTs) improves photosynthetic efficiency and metabolite accumulation (Raliya et al., 2015). These materials also act as stimulants or transporters, enhancing productivity without affecting cell viability. Ecotoxicological assessment and the use of green nanomaterials are crucial to ensure sustainability (Jeevanandam et al., 2022).

13.3.3 Increased Lipids and Sugars with Advanced Interventions

Nanoparticles of silver, iron oxide, carbon and magnesium have shown increases in lipid production in algae (Hasnain et al., 2022; P. Kumar et al., 2023). Genetic engineering has also improved triacylglyceride and oil accumulation in model species such as Arabidopsis and tobacco (Sanjaya et al., 2011). The use of microRNAs can reduce cellular recalcitrance and improve bioenergetic yield (Trumbo et al., 2015). These advanced strategies consolidate lower plants as technological pillars in the transition towards clean energy.

13.4 HIGHER PLANTS IN BIOETHANOL, BIODIESEL AND BIOGAS GENERATION

Plant-based biofuels represent sustainable alternatives to fossil fuels (Malik et al., 2024). The former are derived from food crops (maize, wheat), while the latter are derived from more sustainable lignocellulosic biomass (Cosentino et al., 2008; Joshi & Nookaraju, 2012). Microalgae, the basis of the third generation, offer high efficacy but are resource-intensive (Callegari et al., 2020). Fourth-generation strategies employ genetic engineering to increase CO_2 fixation and biomass productivity (Peña-Castro et al., 2017).

13.4.1 Biofuels Derived from Higher Plants

Bioethanol: Bioethanol is produced from fermentable sugars or polysaccharides. Plants such as *Zea mays* and *Saccharum officinarum* dominate the first generation, while species such as *Miscanthus* and switchgrass contribute biomass for the second generation (Balat & Balat, 2009; Somerville et al., 2010).

Biodiesel: Biodiesel is obtained by transesterification of vegetable oils from soybean, canola, sunflower and *Jatropha curcas*. The lipid profile has an impact on the quality of the product (Knothe, 2005; Gui et al., 2008).

Biogas: Biogas is generated by anaerobic digestion of plant residues (straw, stubble, manure). Pretreatment of biomass improves its conversion (Angelidaki & Sanders, 2004; Taherzadeh & Karimi, 2008).

13.4.2 Nanomaterials and Plant Metabolism

Nanomaterials (NMs) can enhance physiological processes such as photosynthesis, metabolite production and anaerobic digestion (Hatami et al., 2016; Rahman et al., 2016). Their efficacy depends on properties such as size and active surface area (W. Zhao et al., 2023). Although promising, their ecotoxicological risks need to be assessed (Pradhan & Mailapalli, 2017).

13.4.2.1 Physiological Effects of Nanoparticles

Metal nanoparticles, such as AgNPs, ZnO-NPs and CuO-NPs, induce moderate oxidative stress and activate biosynthetic pathways (Faizan et al., 2018) for valuable metabolites production, including phenols and flavonoids (Hasnain et al., 2022). ZnO-NPs enhance photosynthesis by activating Rubisco and carotenoid synthesis (Faizan et al., 2018). TiO_2-NPs strengthen photoprotection, though high doses can be phytotoxic (Chahardoli et al., 2022).

At the antioxidant level, NMs modulate enzymes such as SOD, CAT and APX. At adequate concentrations, they strengthen cellular defences, but in excess, they induce oxidative damage (Jomova et al., 2024). Cyclodextrins and immobilised moulds also increase the production and extraction of secondary metabolites (Martínez-Abaigar & Núñez-Olivera, 2021). Functionalisation of nanomaterials is recommended to minimise environmental risks (Razack et al., 2021).

13.4.3 Genetic Enhancements to Increase Biomass

Genetic engineering seeks to increase biomass yield and TAG accumulation by overexpressing genes such as DGAT, WRI1, and AGPase deletion (Andrianov et al., 2010; Sanjaya et al., 2011). Other strategies include modifying the cell wall or expressing enzymes that enhance biomass conversion (Furtado et al., 2014). Microalgae are also being modified for their photosynthetic efficiency (Radakovits et al., 2010). These improvements aim to optimise the energy density of higher plants and their potential for biofuels (Saha & Ramachandran, 2013).

13.5 APPLICATION OF NANOTECHNOLOGY IN BIOMASS CONVERSION AND PROCESSING

Nanotechnology has revolutionised the conversion of biomass into biofuels by employing nanoparticles with large surface area, reactivity and catalytic capacity. These properties have optimised processes such as transesterification, enzymatic hydrolysis, fermentation and inhibitor removal.

13.5.1 Nanocatalysis in Transesterification and Fermentation

Heterogeneous nanocatalysts improve the speed, selectivity and reusability of key reactions. Examples include Fe_3O_4 nanoparticles functionalised with sulfonic groups, which achieve up to 96% cellobiose conversion (Arya et al., 2021), and Au/Pt nanocatalysts in dendrimers with high electrochemical efficiency. In fermentation, Fe_2O_3 and TiO_2 nanoparticles enhance biohydrogen production (Thanigaivel et al., 2022). Catalysts such as ZnO, NiO and $CaO\text{-}TiO_2$ have achieved biodiesel yields above 96% (Benitha et al., 2021; Ghosh & Halder, 2022).

13.5.2 Nanomaterials in Enzymatic Hydrolysis

Immobilisation on nanomaterials increases enzymatic stability and reusability. Functionalised Fe_3O_4 achieves 94% cellulase immobilisation and up to 58.3% glucose conversion from cane bagasse (Sheikh et al., 2023). Silica nanoparticles and ferrites are also used to enhance enzymatic cycling (Rai et al., 2019). In microalgae, nanoparticles such as CNTs and MgO promote harvesting and lipid accumulation (S. Kumar et al., 2023). These systems reduce operating costs and increase catalytic efficiency (Singhvi & Kim, 2020).

13.5.3 Reduction of Inhibitors During Conversion

Nanomaterials suppress inhibitors such as phenols, aliphatic acids and furan derivatives, improving hydrolysis and fermentation efficiency (Thakur et al., 2022; Srivastava et al., 2023). Magnetic nano-biocatalysts, carbon nanotubes and metal oxides are used to facilitate pretreatment and increase cellulose digestibility (Chandel et al., 2024). These processes are more sustainable and efficient than conventional detoxification methods (Kumar V. et al., 2020).

13.5.3.1 Future Perspectives

The integration of nanomaterials in biorefineries improves yields, sustainability and economic viability. Green nanocatalysts have achieved up to 80% conversion with low energy consumption (Usanova et al., 2024). While toxicological risks need to be assessed, nanotechnology remains a key to the transition towards more sustainable energy systems (Karimi-Maleh et al., 2022).

13.6 NANOMATERIAL-ASSISTED GENETIC ENGINEERING

Nanotechnology has expanded the capabilities of plant genetic engineering, allowing more precise and efficient editing using nanovectors and the induction of metabolic pathways to produce energetic compounds.

13.6.1 Nanovectors for Gene Editing (CRISPR-Cas, RNAi)

Nanoparticles such as carbon nanotubes, gold and mesoporous silica facilitate the delivery of CRISPR-Cas and RNAi with high efficiency and biocompatibility (Demirer et al., 2021; Zhi et al., 2022). These systems allow the transformation of recalcitrant species without the need for genomic integration, yielding transgene-free edited plants (Demirer et al., 2019). In addition, they load ribonucleoprotein (RNP) complexes and siRNA, improving specificity and reducing tissue damage (Vats et al., 2022).

These tools have improved yield, stress resistance and regeneration in crops (Iqbal et al., 2024). Nanosensor monitoring allows greater control of gene expression. Despite their potential, phytotoxicity risks must be assessed and their sustainability ensured (Tandon et al., 2023).

13.6.2 Induction of Metabolic Pathways

Nanomaterials can activate biosynthetic pathways to increase the production of lipids, sugars and secondary metabolites essential for biofuels (Hatami et al., 2016). In microalgae, they induce stress that promotes lipid accumulation (Tan et al., 2024). In higher plants, they enhance photosynthesis and primary metabolism.

The combined use of CRISPR-Cas and nanovectors allows precise redirection of biosynthesis towards bioenergetic compounds (Wu et al., 2023). Nanotubes, metal oxides and polymeric nanoparticles have introduced genes that encode key enzymes to enhance biomass conversion (Arya et al., 2021; Kumar Y. et al., 2021). In addition, these materials can neutralise inhibitors, increasing the overall yield (Thakur et al., 2022).

For massive application, their physicochemical properties must be optimised and adverse effects need to be minimised. The development of eco-friendly, scalable and plant-tissue compatible nanomaterials will be the key to the future of this technology (Sankaran et al., 2021).

Overall, nanomaterial-assisted genetic engineering integrates high-precision gene editing with efficient metabolic regulation, offering a robust approach to improve bioenergy crops (Figure 13.3).

13.7 ENVIRONMENTAL AND RISK ASSESSMENT

The application of nanomaterials in bioenergy has generated remarkable advances, but also poses environmental risks. This assessment focuses on three axes: toxicity in plants and soils, ecological impact and emerging regulations.

13.7.1 Toxicity in Plants and Soils

The effects of nanomaterials (NMs) vary according to their type, concentration and plant species. While some, such as carbon nanotubes, stimulate growth and chlorophyll, others reduce biomass and affect germination or DNA (Liné et al., 2021; Gowtham et al., 2024). The generation of reactive oxygen species (ROS) is related to oxidative stress induced by NMs. Although metal oxides and graphene can improve bioenergy efficiency, their excess is phytotoxic and can enter the food chain (Dietz & Herth, 2011; Gowtham et al., 2024).

13.7.2 Ecological Impact

NMs can alter soil microbial communities, affect aquatic organisms and modify ecosystems. Their persistence depends on physicochemical properties such as size and solubility (Murali et al., 2022). In crops such as maize and oilseed rape, they generate contradictory responses affecting photosynthesis and metabolism. They can also transform the agricultural and natural landscape by modifying the microbiome (Ahmed et al., 2023; Gowtham et al., 2024). In algae, toxicity depends on the type of NM, culture medium and species. Studies of diversity, invasiveness and risk are essential, especially with transgenic species (Wolt, 2009; Kumar S. et al., 2023).

13.7.3 Regulations and Life Cycle

LCA assesses energy and environmental impacts from production to disposal (Ojeda et al., 2015). However, there are methodological limitations and a lack of specific regulations for NMs (Ngarize et al., 2013). More precise regulatory approaches, such as stepwise risk analysis and SSWD, are

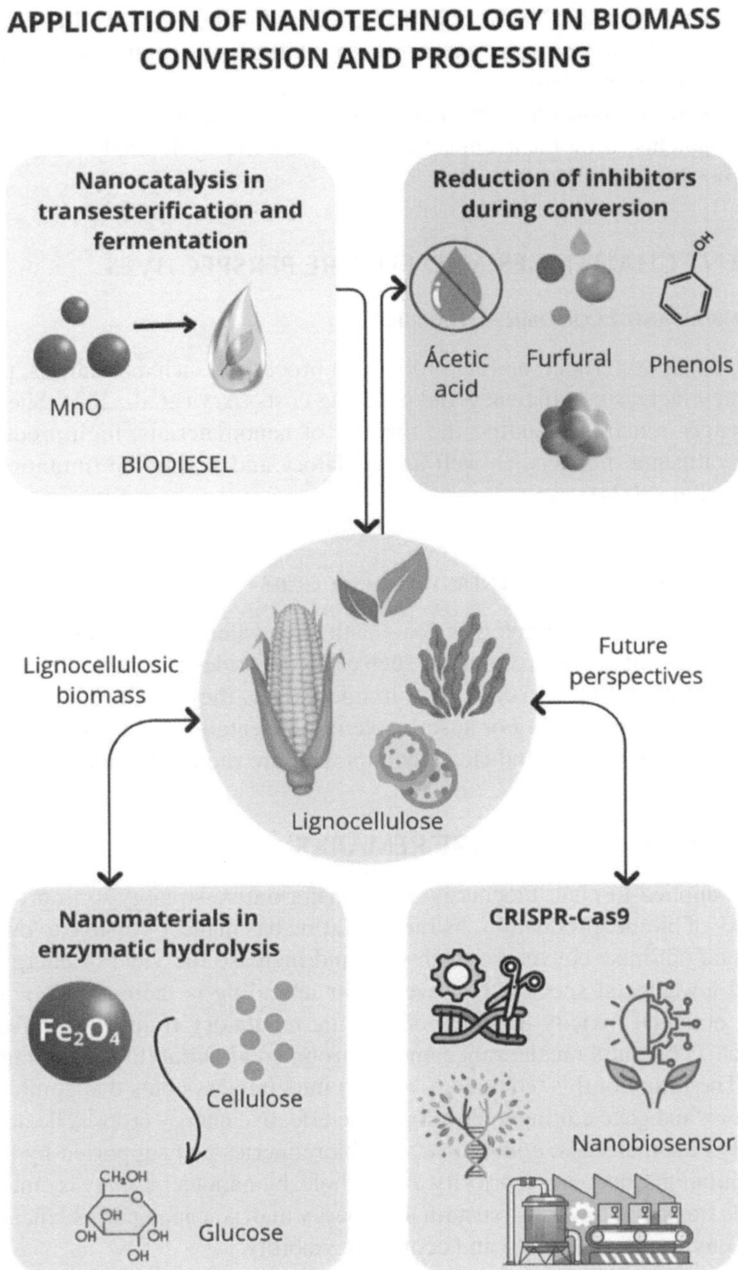

FIGURE 13.3 Applications of nanotechnology in biomass conversion and processing. The illustration highlights the role of nanomaterials in key stages of lignocellulosic and algal biomass utilization: (1) nanocatalysis in transesterification and fermentation processes for biofuel production; (2) the use of nanomaterials in enzymatic hydrolysis to enhance the release of fermentable sugars; (3) suppression of inhibitory compounds such as furfural, acetic acid, and phenolics; and (4) future perspectives including the integration of CRISPR-Cas9 systems and nano-biosensors for genetic enhancement and real-time monitoring. These strategies contribute to a more efficient, selective, and sustainable conversion of biomass into bioenergy products.

required to adequately characterise the effects of NMs on bioenergy crops. In addition, ethical and ecological governance frameworks that include sustainability and social security must be integrated (Semenzin et al., 2015; Stone et al., 2018).

In summary, the responsible and sustainable use of nanotechnology in bioenergy requires rigorous toxicity assessments, ecological impact analyses and life cycle analyses, as well as adaptive regulations (Mehejabin et al., 2024).

13.8 CURRENT CHALLENGES AND FUTURE PERSPECTIVES

13.8.1 Technical and Economical Barriers

Nanotechnology improves crucial biofuel production processes, such as catalysis, pretreatment and separation, thereby increasing efficiency and reducing costs (Arya et al., 2021; Sheikh et al., 2023). However, challenges remain, including the toxicity of nanomaterials, high production costs and health and environmental impacts, as well as regulatory and scaling-up limitations (Ziolkowska, 2018; Mehejabin et al., 2024).

13.8.2 Upscaling of Nano-Bioenergy Technologies

Upscaling of nano-bioenergy technologies deals with the recalcitrance of biomass, which limits the release of fermentable sugars. Nanomaterials such as metal oxides and nanotubes enhance lignocellulosic conversion (Chandel et al., 2024). Within microalgae, they enhance lipid accumulation and harvesting (Mandotra et al., 2018). For sustainable implementation, integrated systems combining genetic engineering, biorefineries and clean technologies are required (Ambaye et al., 2021).

13.9 CONCLUSION AND FUTURE REMARKS

Nanotechnology applied to plant bioenergy is a transformative strategy to improve the efficiency and sustainability of biofuel production. Its incorporation has made it possible to overcome biomass structural barriers, optimise conversion pathways and increase the yield of energy compounds in both higher and lower plant species. However, their upscaling is facing challenges such as high synthesis costs, potential toxicity and lack of specific regulatory frameworks. Technical barriers also remain, such as biomass recalcitrance and environmental variability, affecting the stability of nanomaterials. The future of this technology aims at integrated systems that combine nanotechnology, biotechnology and gene editing to develop more effective energy crops. The design of environmentally friendly nanomaterials, compatible with biorefineries and supported by life cycle studies and adaptive regulations, becomes a priority. As a whole, bionanotechnology is emerging as a major component in the transition towards a sustainable energy matrix, capable of balancing technological innovation with environmental safety and economic viability.

COMPETING INTERESTS

The authors declare that they have no competing interest.

DECLARATION OF GENERATIVE AI AND AI-ASSISTED TECHNOLOGIES IN THE WRITING PROCESS

During the preparation of this work, the book chapter author(s) have not taken any assistance from AI, and the book chapter author(s) reviewed and edited the content as needed and take(s) full responsibility for the content of the publication.

REFERENCES

Abomohra, A. E.-F., El-Naggar, A. H., & Baeshen, A. A. (2018). Potential of macroalgae for biodiesel production: Screening and evaluation studies. *Journal of Bioscience and Bioengineering*, 125(2), 231–237.

Ahmed, S. F., Debnath, J. C., Mehejabin, F., Islam, N., Tripura, R., Mofijur, M., Hoang, A. T., Rasul, M. G., & Vo, D.-V. N. (2023). Utilization of nanomaterials in accelerating the production process of sustainable biofuels. *Sustainable Energy Technologies and Assessments*, 55, 102894.

Akhtar, E. T. (2023). Biofuels: A renewable solution for energy security and climate change mitigation. *SSRN Electronic Journal*, 4.

Alsaiari, N. S., Alzahrani, F. M., Amari, A., Osman, H., Harharah, H. N., Elboughdiri, N., & Tahoon, M. A. (2023). Plant and microbial approaches as green methods for the synthesis of nanomaterials: Synthesis, applications, and future perspectives. *Molecules*, 28(1), 463.

Alvira, P., Tomás-Pejó, E., Ballesteros, M., & Negro, M. J. (2010). Pretreatment technologies for an efficient bioethanol production process based on enzymatic hydrolysis: A review. *Bioresource Technology*, 101(13), 4851–4861.

Ambaye, T. G., Vaccari, M., Bonilla-Petriciolet, A., Prasad, S., Van Hullebusch, E. D., & Rtimi, S. (2021). Emerging technologies for biofuel production: A critical review on recent progress, challenges and perspectives. *Journal of Environmental Management*, 290, 112627.

Andrianov, V., Borisjuk, N., Pogrebnyak, N., Brinker, A., Dixon, J., Spitsin, S., Flynn, J., Matyszczuk, P., Andryszak, K., Laurelli, M., Golovkin, M., & Koprowski, H. (2010). Tobacco as a production platform for biofuel: Overexpression of *Arabidopsis DGAT* and *LEC2* genes increases accumulation and shifts the composition of lipids in green biomass. *Plant Biotechnology Journal*, 8(3), 277–287.

Angelidaki, I., & Sanders, W. (2004). Assessment of the anaerobic biodegradability of macropollutants. *Reviews in Environmental Science and Bio/Technology*, 3(2), 117–129.

Antunes, F. A. F., Gaikwad, S., Ingle, A. P., Pandit, R., Dos Santos, J. C., Rai, M., & Da Silva, S. S. (2017). Bioenergy and biofuels: Nanotechnological solutions for sustainable production. In M. Rai & S. S. Da Silva (Eds.), *Nanotechnology for Bioenergy and Biofuel Production* (pp. 3–18). Springer International Publishing.

Arya, I., Poona, A., Dikshit, P. K., Pandit, S., Kumar, J., Singh, H. N., Jha, N. K., Rudayni, H. A., Chaudhary, A. A., & Kumar, S. (2021). Current trends and future prospects of nanotechnology in biofuel production. *Catalysts*, *11*(11), 1308.

Awogbemi, O., & Kallon, D. V. V. (2024). Recent advances in the application of nanomaterials for improved biodiesel, biogas, biohydrogen, and bioethanol production. *Fuel*, 358, 130261.

Balan, V. (2014). Current challenges in commercially producing biofuels from lignocellulosic biomass. *ISRN Biotechnology*, *2014*, 1–31.

Balat, M., & Balat, H. (2009). Recent trends in global production and utilization of bio-ethanol fuel. *Applied Energy*, 86(11), 2273–2282.

Benitha, V. S., Prabhahar, R. S. S., & Nagarajan, J. (2021). Enhanced yield of biodiesel through nano catalytic transesterification of palm oil. *Materials Today:* Proceedings, 47, 3088–3094.

Callegari, A., Bolognesi, S., Cecconet, D., & Capodaglio, A. G. (2020). Production technologies, current role, and future prospects of biofuels feedstocks: A state-of-the-art review. *Critical Reviews in Environmental Science and Technology*, 50(4), 384–436.

Capareda, S. C. (2011). Biomass energy conversion. In M. Nayeripour & M. Kheshti (Eds.), *Sustainable Growth and Applications in Renewable Energy Sources* (pp. 123–144). IntechOpen.

Chahardoli, A., Sharifan, H., Karimi, N., & Kakavand, S. N. (2022). Uptake, translocation, phytotoxicity, and hormetic effects of titanium dioxide nanoparticles (TiO_2NPs) in *Nigella arvensis* L. *Science of the Total Environment*, 806, 151222.

Chandel, H., Kumar, P., Chandel, A. K., & Verma, M. L. (2024). Biotechnological advances in biomass pretreatment for bio-renewable production through nanotechnological intervention. *Biomass Conversion and Biorefinery*, 14(3), 2959–2981.

Cosentino, S. L., Copani, V., Patanè, C., Mantineo, M., & D'Agosta, G. M. (2008). Agronomic, energetic and environmental aspects of biomass energy crops suitable for Italian environments. *Italian Journal of Agronomy*, 3(2), 81–95.

Demirer, G. S., Silva, T. N., Jackson, C. T., Thomas, J. B., W. Ehrhardt, D., Rhee, S. Y., Mortimer, J. C., & Landry, M. P. (2021). Nanotechnology to advance CRISPR–Cas genetic engineering of plants. *Nature Nanotechnology*, 16(3), 243–250.

Demirer, G. S., Zhang, H., Matos, J. L., Goh, N. S., Cunningham, F. J., Sung, Y., Chang, R., Aditham, A. J., Chio, L., Cho, M.-J., Staskawicz, B., & Landry, M. P. (2019). High aspect ratio nanomaterials enable delivery of functional genetic material without DNA integration in mature plants. *Nature Nanotechnology*, 14(5), 456–464.

Dietz, K.-J., & Herth, S. (2011). Plant nanotoxicology. *Trends in Plant Science*, 16(11), 582–589.

Faizan, M., Faraz, A., Yusuf, M., Khan, S. T., & Hayat, S. (2018). Zinc oxide nanoparticle-mediated changes in photosynthetic efficiency and antioxidant system of tomato plants. *Photosynthetica*, 56(2), 678–686.

Furtado, A., Lupoi, J. S., Hoang, N. V., Healey, A., Singh, S., Simmons, B. A., & Henry, R. J. (2014). Modifying plants for biofuel and biomaterial production. *Plant Biotechnology Journal*, 12(9), 1246–1258.

Ghosh, N., & Halder, G. (2022). Current progress and perspective of heterogeneous nanocatalytic transesterification towards biodiesel production from edible and inedible feedstock: A review. *Energy Conversion and Management*, 270, 116292.

Gouveia, L., & Oliveira, A. C. (2009). Microalgae as a raw material for biofuels production. *Journal of Industrial Microbiology & Biotechnology*, 36(2), 269–274.

Gowtham, H. G., Shilpa, N., Singh, S. B., Aiyaz, M., Abhilash, M. R., Nataraj, K., Amruthesh, K. N., Ansari, M. A., Alomary, M. N., & Murali, M. (2024). Toxicological effects of nanoparticles in plants: Mechanisms involved at morphological, physiological, biochemical and molecular levels. *Plant Physiology and Biochemistry*, 210, 108604.

Gui, M. M., Lee, K. T., & Bhatia, S. (2008). Feasibility of edible oil vs. non-edible oil vs. waste edible oil as biodiesel feedstock. *Energy*, 33(11), 1646–1653.

Gupta, R., Chauhan, H., Garg, V. K., & Kataria, N. (2022). Chemical and physical properties of nanoparticles and hybrid materials. In R. R. Karri, J. R. Koduru, N. M. Mubarak, & E. R. Bandala (Eds.), *Sustainable Nanotechnology for Environmental Remediation* (pp. 199–220). Elsevier.

Hasnain, M., Munir, N., Abideen, Z., Dias, D. A., Aslam, F., & Mancinelli, R. (2022). Applying silver nanoparticles to enhance metabolite accumulation and biodiesel production in new algal resources. *Agriculture*, 13(1), 73.

Hatami, M., Kariman, K., & Ghorbanpour, M. (2016). Engineered nanomaterial-mediated changes in the metabolism of terrestrial plants. *Science of the Total Environment*, 571, 275–291.

Huston, M., DeBella, M., DiBella, M., & Gupta, A. (2021). Green synthesis of nanomaterials. *Nanomaterials*, 11(8), 2130.

Iqbal, N., Ali Akbar, B., Taskeen, N., & Mubashar, M. (2024). Nanoparticles mediated plant genetic engineering: Emerging field with promising applications. *Global Immunological & Infectious Diseases Review*, IX(I), 34–53.

Jeevanandam, J., Kiew, S. F., Boakye-Ansah, S., Lau, S. Y., Barhoum, A., Danquah, M. K., & Rodrigues, J. (2022). Green approaches for the synthesis of metal and metal oxide nanoparticles using microbial and plant extracts. *Nanoscale*, 14(7), 2534–2571.

Jomova, K., Alomar, S. Y., Alwasel, S. H., Nepovimova, E., Kuca, K., & Valko, M. (2024). Several lines of antioxidant defense against oxidative stress: Antioxidant enzymes, nanomaterials with multiple enzyme-mimicking activities, and low-molecular-weight antioxidants. *Archives of Toxicology*, *98*(5), 1323–1367.

Joshi, C. P., & Nookaraju, A. (2012). New avenues of bioenergy production from plants: Green alternatives to petroleum. *Journal of Petroleum & Environmental Biotechnology*, 3(7), 134.

Kamla, M., Sushil, A., Karmal, M., Shweta, & Yadvika. (2021). Nanotechnology: A Sustainable Solution for Bioenergy and Biofuel Production. *Journal of Nanoscience and Nanotechnology*, 21(6), 3481–3494.

Karimi-Maleh, H., Rajendran, S., Vasseghian, Y., & Dragoi, E.-N. (2022). Advanced integrated nanocatalytic routes for converting biomass to biofuels: A comprehensive review. *Fuel*, 314, 122762.

Karp, A., & Shield, I. (2008). Bioenergy from plants and the sustainable yield challenge. *New Phytologist*, 179(1), 15–32.

Khan, N., Sudhakar, K., & Mamat, R. (2021). Role of biofuels in energy transition, green economy and carbon neutrality. *Sustainability*, 13(22), 12374.

Khoo, K. S., Chia, W. Y., Tang, D. Y. Y., Show, P. L., Chew, K. W., & Chen, W.-H. (2020). Nanomaterials utilization in biomass for biofuel and bioenergy production. *Energies*, 13(4), 892.

Knothe, G. (2005). Dependence of biodiesel fuel properties on the structure of fatty acid alkyl esters. *Fuel Processing Technology*, 86(10), 1059–1070.

Kour, D., Rana, K. L., Yadav, N., Yadav, A. N., Rastegari, A. A., Singh, C., Negi, P., Singh, K., & Saxena, A. K. (2019). Technologies for biofuel production: Current development, challenges, and future prospects. In A. A. Rastegari, A. N. Yadav, & A. Gupta (Eds.), *Prospects of Renewable Bioprocessing in Future Energy Systems* (Vol. 10, pp. 1–50). Springer International Publishing.

Kumar, P., Malik, S., & Dubey, K. K. (2023). A review on the use of nanomaterials in agriculture: Benefits and associated health risks. *Current Nanomaterials*, 8(1), 44–57.

Kumar, S., Tirlangi, S., Kumar, A., Imran, M., Pillai Hp, J. S., Kumar Koshariya, A., Sathish, T., Ubaidullah, M., Ayub, R., Minnam Reddy, V. R., AlAbdulaal, T. H., Algarni, H., Maiz, F., & Shkir, Mohd. (2023). A review on the contribution of nanotechnology for biofuel production from algal biomass: A bridge to the reduction of carbon footprint. *Sustainable Energy Technologies and Assessments*, 60, 103498.

Kumar, V., Yadav, S. K., Kumar, J., & Ahluwalia, V. (2020). A critical review on current strategies and trends employed for removal of inhibitors and toxic materials generated during biomass pretreatment. *Bioresource Technology*, 299, 122633.

Kumar, Y., Yogeshwar, P., Bajpai, S., Jaiswal, P., Yadav, S., Pathak, D. P., Sonker, M., & Tiwary, S. K. (2021). Nanomaterials: Stimulants for biofuels and renewables, yield and energy optimization. *Materials Advances*, 2(16), 5318–5343.

Li, Y., Horsman, M., Wu, N., Lan, C. Q., & Dubois-Calero, N. (2008). Biofuels from microalgae. *Biotechnology Progress*, 24(4), 815–820.

Liné, C., Manent, F., Wolinski, A., Flahaut, E., & Larue, C. (2021). Comparative study of response of four crop species exposed to carbon nanotube contamination in soil. *Chemosphere*, 274, 129854.

Lopez-Contreras, A. M., Harmsen, P. F. H., Hou, X., Huijgen, W. J. J., Ditchfield, A. K., Bjornsdottir, B., Obata, O. O., Hreggvidsson, G. O., Van Hal, J. W., & Bjerre, A.-B. (2017). Biorefinery approach to the use of macroalgae as feedstock for biofuels. In L. Pereira (Ed.), *Algal Biofuels* (1st ed., pp. 103–139). CRC Press.

Lu, Y., Eiriksson, F. F., Thorsteinsdóttir, M., & Simonsen, H. T. (2019). Valuable fatty acids in bryophytes—Production, biosynthesis, analysis and applications. *Plants*, 8(11), 524.

Malhotra, R. (2024). Biofuels: Bridging the gap to a greener energy future. *The International Journal of Engineering & Science*, 13(10), 57–68.

Malik, K., Capareda, S. C., Kamboj, B. R., Malik, S., Singh, K., Arya, S., & Bishnoi, D. K. (2024). Biofuels production: A review on sustainable alternatives to traditional fuels and energy sources. *Fuels*, *5*(2), 157–175.

Mandotra, S. K., Kumar, R., Upadhyay, S. K., & Ramteke, P. W. (2018). Nanotechnology: A new tool for biofuel production. In N. Srivastava, M. Srivastava, H. Pandey, P. K. Mishra, & P. W. Ramteke (Eds.), *Green Nanotechnology for Biofuel Production* (Vol. 5, pp. 17–28). Springer International Publishing.

Martínez-Abaigar, J., & Núñez-Olivera, E. (2021). Novel biotechnological substances from bryophytes. In R. P. Sinha & D. P. Häder (Eds.), *Natural Bioactive Compounds* (pp. 233–248). Academic Press.

Mehejabin, F., Musharrat, A., Ahmed, S. F., Kabir, Z., Khan, T. M. Y., & Saleel, C. A. (2024). Sustainable biofuel production utilizing nanotechnology: Challenges and potential solutions. *GCB Bioenergy*, 16(10), e70001.

Mignogna, D., Szabó, M., Ceci, P., & Avino, P. (2024). Biomass energy and biofuels: Perspective, potentials, and challenges in the energy transition. *Sustainability*, 16(16), 7036.

Murali, M., Gowtham, H. G., Singh, S. B., Shilpa, N., Aiyaz, M., Alomary, M. N., Alshamrani, M., Salawi, A., Almoshari, Y., Ansari, M. A., & Amruthesh, K. N. (2022). Fate, bioaccumulation and toxicity of engineered nanomaterials in plants: Current challenges and future prospects. *Science of the Total Environment*, 811, 152249.

Muthumari, P., & Subbaiyan, N. (2024). A comprehensive review on pioneering nanotechnologies in advancing next-generation biofuel production. *International Research Journal of Multidisciplinary Technovation*, *6*(5), 110–133.

Naik, S., Goud, V. V., Rout, P. K., Jacobson, K., & Dalai, A. K. (2010). Characterization of Canadian biomass for alternative renewable biofuel. *Renewable Energy*, 35(8), 1624–1631.

Ngarize, S., Makuch, K. E., & Pereira, R. (2013). The case for regulating nanotechnologies: International, European and national perspectives. *Review of European, Comparative & International Environmental Law*, 22(2), 131–145.

Ojeda, K. A., Herrera, A. P., Sierra, M. J., & Tamayo, K. (2015). Evaluación del impacto ambiental del uso de nanopartículas de alúmina como aditivo de mezclas biodiesel/diésel mediante análisis de ciclo de vida. *Ingeniería y Competitividad*, 17(1), 133–142.

Osman, A. I., Zhang, Y., Farghali, M., Rashwan, A. K., Eltaweil, A. S., Abd El-Monaem, E. M., Mohamed, I. M. A., Badr, M. M., Ihara, I., Rooney, D. W., & Yap, P.-S. (2024). Synthesis of green nanoparticles for energy, biomedical, environmental, agricultural, and food applications: A review. *Environmental Chemistry Letters*, 22(2), 841–887.

Palaniappan, K. (2017). An overview of applications of nanotechnology in biofuel production. *World Applied Sciences Journal*, 35(8), 1305–1311.

Parveen, K., Banse, V., & Ledwani, L. (2016). *Green synthesis of nanoparticles: Their advantages and disadvantages. AIP Conference Proceedings, 1724*, 020048.

Peña-Castro, J. M., Del Moral, S., Núñez-López, L., Barrera-Figueroa, B. E., & Amaya-Delgado, L. (2017). Biotechnological strategies to improve plant biomass quality for bioethanol production. *BioMed Research International*, *2017*, 1–10.

Pradhan, S., & Mailapalli, D. R. (2017). Interaction of engineered nanoparticles with the agri-environment. *Journal of Agricultural and Food Chemistry*, 65(38), 8279–8294.

Radakovits, R., Jinkerson, R. E., Darzins, A., & Posewitz, M. C. (2010). Genetic engineering of algae for enhanced biofuel production. *Eukaryotic Cell*, 9(4), 486–501.

Rahman, K. M., Melville, L., Huq, S. M. I., & Khoda, S. K. (2016). Understanding bioenergy production and optimisation at the nanoscale – A review. *Journal of Experimental Nanoscience*, 11(10), 762–775.

Rai, M., Ingle, A. P., Pandit, R., Paralikar, P., Biswas, J. K., & Da Silva, S. S. (2019). Emerging role of nanobiocatalysts in hydrolysis of lignocellulosic biomass leading to sustainable bioethanol production. *Catalysis Reviews*, 61(1), 1–26.

Raliya, R., Nair, R., Chavalmane, S., Wang, W.-N., & Biswas, P. (2015). Mechanistic evaluation of translocation and physiological impact of titanium dioxide and zinc oxide nanoparticles on the tomato (*Solanum lycopersicum L.*) plant. *Metallomics*, 7(12), 1584–1594.

Razack, S. A., Duraiarasan, S., Ramakrishnan, G., & Hussain, C. M. (2021). Functionalized nanomaterials and the environment. In V. Kumar & C. M. Hussain (Eds.), *Handbook of Functionalized Nanomaterials* (pp. 29–54). Elsevier.

Rial, R. C. (2024). Biofuels versus climate change: Exploring potentials and challenges in the energy transition. *Renewable and Sustainable Energy Reviews*, 196, 114369.

Saha, S., & Ramachandran, S. (2013). Genetic improvement of plants for enhanced bio-ethanol production. *Recent Patents on DNA & Gene Sequences*, 7(1), 36–44.

Sanjaya, Durrett, T. P., Weise, S. E., & Benning, C. (2011). Increasing the energy density of vegetative tissues by diverting carbon from starch to oil biosynthesis in transgenic Arabidopsis. *Plant Biotechnology Journal*, 9(8), 874–883.

Sankaran, R., Markandan, K., Khoo, K. S., Cheng, C. K., Ashokkumar, V., Deepanraj, B., & Show, P. L. (2021). The expansion of lignocellulose biomass conversion into bioenergy via nanobiotechnology. *Frontiers in Nanotechnology*, 3, 793528.

Semenzin, E., Lanzellotto, E., Hristozov, D., Critto, A., Zabeo, A., Giubilato, E., & Marcomini, A. (2015). Species sensitivity weighted distribution for ecological risk assessment of engineered nanomaterials: The n-TiO_2 case study. *Environmental Toxicology and Chemistry*, 34(11), 2644–2659.

Sheikh, Z. U. D., Bajar, S., Devi, A., Rose, P. K., Suhag, M., Yadav, A., Yadav, D. K., Deswal, T., Kaur, J., Kothari, R., Pathania, D., Rani, N., & Singh, A. (2023). Nanotechnology based technological development in biofuel production: Current status and future prospects. *Enzyme and Microbial Technology*, 171, 110304.

Singhvi, M., & Kim, B. S. (2020). Current developments in lignocellulosic biomass conversion into biofuels using nanobiotechology approach. *Energies*, 13(20), 5300.

Somerville, C., Youngs, H., Taylor, C., Davis, S. C., & Long, S. P. (2010). Feedstocks for lignocellulosic biofuels. *Science*, 329(5993), 790–792.

Srivastava, N., Singh, R., Srivastava, M., Mohammad, A., Harakeh, S., Pratap Singh, R., Pal, D. B., Haque, S., Tayeb, H. H., Moulay, M., & Kumar Gupta, V. (2023). Impact of nanomaterials on sustainable pretreatment of lignocellulosic biomass for biofuels production: An advanced approach. *Bioresource Technology*, *369*, 128471.

Stone, V., Führ, M., Feindt, P. H., Bouwmeester, H., Linkov, I., Sabella, S., Murphy, F., Bizer, K., Tran, L., Ågerstrand, M., Fito, C., Andersen, T., Anderson, D., Bergamaschi, E., Cherrie, J. W., Cowan, S., Dalemcourt, J., Faure, M., Gabbert, S., … Poortvliet, P. M. (2018). The essential elements of a risk governance framework for current and future nanotechnologies. *Risk Analysis*, 38(7), 1321–1331.

Taherzadeh, M. J., & Karimi, K. (2008). Pretreatment of lignocellulosic wastes to improve ethanol and biogas production: A review. *International Journal of Molecular Sciences*, 9(9), 1621–1651.

Tan, C. H., Low, S. S., Cheah, W. Y., Singh, J., Chai, W. S., Tiong, S. K., & Show, P. L. (2024). Futuristic opportunities for pretreatment processes in biofuel production from microalgae. *GCB Bioenergy*, 16(5), e13136.

Tandon, A., Singh, A., Thakur, A., & Sharma, V. (2023). Nanomaterial mediated genome engineering for sustainable food production: Current status and future prospects. *Biocatalysis and Agricultural Biotechnology*, 54, 102891.

Thakur, A., Kumar, A., Kaya, S., Vo, D.-V. N., & Sharma, A. (2022). Suppressing inhibitory compounds by nanomaterials for highly efficient biofuel production: A review. *Fuel*, 312, 122934.

Thanigaivel, S., Priya, A. K., Dutta, K., Rajendran, S., Sekar, K., Jalil, A. A., & Soto-Moscoso, M. (2022). Role of nanotechnology for the conversion of lignocellulosic biomass into biopotent energy: A biorefinery approach for waste to value-added products. *Fuel*, 322, 124236.

Trumbo, J. L., Zhang, B., & Stewart, C. N. (2015). Manipulating micro RNAs for improved biomass and biofuels from plant feedstocks. *Plant Biotechnology Journal*, 13(3), 337–354.

Usanova, K. I., Dhall, H., Chandna, M., Pandey, N., Mouli, K. C., & Vyas, A. (2024). Catalytic conversion of biomass to biofuels using green nanocatalysts. *E3S Web of Conferences*, 537, 07011.

Vats, S., Kumawat, S., Brar, J., Kaur, S., Yadav, K., Magar, S. G., Jadhav, P. V., Salvi, P., Sonah, H., Sharma, S., & Deshmukh, R. (2022). Opportunity and challenges for nanotechnology application for genome editing in plants. *Plant Nano Biology*, 1, 100001.

Wolt, J. D. (2009). Advancing environmental risk assessment for transgenic biofeedstock crops. *Biotechnology for Biofuels*, 2(1), 27.

Wu, K., Xu, C., Li, T., Ma, H., Gong, J., Li, X., Sun, X., & Hu, X. (2023). Application of nanotechnology in plant genetic engineering. *International Journal of Molecular Sciences*, 24(19), 14836.

Zhao, L., Lu, L., Wang, A., Zhang, H., Huang, M., Wu, H., Xing, B., Wang, Z., & Ji, R. (2020). Nano-biotechnology in agriculture: Use of nanomaterials to promote plant growth and stress tolerance. *Journal of Agricultural and Food Chemistry*, 68(7), 1935–1947.

Zhao, W., Zhou, P., Lou, B., Jiang, Y., Li, Y., Li, M., Shakoor, N., & Rui, Y. (2023). Translocation and transformation of engineered nanomaterials in plant cells and their effect on metabolism. *Biocell*, 47(3), 493–502.

Zhi, H., Zhou, S., Pan, W., Shang, Y., Zeng, Z., & Zhang, H. (2022). The promising nanovectors for gene delivery in plant genome engineering. *International Journal of Molecular Sciences*, 23(15), 8501.

Ziolkowska, J. R. (2018). Introduction to biofuels and potentials of nanotechnology. In N. Srivastava, M. Srivastava, H. Pandey, P. K. Mishra, & P. W. Ramteke (Eds.), *Green Nanotechnology for Biofuel Production* (Vol. 5, pp. 1–15). Springer International Publishing.

14 *Jatropha* as an Efficient Biofuel Producer

A Critical Analysis for Sustainable Uses

Tesfaye Kassaw Bedru, Abayneh Getachew Demesa, Samuel Bernardo Perez Vega, Mani Jayakumar, and Selvakumar Periyasamy

14.1 INTRODUCTION

As the global energy system moves steadily toward net-zero emissions, there is growing interest in biofuels derived from non-food oilseed crops. In this context, Jatropha curcas has once again attracted attention. It is a perennial shrub with a relatively high oil yield and a strong ability to tolerate drought and low-quality soils, which makes it suitable as a second-generation biofuel feedstock that does not directly compete with food crops (Yang et al., 2024). Its capacity to grow on marginal and degraded land is especially relevant for countries in the Global South, where pressures around land use and food security are both sensitive and politically charged.

Recent advances have confirmed that *Jatropha* can yield up to 40% oil by seed weight with favorable biodiesel properties that meet main fuel standards (Gizaw et al., 2025; Riayatsyah et al., 2022). Biodiesel from *Jatropha* has been shown to possess superior cetane number, minimal sulfur emissions, and adequate oxidative stability, and it is suitable to be used as a substitute for fossil diesel in transportation and power generation sectors (Aguilar-Aguilar et al., 2025). Moreover, life cycle assessment (LCA) indicate that there is a significant reduction in greenhouse gas (GHG) emissions (up to 60–70%) when *Jatropha* biodiesel produced in sustainably way (Yang et al., 2024).

However, despite its clear agronomic flexibility, Jatropha curcas is constrained by several biological and economic limitations that must be addressed before it can be adopted at scale. These include poor and variable seed yield through virtue of genetic heterogeneity, absence of elite domesticated varieties, and lack of region-specific best farming practices (Nahar, 2025). In addition, socioeconomic studies in countries like Mexico and Brazil reveal that institutional and policy frameworks are often not well-designed to support large-scale *Jatropha* biodiesel initiatives without inducing unwanted socio-environmental trade-offs (Aguilar-Aguilar et al., 2025; Eduardo de Oliveira Aparecido et al., 2025).

Environmental vulnerability research has also revealed that climate change-induced stressors, such as rising salinity and weather extremes, could affect the productivity of *Jatropha*. Latest research on genetic and metabolomics adaptations, however, promises to make the plant more salt- and drought-tolerant. MALDI-ToF/ToF-MS proteomics has opened up new areas of research with implications for improving the stress tolerance and oil yields of *Jatropha* (Pompelli et al., 2025).

This chapter provides a comprehensive assessment of *J. curcas* for sustainable biofuel production, ranging from regional differences in oil content to conversion technologies like transesterification and pyrolysis, and agronomic and environmental constraints. It contrasts the economic, environmental, and life cycle performance of Jatropha fuels with conventional fuels and points out future directions like genetic enhancement, climate adaptability, and policy reforms in emerging economies.

DOI: 10.1201/9781003600169-14

14.2 PRACTICES UTILIZED FOR *JATROPHA* OIL EXTRACTION

Efficient oil extraction is essential to realizing the biofuel potential of Jatropha curcas, an inedible oilseed crop cultivated at large scale for biodiesel production. The oil content in the seeds of *Jatropha* is 30–40% by dry weight, but extraction efficiency and oil quality are prone and responsive to technological, environmental, and agricultural conditions (Yang et al., 2024). The past 5 years have seen significant improvement in mechanical, chemical, enzymatic, and hybrid extraction technologies, enhancing substantially both the yield and sustainability of *Jatropha* oil production. Table 14.1 shows the variations in oil content between *Jatropha* seeds cultivated in various locations and using various methods for extracting oil.

14.2.1 Mechanical Extraction Techniques

Mechanical extraction remains the most prevalent method of extracting oil from *J. curcas* seeds, especially in rural and resource-poor regions of Asia and Africa. The process involves the application of physical pressure to force oil from ground kernels with the use of mechanical equipment such as screw presses or hydraulic presses. Of these, the screw press is especially preferred due to ease of operation, minimal maintenance requirements, and applicability to small capacity and decentralized oil plants. However, extraction efficiency in this process is relatively low, and average recoveries of oil vary between 25% and 35% based on equipment design, seed quality, and operating conditions (Khater et al., 2024; Riayatsyah et al., 2022).

Among the significant reasons for this poor efficiency is the incomplete bursting of oil-contained cells during pressing. High fibrous density of *Jatropha* seed kernel, combined with natural viscosity of oil, offers resistance to expression. To overcome this, some recent studies pointed toward pre-conditioning treatments to increase oil release. For example, preheating the seeds to 60–70 °C has been reported to reduce oil viscosity, improve flow behavior, and soften the seed matrix, thereby promoting more effective cell rupture during mechanical pressing. Yuan et al. (2021) observed that thermally preconditioned seeds had up to a 7–12% gain in oil recovery compared to non-treated seeds without much alteration to oil quality (Yuan et al., 2021).

Another significant mechanical extraction parameter is seed moisture content. The ideal moisture content for seeds is 6–8%. Over-dryness causes the kernel to crack rather than compress, while high moisture causes poor oil flow and may lead to emulsion formation during pressing. Optimizing the moisture is required to obtain the maximum yield and minimize mechanical energy loss (Lavenburg et al., 2021; Mursalykova et al., 2023).

In terms of equipment, screw presses are more suitable than hydraulic presses for continuous operation. A screw press operates by the application of a rotating screw shaft to pressurize dehulled or partially ground seeds contained in a metal barrel. Perforated fine holes enable the expressed oil to exit, while the remaining press cake is expelled through the outlet. Despite the potential of hydraulic presses to apply higher force per stroke, they operate in batch mode, which is less effective and slower for industrial operations. The residual oil content in mechanically pressed cake remains relatively high, typically in the range of 10–15%, and therefore requires additional processing when complete oil recovery is desired(Khater et al., 2024).

For performance improvement, dual-stage pressing systems are designed. In this process, seeds undergo two continuous stages of pressing, initially low pressure for primary oil extraction, followed by high-pressure extraction for residual oil recovery. This process has led to an improvement in oil recovery up to 15% over single-stage pressing (Riayatsyah et al., 2022). In addition, innovations such as conditioning drums, variable pitch screws, and temperature-controlled press chambers have enhanced efficiency and consistency of output, especially in semi-commercial operations.

Despite all these technological advances, mechanical pressing has its drawbacks, including standardization of presses, maintenance requirements, and management of residue. Presses manufactured locally are unlikely to be standardized in design and sometimes without pressure control, leading to non-uniform yields. Furthermore, frictional heat generated during pressing can degrade susceptible components of oil such as unsaturated fatty acids and antioxidants unless controlled.

14.2.2 Solvent Extraction Techniques

Solvent extraction has been the most efficient and feasible means of oil extraction from *J. curcas* seeds, especially higher oil recovery and yields needed. Solvent extraction is a method that employs the principle of solubility, where non-polar solvents such as n-hexane invade the cell microstructures of the seed and solubilize lipophilic material, allowing oil extraction. Of all solvent chemicals, n-hexane is still the industrial choice owing to its low boiling point, high triglyceride selectivity, and relative cheapness. Current experimental studies indicate that the extraction of the oil by n-hexane extraction is around 35–45%, far higher compared to mechanical extraction (Bedru et al., 2024; Garuma et al., 2025; Nde and Foncha, 2020; Riayatsyah et al., 2022).

However, the widespread use of *n*-hexane has generated additional controversy due to toxicity, flammability, and environmental persistence. In response to these concerns, researchers have turned to green solvents like ethanol and isopropanol, which offer a greener and biodegradable alternative. Even though their polarity makes them less efficient at lipid solubilization compared to *n*-hexane, advances in co-solvent systems and intensification of the process, such as heating and agitation, have improved their extraction capability. For example, Yusuff (2021) demonstrated that a 70:30 ethanol–water mixture achieved an oil recovery of nearly 36%, comparable to conventional n-hexane extraction, while significantly reducing environmental and occupational hazards(Yusuff, 2021).

More recent research has increasingly focused on facilitated extraction methods to improve oil recovery efficiency. Among the most promising of these are ultrasound-assisted extraction (UAE) and microwave-assisted extraction (MAE). These enhance penetration by the solvent and mass transfer by destroying seed tissue cellular walls through physical energy. With UAE, acoustic cavitation provides for entry of the solvent into intracellular oil reservoirs, with reduced extraction time and consumption of solvent. Microwave extraction, on the other hand, exploits the immediate internal heating that shatters oil bodies within seeds. Studies by Lima-Pereira et al. (2025) and Aziz et al. (2021) show that UAE and MAE, with green solvents, are able to enhance yields by up to 20% compared to conventional soaking operations and without the degradation of the oil's oxidative stability or nutritional quality (Lima-Pereira et al., 2025; Mohamad Aziz et al., 2021).

The other key area in solvent extraction is the efficiency of solvent recovery. Solvent recovery systems, typically by rotary evaporators or vacuum distillation, are necessary in order to recycle and reuse solvents, especially in large or commercial-scale systems. Both the cost and energy demand of this step can be appreciable, and the optimization of recovery systems is one of the main factors for the economic success of solvent extraction. Coupling closed-loop vapor recovery and solvent condensation loops has the potential to increase the efficiency of solvent recycling by over 80%, significantly reducing operating costs (Keneni et al., 2021).

While solvent extraction is a better process relative to oil yield, solvent extraction is still plagued by several challenges for wide-scale application in low-income or rural communities. Explosion-proof solvent handling facilities and the need for continuous thermal control render the process less economic for decentralized or cottage-scale utilization. Research efforts notwithstanding, refinements are still made to improve the safety, environmental compatibility, and cost of solvent-based processes. Hybrid systems that combine mechanical pre-pressing with solvent finishing are also gaining prominence, allowing partial oil recovery through pressing and following solvent extraction of oil from the residue seed cake, thus combining the advantages of the two systems (Bakhshabadi et al., 2025).

14.2.3 Enzyme-Assisted Extraction

Enzyme-Assisted Extraction (EAE) is a recent green technology that has also been favored for oil recovery from *J. curcas* seeds, particularly where the focus is on sustainability, reduced solvent usage, and improved oil quality. This method uses hydrolytic enzymes such as cellulase, pectinase, and protease to specifically break down the structural components of the seed cell wall, including cellulose, hemicellulose, lignin, and bound polysaccharides, which helps to improve oil release. These cellular structures form rigid barriers to oil release; oil is more readily released from intracellular compartments through enzymatic breakdown (Liu et al., 2016; Martínez Herrera et al., 2019).

EAE typically entails the soaking of ground or milled *Jatropha* seeds or seedcake in a buffer water solution and subsequent addition of a proper enzyme or combination of enzymes. Extraction efficiency is sensitive to multiple parameters, including enzyme concentration, reaction temperature, pH, and incubation time. Experiments have shown that maximum oil recovery occurs in moderate temperatures (40–55°C) and moderately acidic to close to neutral pH levels (pH 5.0–7.0), which support maximum enzyme stability and activity (Lavenburg et al., 2021). For instance, in the course of enzyme tests in Brazil and Bangladesh, researchers attained oil recovery yields above 40%, a similar or superior rate compared to solvent extraction under similar conditions (Rollin et al., 2023).

Cellulase and pectinase break the polysaccharide structure of the seed coat, whereas protease breaks protein–lipid complexes, opening the way to imbedded lipids more easily. Synergistic use of mixtures of enzymes has been shown to yield much higher oil extraction compared to treatment with an individual enzyme (Bakhshabadi et al., 2025). Besides, this process preserves susceptible oil components, such as unsaturated fatty acids and antioxidants, because the extraction occurs at relatively low temperatures and without the use of organic solvents, minimizing oxidative deterioration.

Despite these advantages, EAE is not yet significantly commercialized due to some major drawbacks. First, the enzyme cost remains expensive, particularly when acquired at large-scale production capacities. Second, the incubation time is relatively long, typically ranging from 4 to 24 hours, which can reduce throughput and complicate batch planning in industrial settings. Third, the activity of enzymes is sensitive to pH and temperature, which may fluctuate due to variability among seed lots or operational inconsistencies in remote processing plants(Martínez Herrera et al., 2019).

14.2.4 Regional Variability in Oil Yield

The oil productive potential of *J. curcas* is heavily influenced not only by extraction processes but also by local environmental conditions, farming practices, and genetic diversity. Oil content has been shown in numerous studies to vary widely between geographical locations due to climatic conditions, altitude, soil fertility, rainfall pattern, and temperature variations, all of which affect seed development and lipid biosynthesis (Neupane et al., 2021; Yusuff, 2021).

In the subtropics and tropics, e.g., southern India, Thailand, and parts of Brazil, where *Jatropha* cultivation has fairly consistent rainfall and moderate temperatures, seed oil content consistently exceeds 38–40% dry weight. In contrast, plants grown in semi-arid or drought-prone conditions, such as in northern Ethiopia, Botswana, or central Nigeria, tend to produce seeds with relatively lower oil content, often between 2530% (Malik and Tripathi, 2019; Tadano et al., 2021). The reduction is essentially caused by water stress during the flowering and seed filling stages, which limits lipid accumulation in seed endosperm tissue.

Soil quality also matters to the oil yield. Research in Pakistan and West Africa has shown that higher oil output is associated with loamy or freely drained sandy soils that are most productive in potassium and phosphorus (Ramkat et al., 2011). Agronomic practices such as supplemental irrigation and pruning have been shown to improve Jatropha curcas growth and yield by enhancing branching, inflorescence density, and seed production, which can ultimately increase oil yield (Arunyanark et al., 2025; Wasnik & Singh, 2024).

Genetic variation among *Jatropha curcas* ecotypes is a major factor driving differences in oil yield across geographical regions. Landraces and wild populations from Africa and Central America show high genetic and biochemical diversity, particularly in oil content and composition, whereas elite breeding lines developed in Brazil, China, and India tend to be more uniform in their oil characteristics and generally exhibit greater stability under environmental stress(de Souza Carneiro et al., 2022; Laviola et al., 2019). These hybrid cultivars are typically selected for their kernel-to-husk ratio, oil biosynthetic enzyme activity, and early maturation. Marker-assisted breeding and QTL mapping are also employed to select genotypes that give high similar oil content, even under local climate fluctuations (Kancharla et al., 2019).

Efforts have been made to systematically align Jatropha curcas genotypes with appropriate agro-climatic regions. For example, Iannella et al. (2021) examined spatial connectivity patterns in Sub-Saharan Africa and proposed zonation models in which genotype performance is matched with region-specific factors such as pest pressure, soil salinity, and rainfall regimes(Iannella et al., 2021). In Mexico and Central America, Jatropha curcas germplasm exhibits high genetic diversity that reflects its role as a center of origin and early domestication, with southern Mexican populations showing distinct genetic structure likely shaped by local environmental gradients and human selection (Li et al., 2017).

National policy frameworks are increasingly incorporating land-use zoning based on agro-ecological compatibility to guide sustainable biofuel development. For example, agro-climatic suitability models in Brazil have matched climatic requirements with regional conditions to identify areas suitable for Jatropha curcas cultivation, helping avoid unsuitable environments and maximize productivity. Similarly, GIS-based land suitability analyses in Mexico have combined climate, soil, and land-use data to delineate potential Jatropha production zones that minimize competition with food crops and support strategic biofuel planning (Yamada & Sentelhas, 2014). Despite its biological potential, one of the main constraints to large-scale adoption is the pronounced year-to-year and regional variability in Jatropha oil yield. Such variability undermines investor confidence and complicates supply chain planning for biofuel projects. Addressing this challenge requires an integrated approach that combines careful genotype selection with improved agronomic management and site-specific adaptation to climatic variability.

14.2.5 Agronomic Interventions for Improved Yield

Optimization of agronomic practice is necessary in order to increase the oil content and yield of *J. curcas* under marginal soils and climatic constraints. As oil content and seed yield are highly sensitive to cultivation environmental conditions, several management intervention experiments have been conducted and validated in diverse agro-ecological conditions.

Organic fertilizer and manure are great for encouraging vegetative development and oil biosynthesis. Composting of neem cake and farmyard manure improved the root elongation, microbial counts, and availability of nutrients that have led to better pod formation and flowering synchrony (Pagani et al., 2024). Particularly in sub-Saharan Africa and South Asia, replacing synthetics with organic fertilizers not only is cost-saving but also increases long-term soil structure and fertility. In addition, the use of phosphorus and potassium at the seedling stages has been linked to increased oil content in seeds due to the roles they play in lipid biosynthesis and energy metabolism (Kakade et al., 2025).

Micro-irrigation management is essential for plant physiology management and optimization of productivity. Drip irrigation under management in semi-arid environments such as India and Ethiopia doubled fruit-to-flower ratio and raised seed oil content by 10–15% compared to rain-fed situations (Safdar et al., 2023). This targeted approach ensures proper water application timing and prevents plant stress, especially at critical phenological stages such as flowering and seed formation.

Intercropping with legumes like pigeon pea (*Cajanus cajan*) and cowpea (*Vigna unguiculata*) has picked up steam because they can fix atmospheric nitrogen, prevent weeds, and enhance soil moisture. Current experiments verified that *Jatropha* when intercropped with legumes yielded up to 18% increase in oil yield compared to monoculture systems (Yadav et al., 2023). The advantage comes from enhanced rhizosphere health and less competition in nutrient absorption. Camelina intercropping with Jatropha plantations was also useful in improving land use efficiency and biomass production without a reduction in oil quality from seeds (Pagani et al., 2024).

TABLE 14.1
The Variations in Oil Content between *Jatropha* Seeds Cultivated in Various Locations and Using Various Methods for Extracting Oil

S. No.	Cultivating Nation	1. Extraction Method	2. Average Oil Yield (% by wt.)	3. Key Reference
1	Botswana	Solvent	42.25	Jonas et al. (2020)
2	Ethiopia	Solvent	47.04	Keneni et al. (2021)
3	Tanzania	Solvent	47.5	Ntalikwa (2021)
4	South Africa	Solvent	56.69	Yusuff (2021)
5	Brazil	Solvent	60.30	Santos et al. (2015)
8	China	Subcritical fluid	40.28	J. Liu et al. (2014)
9	Indonesia	Solvent extraction	39.00	Amalia Kartika et al. (2013)
10	Bangladesh	Mechanical	24.87	Prodhan et al. (2025)
11	Tanzanian	Mechanical	26.02	Aboubakar et al. (2017)
12	Colombia	Mechanical	23.71	Yate et al. (2020)
13	Germany	Mechanical	35.55	Karaj and Müller (2011)
14	Indonesia	Mechanical	32.5	Subroto et al. (2015a,b)
15	Malaysia	Mechanical	27.65	Tambunan et al. (2012)
16	Malaysia	Mechanical	30.8	Subroto et al. (2015a,b)
17	Austria	Enzymatic	47.5	Winkler et al., 1997)
18	Malaysia	Enzymatic	74–75	Shah et al., 2005)

Soil salinity and pH management have also been examined as further strategies in the context of reclaimed or degraded land. Gypsum, biochar, and municipal solid waste compost have been reported to restore soil balance and enable nutrient uptake in saline-affected lands, with promising experiences from northern India and coastal Brazil (Elshamly et al., 2025). These products not only act as buffers against pH but also enhances water-holding capacity and carbon sequestration, aligning with climate-resilient agriculture models.

Finally, agronomic zoning is increasingly well-known as a policy-relevant tool for scaling up *Jatropha* production. With the integration of edaphic, climatic, and market access data, zoning becomes possible to map out priority areas for cultivation. Zoning has already been implemented in some areas of Brazil and Kenya to reduce regional crop-site mismatches (Ogunwole et al., 2019). The variations in oil content between *Jatropha* seeds cultivated in various locations and using various methods for extracting oil are illustrated in Table 14.1.

14.3 *JATROPHA CURCAS* AS A POSSIBLE SUSTAINABLE ENERGY SOURCE

The need to reduce global dependence on fossil fuels and address climate change challenges has breathed new life into the use of renewable and non-food oil crops for biofuel. Of all these, *J. curcas* leads the list as it is oil-rewarding, drought tolerant, and can be cultivated on poor soil that is impossible for food crop production. This unique set of agronomic characteristics identifies *Jatropha* as a biofuel crop that can be utilized to enhance rural development and ecological rehabilitation without aggravating food insecurity. The seeds of *Jatropha* contain approximately 30–40% oil by weight, and the oil has the potential to be efficiently converted into biofuel via transesterification or thermal means such as catalytic pyrolysis (Guedes et al., 2025).

Previous studies point to *Jatropha*'s oil promise to produce renewable hydrocarbons by using catalytic process. Scientists used emerging mesoporous catalysts like Co/KIT-6 to achieve high deoxygenation of *Jatropha* oil, resulting in hydrocarbons of C_{10}–C_{21}, which is suitable for green diesel and bio-kerosene (Guedes et al., 2025). In this process, the proportion of fuel-grade compounds is notably high, while the required reaction temperature is substantially lower than that of conventional thermal cracking, leading to reduced overall energy input. Similarly, Nahar (2025) probed the utilization of *Jatropha* in Bangladesh as a second-generation (2G) bioenergy crop, whereby its application was revealed to address both energy scarcity and climate resilience demands. They further demonstrated that with medium agronomic inputs, *Jatropha* plantations were capable of offering stable biodiesel yields in arid and semi-arid areas while concurrently addressing soil erosion and sequestering atmospheric carbon (Nahar, 2025).

Policy analysis also confirms the socioeconomic viability of *J. curcas*. Aguilar-Aguilar and Mena-Cervantes (2025) evaluated Mexican national biofuel policies and indicated that *J. curcas* is a prominent frontier non-food feedstock with greater similarity to renewable energy objectives and greater rural employment schemes. Their findings indicate the institutional support, seed system enhancement, and value-chain integration to maximize the potential of the crop to the desired extent (Aguilar-Aguilar et al., 2025).

Biophysically, *Jatropha* is also very resilient in unfavorable climatic conditions. de Oliveira Aparecido et al. (2025) used agroecological modeling to show that *Jatropha* would be capable of maintaining stable yields in several South American dryland ecosystems, as well as being more robust and water-frugal than a majority of other energy crops (Eduardo de Oliveira Aparecido et al., 2025). Besides, technical research indicates that the physicochemical properties of *Jatropha* biodiesel satisfy international fuel standards. Desta et al. (2025) determined that *J. curcas* seed oil biodiesel, by using fly ash as a catalyst, has good combustion characteristics in the form of high cetane number and low sulfur content. Their findings suggest that, through localized processing and environmentally sustainable supply chains, Jatropha biodiesel could be a key sector for energy transition policy, most notably in the Global South where access to energy and ecological regeneration are typical policy priorities (Desta et al., 2025). Together, these new results reflect growing consensus among scientists that *Jatropha curcas* offers a technically, economically, and environmentally viable path to decarbonized energy systems.

14.4 *JATROPHA CURCAS* LIMITATIONS AS A CROP FOR BIOFUEL

Despite its promising role as a sustainable source of biofuel, *J. curcas* has several agronomic, environmental, and socioeconomic constraints that limit its broad application. The most critical one is the variability in seed yield, which is primarily driven by genetic heterogeneity, climatic factors, and soil fertility. In contrast to staple crops with matured cultivars, *Jatropha* is underdeveloped as a domesticated crop with focused breeding, which results in unstable and often below-potential yields (Mundale and Malgundkar, 2024).

Moreover, the crop is highly vulnerable to frost and excessive moisture, restricting its growth to some agro-climatic areas. Even though it has been promoted as a drought-tolerant crop, studies have shown that extended water stress significantly reduces oil yield and biomass production (Gomes et al., 2024). In addition, under water-limited conditions, *Jatropha* may prove to be an inferior competitor to other perennial bioenergy crops possessing better water use efficiency (WUE).

Another key weakness of *Jatropha* is vulnerability to disease and pests. Leaf spot, root rot, and insect pests such as the *Scutellera nobilis* bug have been reported to have a significant impact on plant fitness and yield, as well as necessitate chemical controls contradicting its image as an environmentally friendly crop (Anitha and Varaprasad, 2012). Furthermore, problem areas in propagation persist, most specifically regarding seed dormancy as well as poor germination levels. Vegetative propagation by stem cutting, although valuable for clonal uniformity, will result in shallow root and lowered drought tolerance (Mundale and Malgundkar, 2024).

Socioeconomic limitations equally affect the commercialization of *Jatropha*. Early large-scale production, particularly in India and Africa, failed due to overestimated harvests, lack of incentives for farmers, and inefficient processing and distribution networks (Nahar, 2025). Land use conflict is another pressing problem. Although repeatedly promoted to cultivate on "marginal lands," experiences indicate that *Jatropha* yields are not fulfilling large inputs without irrigation, fertilization, and pest control. Hence, for economically viable yields, farmers may divert production to crop land, thereby increasing food crop competition (Slingerland and Schut, 2014).

According to energy economics, *Jatropha's* life cycle energy return on investment (EROI) is variable and may be negative in inefficient systems. Moreover, when oil is extracted, seed cake is useful as organic fertilizer or biogas substrate, and it remains toxic with phorbol esters, making safe usage without costly detoxification processes (Nahar, 2025).

To prevail over these challenges, collaborative research on high-yielding, pest-resistant varieties, improved agro-management, and decentered value-chain development is imperative. Unless these biological, ecological, and economic barriers are surmounted, *Jatropha curcas* will fall short of its promise as a world biofuel crop of sustainability.

14.5 BIOFUEL CONVERSION PRACTICES: JATROPHA OIL AND SEED CAKE

14.5.1 Transesterification

Transesterification is the most prevalent chemical reaction to produce biodiesel from Jatropha oil. During transesterification, triglycerides of the oil undergo a reaction with an alcohol (generally methanol) in the presence of a catalyst (sodium or potassium hydroxide) to produce fatty acid methyl esters (FAME) and glycerol as a byproduct. The resulting FAME is used as a biodiesel that is chemically equivalent to petro-diesel and can be directly used in standard diesel engines with minimal alteration (Asfaw et al., 2025).

The free fatty acid (FFA) content of *Jatropha* oil plays an important role in the efficiency of transesterification. The crude *Jatropha* oil contains typically over 1% FFA, which necessitates a two-stage process of acid esterification followed by base transesterification to prevent soap generation (Majamo et al., 2023). Research by Atabani et al. (2013) attested to the effectiveness of the dual-step process in ensuring yields of over 90% (Atabani et al., 2013). Most recently, the reaction has been optimized by using heterogeneous catalysts and ultrasound-assisted methods, which maximize yield and reaction rate and reduce energy input.

Solid acid catalysts like sulfonated biochar from *Jatropha* seed husk have been capable of reducing FFA content in a pretreatment process, thus improving overall transesterification efficiency (Jayakumar et al., 2022; Corro et al., 2010). Nanocatalysts and enzyme-based systems have also been investigated to make biodiesel production more sustainable and environmentally friendly (Zhao et al., 2016).

Though promising, however, transesterification of *Jatropha* oil is not a problem-free process. Toxic phorbol esters in the oil require proper treatment and waste disposal, especially when one is considering glycerol byproducts. Nonetheless, transesterification still remains a commercially viable route, especially when combined with pretreatment systems and using renewable sources of alcohol. Figure 14.1 shows the conversion process of *Jatropha* seed oil seed cake and biomass to biofuel.

14.5.2 Pyrolysis/Thermal Cracking

Thermal cracking or pyrolysis is another thermochemical process in which biomass (*Jatropha* oil, seed cake, or husks) is heated in the absence of oxygen to produce bio-oil, syngas, and biochar. Pyrolysis, in contrast to transesterification, can handle both lipid and lignocellulosic fractions, making it amenable to handle residual seed cake after oil extraction.

The bio-oil yielded by *Jatropha* biomass and seed cake pyrolysis is complex and contains oxygenated species, phenols, and acids, all of which must be upgraded (e.g., by hydrodeoxygenation) often to improve stability and calorific value. Pyrolysis at 500°C under inert atmosphere can yield up to 65% bio-oil from *Jatropha* biomass. Recent studies on catalytic pyrolysis by utilizing zeolites

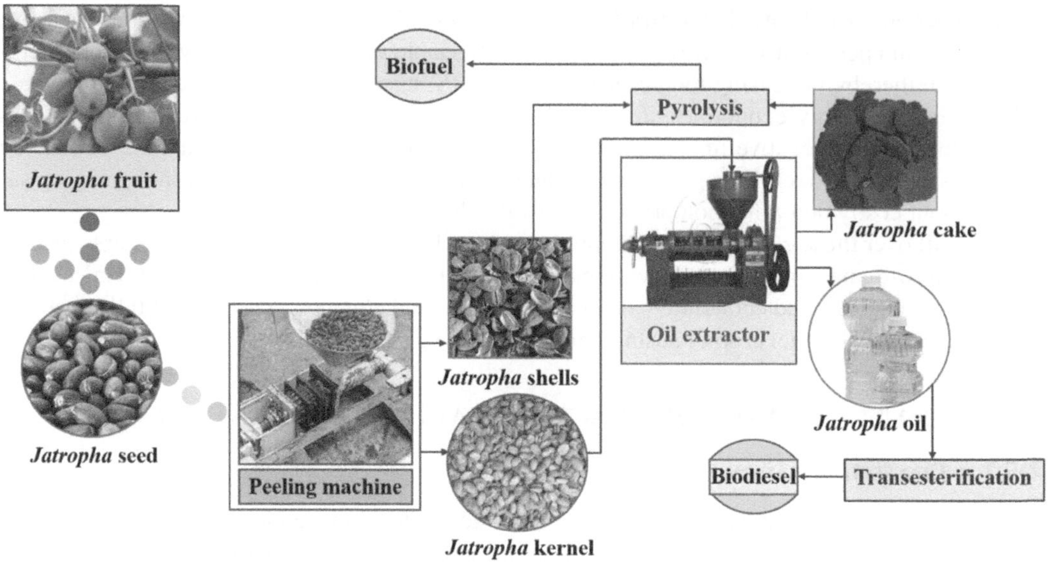

FIGURE 14.1 Biofuel conversion from *Jatropha* seed oil, seed cake, and biomass.

and metal oxides have been discovered to improve the composition and reduce the acidity of bio-oil, thereby rendering it more compatible with combustion engines (Renny et al., 2016).

Pyrolysis of *Jatropha* seed cake using fluidized bed reactors produced bio-oils of improved quality with lower viscosity and higher heating value when catalyzed with CaO and MgO. A solid byproduct biochar has the potential for use as a soil conditioner and as a carbon sequestration material (Rajai et al., 2020).

However, pyrolysis requires more advanced infrastructure than transesterification, i.e., controlled heating, inert gas streams, and downstream units. Pyrolysis energy efficiency and economy largely depend on seed cake preparation, reactor design, and scale-up. Hybrid approaches combining transesterification of oil and pyrolysis of seed cake are a good step toward full-plant utilization with circular bioeconomy objectives (Ruatpuia et al., 2024).

14.6 ECONOMIC AND ENVIRONMENTAL IMPACT

14.6.1 Economic Impact: Promise versus Reality

Initially, Jatropha was lauded for its inedibility and 30–40% oil content. It was a golden promise for the production of biofuels without compromising food security. All African, Asian, and Latin American nations promoted *Jatropha* with huge plantation development plans and small farmer contracts. At the farm level, theoretical models did not materialize, as severe deficiencies in actual yield performance, input costs, and marketability were reported.

In Southern Africa, Kgathi et al. (2017) investigated the performance of *Jatropha* production projects in Botswana and found that most of the smallholder enterprises were economically not sustainable due to the fact that they faced high input costs, low productivity, and a lack of secured market linkages (Kgathi et al., 2017). In India also, Ariza-Montobbio and Lele (2010) illustrated how the yield averaged much below the potential 2 tons per hectare, rendering most plantations economically not feasible (Ariza-Montobbio and Lele, 2010).

Despite these shortcomings, more recent techno-economic assessments are more balanced. Liu et al. (2021) had carried out a China-based life cycle techno-economic analysis and concluded that under optimized conditions (e.g., silvi-pastoral integration, local processing units), *Jatropha*-based biofuel can be as competitive as fossil fuels (Liu et al., 2021). But this success depends enormously on the availability of capital, technology inputs, and land tenure; security.

A significant consideration is the opportunity cost of land. When *Jatropha* takes up productive land, it may directly compete with food crops and hence jeopardize rural livelihood and food security. In addition, implicit subsidies such as government subsidies or zero-cost land leases often skew economic viability estimates. Thus, a truly sustainable economic model must consider land use efficiency, productivity consistency, and end-to-end life cycle costing.

14.6.2 Environmental Impact: Carbon Mitigation or Ecological Burden?

Jatropha's environmental charm is largely owed to its perceived ability to grow on marginal land, sequester carbon, and produce low-carbon biofuel. However, field-based life cycle assessments (LCA) indicate a much more complex environmental reality. Studies such as Gmünder et al. (2012) and Khanam et al. (2021) evaluated the environmental performance of *Jatropha* biodiesel using detailed LCAs. While *Jatropha* biodiesel produced fewer GHG emissions than petro-diesel, its overall environmental benefit was extremely sensitive to farming practices, fertilizer use, and land conversion (Gmünder et al., 2012; Khanam et al., 2021).

`One of the biggest worries is indirect land use change (ILUC). Grasslands or forests conversion into *Jatropha* plantations may lead to net carbon emissions instead of savings. For example, Bailis and Kavlak (2013) had stated that in Brazil's Mato Grosso state, *Jatropha* expansion onto cleared lands could offset any carbon benefit for 30 years (Bailis and Kavlak, 2013). Moreover, *Jatropha* cultivation often involves high chemical and water inputs in semi-arid areas to maintain yield, thereby eliminating its supposed low-input crop status. Van Eijck et al. (2014) noted that pesticide and fertilizer use had a high cost on eutrophication and eco-toxicity potential, especially in supply chains based on plantations (van Eijck et al., 2014). In particular, integrated multi-output systems (like seedcake biogas with biodiesel or biochar for land conditioning) can significantly enhance environmental performance. Abdudeen et al. (2023) promoted such circular bioeconomy ideas to transform *Jatropha's* environmental footprint from linear deterioration toward regenerative potential (Abdudeen et al., 2023).

14.7 LIFE CYCLE ASSESSMENT (LCA)

Life cycle assessment (LCA) is a robust comparative measure of environmental performance of *J. curcas* biodiesel, a non-food oilseed promoted for sustainable fuel production potential. LCA makes explicit genuine benefits and trade-offs in terms of GHG emissions, land use, water use, energy input, and biodiversity impacts by considering the whole life cycle, from crop plant to end use of fuel. Previous study proves that *Jatropha* biodiesel has the potential to produce 60%–80% less GHGs than fossil diesel when it is cultivated on degraded land with low inputs (Rattanaphra et al., 2024). Deforestation or grassland conversion for land use for *Jatropha* plantation, on the other hand, would incur a carbon debt that would be repaid in decades (Portugal-Pereira et al., 2016).

While *Jatropha* is intended to be grown on marginal land, commercially viable yields typically require irrigation and fertilizers. LCA by Pérez and Islas-Samperio (2021) revealed that sustainable GHG savings are only achievable under optimized input and water management. In water-scarce areas, though, the crop's high water demand can negate its advantages, as seen in comparative studies with palm oil in Malaysia (Lam et al., 2009). Soil degradation and biodiversity loss are also concerns, especially in monoculture plantations. Pandey et al. (2011) found deteriorating soil health under high-input systems, whereas integrated systems with intercrops or livestock had more favorable ecological outcomes (Pandey et al., 2011).

A concurrent LCA–LCC approach by Liu et al. (2024) indicated that, while *Jatropha* biodiesel is not commercially viable in the absence of policy support, it leads on environmental metrics including non-renewable energy consumption and acidifying potential (Liu et al., 2024).

Such findings highlight that sustainability outcomes are highly context-specific, affected by factor like land use history, cultivation practices, input intensity, and processing systems. Furthermore, the majority of LCA researches overlook temporal attributes such as carbon payback times, demanding dynamic models incorporating long-term as well as social dimensions of *Jatropha*-based biodiesel systems.

14.8 FUTURE OUTLOOK AND STRATEGIC RESEARCH DIRECTIONS

J. curcas was marketed on its bioenergy crop potential in terms of adaptability, oil yields, and non-food status, but functional limits, low yields, susceptibility to pests, and socioeconomic factors, have seen its contribution then re-appraised. A shift is occurring from large-scale promotion to local, evidence-led deployment strategies. Abdudeen et al. (2023) emphasize that *Jatropha* viability is now dependent on region-specific approaches with respect to Genotype-–environment interactions, management of the cost of inputs, and full value-chain integration. Instead of monocultures, agro-ecological systems like intercropping and integration of biorefineries are considered more sustainable. The application of the *Jatropha* is also expanding on the application of bioplastics, biochar, and green chemicals, thereby making it more relevant in a circular bioeconomy (Abdudeen et al., 2023).

Strategic research now includes breeding for yield stability and stress tolerance by utilizing QTL mapping and molecular markers. Arockiasamy et al. (2021) aligned stress- and oil content-enhancing QTLs for accurate breeding. Johnson et al. (2011) studied biotech technologies such as CRISPR and RNAi that are aimed at higher oil yield, phorbol ester pathway silencing resulting in detoxification of seedcake, and increased pest resistance. However, the lack of domestication is a significant concern for high-yielding varieties. Montes and Melchinger (2016) highlight germplasm collection, genetic diversity preservation, and gene bank development as the center of their attention.

Agronomic models specific to climate and soil type are required. Intercropping, dryland farming, and organic manure can enhance sustainability. Liu et al. (2024) suggest the incorporation of *Jatropha* in agroforestry systems for maximal land use and enhancing ecosystem services (Liu et al., 2024). Socioeconomically, inclusive policy structures ensuring land rights, making smallholder inclusion possible, and implementing certification schemes like ISCC and RSB are required for equitable and sustainable value chains (Bailis et al., 2014).

Lastly, integrated biorefineries are a promising opportunity. By producing biodiesel, biochar, fertilizers, and biogas together, these systems can minimize environmental footprints while enhancing rural access to energy. Life cycle cost-benefit analysis, e.g., by Abdudeen et al. (2023), suggests that such models, especially those with incorporation of waste valorization and energy recovery, can offer scalable solutions to climate resilience and local development (Abdudeen et al., 2023).

14.9 CONCLUSION

Jatropha curcas shows strong promise as a sustainable biofuel source, but its success depends on solutions tailored to specific contexts. Its high non-food oil content and resilience make it a viable alternative to edible oils for biofuel, yet practical experience highlights challenges such as inconsistent yields, pest susceptibility, and economic constraints. Conversion technologies like transesterification and pyrolysis are effective for fuel production, but careful management of environmental impacts, particularly land use and greenhouse gas emissions, is essential. Life cycle assessments indicate that cultivating Jatropha on degraded, low-input lands can achieve net climate benefits. Future research should prioritize genetic improvement, integrated farming practices, and the development of robust value chains. With sound policies and strategic investment, Jatropha curcas could play a meaningful role in advancing low-carbon energy transitions and supporting rural development.

COMPETING INTERESTS

The authors declare that they have no competing interest.

DECLARATION OF GENERATIVE AI AND AI-ASSISTED TECHNOLOGIES IN THE WRITING PROCESS

During the preparation of this work, the book chapter author(s) have not taken any assistance from AI, and the book chapter author(s) reviewed and edited the content as needed and take(s) full responsibility for the content of the publication.

REFERENCES

Abdudeen, A., Selim, M. Y. E., Sekar, M., & Elgendi, M. (2023). Jatropha's rapid developments and future opportunities as a renewable source of biofuel—A review. *Energies*, *16*(2). https://doi.org/10.3390/en16020828

Aboubakar, X., Goudoum, A., Bébé, Y., & Mbofung, C. M. F. (2017). Optimization of *Jatropha curcas* pure vegetable oil production parameters for cooking energy. *South African Journal of Chemical Engineering*, 24, 196–212. https://doi.org/10.1016/j.sajce.2017.09.002

Aguilar-Aguilar, F. A., Mena-Cervantes, V. Y., & Hernández-Altamirano, R. (2025). Analysis of public policies and resources for biodiesel production in México. *Biomass and Bioenergy*, 196, 107762. https://doi.org/10.1016/j.biombioe.2025.107762

Amalia Kartika, I., Yani, M., Ariono, D., Evon, Ph., & Rigal, L. (2013). Biodiesel production from jatropha seeds: Solvent extraction and in situ transesterification in a single step. *Fuel*, 106, 111–117. https://doi.org/10.1016/j.fuel.2013.01.021

Anitha, K., & Varaprasad, K. S. (2012). Jatropha pests and diseases: An overview. In B. Bahadur, M. Sujatha, & N. Carels (Eds.), *Jatropha, Challenges for a New Energy Crop: Volume 1: Farming, Economics and Biofuel* (pp. 175–218). Springer New York. https://doi.org/10.1007/978-1-4614-4806-8_10

Ariza-Montobbio, P., & Lele, S. (2010). Jatropha plantations for biodiesel in Tamil Nadu, India: Viability, livelihood trade-offs, and latent conflict. *Ecological Economics*, 70(2), 189–195. https://doi.org/10.1016/j.ecolecon.2010.05.011

Arockiasamy, S., Kumpatla, J., Hadole, S., Yepuri, V., Patil, M., Shrivastava, V., Rao, C., Kancharla, N., Jalali, S., Varshney, A., Madan, N., Pothakani, S., Nair, V., Peyyala, S., Mudunuri, V., Gopal, A., Kumar, N. S., Pachiyannan, J., Seelamanthula, S., … Dasgupta, S. (2021). Breeding and biotechnological efforts in *Jatropha curcas* L. for sustainable yields. *Oil Crop Science*, 6(4), 180–191. https://doi.org/10.1016/j.ocsci.2021.10.004

Arunyanark, A., Foytong, K., Jompuk, C., Srinives, P., & Tanya, P. (2025). Correlation and effect of pruning and spacing on jatropha flower, fruit, and seed performances. Agriculture and Natural Resources, 59(2), 590212. https://doi.org/10.34044/j.anres.2025.59.2.12

Asfaw, B. T., Gari, M. T., & Jayakumar, M. (2025). Transesterification of biodiesel from non-edible oils using heterogeneous base catalysts: A comprehensive review of potential renewable biomass feedstocks. *Chemical Engineering Journal*, 511, 162028. https://doi.org/10.1016/j.cej.2025.162028

Atabani, A. E., Silitonga, A. S., Ong, H. C., Mahlia, T. M. I., Masjuki, H. H., Badruddin, I. A., & Fayaz, H. (2013). Non-edible vegetable oils: A critical evaluation of oil extraction, fatty acid compositions, biodiesel production, characteristics, engine performance and emissions production. *Renewable and Sustainable Energy Reviews*, 18, 211–245. https://doi.org/10.1016/j.rser.2012.10.013

Aziz, N. A. M., Yunus, R., Kania, D., & Abd Hamid, H. (2021). Prospects and challenges of microwave-combined technology for biodiesel and biolubricant production through a transesterification: A review. *Molecules*, *26*(4). https://doi.org/10.3390/molecules26040788

Bailis, R., & Kavlak, G. (2013). Environmental implications of Jatropha biofuel from a Silvi-pastoral production system in Central-West Brazil. *Environmental Science & Technology*, 47(14), 8042–8050. https://doi.org/10.1021/es303954g

Bailis, R., Solomon, B. D., Moser, C., & Hildebrandt, T. (2014). Biofuel sustainability in Latin America and the Caribbean – A review of recent experiences and future prospects. *Biofuels*, 5(5), 469–485. https://doi.org/10.1080/17597269.2014.992001

Bakhshabadi, H., Ganje, M., Gharekhani, M., Mohammadi-Moghaddam, T., Aulestia, C., & Morshedi, A. (2025). A review of new methods for extracting oil from plants to enhance the efficiency and physicochemical properties of the extracted oils. *Processes*, *13*(4). https://doi.org/10.3390/pr13041124

Bedru, T. K., Meshesha, B. T., & Mohammed, S. A. (2024). Extraction of *Argemone ochroleuca* seeds oil and parametric optimization for biodiesel and epoxy oil production. *Biomass Conversion and Biorefinery*. https://doi.org/10.1007/s13399-024-05874-x

Corro, G., Tellez, N., Ayala, E., & Marinez-Ayala, A. (2010). Two-step biodiesel production from *Jatropha curcas* crude oil using SiO_2·HF solid catalyst for FFA esterification step. *Fuel*, 89(10), 2815–2821. https://doi.org/10.1016/j.fuel.2010.04.023

de Souza Carneiro, A., dos Santos, A., Laviola, B. G., Teodoro, L. P. R., Teodoro, P. E., & Rodrigues, E. V. (2022). Genetic diversity and population structure in Jatropha (*Jatropha curcas* L.) based on molecular markers. *Genetic Resources and Crop Evolution*, 69(1), 245–254. https://doi.org/10.1007/s10722-021-01224-2

Desta, T. G., Gebresilasie, G. G., Meressa, G. G., Abraha, S. K., Weldeslassie, M. W., Abdu, K. Y., Endris, Y. A., & Shah, M. A. (2025). Production and characterization of biodiesel from *Jatropha curcas* seed oil by using fly ash as a catalyst. *ACS Omega*, 10(24), 25498–25505. https://doi.org/10.1021/acsomega.5c00712

Eduardo de Oliveira Aparecido, L., Torsoni, G. B., Costa Baratti, A. C., Ferreira de Melo Rossi, M., Lorençone, J. A., Lorençone, P. A., & Castro, R. P. (2025). Agroecological vulnerability and climate resilience of *Jatropha curcas* under anthropogenic climate extremes in South America. *Biomass and Bioenergy*, 194, 107654. https://doi.org/10.1016/j.biombioe.2025.107654

Elshamly, A. M. S., Abaza, A. D. S., Mustafa, A. E.-Z. M. A., Rizwana, H., Iqbal, J., Ahmad, S., Iqbal, R., & Gruda, N. S. (2025). Synergistic effect of ammonium and potassium on carrot growth, physio-biochemical mechanisms, and water use efficiency under varying irrigation regimes. *Scientific Reports*, 15(1), 16151. https://doi.org/10.1038/s41598-025-00690-3

Garuma, W. B., Bedru, T. K., Gindaba, G. T., Jayakumar, M., Edae, I. T., Meshesha, B. T., Mohammed, S. A., & Demesa, A. G. (2025). Optimization and characterization of ultrasonically extracted oil from watermelon seeds using response surface methodology: A pathway to biodiesel and epoxy oil production. *Biomass Conversion and Biorefinery*. https://doi.org/10.1007/s13399-025-06746-8

Gizaw, D. G., Periyasamy, S., Beula Isabel, J., Hunegnaw, B. M., Jayakumar, M., Asaithambi, P., Gupta, S. K., & Sivashanmugam, P. (2025). Potential oil sources, conversion routes and process limiting factors for sustainable biodiesel production – A comprehensive outlook. *Journal of the Taiwan Institute of Chemical Engineers*, 106240. https://doi.org/10.1016/j.jtice.2025.106240

Gmünder, S., Singh, R., Pfister, S., Adheloya, A., & Zah, R. (2012). Environmental impacts of Jatropha curcas biodiesel in India. *Journal of Biomedicine and Biotechnology*, *2012*. https://doi.org/10.1155/2012/623070

Gomes, F. P., Griffiths, H., Gomes, L. M. C., & Laviola, B. G. (2024). Carbon isotope composition of biomass and water use efficiency in young plants of *Jatropha curcas* L. under irrigated or water deficit conditions. *South African Journal of Botany*, 174, 624–633. https://doi.org/10.1016/j.sajb.2024.09.037

Guedes, J. M., Garcia, K. G. V., Araújo, A. M. de M., Gondim, A. A. D., Fernandes, F. R. D., Dos Santos, A. P. B., Caldeira, V. P. da S., Dos Santos, H. S., & Santos, A. G. D. (2025). Renewable hydrocarbons obtained by catalytic pyrolysis of *Jatropha curcas* L. oil. *Quimica Nova*, *48*(4). https://doi.org/10.21577/0100-4042.20250111

Iannella, M., De Simone, W., Cerasoli, F., D'Alessandro, P., & Biondi, M. (2021). A continental-scale connectivity analysis to predict current and future colonization trends of biofuel plant's pests for sub-Saharan African countries. *Land*, *10*(11). https://doi.org/10.3390/land10111276

Jayakumar, M., Gebeyehu, K. B., Selvakumar, K. V., Parvathy, S., Kim, W., & Karmegam, N. (2022). Waste Ox bone based heterogeneous catalyst synthesis, characterization, utilization and reaction kinetics of biodiesel generation from Jatropha curcas oil. *Chemosphere*, 288, 132534. https://doi.org/10.1016/j.chemosphere.2021.132534

Johnson, T. S., Eswaran, N., & Sujatha, M. (2011). Molecular approaches to improvement of *Jatropha curcas* Linn. as a sustainable energy crop. *Plant Cell Reports*, 30(9), 1573–1591. https://doi.org/10.1007/s00299-011-1083-1

Jonas, M., Ketlogetswe, C., & Gandure, J. (2020). Variation of *Jatropha curcas* seed oil content and fatty acid composition with fruit maturity stage. *Heliyon*, 6(1), e03285. https://doi.org/10.1016/j.heliyon.2020.e03285

Kakade, P. S., Suryawanshi, Y., & Patil, R. B. (2025). Sustainable era of oilseed plants. In *Oilseed Crops* (pp. 1–24). John Wiley & Sons, Ltd. https://doi.org/10.1002/9781394186426.ch1

Kancharla, N., Jalali, S., Narasimham, J. V, Nair, V., Yepuri, V., Thakkar, B., Reddy, V. B., Kuriakose, B., Madan, N., & S, A. (2019). De novo sequencing and hybrid assembly of the biofuel crop *Jatropha curcas* L.: Identification of quantitative trait loci for geminivirus resistance. *Genes*, *10*(1). https://doi.org/10.3390/genes10010069

Karaj, S., & Müller, J. (2011). Optimizing mechanical oil extraction of *Jatropha curcas* L. seeds with respect to press capacity, oil recovery and energy efficiency. *Industrial Crops and Products*, 34(1), 1010–1016. https://doi.org/10.1016/j.indcrop.2011.03.009

Keneni, Y. G., Bahiru, L. A., & Marchetti, J. M. (2021). Effects of different extraction solvents on oil extracted from Jatropha seeds and the potential of seed residues as a heat provider. *BioEnergy Research*, 14(4), 1207–1222. https://doi.org/10.1007/s12155-020-10217-5

Kgathi, D. L., Mmopelwa, G., Chanda, R., Kashe, K., & Murray-Hudson, M. (2017). A review of the sustainability of Jatropha cultivation projects for biodiesel production in southern Africa: Implications for energy policy in Botswana. *Agriculture, Ecosystems & Environment, 246*, 314–324. https://doi.org/10.1016/j.agee.2017.06.014

Khanam, T., Khalid, F., Manzoor, W., Rashedi, A., Hadi, R., Ullah, F., Rehman, F., Akhtar, A., Karthik Babu, N. B., & Hussain, M. (2021). Environmental sustainability assessment of biodiesel production from *Jatropha curcas* L. seeds oil in Pakistan. *PLoS ONE, 16*(11 November). https://doi.org/10.1371/journal.pone.0258409

Khater, E.-S. G., AbdAlla, S. A., Bahnasawy, A. H., & AbuHashish, H. M. (2024). Improvement of the production of bio-oil and biodiesel from Egyptian Jatropha seeds by using microwave and ultrasonic. *Scientific Reports, 14*(1), 1882. https://doi.org/10.1038/s41598-024-51579-6

Lam, M. K., Lee, K. T., & Mohamed, A. R. (2009). Life cycle assessment for the production of biodiesel: A case study in Malaysia for palm oil versus jatropha oil. *Biofuels, Bioproducts and Biorefining*, 3(6), 601–612. https://doi.org/10.1002/bbb.182

Lavenburg, V. M., Rosentrater, K. A., & Jung, S. (2021). Extraction methods of oils and phytochemicals from seeds and their environmental and economic impacts. *Processes, 9*(10). https://doi.org/10.3390/pr9101839

Laviola, B. G., Rodrigues, E., Ribeiro, L. P., da Silva, L. Q. M., de Azevedo Peixoto, L., & Bhering, L. L. (2019). Strategies in the genetic breeding of *Jatropha curcas* for biofuel production in Brazil. In B. Bahadur, M. Sujatha, & N. Carels (Eds.), *Jatropha, Challenges for a New Energy Crop: Volume 3: A Sustainable Multipurpose Crop* (pp. 45–62). Springer Singapore. https://doi.org/10.1007/978-981-13-3104-6_3

Li, H., Tsuchimoto, S., Harada, K., Yamasaki, M., Sakai, H., Wada, N., Alipour, A., Sasai, T., Tsunekawa, A., Tsujimoto, H., Ando, T., Tomemori, H., Sato, S., Hirakawa, H., Quintero, V. P., Zamarripa, A., Santos, P., Hegazy, A., Ali, A. M., & Fukui, K. (2017). Genetic Tracing of Jatropha curcas L. from Its Mesoamerican Origin to the World. *Frontiers in Plant Science, 8*, 1539. https://doi.org/10.3389/fpls.2017.01539

Lima-Pereira, Y., de Souza, E. M. O., dos Reis, D. S., Barcellos-Silva, I. G. C., Miki, K. S. L., Veiga-Júnior, V. F., & Teixeira-Costa, B. E. (2025). Current perspectives on the extraction, isolation, and identification of fats and fatty acids using conventional and green methods. *Separations, 12*(6). https://doi.org/10.3390/separations12060160

Liu, H., Zhang, C., Tian, H., Li, L., Wang, X., & Qiu, T. (2021). Environmental and techno-economic analyses of bio-jet fuel produced from jatropha and castor oilseeds in China. *The International Journal of Life Cycle Assessment*, 26(6), 1071–1084. https://doi.org/10.1007/s11367-021-01914-0

Liu, J., Chen, P., He, J., Deng, L., Wang, L., Lei, J., & Rong, L. (2014). Extraction of oil from *Jatropha curcas* seeds by subcritical fluid extraction. *Industrial Crops and Products*, 62, 235–241. https://doi.org/10.1016/j.indcrop.2014.08.039

Liu, J., Gasmalla, M. A. A., Li, P., & Yang, R. (2016). Enzyme-assisted extraction processing from oilseeds: Principle, processing and application. *Innovative Food Science & Emerging Technologies*, 35, 184–193. https://doi.org/10.1016/j.ifset.2016.05.002

Liu, Y., Zhu, Z., Zhang, R., & Zhao, X. (2024). Life cycle assessment and life cycle cost analysis of Jatropha biodiesel production in China. *Biomass Conversion and Biorefinery*, 14(22), 28635–28660. https://doi.org/10.1007/s13399-022-03614-7

Majamo, S. L., Amibo, T. A., & Bedru, T. K. (2023). Synthesis and application of biomass-derived magnetic biochar catalyst for simultaneous esterification and trans-esterification of waste cooking oil into biodiesel: Modeling and optimization. *Materials for Renewable and Sustainable Energy*, 12(2), 147–158. https://doi.org/10.1007/s40243-023-00236-5

Martínez Herrera, J., Sánchez-Chino, X., Corzo-Ríos, L. J., Dávila-Ortiz, G., & Jiménez Martínez, C. (2019). Comparative extraction of *Jatropha curcas* L. lipids by conventional and enzymatic methods. *Food and Bioproducts Processing*, 118, 32–39. https://doi.org/https://doi.org/10.1016/j.fbp.2019.08.013

Montes, J. M., & Melchinger, A. E. (2016). Domestication and breeding of *Jatropha curcas* L. *Trends in Plant Science*, 21(12), 1045–1057. https://doi.org/10.1016/j.tplants.2016.08.008

Mundale, P. R., & Malgundkar, H. K. (2024). A comprehensive overview on *Jatropha curcas* Linn. *International Journal of Pharmacognosy*, 11(12), 661 674. https://doi.org/10.13040/IJPSR.0975-8232.IJP.11(12).664-74

Mursalykova, M., Kakimov, M., Kassenov, A., Iskakov, B., Sergibayeva, Z., Kaspakov, E., Zhumadilova, G., Shulenova, A., Kokayeva, G., & Suychinov, A. (2023). Mathematical modeling of screw press configuration for processing safflower oil. *Applied Sciences, 13*(5). https://doi.org/10.3390/app13053057

Nahar, K. (2025). Biofuel crop to combat with climate change. In *Jatropha curcas L: A Potential 2G Energy Crop to Produce Biofuel in Bangladesh: Agronomy, Biotechnology, Biodiesel and Byproducts* (pp. 81–89). Springer Nature Switzerland. https://doi.org/10.1007/978-3-031-77642-7_6

Nde, D. B., & Foncha, A. C. (2020). Optimization methods for the extraction of vegetable oils: A review. *Processes*, *8*(2). https://doi.org/10.3390/pr8020209

Neupane, D., Bhattarai, D., Ahmed, Z., Das, B., Pandey, S., Solomon, J. K. Q., Qin, R., & Adhikari, P. (2021). Growing Jatropha (*Jatropha curcas* L.) as a potential second-generation biodiesel feedstock. *Inventions*, *6*(4). https://doi.org/10.3390/inventions6040060

Ntalikwa, J. W. (2021). Solvent extraction of Jatropha oil for biodiesel production: Effects of solvent-to-solid ratio, particle size, type of solvent, extraction time, and temperature on oil yield. *Journal of Renewable Energy*, 2021(1), 9221168. https://doi.org/10.1155/2021/9221168

Ogunwole, J. O., Alabi, O., Ugbabe, O., & Birhanu, B. Z. (2019). Promoting Jatropha agriculture for sustainable soil capital improvement: A win-win technology for rehabilitating degraded lands in Africa. In N. F. Unuigbe & E. T. Ayuk (Eds.), *New Frontiers in Natural Resources Management in Africa* (pp. 27–39). Springer International Publishing. https://doi.org/10.1007/978-3-030-11857-0_3

Pagani, E., Zanetti, F., Ferioli, F., Facciolla, E., & Monti, A. (2024). Camelina intercropping with pulses a sustainable approach for land competition between food and non-food crops. *Agronomy*, *14*(6). https://doi.org/10.3390/agronomy14061200

Pandey, K. K., Pragya, N., & Sahoo, P. K. (2011). Life cycle assessment of small-scale high-input Jatropha biodiesel production in India. *Applied Energy*, 88(12), 4831–4839. https://doi.org/10.1016/j.apenergy.2011.06.026

Pérez, G., & Islas-Samperio, J. M. (2021). Sustainability evaluation of non-toxic *Jatropha curcas* in rural marginal soil for obtaining biodiesel using life-cycle assessment. *Energies*, *14*(10). https://doi.org/10.3390/en14102746

Pompelli, M. F., Seleiman, M. F., Dey, P., Suarez-Padrón, I. E., Zumaqué, L. E. O., Jarma-Orozco, A., Jaraba-Navas, J., Pineda-Rodríguez, Y. Y., Rodríguez-Páez, L. A., & Kido, É. A. (2025). MALDI-ToF/ToF-MS detection of differential protein expression and metabolomic profiles in *Jatropha curcas* under salinity: Advancing the understanding of salt stress mechanisms. *Journal of Agronomy and Crop Science*, 211(3), e70058. https://doi.org/10.1111/jac.70058

Portugal-Pereira, J., Nakatani, J., Kurisu, K., & Hanaki, K. (2016). Life cycle assessment of conventional and optimised Jatropha biodiesel fuels. *Renewable Energy*, 86, 585–593. https://doi.org/10.1016/j.renene.2015.08.046

Prodhan, A., Hasan, Md. I., Sujan, S. M. A., Hossain, M., Quaiyyum, Md. A., & Ismail, M. (2025). Production and characterization of biodiesel from Jatropha (*Jatropha curcas*) seed oil available in Bangladesh. *Energy Sources, Part A: Recovery, Utilization, and Environmental Effects*, 47(1), 3343–3358. https://doi.org/10.1080/15567036.2020.1851819

Rajai, V., Shah, H., Patel, D., Himanshu, P., & Maiti, S. (2020). Generation and characterization of bio-oil through slow pyrolysis process from *Jatropha curcas* shell. In A. K. Parwani, P. L. Ramkumar, K. Abhishek, & S. K. Yadav (Eds.), *Recent Advances in Mechanical Infrastructure* (pp. 143–151). Springer Singapore.

Ramkat, R. C., Calari, A., Maghuly, F., & Laimer, M. (2011). Biotechnological approaches to determine the impact of viruses in the energy crop plant *Jatropha curcas*. *Virology Journal*, 8(1), 386. https://doi.org/10.1186/1743-422X-8-386

Rattanaphra, D., Tawkaew, S., Chuichulcherm, S., Kingkam, W., Nuchdang, S., Kitpakornsanti, K., & Suwanmanee, U. (2024). Evaluation of life cycle assessment of Jatropha biodiesel processed by esterification of Thai domestic rare earth oxide catalysts. *Sustainability*, *16*(1). https://doi.org/10.3390/su16010100

Renny, A., Santhosh, V., Somkuwar, N., Gokak, D. T., Sharma, P., & Bhargava, S. (2016). Pyrolysis of de-oiled seed cake of *Jatropha curcas* and catalytic steam reforming of pyrolytic bio-oil to hydrogen. *Bioresource Technology*, 220, 151–160. https://doi.org/10.1016/j.biortech.2016.08.052

Riayatsyah, T. M. I., Sebayang, A. H., Silitonga, A. S., Padli, Y., Fattah, I. M. R., Kusumo, F., Ong, H. C., & Mahlia, T. M. I. (2022). Current progress of *Jatropha curcas* commoditisation as biodiesel feedstock: A comprehensive review. *Frontiers in Energy Research*, *9–2021*. https://doi.org/10.3389/fenrg.2021.815416

Rollin, S., Gupta, A., Franco, C. M. M., Singh, S., & Puri, M. (2023). Development of sustainable downstream processing for nutritional oil production. *Frontiers in Bioengineering and Biotechnology*, 11*–2023*. https://doi.org/10.3389/fbioe.2023.1227889

Ruatpuia, J. V. L., Halder, G., Vanlalchhandama, M., Lalsangpuii, F., Boddula, R., Al-Qahtani, N., Niju, S., Mathimani, T., & Rokhum, S. L. (2024). *Jatropha curcas* oil a potential feedstock for biodiesel production: A critical review. *Fuel*, 370, 131829. https://doi.org/10.1016/j.fuel.2024.131829

Safdar, M. E., Qamar, R., Javed, A., Nadeem, M. A., Javeed, H. M. R., Farooq, S., Głowacka, A., Michałek, S., Alwahibi, M. S., Elshikh, M. S., & Ahmed, M. A. A. (2023). Combined application of boron and zinc improves seed and oil yields and oil quality of oilseed rape (*Brassica napus* L.). *Agronomy*, *13*(8). https://doi.org/10.3390/agronomy13082020

Santos, S. B. dos, Martins, M. A., Caneschi, A. L., Aguilar, P. R. M., & Coimbra, J. S. dos R. (2015). Kinetics and thermodynamics of oil extraction from *Jatropha curcas* L. using ethanol as a solvent. *International Journal of Chemical Engineering*, 2015(1), 871236. https://doi.org/10.1155/2015/871236

Shah, S., Sharma, A., & Gupta, M. N. (2005). Extraction of oil from *Jatropha curcas* L. seed kernels by combination of ultrasonication and aqueous enzymatic oil extraction. *Bioresource Technology*, 96(1), 121–123. https://doi.org/10.1016/j.biortech.2004.02.026

Slingerland, M., & Schut, M. (2014). Jatropha developments in mozambique: Analysis of structural conditions influencing niche-regime interactions. *Sustainability*, 6(11), 7541–7563. https://doi.org/10.3390/su6117541

Subroto, E., Manurung, R., Heeres, H. J., & Broekhuis, A. A. (2015a). Mechanical extraction of oil from *Jatropha curcas* L. kernel: Effect of processing parameters. *Industrial Crops and Products*, 63, 303–310. https://doi.org/10.1016/j.indcrop.2014.06.018

Subroto, E., Manurung, R., Heeres, H. J., & Broekhuis, A. A. (2015b). Optimization of mechanical oil extraction from *Jatropha curcas* L. kernel using response surface method. *Industrial Crops and Products*, 63, 294–302. https://doi.org/10.1016/j.indcrop.2014.08.050

Tadano, S., Chiyapo, G., Ishimoto, Y., Konaka, T., Mazereku, C., Tsujimoto, H., & Akashi, K. (2021). Multivariate analysis of seed chemical diversity among *Jatropha curcas* in Botswana. *Agronomy*, *11*(8). https://doi.org/10.3390/agronomy11081570

Tambunan, A. H., Situmorang, J. P., Silip, J. J., Joelianingsih, A., & Araki, T. (2012). Yield and physicochemical properties of mechanically extracted crude *Jatropha curcas* L. oil. *Biomass and Bioenergy*, 43, 12–17. https://doi.org/10.1016/j.biombioe.2012.04.004

van Eijck, J., Romijn, H., Smeets, E., Bailis, R., Rooijakkers, M., Hooijkaas, N., Verweij, P., & Faaij, A. (2014). Comparative analysis of key socio-economic and environmental impacts of smallholder and plantation based jatropha biofuel production systems in Tanzania. *Biomass and Bioenergy*, 61, 25–45. https://doi.org/10.1016/j.biombioe.2013.10.005

Wasnik, G., & Singh, K. (2024). Effect of plantation season on growth of different genotypes of Jatropha curcas as controlled irrigation and Mycorrhizal association under semi-arid region. *International Journal of Research in Agronomy*, *7*(5), 244–250. https://doi.org/10.33545/2618060X.2024.v7.i5d.683

Winkler, E., Foidl, N., Gübitz, G. M., Staubmann, R., & Steiner, W. (1997). Enzyme-supported oil extraction from *Jatropha curcas* Seeds. *Applied Biochemistry and Biotechnology*, 63(1), 449–456. https://doi.org/10.1007/BF02920445

Yadav, R. K., Kumar, R., Singh, A., & Dagar, J. C. (2023). Utilization of saline and other poor-quality waters to sustain agroforestry production. In J. C. Dagar, S. R. Gupta, & G. W. Sileshi (Eds.), *Agroforestry for Sustainable Intensification of Agriculture in Asia and Africa* (pp. 243–271). Springer Nature Singapore. https://doi.org/10.1007/978-981-19-4602-8_8

Yamada, E. S. M., & Sentelhas, P. C. (2014). Agro-climatic zoning of Jatropha curcas as a subside for crop planning and implementation in Brazil. International Journal of Biometeorology, 58(9), 1995–2010. https://doi.org/10.1007/s00484-014-0803-y

Yang, L., Zhou, J., Pan, C., Tu, Q., Wang, G., & Duan, J. (2024). Environmental, energy, and economic assessment of *Jatropha curcas* L. biodiesel production in China. *Journal of Cleaner Production*, 470, 143286. https://doi.org/10.1016/j.jclepro.2024.143286

Yate, A. V, Narváez, P. C., Orjuela, A., Hernández, A., & Acevedo, H. (2020). A systematic evaluation of the mechanical extraction of *Jatropha curcas* L. oil for biofuels production. *Food and Bioproducts Processing*, 122, 72–81. https://doi.org/10.1016/j.fbp.2020.04.001

Yuan, M.-H., Chang, C.-C., Hsu, T.-C., Shie, J.-L., Chen, Y.-H., Chang, C.-Y., Lin, C.-F., Yu, C.-P., Wu, C.-H., Do, M. Van, Lin, F.-C., Lee, D.-J., Liu, B.-L., Chen, Y.-H., & Huang, M. (2021). A technical analysis of solid recovered fuel from torrefied Jatropha seed residue via a two-stage mechanical screw press and solvent extraction process. *Energies*, *14*(23). https://doi.org/10.3390/en14237876

Yusuff, A. S. (2021). Parametric optimization of solvent extraction of Jatropha curcas seed oil using design of experiment and its quality characterization. *South African Journal of Chemical Engineering*, 35(November 2020), 60–68. https://doi.org/10.1016/j.sajce.2020.11.006

Zhao, K., Di, Q., Cao, X., Wang, M., Deng, L., & Wang, F. (2016). Production of biodiesel using immobilized lipase and the characterization of different co-immobilizing agents and immobilization methods. *Sustainability*, *8*(9). https://doi.org/10.3390/su8090764

15 Palm as an Efficient Biofuel Producer

A Critical Analysis for Sustainable Uses

Desta Getachew Gizaw, Zinnabu Tassew Redda, Leta Lemma Sharbe, Andualem Bahiru Abera, and Selvakumar Periyasamy

15.1 INTRODUCTION

It is evident that having access to reliable and efficient energy supplies is essential to both social growth and national well-being, given the close relationship involving energy use and advancement in society as well as the crucial part that energy plays in national development (Mignogna et al., 2024). The achievement of the 2030 Agenda for Sustainable Development depends on having reliable and reasonably priced energy, which makes it possible to reduce food insecurity and poverty, promote healthcare, boost communication and education, and mitigate the effects of climate change (Gaurav, 2017). Nowadays, fuels made from petroleum dominate the world's energy supply, accounting for about 80% of it due to the rising worldwide need for energy (Gaurav, 2017). The erratic and frequently unforeseen character of fuel demand in our quickly modernizing world, which is fueled by industry and extensive transportation, makes the dependency on fossil fuels, especially for engines with internal combustion (Gaurav, 2017).

These limited resources are being rapidly depleted as a result of the world's continued dependence on conventional fossil fuels (Kaloudas, 2021). One of the main causes of this growing energy consumption is the growing human population. A growing human population is one factor contributing to this increased energy demand. Although fossil fuels are presently used to supply this need because of their technological advances and affordable prices, this strategy is not sustainable. Fossil fuels are non-renewable and are projected to be depleted within approximately 50 years (Aron, 2020). Moreover, their combustion releases greenhouse gases (GHGs), contributing significantly to global warming and negatively impacting ecosystems (Aron, 2020). The burning of fossil fuels raised the release of carbon dioxide by 1.6% in 2017, and more hikes are expected (Aron, 2020). These elements—environmental adverse effects, declining resources, and rising demand—highlight the pressing need for an ideological shift that prioritizes the investigation and advancement of a variety of sources of clean energy (Mignogna et al., 2024).

In response to the escalating global challenges of climate change as well as diminishing availability for environmentally and politically risky petroleum resources, interest in large-scale biofuel applications has grown considerably (Solomon, 2010). This growing interest in biofuels reflects a wider global push for sustainable energy solutions. Governments worldwide are actively promoting the development and deployment of diverse renewable resources—including biomass, wind, solar, and hydropower—to meet rising energy demands while simultaneously decreasing dependence on fossil fuels as well as lowering carbon dioxide emissions (Mignogna et al., 2024). Bioenergy, derived

DOI: 10.1201/9781003600169-15

from the energy stored in organic plant and animal matter, offers a compelling option within this renewable energy landscape. Most importantly, it offers a way to minimize negative consequences of energy generation on the ecosystem by lowering hazardous gases released in comparison to conventional fuels (Abdel-Basset et al., 2021). In the end, shifting to other forms of energy like bioenergy is crucial to attaining environmental and energy sustainability, which calls for cooperation from organizations, corporations, governments, and individuals (Simon, 2022).

The main industrial source of palm seed oil is the oil palm, *Elaeis guineensis* (Arecaceae family). Even though *Elaeis oleifera* and other *Elaeis* species help with production, but their contribution is not as substantial (Mahlia, 2019). This extremely profitable crop has a variety of uses, using every part of the plant, from fruit harvesting to wastes from subsequent processing, with the possibility of producing environmentally friendly energy and biofuel (Kaniapan, 2021). Nearly 60% of the world's oilseed supplies are palm oil, demonstrating its control in the vegetable oil industry (Ayompe, 2021). The amount of land in oil palm farming as well as the output of palm oil have grown rapidly as a consequence of this trade boom, which has been stimulated by rising worldwide consumption (Ayompe, 2021).

This chapter provides a comprehensive overview of the biology, cultivation, processing, and environmental and socioeconomic effects associated with the application of palm oil in production for biofuels. It investigates an extensive examination of the environmental implications of palm oil being used as biofuel, particularly deforestation and emissions of GHGs. The chapter goes on to discuss the intricate socioeconomic issues of palm oil production, such as how it affects food availability, land tenure, and rural area development. Current sustainability activities, such as accreditation systems like the Roundtable on Sustainable Palm, are assessed keeping these complex issues in view. The conclusion addresses research needs and potential avenues with regard to innovative technology, the best conversion process, and the linkage between palm oil biofuel production and overall sustainable development goals.

15.2 PALM BIOLOGY AND CULTIVATION

15.2.1 Botanical Characteristics of Oil Palm (*Elaeis guineensis*)

Oil palm is a tropical perennial crop belonging to the family Arecaceae. Native to West Africa, it is widely cultivated for its high-yielding oil-rich fruits, which serve as a primary feedstock for biofuel production (Waisundara and Grichar, 2024). The plant exhibits a monopodial growth habit, featuring a single, unbranched trunk with a crown of large, pinnate leaves. It can grow up to 20–30m in height and has an economic lifespan of approximately 25–30 years. Oil palm fruit consists of a fibrous mesocarp, a seed (kernel), and a hard endocarp (Ahmad, 2025). Both the mesocarp and kernel yield different types of oil: palm oil and palm kernel oil, respectively, with palm oil being the main source for biofuel production as a result of its high fatty acid content.

Oil palm is a C_3 photosynthetic plant but demonstrates efficient carbon assimilation, contributing to its high productivity (Qin et al., 2025). The tree produces inflorescences, which can be either male or female, and pollination primarily occurs through insect vectors such as *Elaeidobius kamerunicus* weevils. Fruit bunches mature in about 5 to 6 months, allowing for continuous harvesting throughout the year.

15.2.2 Global Distribution and Cultivation Practices

Oil palm is primarily cultivated in tropical regions between 10° north and south of the equator, where warm temperatures (24°C–32°C) and high annual rainfall (1,500–2,500mm) support optimal growth (Varma et al., 2025). The leading oil palm-producing countries include Indonesia and Malaysia, accounting for 85% of global production, followed by Thailand, Nigeria, Colombia, and Ecuador.

Cultivation practices vary based on regional agro-climatic conditions and land availability. Traditional smallholder farming coexists with large-scale commercial plantations. Standard agronomic practices for oil palm include:

- **Land Preparation and Nursery Management:** Seeds are germinated and grown in nurseries for 10–12 months before field transplanting.
- **Spacing and Planting Density:** Typical planting densities range from 136 to 160 palms per hectare to maximize yield while ensuring adequate sunlight penetration.
- **Fertilization and Soil Management:** Nutrient supplementation, particularly nitrogen, phosphorus, potassium, and magnesium, is essential for optimal fruit development.
- **Weed and Pest Control:** Integrated pest management (IPM) strategies, including biological control and agrochemical applications, help mitigate losses.
- **Harvesting Practices:** Fruit bunches are harvested manually or mechanically when ripe, generally at 10–15-day intervals.

15.2.3 Oil Palm Yield and Productivity: Factors Affecting Production

Oil palm is the world's top-producing oilseed crop, with an average oil yield of 3.5–4.0 tons per hectare per year, significantly surpassing other oil crops such as soybean (0.4 t/ha), sunflower (0.6 t/ha), and rapeseed (1.0 t/ha) (Abrol, 2024). Several factors influence oil palm yield and productivity:

- **Climatic Conditions:** Optimal growth requires high rainfall, consistent temperature, and ample sunlight. Prolonged drought or excessive flooding can negatively impact yields.
- **Soil Quality:** Fertile, well-drained soils with a pH of 4.0–6.0 are ideal for oil palm cultivation.
- **Planting Material:** Utilizing high-yielding hybrid varieties and improved cultivars enhances fruit production and oil extraction rates.
- **Agronomic Practices:** Proper pruning, pest and disease control, and efficient fertilization improve tree health and productivity.
- **Harvesting Efficiency:** Proper timing and frequency of harvesting significantly affect oil content and quality.

15.2.4 Genetic Improvement and Breeding Programs for Enhanced Biofuel Traits

Advancements in genetic improvement and breeding programs have played a crucial role in enhancing oil palm traits for biofuel production (Abubakar et al., 2023). Breeding efforts focus on increasing oil yield, improving disease resistance, and optimizing fatty acid composition for superior biofuel characteristics.

- **Hybrid Breeding Programs:** Conventional breeding methods have led to the development of *Tenera* hybrids (a cross between *Dura* and *Pisifera* varieties), which produce higher oil content and thinner shells, improving extraction efficiency.
- **Molecular Breeding and Genomic Selection:** Recent developments in genome sequencing and marker-assisted selection (MAS) enable targeted breeding for specific traits, such as high oleic acid content, which is preferable for biodiesel stability.
- **Genetic Modification (GM) and CRISPR Technology:** Genetic engineering techniques, including CRISPR-Cas9, are being explored to enhance drought tolerance, reduce disease susceptibility, and optimize oil composition for renewable energy applications.
- **Clonal Propagation and Tissue Culture:** Micropropagation techniques are being developed to mass-produce elite planting materials with uniform, high-yielding traits.

These genetic advancements not only improve oil palm productivity but also contribute to sustainability by reducing the land footprint required for high yields, making oil palm a viable and efficient biofuel source.

15.3 PALM OIL PROCESSING AND BIOFUEL PRODUCTION

Per hectare oil production, oil palm (*E. guineensis*) leads all oilseed crops ahead of major oil producers like soybean, sunflower, and rapeseed (Nair, 2021). Palm oil production now exceeds 58 million metric tons annually worldwide, making it the top vegetable oil, surpassing soybean oil (Murphy et al., 2021). Oil palm fruit consists of three layers—exocarp, mesocarp, and endocarp—and yields two distinct oils: The mesocarp of the oil palm fruit produces crude palm oil (CPO), while crude palm kernel oil is derived using the kernel according to Jeguirim et al. (2023). Palm oil receives growing attention for biofuel production because its exceptional yield reaches up to 4,000 kg/ha while providing a renewable energy source that replaces fossil fuels. Environmental sustainability for palm oil-based biofuels relies on extraction and purification processes that reduce negative environmental consequences.

Understanding these processes is essential for maximizing palm oil's viability and efficiency for use in biofuel applications.

15.3.1 Palm Oil Extraction and Refining Processes

Reception and quality assessment of fresh fruit bunches are among the first steps in palm oil extraction, and it is defined that OER (oil extraction rate) and CPO quality depend on ripeness and physical condition. Fruit ripeness, as determined by fruit detachment standards established by the Malaysian Palm Oil Board, results in the maximum oil yields because full triacylglycerol (TAG) production has taken place. Optical methods, such as RGB imaging and near-infrared (NIR) sensors, are being developed in addition to traditional visual inspection, which is often inconsistent.

Sterilization is then performed at >55°C to inactivate lipase enzymes that can degrade the oil and allow the fruit to be removed after transportation. Fruits are mechanically separated from bunch stalks by a process called threshing, while empty fruit bunches (EFB) are repurposed either for energy or fertilizers. The oil in the mesocarp must therefore be isolated without breaking the nuts so that the oil in the kernel is not introduced into the oil. Finally, oil recovery steps are taken, including clarification, purification, and drying to separate CPO from impurities, thereby guaranteeing product quality and sustainability of the entire process.

15.3.2 Biodiesel Production from Palm Oil: Chemical Pathways and Technologies

Taking into consideration, biodiesel production from palm oil majority uses conventional methods such as transesterification (oil and fat can be transformed by alkaline and acid-catalyzed transesterification) and catalytic cracking. The following paragraphs elaborate on these technologies, exploring their benefits, drawbacks, and latest advancements in biodiesel synthesis.

15.3.2.1 Transesterification with Alkali Catalyst

Because it is the most common and well-established, palm oil transesterification to biodiesel is primarily accelerated by alkali. The transesterification process produces fatty acid methyl/ethyl esters and glycerol by reacting alcohol with triglycerides in the presence of a base catalyst (usually NaOH and KOH) (Olatundun et al., 2024). Furthermore, it is commonly recognized that the yield and efficiency of the process are largely affected by reaction conditions, which include temperature, the alcohol-to-oil ratio, the intensity of mixing, catalyst loading, and quality of the feedstock (Tran-Nguyen et al., 2020). This is a vast method as it is cost-effective and easy to conduct.

15.3.2.2 Acid-Catalyzed Esterification (for High FFA Feedstock)

Acid-catalyzed esterification is used for feeds with high free fatty acid (FFA), such as waste oil or palm oil of lower grade. FFAs here will undergo catalytic esterification with the help of strong acid catalysts (i.e., H_2SO_4) before it is transesterified using alcohol (Anguebes-Franseschi et al., 2017). It is more time- and temperature-demanding than alkaline transesterification, but a necessity in processing high-FFA total feedstock that cannot be processed by alkaline-based catalysts due to soap formation and low biodiesel yield.

15.3.2.3 Other New Technologies/Techniques

The majority of new technologies are under investigation to bypass the constraints of conventional approaches. Because of its specificity toward triglycerides and favorable reaction conditions, lipase or whole-cell catalyst catalysis (e.g., Pichia pastoris) has been the center of attention (Lv et al., 2021). Supercritical methanol transesterification is also performed under elevated temperature and pressure, which helps to achieve higher reaction conversions and product yields by excluding the use of catalysts and pretreatment processes (Saha et al., 2017). Although this process facilitates efficiency improvements, it also has high-energy demands, in addition to complexity in our reactor design. Despite these issues, these technologies have the potential for enhancing the economics and the sustainability of future biodiesel production.

15.3.3 Palm Oil Feedstock for Sustainable Aviation Fuel (SAF) Production

Due to the potential for decreasing emissions and the utilization of fossil fuels, the production of Sustainable Aviation Fuel (SAF) from palm oil has become highly sought after. Different catalytic and process optimization techniques have been pursued to deliver improved conversion efficiency and improved fuel quality. As much as 78.33% (bio-aviation fuel inclusive) of the biofuels might be yielded in optimized conditions via hydrocracking palm oil utilizing molybdenum nitride-bentonite catalysts (Hasanudin et al., 2022). A hydrogenic and hydrocracking/isomerization procedure has also been utilized for producing jet fuel-range hydrocarbons from alkanes of palm origin through Ni-Ag-catalysis with optimized reaction conditions to increase the isomer-to-normal alkane ratio, along with aromatics content (Lin et al., 2020). Hydrodeoxygenation of palm sources feedstocks using different heterogeneous catalysts, such as Pd/C, under apparently ideal conditions produced good jet paraffin selectivity (73%) and fuel properties to meet the ASTM Jet A-1 standard (Why et al., 2022). Correspondingly, a new hydroprocessing catalyst generation, tantalum phosphate, led to the yield of bio-jet from date palm seed by up to 53.6%, determining the viability of a one-pot reaction under mild conditions (Rambabu et al., 2023). Apart from that, Cican et al. (2023) recently successfully altered the characteristics of micro-turbo combustion for use on the airside, as the combustion characteristics were stable when a micro-turbine engine was powered using the aviation fuel blended with biodiesel from waste palm oil. Palm kernel oil-based FAME-like biofuels have also been studied concerning their possible mixing with Jet-A1 fuel, enhancing cold flow properties and combustion characteristics (Mayorga et al., 2020). Furthermore, transesterification of palm kernel, jatropha, and babassu oils yielded bio-jet fuel mixtures whose quality conformed to commercial aviation fuel standards (Ranucci et al., 2018). This is one additional step demonstrating that SAF made using palm oil is a sustainable substitute for conventional jet fuel.

15.4 ENVIRONMENTAL IMPACTS OF PALM OIL BIOFUEL PRODUCTION

Palm oil plays a significant role in contributing to the economic development of a nation by creating job opportunities and providing alternative sources of energy. Palm oil is an attractive commodity for biofuel production because of the high yield of oil produced from oil palm per hectare of land used and the low production cost. Biofuel produced from palm oil is a potential source of clean energy, which reduces the consumption of fossil fuels. This potentiality fuels an excitement for large-scale feedstock cultivation of oil palm, followed by production of biofuels. However, increasing concerns have emerged regarding the potential environmental and socioeconomic impact of deforestation and the loss of wet and peatland habitats for oil palm cultivation purposes.

15.4.1 Deforestation and Habitat Loss: Biodiversity Concerns

Deforestation and habitat losses are issues faced by the oil palm industry, especially in developing countries where rainforests contain a large volume of biodiversity (Hoffmann et al., 2010; Leadley, 2010). Oil palm is a main reason for deforestation and leads to substantial habitat losses, especially in tropical regions. According to the Food and Agriculture Organization (FAO), the palm oil industry has contributed to an estimated 5% of tropical deforestation in tropical regions and 2.3% of global deforestation. In Indonesia, the biofuel industry accounts for 6.5% of deforestation, and around 24 million hectares of rainforest were razed between 1990 and 2015 (Obidzinski et al., 2012).

Oil palm only grows in biodiversity-rich tropics and requires light, and this requirement leads to the removal of all other vegetation. Thus, one plant species predominates on oil palm plantations (Danielsen et al., 2009; Gillison and Liswanti, 1999) and supports a much lower biodiversity compared to natural forests. The deforestation for this purpose mostly includes both primary and secondary forests (Figure 15.1). These forests are rich in biodiversity and home to numerous animals and plants. Degradation of these forests, followed by monoculture plantations, leads to unbalanced ecological interaction compared to diverse and complex forest ecosystems.

The conversion of forestland into oil palm farms caused a loss of biodiversity, endangered species, and a substantial increase in GHG emissions. The decline and extinction of numerous species like orangutans, tigers, and elephants indicate a significant impact of the palm oil industry on the environment. Besides, oil palm plantations resulted in conflict between humans and other animals, such as human-orangutan and human-tiger conflict (Azlan and Sharma, 2006; Meijaard et al., 2011, 2012). Due to the expansion of palm oil production, 193 species classified as endangered and vulnerable are threatened globally (Meijaard et al., 2018). Over the past year, the situation elevated

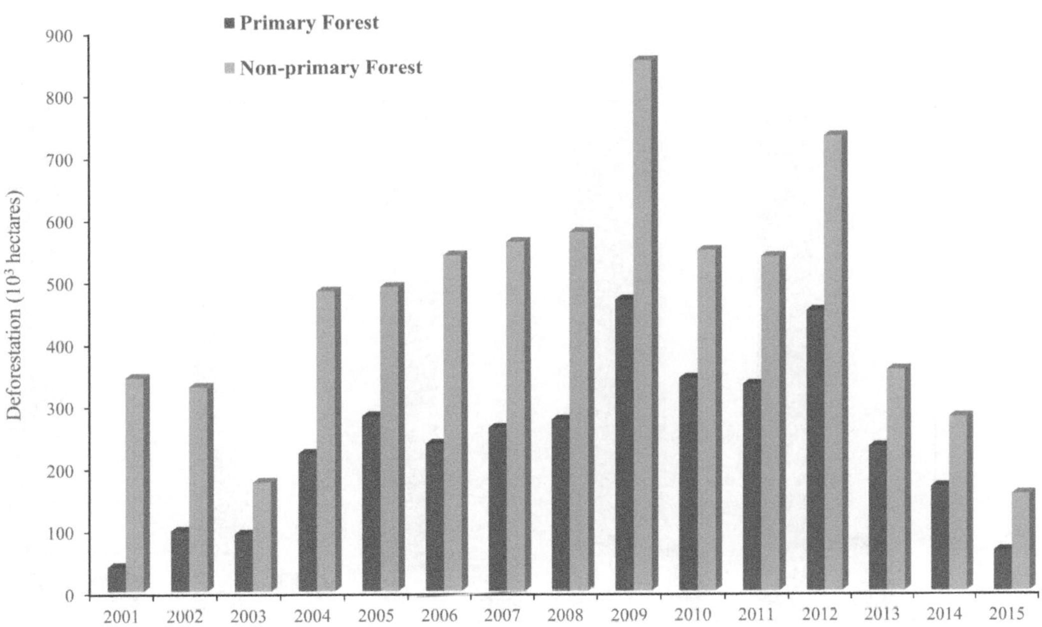

FIGURE 15.1 Total forest area replaced by oil palm globally (directly and indirectly), 2001–2015. (Source: World Resource Institute, 2021.)

due to the utilization of palm oil as a feedstock in biofuel synthesis in addition to its use for food, cosmetics, and cleaning products.

Since 2017, the deforestation linked to palm oil in Indonesia, Malaysia, and Papua New Guinea has been decreasing (Idris et al., 2025). This decrease is linked to sustainable land use practices, protection, and restoration of critical habitats. This progress shows that the connection between deforestation and the expansion of oil palm plantations could be broken eventually. The decline is attributed to strengthened law enforcement, ensuring national policies implementation, certification of palm oil plantations, and drops in prices.

15.4.2 Greenhouse Gas Emissions: Carbon Footprint Analysis

Biodiesel, a liquid biofuel produced from palm oil and other oils, is a renewable alternative that can replace petrodiesel (Atabani et al., 2012). Biofuel can reduce GHG emissions to the surrounding environment by 46%–73% when compared to diesel from fossil fuels (Silalertruksa and Gheewala, 2012). Biofuel is a carbon-neutral energy source, but deforestation and peatland destruction for oil palm plantations are major sources of GHG emissions that contribute to climate change (Jon et al., 2010; Tan et al., 2009). The GHG emissions increased due to the decomposition of organic materials and carbon stock loss in peatlands. Over the last two decades, Indonesia and Malaysia together were responsible for 50% loss of forestland for oil palm activities (Wicke et al., 2011), and this led to a huge amount of CO_2 released into the environment. For instance, 756–1,043 metric tons (MT) of CO_2 have been released into the surrounding environment due to deforestation for oil palm plantations in Sumatra, Indonesia (Lee et al., 2014). Besides, peatland subsides and fires on drained peatlands account for 92% of average annual GHG emissions in Indonesia from 2015 to 2022 (Benedict and Heilmayr, 2024). While GHG emissions from peat subsidence (million tonnes CO_2 eq) were increasing, they decreased from industrial palm oil land conversion and peat burning from 2018 to 2022 (Figure 15.2) (Benedict and Heilmayr, 2024; Trase, 2024). The amount of GHG emission (Kg CO_2 eq. per liter of biodiesel,

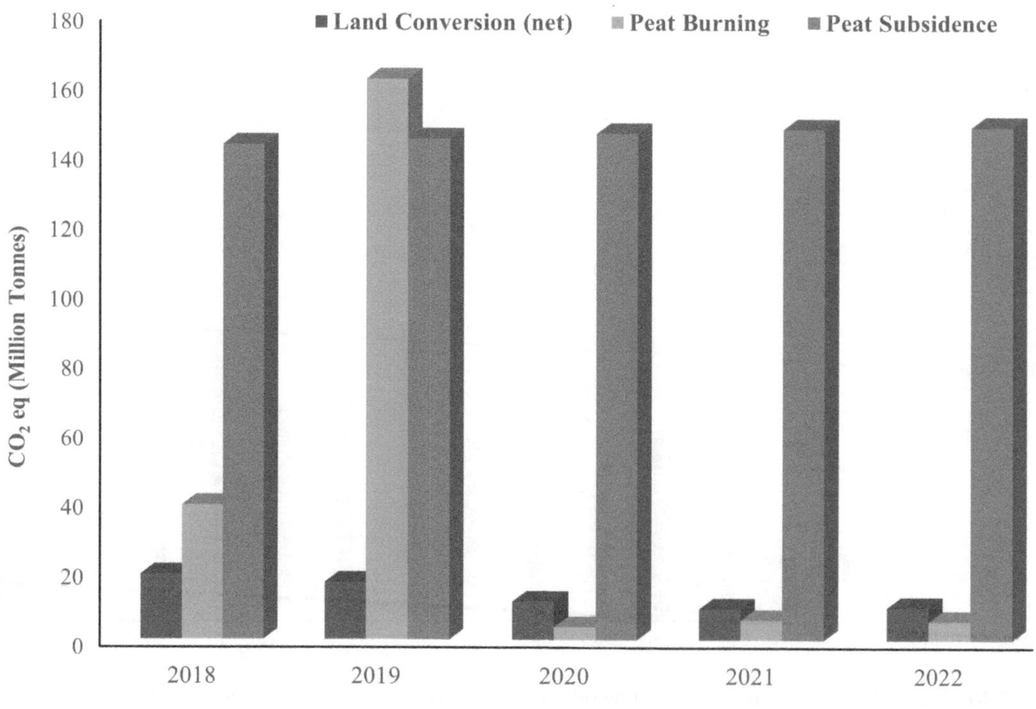

FIGURE 15.2 Greenhouse gas emission estimates for the palm oil sector's land use by source, 2018–2022. (GHG Indicator.)

production and use) due to the conversion of forestland to oil palm is significantly higher than that of petro diesel (Silalertruksa and Gheewala, 2012). The largest overall environmental impact on global warming was caused by the manufacture of biofuel from palm oil (Siregar et al., 2015).

15.4.3 Water Consumption and Pollution: Impacts on Water Resources

The effect of deforestation for oil palm cultivation purposes not only increases CO_2 emissions but also decreases the amount and quality of water. For oil palm to be grown with an effective yield, high water levels are necessary. During its growth, the oil palm drastically reduces soil fertility by absorbing a significant amount of water and soil nutrients. In the monsoon months, 280–350 L/day water is required for a single oil palm tree for optimum yield (Carr, 2011). The plant requires rainfalls of 2500–4000 mm per year under evenly distributed situations. In uneven distribution or a lack of rainfall, it is necessary to ensure a water supply with irrigation.

15.4.4 Soil Degradation and Nutrient Depletion

Oil palm has a greater ability to sequester carbon and nitrogen content in the soil compared to other plants, such as rubber, resulting in soil degradation (Guillaume et al., 2016). Marginal areas, such as peatlands, which have high carbon content and stock, were preferred for oil palm plantations. Gradually, the carbon content decreases and the bulk density of the soil increases as the plant gets older, leading to extensive soil degradation (low carbon content and stock, and high bulk density). Furthermore, rather than converting natural ecosystems to agroecosystems, most of the oil palm production was done through land use intensification (Gibbs et al., 2010). This leads to the application of fertilizer to increase the carbon content and to improve plant yield. Application of fertilizers leads to degradation of soil quality and reduction in food availability for animals in soils (Goh et al., 2016; Tao et al., 2016).

15.4.5 Peatland Conversion and Its Consequences

The expansion of palm oil industries is a major driver of the conversion of peatlands to oil palm cultivation. Peatlands are rich in organic matter content and suitable for oil palm cultivation. Annually, 2.2% of peat swamp forests are being converted to oil palm plantations in Southeast Asia (Miettinen et al., 2012), and this leads to the extinction of endemic species, GHG emissions, and climate change (Adila et al., 2017). Moreover, peatland subsidence and fires on drained peatlands increase the emission of GHG. Peatlands converted to oil palm fields were characterized by strong loss of water storage and infiltration, carbon content, and nutrient cycling in soil composition (Guillaume et al., 2016). This leads to substitutional degradation of soil fertility and increases the application of fertilizers. Soil properties such as carbon-to-nitrogen ratio, moisture content, pH, and soil-surface temperature were affected by the establishment of vegetation (Anamulai et al., 2019).

15.5 SOCIOECONOMIC CONSIDERATIONS

15.5.1 Economic Benefits and Rural Development: Job Creation and Income Generation

The growth of oil palm, particularly in tropical nations, is seen as a major force behind economic progress and is probably an excellent place to create jobs. Oil palm plantations are considered a commodity that has a significant role in state development, contributing to state revenues and the provision of employment opportunities. Moreover, palm oil plays a significant role in the contribution to the GDP of the nation, especially in Southeast Asia and West and Central Africa (Meijaard et al., 2018). On the other side, searching for land for oil palm plantations leads to deforestation, biodiversity loss, and other social issues (Sodhi et al., 2010). Palm oil can be used as an alternative energy source in various sectors, including transportation, power plants, and households, and it has a significant role in economic development (Atabani et al., 2012). The manufacture of biofuel from palm oil is seen as a high-productivity commodity that is significant to society's economic standing. The trade-off between economic development and environmental impact should be considered in the palm oil industry for sustainable development.

15.5.2 Land Tenure and Community Rights: Social Justice Issues

One of the main forces behind the economic growth of tropical nations was the palm oil sector. However, the development affects the availability of land for agriculture, often causes land use conflict, and increases the land prices (Afrizal et al., 2024; Obidzinski et al., 2012). The increases in land prices led to conflict between the communities, especially indigenous communities and companies (Mukherjee and Sovacool, 2014). The companies have abused and manipulated the locally based land title system to get more land for cultivation. The landowners mostly lose their lands due to the limitation of access to conflict resolution mechanisms (Afrizal et al., 2024).

15.5.3 Food Security and Competition for Resources: Balancing Food and Fuel Needs

Palm oil is the largest edible oil globally and the most promising feedstock for the production of biofuel. It is one of the most productive edible oils compared to soybeans, rapeseed, or sunflower, and an instrumental and key driver in maintaining global food security. While 90% of palm oil produced is used for food consumption, 10% is used for other purposes such as cosmetic production and biofuel purposes (Adade, 2022). Oil palm, which is grown only on 5% of the world's agricultural lands used for edible oil production, accounts for 25% of the global market share. Some study shows that palm oil is one of the most economical and sustainable sources of both food and energy among other edible oils (Lam et al., 2009). However, the unsustainable cultivation of oil palm and the extensive conversion of agricultural land to oil palm fields raise questions about food security. The use of palm oil as feedstock for biofuel production significantly reduces the amount of edible palm oil in the market, and it may have an impact on the prices of other edible oils, such as soybean. Oil palm-producing tropical countries face food security problems due to the implementation of oil palm production policies that do not support food production, the conversion of smallholder farmers' land to oil palm cultivation, and increased land use pressure from expanding palm oil industries (Adade, 2022; Yengoh and Armah, 2015).

15.5.4 Labor Practices and Working Conditions: Ethical Considerations

In tropical areas, particularly in Indonesia and Malaysia, where 85% of the world's oil palm is grown, the palm oil business is one of the main employers. The working conditions in palm oil industries raise labor-related issues, such as low wages, forced labor, child labor, gender discrimination, and harsh environmental working conditions (Amnesty International, 2016). Due to its social interaction complexity, it is difficult to audit exactly the social issues related to the palm oil industry. This puts pressure on the workers and leads to working in harsh environments at low wages. It darkens the importance of social issues, including labor conditions and human trafficking. Many of the workers working on oil palm have been migrants, economic refugees, or undocumented people from other regions who have no access to social rights.

15.6 SUSTAINABILITY CRITERIA AND CERTIFICATION SCHEMES

The sustainability of palm oil as a biofuel feedstock continues to be a concern due to the effects of its exploitation on the economy, society, and the environment. Various certification systems and standards have been established to contribute to sustainable production and consumption. In a manner that will not compromise economic feasibility, the programs can ensure reduced deforestation, conserve biodiversity, and improve decent work practices. This section takes into account the leading sustainability certification schemes that regulate palm oil production, the justification for them, and the possibilities and challenges of increasing sustainable levels.

15.6.1 Sustainable Palm Oil: Principles and Criteria

The Roundtable on Sustainable Palm Oil (RSPO) is the most recognized certification scheme for sustainable palm oil production. RSPO is an organization that was formed in 2004 and formulates guidelines to guarantee environmentally and socially sustainable palm oil production. RSPO certification relies upon a series of underlying principles and criteria, which are:

- **No Deforestation and Environmental Protection**: The RSPO prohibits the destruction of primary forests, high conservation value (HCV) areas, and peatlands for palm oil cultivation. This assists in the preservation of biodiversity and protects ecosystem processes.
- **Transparent and Legal Operations**: RSPO-certified producers must comply with national and international laws, ensuring legal land acquisition and operational transparency.
- **Sustainable Agricultural Practices**: Certification requires responsible land management, minimizing chemical use, improving soil fertility, and optimizing water management to reduce negative environmental impacts.

15.6.2 Other Certification Schemes and Standards for Sustainable Palm Oil Production

Aside from RSPO, a number of other certification programs and standards for sustainability govern palm oil production to achieve social and environmental responsibility:

- **International Sustainability & Carbon Certification (ISCC):** Focusing on social responsibility, sustainable land management, and reducing GHGs, the ISCC certification scheme ensures compliance with the EU Renewable Energy Directive (RED) (Mai-Moulin et al., 2021).
- **Malaysian Sustainable Palm Oil (MSPO) and Indonesian Sustainable Palm Oil (ISPO):** Employed to encourage sustainable operations and national compliance, the domestic certification schemes will assist in regulating the palm oil industry in Malaysia and Indonesia, two of the world's largest palm oil producers.
- **Rainforest Alliance Certification**: This scheme promotes biodiversity conservation and ethical labor practices in palm oil production.
- **Palm Oil Innovation Group (POIG)**: A voluntary initiative that extends beyond RSPO requirements, emphasizing stricter environmental and social governance practices.
- **Fair Trade Certification**: Ensures that palm oil producers receive fair compensation, promoting social equity and economic sustainability.

For sustainable palm oil production to be a reality and for safe use of palm oil in the biofuel sector, all these schemes of certification are required.

15.6.3 Challenges and Opportunities for Improving Sustainability Practices

The total application of sustainable palm oil production is hindered by various problems despite the improvement in sustainability certification:

- **Weak Enforcement and Compliance**: In some regions, lax enforcement of sustainability standards results in continued deforestation, land conflicts, and unethical labor practices.

- **High Certification Costs**: Smallholder farmers, who account for a significant portion of global palm oil production, often face financial barriers in obtaining certification.
- **Greenwashing and Transparency Issues**: Some companies claim sustainability without genuine compliance, undermining the credibility of certification schemes.
- **Indirect Land Use Change (ILUC)**: The expansion of palm oil plantations can indirectly lead to deforestation and biodiversity loss elsewhere.

15.7 FUTURE DIRECTIONS AND RESEARCH NEEDS

Future palm biofuel production depends on the advancement and introduction of new technologies that improve sustainability and meet social and environmental issues. Some examples of new technologies that can improve palm oil productivity significantly, while reducing land use and deforestation, are precision agriculture, remote sensing, and automated monitoring systems (Abubakar et al., 2025). Besides this, further research must be carried out to adopt sustainable processing techniques, such as catalytic and enzymatic conversion processes, for reducing GHG emissions, improving efficiency, and eliminating waste. Modern biorefinery technologies can enhance resource efficiency and circularity through the use of residual palm biomass for biofuel and bio-based chemicals production. This will render the bioeconomy more sustainable.

Enhancing transesterification, hydroprocessing, and catalytic cracking process effectiveness is one of the highest research priorities to optimize biofuel conversion processes. Researchers should explore new catalysts, intensification strategies, and hybrid conversion systems to reduce energy usage and enhance fuel output. To ensure that the innovations align with global climate objectives, life cycle assessment (LCA) research must be expanded to include the overall environmental burden of various biofuel production routes. Furthermore, technology improvement efforts must address ILUC concerns and reduce the carbon footprint of palm oil.

In the view of Das et al. (2024), an integrated approach to palm oil biofuel production involves alignment with wider sustainable development goals (SDGs) like eradicating poverty, ensuring energy security, and climate change mitigation. Sustainable palm growth and biofuel production community models can assist in fair trade practices and livelihood support to rural communities. Besides, policy guidelines toward the responsible promotion of palm oil expansion under the protection of economic and environmental interests need interdisciplinary research. Mixing palm oil biofuel with other renewable fuels like solar or wind energy can yet further enhance sustainability and energy diversification.

Some are areas of research deficits, which still need to be bridged, though there are sweet achievements. There is an urgent need to investigate genetically engineered palm types for increased productivity and disease resistance and to minimize the environmental impact of agriculture. Additional studies are also needed to innovate low-cost and scalable waste valorization technologies, including the production of biofuels and bioproducts from palm oil mill effluent (POME) and EFB (Siagian et al., 2024). It is also required to carry out social and economic impact assessments of palm-based biofuels to deliver fair benefits across supply chains. All these research requirements need to be fulfilled through combined interdisciplinary approaches in a bid to make palm-based biofuels a viable and sustainable source of energy in the future.

15.8 CONCLUSIONS

This chapter is a detailed account of palm oil application in the biofuel industry based on its potential as a renewable energy source and meeting environmental and socioeconomic challenges. The chapter discusses the biology of oil palm, cultivation, processing, and biofuels, such as biodiesel and green airplane fuel. The discourse critically assesses the environmental considerations, such as deforestation and emissions, as well as socioeconomic considerations, such as rural development and food security. Finally, the chapter concludes by highlighting the cutting-edge areas of research and innovation in technology needed in a sustainable palm oil biofuels industry, such as advanced production technologies, maximization of biofuel conversion, and integration into wider sustainable development. This chapter calls for evidence-based policy driven by focused research to facilitate a sustainable and responsible industry.

COMPETING INTERESTS

The authors declare that they have no competing interest.

DECLARATION OF GENERATIVE AI AND AI-ASSISTED TECHNOLOGIES IN THE WRITING PROCESS

During the preparation of this work, the book chapter author(s) have not taken any assistance from AI, and the book chapter author(s) reviewed and edited the content as needed and take(s) full responsibility for the content of the publication.

REFERENCES

Abdel-Basset, M., Gamal, A., Chakrabortty, R. K., & Ryan, M. (2021). Development of a Hybrid Multi-Criteria Decision-Making Approach for Sustainability Evaluation of Bioenergy Production Technologies: A Case Study. *Journal of Cleaner Production*, 290. https://doi.org/10.1016/j.jclepro.2021.125805

Abrol, D. P. (2024). *Pollination Biology of Cultivated Oil Seeds and Pulse Crops*. CRC Press.

Abubakar, A. M., Zakarya, I. A., Hasnain, M., Sarkinbaka, Z. M., Mukwana, K. C., & Abdo, A. (2025). Potential Breakthroughs in Environmental Monitoring and Management. In F. D. Mobo (Ed.), *Harnessing AI in Geospatial Technology for Environmental Monitoring and Management* (pp. 239–282). IGI Global Scientific Publishing.

Abubakar, A., Kasim, S., Ishak, M. Y., & Uddin, M. K. (2023). Maximizing Oil Palm Yield: Innovative Replanting Strategies for Sustainable Productivity. *Journal of Environmental & Earth Sciences*, 5(02), 61–75.

Adade, F. B. (2022). Oil Palm (*Elaeis guineensis*) Cultivation and Food Security in the Tropical World. In K. Hesam (Ed.), *Elaeis Guineensis* (p. 171). IntechOpen.

Adila, N., Sasidhran, S., Kamarudin, N., Puan, C. L., Azhar, B., & Lindenmayer, D. B. (2017). Effects of Peat Swamp Logging and Agricultural Expansion on Species Richness of Native Mammals in Peninsular Malaysia. *Basic and Applied Ecology*, 22, 1–10.

Afrizal, E. V. P., & Elida, L. (2024). Palm Oil Expansion, Insecure Land Rights, and Land-Use Conflict: A Case of Palm Oil Centre of Riau, Indonesia. *Land Use Policy*, 146, 107325. https://doi.org/10.1016/j.landusepol.2024.107325.

Ahmad, H. (2025). An Effectiveness of Mycorrhia and Lime Application on the Growth of Oil Palm Seedlings (*Elaeis guineensis* Jacq) in peat soil. *INFOKUM*, 13(01), 60–71.

Anamulai, S., Sanusi, R., Zubaid, A., Lechner, A. M., Ashton-Butt, A., & Azhar, B. (2019). Land Use Conversion from Peat Swamp Forest to Oil Palm Agriculture Greatly Modifies Microclimate and Soil Conditions. *PeerJ*, 7, e7656.

Anguebes-Franseschi, F., Abatal, M., Bassam, A., Escalante Soberanis, M. A., May Tzuc, O., Bucio-Galindo, L., Cordova Quiroz, A. V., Aguilar Ucan, C. A., & Ramirez-Elias, M. A. (2017). Esterification Optimization of Crude African Palm Olein Using Response Surface Methodology and Heterogeneous Acid Catalysis. *Energies*, 11(1), 157. https://doi.org/10.3390/en11010157

Aron, N. S. M., Khoo, K. S., Chew, K. W., Show, P. L., Chen, W.-H., & Nguyen, T. H. P. (2020). Sustainability of the Four Generations of Biofuels – A Review. *International Journal of Energy Research*. https://doi.org/10.1002/er.5557

Atabani, A. E., Silitonga, A. S., Badruddin, I. A., Mahlia, T. M. I., Masjuki, H. H., & Mekhilef, S. (2012). A Comprehensive Review on Biodiesel as an Alternative Energy Resource and its Characteristics. *Renewable and Sustainable Energy Reviews*, 16(4), 2070–2093. https://doi.org/10.1016/j.rser.2012.01.003.

Ayompe, L. M., Schaafsma, M., & Egoh, B. N. (2021). Towards Sustainable Palm Oil Production: The Positive and Negative Impacts on Ecosystem Services and Human Wellbeing. *Journal of Cleaner Production*, 276. https://doi.org/10.1016/j.jclepro.2020.123914

Azlan, J. M., & Sharma, D. S. K. (2006). The Diversity and Activity Patterns of Wild Felids in a Secondary Forest in Peninsular Malaysia. *Oryx*, 40(1), 36–41.

Benedict, J., & Heilmayr, R. (2024). Indonesian Palm Oil Exports and Deforestation. *Trase*.

Carr, M. K. V. (2011). The Water Relations and Irrigation Requirements of Oil Palm (*Elaeis guineensis*): A Review. *Experimental Agriculture*, 47(4), 629–652.

Cican, G., Crunteanu, D. E., Mirea, R., Ceatra, L. C., & Leventiu, C. (2023). Biodiesel from Recycled Sunflower and Palm Oil—A Sustainable Fuel for Microturbo-Engines Used in Airside Applications. *Sustainability*, 15(2), 987.

Danielsen, F., Beukema, H., Burgess, N. D., Parish, F., Brühl, C. A., Donald, P. F., Murdiyarso, D., Phalan, B. E. N., Reijnders, L., & Struebig, M. (2009). Biofuel Plantations on Forested Lands: Double Jeopardy for Biodiversity and Climate. *Conservation Biology*, 23(2), 348–358.

Das, P., Jha, C. K., Saxena, S., & Ghosh, R. K. (2024). Can Biofuels Help Achieve Sustainable Development Goals in India? A Systematic Review. *Renewable and Sustainable Energy Reviews*, 192, 114246.

Gaurav, N., Sivasankari, S., Kiran, G. S., Ninawe, A., & Selvin, J. (2017). Utilization of Bioresources for Sustainable Biofuels: A Review. *Renewable and Sustainable Energy Reviews*, 205–214. https://doi.org/10.1016/j.rser.2017.01.070

Gibbs, H. K., Ruesch, A. S., Achard, F., Clayton, M. K., Holmgren, P., Ramankutty, N., & Foley, J. A. (2010). Tropical Forests Were the Primary Sources of New Agricultural Land in the 1980s and 1990s. *Proceedings of the National Academy of Sciences*, 107(38), 16732–16737.

Gillison, A., & Liswanti, N. (1999). *Impact of Oil Palm Plantations on Biodiversity in Jambi, Central Sumatra, Indonesia*. Center for International Forestry Research, Bogor.

Goh, C. S., Wicke, B., Verstegen, J., Faaij, A., & Junginger, M. (2016). Linking Carbon Stock Change from Land-Use Change to Consumption of Agricultural Products: A Review with Indonesian Palm Oil as a Case Study. *Journal of Environmental Management*, 184, 340–352.

Guillaume, T., Holtkamp, A. M., Damris, M., Brümmer, B., & Kuzyakov, Y. (2016). Soil Degradation in Oil Palm and Rubber Plantations under Land Resource Scarcity. *Agriculture, Ecosystems & Environment*, 232, 110–118. https://doi.org/10.1016/j.agee.2016.07.002.

Hasanudin, H., Asri, W. R., Said, M., Hidayati, P. T., Purwaningrum, W., Novia, N., & Wijaya, K. (2022). Hydrocracking Optimization of Palm Oil to Biogasoline and Bio-Aviation Fuels Using Molybdenum Nitride-Bentonite Catalyst. *RSC Advances*, 12(3), 2547–2561.

Hoffmann, M., Hilton-Taylor, C., Angulo, A., Böhm, M., Brooks, T. M., Butchart, S. H. M., Carpenter, K. E., Chanson, J., Collen, B., & Cox, N. A. (2010). The Impact of Conservation on the Status of the World's Vertebrates. *Science*, 330(6010), 1503–1509.

Idris, M. S., Nor, M. S. Md., Ali, S., Ali, N. M., Yoong, S. W., Nor, M. N. M., Muhili, M. S. M., & Hassim, M. F. N. (2025). Big Data-Driven Risk Analysis of Environmental Grievances in the Palm Oil Industry for Sustainable Decision-Making. *Jurnal Manajemen Hutan Tropika*, 31(1), 72–88.

Jeguirim, M., Khiari, B., & Jellali, S. (2023). Identification, Quantification, and Characterization of Palm-Tree and Fruit Wastes. In M. Jeguirim, B. Khiari, & S. Jellali (Eds.), *Palm Trees and Fruits Residues: Recent Advances for Integrated and Sustainable Management* (Vol. 1, pp. 1–58). Academic Press.

Jon, D., Usha, M., Sarala, A., Nyon, Y. C., & Gabriel, C. (2010). A Quick Scan of Peatlands in Malaysia. *Wetlands International Malaysia*, 1.

Kaloudas, D., Pavlova, N., & Penchovsky, R. (2021). Lignocellulose, Algal Biomass, Biofuels, and Biohydrogen: A Review. *Environmental Chemistry Letters*, 19, 2809–2824. https://doi.org/10.1007/s10311-021-01213-y

Kaniapan, S., Hassan, S., Ya, H., Patma Nesan, K., & Azeem, M. (2021). The Utilisation of Palm Oil and Oil Palm Residues and the Related Challenges as a Sustainable Alternative in Biofuel, Bioenergy, and Transportation Sector: A Review. *Sustainability*, 13(6). https://doi.org/10.3390/su13063110

Lam, M. K., Tan, K. T., Lee, K. T., & Mohamed, A. R. (2009). Malaysian Palm Oil: Surviving the Food versus Fuel Dispute for a Sustainable Future. *Renewable and Sustainable Energy Reviews*, 13(6), 1456–1464. https://doi.org/10.1016/j.rser.2008.09.009.

Leadley, P. (2010). *Biodiversity Scenarios: Projections of 21st Century Change in Biodiversity, and Associated Ecosystem Services: A Technical Report for the Global Biodiversity Outlook 3*. UNEP/Earthprint.

Lee, J. S. H., Abood, S., Ghazoul, J., Barus, B., Obidzinski, K., & Koh, L. P. (2014). Environmental Impacts of Large-Scale Oil Palm Enterprises Exceed That of Smallholdings in Indonesia. *Conservation Letters*, 7(1), 25–33.

Lin, C. H., Chen, Y. K., & Wang, W. C. (2020). The Production of Bio-Jet Fuel from Palm Oil-Derived Alkanes. *Fuel*, 268, 117198.

Lv, L., Dai, L., Du, W., & Liu, D. (2021). Progress in Enzymatic Biodiesel Production and Commercialization. *Processes*, 9(2), 355.

Mahlia, T. M. I., Ismail, N., Hossain, N., Silitonga, A. S., & Shamsuddin, A. H. (2019). Palm Oil and its Wastes as Bioenergy Sources: A Comprehensive Review. *Environmental Science and Pollution Research*. https://doi.org/10.1007/s11356-019-04563-x

Mai-Moulin, T., Hoefnagels, R., Grundmann, P., & Junginger, M. (2021). Effective Sustainability Criteria for Bioenergy: Towards the Implementation of the European Renewable Directive II. *Renewable and Sustainable Energy Reviews*, 138, 110645.

Mayorga, M. A., López, M., López, C. A., Bonilla, J. A., Silva, V., Talero, G. F., Correa, F., & Noriega, M. A. (2020). Production of Aviation Biofuel from Palm Kernel Oil. *Chemical Engineering Transactions*, 82, 25–30.

Meijaard, E., Buchori, D., Hadiprakarsa, Y., Utami-Atmoko, S. S., Nurcahyo, A., Tjiu, A., Prasetyo, D., Nardiyono, L. C., & Ancrenaz, M. (2011). Quantifying Killing of Orangutans and Human-Orangutan Conflict in Kalimantan, Indonesia. *PloS One*, 6(11), e27491.

Meijaard, E., Carlson, K. M., Garcia, U., Sheil, D., Juffe-Bignoli, D., & Wich, S. (2018). *Oil palm and biodiversity: A situation analysis by the IUCN Oil Palm Task Force*. International Union for Conservation of Nature - IUCN. https://catalogo.fedepalma.org/cgi-bin/koha/opac-detail.pl?biblionumber=29970

Meijaard, E., Wich, S., Ancrenaz, M., & Marshall, A. J. (2012). Not by Science Alone: Why Orangutan Conservationists Must Think Outside the Box. *Annals of the New York Academy of Sciences*, 1249(1), 29–44.

Miettinen, J., Shi, C., & Liew, S. C. (2012). Two Decades of Destruction in Southeast Asia's Peat Swamp Forests. *Frontiers in Ecology and the Environment*, 10(3), 124–128.

Mignogna, D., Szabó, M., Ceci, P., & Avino, P. (2024). Biomass Energy and Biofuels: Perspective, Potentials, and Challenges in the Energy Transition. *Sustainability*, 16(16). https://doi.org/10.3390/su16167036

Mukherjee, I., & Sovacool, Benjamin K. (2014). Palm Oil-Based Biofuels and Sustainability in Southeast Asia: A Review of Indonesia, Malaysia, and Thailand. *Renewable and Sustainable Energy Reviews*, 37, 1–12.

Murphy, D. J., Goggin, K., & Paterson, R. R. M. (2021). Oil Palm in the 2020s and Beyond: Challenges and Solutions. *CABI Agriculture and Bioscience*, 2(39), 1–10.

Nair, K. P. (2021). *Oil Palm (Elaeis guineensis Jacquin). Tree Crops*. Springer, Cham.

Obidzinski, K., Andriani, R., Komarudin, H., & Andrianto, A. (2012). Environmental and Social Impacts of Oil Palm Plantations and Their Implications for Biofuel Production in Indonesia. *Ecology and Society*, 17(1), 1–19.

Olatundun, T. O., Popoola, V. A., Fakoyede, P. D., Adebayo, D. O., Kehinde, E. D., Adetoro, Q. A., Akhabue, O. B., Ewemade, C. E., Okpako, O., & Anyalebechi, P. E. (2024). Production of Biodiesel from Palm Kernel Oil Through Base-Catalyzed Transesterification Process. *World Journal of Advanced Research and Reviews*, 23(1), 2054–2081.

Pavičić, J., Novak Mavar, K., Brkić, V., & Katarina, S. (2022). Biogas and Biomethane Production and Usage: Technology Development, Advantages and Challenges in Europe. *Energies*, 15(8). https://doi.org/10.3390/en15082940

Qin, K., Ye, X., Luo, S., Fernie, A. R., & Zhang, Y. (2025). Engineering Carbon Assimilation in Plants. *Journal of Integrative Plant Biology*, 67(4), 926–948.

Rambabu, K., Bharath, G., Sivarajasekar, N., Velu, S., Sudha, P. N., Wongsakulphasatch, S., & Banat, F. (2023). Sustainable Production of Bio-Jet Fuel and Green Gasoline from Date Palm Seed Oil via Hydroprocessing over Tantalum Phosphate. *Fuel*, 334, 127157.

Ranucci, C. R., Alves, H. J., Monteiro, M. R., Kugelmeier, C. L., Bariccatti, R. A., de Oliveira, C. R., & da Silva, E. A. (2018). Potential Alternative Aviation Fuel from Jatropha (*Jatropha curcas* L.), Babassu (*Orbignya phalerata*), and Palm Kernel (*Elaeis guineensis*) as Blends with Jet-A1 Kerosene. *Journal of Cleaner Production*, 172, 3089–3098.

Saha, B., Aboelazayem, O., & Gadalla, M. (2017). Biodiesel Production from Waste Cooking Oil via Supercritical Methanol: Optimisation and Reactor Simulation. *Renewable Energy*, 124, 144–154.

Siagian, U. W. R., Wenten, I. G., & Khoiruddin, K. (2024). Circular Economy Approaches in the Palm Oil Industry: Enhancing Profitability Through Waste Reduction and Product Diversification. *Journal of Engineering and Technological Sciences*, 56(1), 25–49.

Silalertruksa, T., & Gheewala, S. H. (2012). Environmental Sustainability Assessment of Palm Biodiesel Production in Thailand. *Energy*, 43(1), 306–314. https://doi.org/10.1016/j.energy.2012.04.025.

Siregar, K., Tambunan, A. H., Irwanto, A. K., Wirawan, S. S., & Araki, T. (2015). A Comparison of Life Cycle Assessment on Oil Palm (*Elaeis guineensis* Jacq.) and Physic Nut (*Jatropha curcas* Linn.) as Feedstock for Biodiesel Production in Indonesia. *Energy Procedia*, 65, 170–179.

Sodhi, N. S., Koh, L. P., Clements, R., Wanger, T. C., Hill, J. K., Hamer, K. C., Clough, Y., Tscharntke, T., Posa, M. R. C., & Lee, T. M. (2010). Conserving Southeast Asian Forest Biodiversity in Human-Modified Landscapes. *Biological Conservation*, 143(10), 2375–2384.

Solomon, B. D. (2010). Biofuels and Sustainability. *Annals of the New York Academy of Sciences*. https://doi.org/10.1111/j.1749-6632.2009.05279.x

Tan, K. T., Lee, K. T., Mohamed, A. R., & Bhatia, S. (2009). Palmoil: Addressing Issues and Towards Sustainable Development. *Renewable and Sustainable Energy Reviews*, 13(2), 420–427.

Tao, H.-H., Slade, E. M., Willis, K. J., Caliman, J.-P., & Snaddon, J. L. (2016). Effects of Soil Management Practices on Soil Fauna Feeding Activity in an Indonesian Oil Palm Plantation. *Agriculture, Ecosystems & Environment*, 218, 133–140.

Tran-Nguyen, P. L., Ong, L. K., Go, A. W., Ju, Y. H., & Angkawijaya, A. E. (2020). Non-Catalytic and Heterogeneous Acid/Base-Catalyzed Biodiesel Production: Recent and Future Developments. *Asia-Pacific Journal of Chemical Engineering*, 15(4), e2490.

Trase. (2024). SEI-PCS Indonesia Palm Oil v1.2.2 Supply Chain Map: Data Sources and Methods. *Trase*.

Varma, V., Mosedale, J. R., Alvarez, J. A. G., & Bebber, D. P. (2025). Socio-Economic Factors Constrain Climate Change Adaptation in a Tropical Export Crop. *Nature Food*, 6, 343–352.

Waisundara, V. Y., & Grichar, W. J. (2024). *Elaeis guineensis*: New Insights. BoD–Books on Demand.

Why, E. S. K., Ong, H. C., Lee, H. V., Chen, W. H., Asikin-Mijan, N., Varman, M., & Loh, W. J. (2022). Single-Step Catalytic Deoxygenation of Palm Feedstocks for the Production of Sustainable Bio-Jet Fuel. *Energy*, 250, 123456.

Wicke, B., Sikkema, R., Dornburg, V., & Faaij, A. (2011). Exploring Land Use Changes and the Role of Palm Oil Production in Indonesia and Malaysia. *Land Use Policy*, 28(1), 193–206.

Yengoh, G. T., & Armah, F. A. (2015). Effects of Large-Scale Acquisition on Food Insecurity in Sierra Leone. *Sustainability*, 7(7), 9505–9539.

16 Soybean as an Efficient Biofuel Producer

A Critical Analysis for Sustainable Uses

Karen Christina, Suresh Kumar Krishnan, Kavitha Subbaih, Karthik Velusamy, and Selvakumar Periyasamy

16.1 INTRODUCTION

Sustainable energy is vital in the 21st century for meeting current needs without compromising future generations, primarily through renewable sources like solar, wind, hydro, geothermal, and biomass (Fathima & Palanisamy, 2018; Rosen & Farsi, 2022). These alternatives reduce reliance on fossil fuels and mitigate greenhouse gas emissions driving climate change. The goal of "zero pollution" emphasizes keeping pollutant levels below environmentally harmful thresholds, aligning with ecological sustainability rather than eliminating emissions entirely (Awewomom et al., 2024). Advancing clean energy innovations enhances energy security, supports green jobs, and promotes economic growth, making the shift to renewables essential amid rising global energy demands and environmental threats (Qu & Chen, 2024).

Biofuels are becoming alternatives or complements to fossil fuels, especially plant or animal wastes, with bioethanol and biodiesel being the main biofuels used in transport (El-Araby, 2024). To reduce greenhouse gas emissions, improve energy diversification, and increase agribusiness and industrial development, governments worldwide promote biofuel blending mandates (Scott, 2025). They are necessary to get to a low-carbon future and to foster the investment in renewable technologies (Altawell, 2021). The global biodiesel demand is expected to rise from 2023 to 2028 by 38 billion liters (an increase of 30%), up to 200 billion liters by 2028, with ethanol and renewable diesel being main drivers of the growth, according to the IEA (2023) (Figure 16.1).

Soybean oil has been one of the primary feedstocks through which biodiesel can be produced due to its high oil yield, ideal fatty acid composition, and widespread availability. Soybean is one of the most widely grown oilseed crops and it is a known and stable source of triglycerides, which can be transesterified easily into fatty acid methyl esters (FAMEs). We have studied the following improvements in the enzymatic and heterogeneous catalytic techniques for biosynthesis from soybean oil, which also met the high efficiency and sustainability requirements for biodiesel in recent decades. In addition, the soybean crop has been bred to higher oil content and also more resistant to pests, supporting the possibility of using soybeans on a large scale in biofuel applications. It can be said that soybean biodiesel is very beneficial from the environmental point of view, as it can reduce greenhouse gas emissions by up to 78% compared to conventional diesel.

Currently, soybean oil is the major feedstock for biodiesel production because it is readily available and has good properties. As of early 2024, roughly 42% of the feedstock is used in the production of biodiesel and renewable diesel in the United States came from soybean oil, retaining supremacy against other inputs such as used cooking oil and tallow (Li et al., 2023). Since soybean oil can be converted to biodiesel, advancements in processing technologies like microreactor use have increased production yield and decreased reaction time (Muranaka et al., 2023). Additionally,

DOI: 10.1201/9781003600169-16

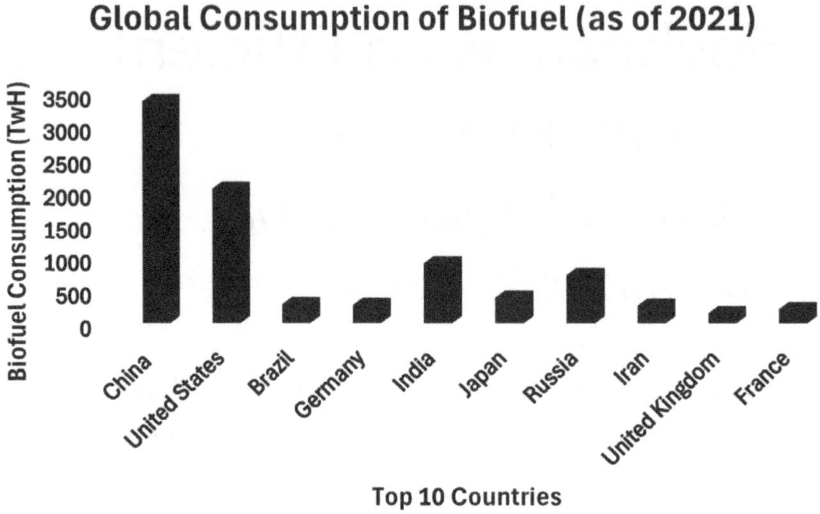

FIGURE 16.1 Global consumption of biofuel (as of 2021).

the carbon footprint of producing soybean oil in the United States is reduced dramatically in a life cycle assessment. These advancements prove that soybean oil is a critical player in the efforts to promote renewable energy and clean up our air of greenhouse gas emissions (Pradhan et al., 2011).

However, as research continues to demonstrate soybean oil as a suitable feedstock for biodiesel, there is a wide variety of gaps in building sustainable production and use for the oil (Dhairiyasamy et al., 2025). While current studies on potential yield and the basic transesterification process are rampant, primarily focusing on the environmental trade-offs, business integration of crop-based renewable systems (CoRS) requires a life cycle assessment, integration of environmental trade-offs, socioeconomic implications, different agro-climatic regions, and the synergy of multifaceted sectors. Also, conventional literature often skews the land use competition, dependence on fertilizer, as well as the waste management after crop has taken place (Silitonga et al., 2025). The purpose of this study is to critically discuss the role of soybean as a source of biofuel by assessing these under-explored issues. The main objectives are to evaluate the efficiency and environmental performance of soybean biodiesel produced in various production systems; to identify technological innovations of oil extraction and transesterification efficiency; to develop integrated sustainability frameworks that consider the energy output, ecological and economic viability at the same time. The outcomes of this analysis will bear further toward the identification of more robust, region-specific approaches to grow soybean as a lynchpin for sustainable biofuel production.

16.2 SOYBEAN AND ITS CULTIVATION

The East Asian legume variety known as soybean exists as an annual bushy plant type that colonizes various soil types and temperature zones to generate nitrogen-enriching seeds that support food production and animal feed, together with industrial applications (Ramlal et al., 2023). Soybeans have become one of the biggest food oil providers and serve as a widely used animal feed protein, yet their key production centers exist in the United States, Brazil, Argentina, China, and India. Modern research in plant biotechnology, combined with scientific advances in food science and medical applications, creates conditions for soybeans to become the primary plant protein consumed by humans throughout the world (Hamza et al., 2024). The rhizosphere contains symbiotic relationships between soybeans and arbuscular mycorrhizal fungi, rhizobia and other biological control organisms that operators use for biofertilization to increase nutrient efficiency, together with disease protection and improved plant health. Sustainable soybean cultivation requires proper management of abiotic soil factors such as pH level, temperature, nutrient composition, and water content as these determine yield quantity and development consistency (Pagano & Miransari, 2016).

Biodiesel production using soybean oil brings several advantages because soybeans have extensive cultivation, while existing processing assets and soybean residue act as crucial animal feed. Soybeans have developed a big framework of machinery infrastructure alongside storage, in addition to railway-cargo shipping and ground transportation networks. Like most agricultural crops, soybeans endure attacks from pests, together with diseases and face conditions from adverse weather conditions (Dreoni et al., 2022). Soybean rust has become a prominent focus of research during the past few years. The fungal disease soybean rust originated in Asia before it moved onto South American soybean fields and reached the US soybean crops. The control of rust requires budget-intensive fungicide applications, while severe yield reduction becomes a possible consequence.

The widespread use of soybean oil creates a major obstacle for soybean cultivation because diversified industries require the same product. The product finds multiple applications throughout the food industry for cooking oil and industrial purposes. The consumption of edible fats and oils in the United States relies on soybeans to the extent of 80% (Kheam et al., 2024). The record-breaking market demand between multiple sectors caused soybean oil prices to escalate throughout the market, which affects the economic viability of soybean-based biodiesel (Miao et al., 2025).

16.3 SOYBEAN OIL EXTRACTION AND OIL CHARACTERIZATION

The extraction of seed oil has been a traditional practice of indigenous people for several centuries until mechanical soybean pressing technology became available in the 1940s. Lots of variations exist in oil-extraction equipment between expeller presses and extruders, yet expeller presses mainly serve small-scale seed-grinding activities on the farm. The solvent extraction technique stands as the primary method to extract soybean oil. The oilseed extraction industry selects n-hexane (hexane) as its main organic solvent since it absorbs oil effectively while being economical and recyclable and nonpolar, and it has a low heat of vaporization and boiling temperature of 63°C–67°C (Lavenburg et al., 2021). The general soybean oil extraction stages are represented in Figure 16.2.

Soybeans are cleaned, cracked, dehulled, and then conditioned into thin flakes for extraction. The solvent field flakes are then transported through a sealed belt conveyor to enclosed vessels,

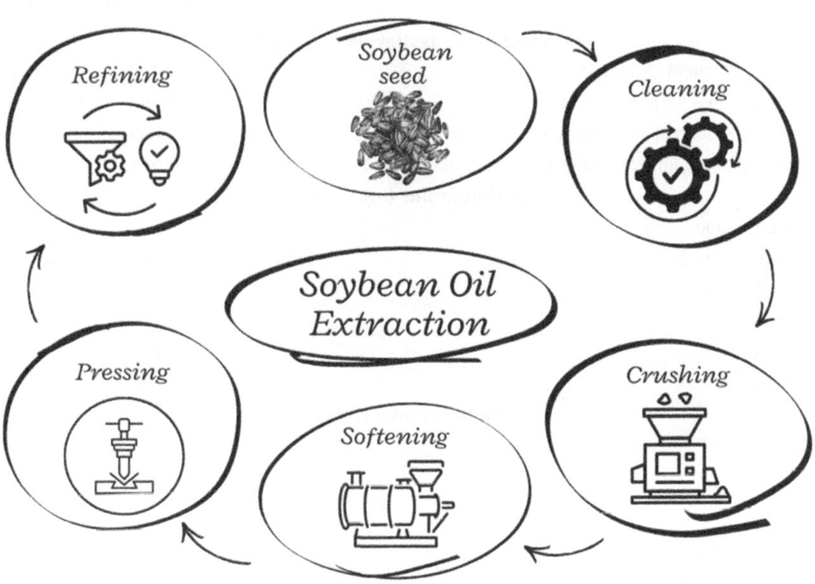

FIGURE 16.2 Extraction flow of soybean oil.

where jacket and sparge steam desolventize the flakes. The oil is then separated from hexane using rising film evaporators and final removal via vacuum distillation. Hexane recovered from both the oil and the meal is condensed using atmospheric condensers (Cheng & Rosentrater, 2017).

Fatty acid composition, oxidative stability, and the rheological behavior of its physicochemical properties are recent focuses of studies. For example, the soybean oil body extract stability and rheological properties under different denaturing conditions are examined in research (Liao et al., 2025). In addition, technological advances in the areas of analytical techniques, such as hyperspectral imaging, have allowed the rapid and nondestructive determination of oil content or fatty acid profiles in soybeans, thereby aiding the quality assessment and breeding programs. Knowledge of these characteristics is necessary for the development of optimized use of soybean oil in food products and industrial applications (Li et al., 2025).

16.3.1 Phytochemicals in Soybean

The bioactive substances present in soybeans include the compounds isoflavones and anthocyanins and saponins and phenolic acids. Scientific analysis shows that soybeans contain twelve isoflavones combined with six types of saponins and one phenol unit and fourteen phenolic acids, together with nine anthocyanins. The levels of saponin substances show higher concentrations in colored varieties rather than black types of soybeans (Rahman et al., 2024). The anthocyanin content stands much higher in black soybeans when compared to alternative colored soybean varieties. The main anthocyanin compound detected in black soybean seed coats exists as cyanidin-3-O-glucoside because it represents 75% of all detected anthocyanins (Kumar et al., 2023; Li et al., 2024). The phytochemical profile of soybean seeds is presented in Table 16.1.

It was identified from Table 16.1 that isoflavone malonylglycosides, including malonyldaidzin, malonylglycitin, and malonylgenistin, exist at the highest content level of 2.591 mg/g, whereas β-glycosides, aglycones, and acetylglycosides follow with respective amounts of 0.467, 0.198, and

TABLE 16.1
Phytochemical Profile of Soybean Seeds: Composition and Key References

Phytochemicals	Composition	Key Reference
Flavonoids—isoflavones (mg/g)		
Aglycones—daidzein, glycitein, genistein	0.198	Choi et al. (2020), Kim et al. (2025)
β-glycosides—daidzein, glycitein, genistein	0.467	
Acetylglycosides—acetyldaidzin, acetylglycitin, acetylgenistin	0.063	
Malonylglycosides—malonyldaidzin, malonylglycitin, malonylgenistin	2.591	
Anthocyanins (mg/100 g)		
Cyanidin 3-O-β-d-glucoside	333.6	Lee et al. (2016a), Choi et al. (2020), Yuui et al. (2025)
Delphinidin-3-O-β-d-glucoside	63.9	
Petunidin 3-O-β-d-glucoside	12.0	
Malvidin-3-*O*-glucoside	7.7	
Peonidin-3-*O*-glucoside	14.6	
Cyanidin-3-*O*-galactoside	8.03	
Saponins (mg/100 g)		
Soyasaponin Af (Acetyl A2)	15.67	Chávez-Mendoza and Sánchez (2017), Fujimatsu et al. (2020)
Deacetyl soyasaponin Af (A2)	3.13	
Soyasaponin Ba (V) ME	3.81	
Soyasaponin αg	11.03	
Soyasaponin βg	5.99	
Soyasaponin γg	2.65	

(*Continued*)

TABLE 16.1 (*Continued*)
Phytochemical Profile of Soybean Seeds: Composition and Key References

Phytochemicals	Composition	Key Reference
Phenolic acids (μg/g)		
Naringin	144.9	Li et al. (2024), Tian et al. (2017)
Protocatechuic acid	86.39	
Phloretic acid	70.0	
Naringenin	31.44	
Salicylic acid	16.48	
Catechin	7.20	
P-coumaric acid	6.19	
Veratric acid	3.29	
Quercetin	2.46	
Chlorogenic acid	2.30	
Ferulic acid	2.14	
Gallic acid	1.11	
Biochanin	1.05	
Caffeic acid	0.38	

0.063 mg/g (Jung et al., 2020). The main anthocyanin compound found in black soybeans is cyanidin 3-O-β-d-glucoside, which comprises 75% of the total anthocyanin content and reaches 333.6 mg/100g concentration levels. The anthocyanin compounds present in the extract contain delphinidin alongside petunidin, malvidin, peonidin, and cyanidin derivatives (Lee et al., 2016b). The highest amount of health-promoting saponins in black soybeans appears as soyasaponin Af with a concentration of 15.67 mg/100g and continues with soyasaponin αg and βg along with other saponin compounds (Lee et al., 2018). The phenolic acid content of soybeans represents naringin at 144.9 μg/g as their major compound, with additional minor amounts of protocatechuic, phloretic, naringenin, salicylic acid, and other similar compounds. Soybean variety determines both the types and proportions of phytochemicals, and colored beans tend to contain higher amounts of specific compounds (Li et al., 2024).

16.4 KEY ASPECTS OF SOYBEAN BIODIESEL PRODUCTION

Biodiesel production encompasses various methods, each with distinct advantages and challenges. The most common technique is yet transesterification, i.e., reaction of triglycerides with alcohols in the presence of catalysts to form biodiesel and glycerol. Following this, heterogenized catalysts have been investigated in which the catalyst can be more easily separated and reused than traditional homogeneous catalysts (Anil et al., 2024). Enzymatic transesterification, catalyzed by lipase enzymes, offers a green alternative under mild conditions, eliminates byproducts, and has lower product costs; however, the enzyme cost is high and the reaction time is longer. Although catalysts are not necessary for supercritical methanol transesterification due to the high temperatures and pressures of the process, reaction rates are improved and feedstocks with high free fatty acid content can also be accommodated. A third approach, biomass thermally decomposed without oxygen in pyrolysis, generates bio-oil from which biodiesel can be upgraded. This method is applicable to waste material use, but it requires the addition of complex upgrading. Reactor design to improve reaction efficiency and reduce the processing time has also been explored with innovations in the design, including ultrasonic- and microwave-assisted systems. The diverse methodologies mentioned are all about efforts to make biodiesel production more sustainable, efficient, and financially viable (Sundaramahalingam et al., 2021).

Soybean oil is extensively used as an alternative fuel. High oil output per hectare enables soybean biodiesel to function as a clean replacement for diesel fuel throughout standard engines (Anil et al., 2024).

FAMEs with higher than 98% purity result from high-oil soybean plant extracts that have been subjected to technical management through catalyst selection and reaction condition control during transesterification when using mechanical extraction or solvent treatment methods (Farouk et al., 2024). The identical procedure of biodiesel manufacturing and distribution achieves 78% lower greenhouse gas outputs than petroleum diesel fuel tools, as demonstrated by life cycle analysis results. The successful fulfillment of ASTM D6751 standards together with EN 14214 regulations allows regulators to enforce rigorous biodiesel standards, thus accelerating market entry with blending requirements (Gbadeyan et al., 2024).

There are several methods to make biodiesel from soybean oil that vary in efficiency, cost, and environmental impact. Transesterification (using alcohol, ordinarily methanol, under alkaline, acid, or enzymatic conditions and with a catalyst to give FAME and glycerol as the byproduct) is the most commonly employed technique. Sodium or potassium hydroxide is one of the alkaline catalysts widely used for its fast reaction rates and costs (Rajak et al., 2024). But crude soybean oil with high free fatty acid (FFA) might need a two-step process with pre-acid treatment to avoid soap formation.

However, due to the fact that enzymatic transesterification is a more environmentally friendly option, it is currently gaining interest. While it involves lipase enzymes catalyzing the reaction in milder conditions, waste and energy use get minimized at a higher cost and slower rate (Awogbemi & Desai, 2025a,b). Supercritical methanol transesterification is another emerging method, which obviates a catalyst and increases reaction speed and product purity. However, greater energy consumption is ultimately needed to fulfill this process, which is called high temperature and pressure (Ağbulut et al., 2024).

Pyrolysis of soybean oil creates a downwardly pyrolysis oil with diesel-like properties, but it is a complex mixture that requires further refining. Recently, ultrasound- and microwave-assisted transesterification are recent innovations in reducing reaction time and energy consumption, thus improving overall process efficiency (Milano et al., 2018). Evolving methods of how biodiesel can be produced highlight ongoing efforts to improve the economic and environmental sustainability of producing biodiesel. Table 16.2 presents a concise summary of how to produce biodiesel from soybeans. The analysis of soybean biodiesel production starts with essential information extending to

TABLE 16.2
Key Parameters and Performance Metrics of Soybean Biodiesel Production

Parameter	Details/Findings	Remarks	Key References
Soybean variety	NRC 157, NRC 150, JS 21–72, etc.	High oil-yielding and pest-resistant	Gai et al. (2025)
Oil content (%)	~18%–22% in typical varieties; up to 33% in select lines	Higher oil content = more biodiesel yield	Adie and Krisnawati (2014)
Biodiesel yield (L/ton)	Approx. 190–240 L/ton of soybean	Varies with processing technology	Ağbulut et al. (2024)
GHG emission reduction	Up to 78% lower than petroleum diesel	Environmentally sustainable fuel option	Wan Osman et al. (2024)
Cultivation regions	Central and Southern India, US Midwest, Brazil	Influences oil yield and processing suitability	Yan et al. (2023)
Economic viability	High in regions with subsidies; market-driven elsewhere	Depends on oil prices, demand, and policy support	Dougher et al. (2023)
Processing technology	Transesterification (chemical & enzymatic), pyrolysis (experimental)	Affects fuel quality and cost-efficiency	Rajak et al. (2024)
Byproducts utilized	Soybean meal (animal feed), glycerol (chemical industry)	Enhances overall economic returns	Awogbemi and Desai (2025b)
Challenges	Land competition with food crops, high input costs, policy inconsistencies	Needs integrated agri-energy policy	El-Araby (2024)
Recent trends (2023–2024)	Increased use in renewable diesel, global oil demand shifts	Expanding crushing capacity and oil trade implications	Pratama et al. (2024)

recent market trends, as show in the table. Soybean proves itself as a superior material for biodiesel production based on technical and monetary evaluation.

Biodiesel producers select soybean lines specifically with a high oil content that meets a range of 18%–22% (including 33% in particular types) and biological resistance to create 190–240 L/ton, which reduces greenhouse gas emissions by 78% compared to petroleum diesel. The production of soybean biodiesel takes place mostly in Central and Southern India as well as in the US Midwest and Brazil, while its profitability depends directly on institutional backing and market demand measures and fuel pricing. Biodiesel manufacturers use transesterification for the main process and generate additional benefits by turning soybean meal and glycerol remnants into valuable products. Renewable diesel applications and shifting global petroleum requirements cannot eliminate the main barriers facing the biodiesel industry, which consist of cost efficiency challenges and political instability.

16.5 BLENDING WITH CONVENTIONAL FUELS

Biofuel provides only 1% of global energy needs, but oil and its products make up around 80% of the market. Biodiesel production started its history many years ago through the continuous process of solving numerous operational challenges. High conversion efficiencies do not alleviate the requirement for additional technological development since existing technologies need further refinement. The main drawback of these industrial processes is their inability to recycle homogeneous catalysts. The processing of transesterification makes use of methanol obtained from fossil fuel reserves. A key environmental concern arises due to competition for natural resources since food production overlaps with energy crop cultivation, leading to difficulties between energy generation and food security (Prabhu et al., 2023).

The transition from conventional fuels to soybean-based products such as biodiesel and renewable diesel will happen through three main approaches, including enhancing production levels, incorporating existing fuels, and building the necessary infrastructure.

The production of biodiesel requires one or more types of raw vegetable oils and animal fats or algae, which undergo chemical transesterification to create methyl esters. The production of biodiesel involves primarily soybean oil, which represents 70% of the vegetable raw materials introduced into the process during 2018. Domestic biodiesel production meets environmental standards through U.S. government guidelines, which declare it carbon-neutral because plant feedstock farming removes atmospheric CO_2, according to CME Group (2021). The biodiesel sector remains apprehensive about identifying new glycerol uses in industry, together with locating alternative oil resources that merge affordability with greater operational effectiveness and production scope. The majority of biodiesel byproducts end up being discarded through environmentally harmful disposal methods, which involve waterway disposal or landfilling (Bibin et al., 2023). The fuel can be applied directly as an alternative to petroleum-based diesel fuel in compression ignition (CI) engines. Research has shown that the optimized blend of soybean biofuel works as a replacement for conventional fuels in Indian commercial vehicles (Rajak et al., 2024).

16.6 ADVANCING AND ENDORSING SOYBEAN-DERIVED FUELS

Financial incentives, which also include tax breaks and regulatory requirements from government authorities, enhance the production and adoption of soybean-based fuels. Through government-backed supportive measures, authorities can establish an advantageous framework that boosts the supply and demand of soybean-based renewable fuel products. The incentives established by governments lead to decreased production expenses for alternative fuels alongside better economic viability and transition toward sustainable power sources (Debnath et al., 2025). Through governmental interaction, the renewable energy transition becomes faster while decreasing reliance on traditional fossil fuels and generating sustainability results for the environment (Chen et al., 2018). The benefits of first-generation biofuels are overshadowed by their major disadvantages of interfering with food production and feedstock-related food shortages, as well as their high production expenses and land use challenges from requiring specific plant components. Second-generation biofuels were derived from non-edible lignocellulosic biomass because they avoid food chain conflicts and achieve higher yields on less land through full plant utilization and waste forestry. The manufacturing process of these products needs advanced technological systems with increased financial investments. The production of third- and fourth-generation biofuels requires prolonged development along with lab-based experimentation executed by experts (Mahapatra et al., 2021).

16.7 CURRENT CHALLENGES AND FUTURE OPPORTUNITIES FOR SUSTAINABILITY PRACTICES

Soybean biodiesel stands at the crossroads of opportunity and challenge in the quest for sustainable energy. One of the foremost concerns is the competition between fuel and food uses of soybeans, which has been linked to rising food prices and land use conflicts. The expansion of soybean farming, particularly for biofuel, has contributed to deforestation and habitat destruction in ecologically sensitive regions like the Amazon and the Cerrado. Furthermore, biodiesel production processes can be energy-intensive and environmentally taxing due to the heavy use of chemical fertilizers and pesticides, compounding concerns over soil degradation and water contamination (Nath, 2024). Despite these issues, efforts to improve the sustainability of soybean biodiesel are gaining traction. Advances in agricultural biotechnology have produced soybean cultivars with higher oil content, greater resistance to pests and diseases, and improved stress tolerance traits, which can lead to reduced input requirements and increased yield (Shelare et al., 2023). Innovations such as nano-additives and bio-waste-derived catalysts are also emerging as key tools for enhancing biodiesel production efficiency while minimizing environmental impact. Policy initiatives like Brazil's "Fuel of the Future" program demonstrate how governmental support can promote higher biodiesel blending mandates, further advancing the development of more sustainable biofuel markets (Farouk et al., 2024).

In addition to these biological and technological advances, systemic and process-related challenges continue to affect the overall sustainability of soybean biodiesel. The use of fossil-derived methanol in the transesterification process offsets many of the environmental benefits of biodiesel, prompting a need for renewable methanol alternatives. Moreover, the reuse of catalysts and the energy-intensive nature of the production process still pose sustainability barriers (Mathew et al., 2021). However, there are strong opportunities to improve these practices. Emerging technologies like eco-friendly solvents and reusable reactors are being explored to reduce the environmental footprint of biodiesel manufacturing. Furthermore, integrated biorefinery models can enhance value recovery by converting byproducts such as glycerol and agricultural residues into secondary products or energy inputs, thereby improving both economic and resource efficiency (Wan Osman et al., 2024). These efforts are complemented by policy frameworks that support biofuel mandates, tax incentives, and carbon pricing mechanisms to encourage the adoption of cleaner energy alternatives. Strategic alignment across technology, agronomy, and regulation will be key to transforming soybean biodiesel from a controversial option into a credible, scalable, and sustainable solution for global energy needs. Addressing these multidimensional challenges will ensure that soybean biodiesel not only contributes to energy security but also aligns with long-term environmental and economic goals (Gbadeyan et al., 2024).

16.8 CONCLUSION

Soybean (Glycine max) is a well-known, globally important crop due to its excellent adaptation, nitrogen fixation, and high oil content, providing food and energy resources. Soybeans are widely grown in the United States, Brazil, India, and Argentina, where they can be successfully grown in agroclimatic zones, especially with very little input, and smallholder and industrial scale agriculture are supported. Soybean seeds are mechanically pressed to extract the oil, or the oil is extracted with a solvent and impurities are removed by refining. The oil is mostly composed of unsaturated fatty acids, making it suitable for conversion into biodiesel by transesterification. Efficient catalysts, optimal conditions, and eco-friendly solvents play a vital role in this process, allowing for an increase in yield and lower environmental impact. Soybean-based biodiesel can be used as a blended fuel (B20 or B100) in conventional gasoline direct injection engines and diesel fueled engines, with no requirement for any type of modification of existing engines or the diesel fuel supply and distribution infrastructure. Policies and market incentives, as well as technological innovations, are all in support of soybean-derived biofuels. There are, however, challenges to overcome, including debates of food versus fuel, pests and diseases, reliance on fossil-based methanol, and inconsistent policy frameworks. Application of the concepts in future soybean breeding will develop high-oil, stress-tolerant varieties to create future opportunities for the development of soybean byproducts into biorefinery byproducts and the realization of integrated sustainability practices. The advances suggested here can make soybean biodiesel a foundation stone of the global transition to clean energy.

COMPETING INTERESTS

The authors declare that they have no competing interest.

DECLARATION OF GENERATIVE AI AND AI-ASSISTED TECHNOLOGIES IN THE WRITING PROCESS

During the preparation of this work, the book chapter author(s) have not taken any assistance from AI, and the book chapter author(s) reviewed and edited the content as needed and take(s) full responsibility for the content of the publication.

REFERENCES

Adie, M. M., & Krisnawati, A. (2014). Soybean opportunity as source of new energy in Indonesia. *International Journal of Renewable Energy Development*, 3(1), 37–43. https://doi.org/10.14710/ijred.3.1.37-43

Ağbulut, Ü., Sathish, T., Kiong, T. S., Sambath, S., Mahendran, G., Kandavalli, S. R., Sharma, P., Gunasekar, T., Kumar, P. S., & Saravanan, R. (2024). Production of waste soybean oil biodiesel with various catalysts, and the catalyst role on the CI engine behaviors. *Energy*, 290, 130157. https://doi.org/10.1016/j.energy.2023.130157

Altawell, N. (Ed.). (2021). 5 - Project management in oil and gas. In *Rural Electrification* (pp. 91–107). Academic Press. https://doi.org/10.1016/B978-0-12-822403-8.00005-9

Anil, N., Rao, P. K., Sarkar, A., Kubavat, J., Vadivel, S., Manwar, N. R., & Paul, B. (2024). Advancements in sustainable biodiesel production: A comprehensive review of bio-waste derived catalysts. *Energy Conversion and Management*, 318, 118884. https://doi.org/10.1016/j.enconman.2024.118884

Awewomom, J., Dzeble, F., Takyi, Y. D., Ashie, W. B., Ettey, E. N. Y. O., Afua, P. E., Sackey, L. N. A., Opoku, F., & Akoto, O. (2024). Addressing global environmental pollution using environmental control techniques: A focus on environmental policy and preventive environmental management. *Discover Environment*, 2(1), 8. https://doi.org/10.1007/s44274-024-00033-5

Awogbemi, O., & Desai, D. A. (2025a). Application of computational technologies for transesterification of waste cooking oil into biodiesel. *Biomass and Bioenergy*, 194, 107620. https://doi.org/10.1016/j.biombioe.2025.107620

Awogbemi, O., & Desai, D. A. (2025b). Progress in the conversion of biodiesel-derived crude glycerol into biofuels and other bioproducts. *Bioresource Technology Reports*, 30, 102106. https://doi.org/10.1016/j.biteb.2025.102106

Bibin, C., Sheeja, R., Devarajan, Y., Madhu, S., Ponnappan, V. S., & Natrayan, L. (2023). Environment impact assessment of agricultural diesel engines utilizing biodiesel derived from phoenix sylvestris oil. *Environmental Research*, 224, 115432. https://doi.org/10.1016/j.envres.2023.115432

Chávez-Mendoza, C., & Sánchez, E. (2017). Bioactive compounds from Mexican varieties of the common bean (*Phaseolus vulgaris*): Implications for health. *Molecules*, *22*(8), 8. https://doi.org/10.3390/molecules22081360

Chen, R., Qin, Z., Han, J., Wang, M., Taheripour, F., Tyner, W., O'Connor, D., & Duffield, J. (2018). Life cycle energy and greenhouse gas emission effects of biodiesel in the United States with induced land use change impacts. *Bioresource Technology*, 251, 249–258. https://doi.org/10.1016/j.biortech.2017.12.031

Cheng, M.-H., & Rosentrater, K. A. (2017). Economic feasibility analysis of soybean oil production by hexane extraction. *Industrial Crops and Products*, 108, 775–785. https://doi.org/10.1016/j.indcrop.2017.07.036

Choi, Y.-M., Yoon, H., Lee, S., Ko, H.-C., Shin, M.-J., Lee, M. C., Hur, O. S., Ro, N. Y., & Desta, K. T. (2020). Isoflavones, anthocyanins, phenolic content, and antioxidant activities of black soybeans (*Glycine max* (L.) Merrill) as affected by seed weight. *Scientific Reports*, *10*(1), 19960. https://doi.org/10.1038/s41598-020-76985-4

Debnath, D., Whistance, J., & Thompson, W. (2025). The U.S. biodiesel outlook: A pathway toward sustainability. *Sustainability*, *17*(2), 2. https://doi.org/10.3390/su17020689

Dhairiyasamy, R., Bunpheng, W., Kit, C. C., & Hasan, N. (2025). Comparative performance and emission analysis of soybean and algae biodiesels in low heat rejection engines. *Energy Science & Engineering*, 13(4), 1732–1748. https://doi.org/10.1002/ese3.2090

Dougher, M., Soh, L., & Bala, A. M. (2023). Techno-economic analysis of interesterification for biodiesel production. *Energy & Fuels*, 37(4), 2912–2925. https://doi.org/10.1021/acs.energyfuels.2c04029

Dreoni, I., Matthews, Z., & Schaafsma, M. (2022). The impacts of soy production on multi-dimensional well-being and ecosystem services: A systematic review. *Journal of Cleaner Production*, 335, 130182. https://doi.org/10.1016/j.jclepro.2021.130182

El-Araby, R. (2024). Biofuel production: Exploring renewable energy solutions for a greener future. *Biotechnology for Biofuels and Bioproducts*, 17(1), 129. https://doi.org/10.1186/s13068-024-02571-9

Farouk, S. M., Tayeb, A. M., Abdel-Hamid, S. M. S., & Osman, R. M. (2024). Recent advances in transesterification for sustainable biodiesel production, challenges, and prospects: A comprehensive review. *Environmental Science and Pollution Research*, 31(9), 12722–12747. https://doi.org/10.1007/s11356-024-32027-4

Fathima, A. H., & Palanisamy, K. (2018). 8 - Renewable systems and energy storages for hybrid systems. In A. H. Fathima, N. Prabaharan, K. Palanisamy, A. Kalam, S. Mekhilef, & J. J. Justo (Eds.), *Hybrid-Renewable Energy Systems in Microgrids* (pp. 147–164). Woodhead Publishing. https://doi.org/10.1016/B978-0-08-102493-5.00008-X

Fujimatsu, T., Endo, K., Yazaki, K., & Sugiyama, A. (2020). Secretion dynamics of soyasaponins in soybean roots and effects to modify the bacterial composition. *Plant Direct*, 4(9), e00259. https://doi.org/10.1002/pld3.259

Gai, Y., Liu, S., Zhang, Z., Wei, J., Wang, H., Liu, L., Bai, Q., Qin, Q., Zhao, C., Zhang, S., Xiang, N., & Zhang, X. (2025). Integrative approaches to soybean resilience, productivity, and utility: A review of genomics, computational modeling, and economic viability. *Plants*, *14*(5), 5. https://doi.org/10.3390/plants14050671

Gbadeyan, O. J., Muthivhi, J., Linganiso, L. Z., Mpongwana, N., Dziike, F., & Deenadayalu, N. (2024). Recent improvements to ensure sustainability of biodiesel production. *Biofuels*, 15(8), 1063–1077. https://doi.org/10.1080/17597269.2024.2318852

Hamza, M., Basit, A. W., Shehzadi, I., Tufail, U., Hassan, A., Hussain, T., Siddique, M. U., & Hayat, H. M. (2024). Global impact of soybean production: A review. *Asian Journal of Biochemistry, Genetics and Molecular Biology*, 16(2), 12–20. https://doi.org/10.9734/ajbgmb/2024/v16i2357

International Energy Agency. (2023). *Renewables 2023: Analysis and Forecasts to 2028*. Paris: IEA. Licence: CC BY 4.0.

Jung, Y. S., Rha, C.-S., Baik, M.-Y., Baek, N.-I., & Kim, D.-O. (2020). A brief history and spectroscopic analysis of soy isoflavones. *Food Science and Biotechnology*, 29(12), 1605–1617. https://doi.org/10.1007/s10068-020-00815-6

Kheam, S., Rubene, D., Markovic, D., Ith, S., Uk, O. N., Soung, S., & Ninkovic, V. (2024). The effects of cultivar mixtures on insect pest and natural enemy abundance, diseases, and yield in tropical soybean cropping system. *Biological Control*, 196, 105571. https://doi.org/10.1016/j.biocontrol.2024.105571

Kim, G.-Y., Kwon, Y. S., Gebru, Y. A., Kim, Y.-H., Kang, D. H., Kim, M.-K., & Choi, H.-S. (2025). Dynamic changes of nutrients, isoflavone composition and antioxidant activities during liquid-state fermentation of soybean embryo homogenate by *Ganoderma lucidum*. *Food Chemistry:* X, 26, 102283. https://doi.org/10.1016/j.fochx.2025.102283

Kumar, M., Suhag, R., Hasan, M., Dhumal, S., Radha, Pandiselvam, R., Senapathy, M., Sampathrajan, V., Punia, S., Sayed, A. A. S., Singh, S., & Kennedy, J. F. (2023). Black soybean (*Glycine max* (L.) Merr.): Paving the way toward new nutraceutical. *Critical Reviews in Food Science and Nutrition*, 63(23), 6208–6234. https://doi.org/10.1080/10408398.2022.2029825

Lavenburg, V. M., Rosentrater, K. A., & Jung, S. (2021). Extraction methods of oils and phytochemicals from seeds and their environmental and economic impacts. *Processes*, *9*(10), 10. https://doi.org/10.3390/pr9101839

Lee, K. J., Lee, J.-R., Ma, K.-H., Cho, Y.-H., Lee, G.-A., & Chung, J.-W. (2016a). Anthocyanin and isoflavone contents in Korean black soybean landraces and their antioxidant activities. *Plant Breeding and Biotechnology*, 4(4), 441–452.

Lee, M., Sorn, S. R., Park, Y., & Park, H.-K. (2016b). Anthocyanin rich-black soybean testa improved visceral fat and plasma lipid profiles in overweight/obese Korean adults: A randomized controlled trial. *Journal of Medicinal Food*, 19(11), 995–1003. https://doi.org/10.1089/jmf.2016.3762

Lee, Y. H., Kim, B., Hwang, S.-R., Kim, K., & Lee, J. H. (2018). Rapid characterization of metabolites in soybean using ultra high performance liquid chromatography coupled with electrospray ionization quadrupole time-of-flight mass spectrometry (UPLC-ESI-Q-TOF-MS/MS) and screening for α-glucosidase inhibitory and antioxidant properties through different solvent systems. *Journal of Food and Drug Analysis*, 26(1), 277–291. https://doi.org/10.1016/j.jfda.2017.05.005

Li, S., Chen, J., Hao, X., Ji, X., Zhu, Y., Chen, X., & Yao, Y. (2024). A systematic review of black soybean (*Glycine max* (L.) Merr.): Nutritional composition, bioactive compounds, health benefits, and processing to application. *Food Frontiers*, *5*(3), 1188–1211. https://doi.org/10.1002/fft2.376

Li, X., Wang, D., Gong, J., Yu, L., Ma, F., Wang, X., Zhang, L., & Li, P. (2025). Rapid and nondestructive detection of oil content and fatty acids of soybean using hyperspectral imaging. *Journal of Food Composition and Analysis*, 139, 107033. https://doi.org/10.1016/j.jfca.2024.107033

Li, Y., Xu, H., Northrup, D., & Wang, M. (2023). Effects of soybean varieties on life-cycle greenhouse gas emissions of biodiesel and renewable diesel. *Biofuels, Bioproducts and Biorefining*, 17(3), 449–462. https://doi.org/10.1002/bbb.2462

Liao, Y., Gao, Y., Pei, Y., Ma, L., Yan, S., Qi, B., & Li, Y. (2025). Characterization of the soybean oil body interface: Revealing the mechanism of changes in the interfacial protein composition, structure, and function of oil bodies extracted under different denaturation conditions. *Food Hydrocolloids*, 162, 110930. https://doi.org/10.1016/j.foodhyd.2024.110930

Mahapatra, S., Kumar, D., Singh, B., & Sachan, P. K. (2021). Biofuels and their sources of production: A review on cleaner sustainable alternative against conventional fuel, in the framework of the food and energy nexus. *Energy Nexus*, 4, 100036. https://doi.org/10.1016/j.nexus.2021.100036

Mathew, G. M., Raina, D., Narisetty, V., Kumar, V., Saran, S., Pugazhendi, A., Sindhu, R., Pandey, A., & Binod, P. (2021). Recent advances in biodiesel production: Challenges and solutions. *Science of the Total Environment*, 794, 148751. https://doi.org/10.1016/j.scitotenv.2021.148751

Miao, R., Li, Y., Khanna, M., Clark, C., Burkholder, D., & Chen, L. (2025). Quantifying the land-use change due to soybean-based biodiesel in the United States. *Applied Economic Perspectives and Policy*, 1–28. https://doi.org/10.1002/aepp.70005

Milano, J., Ong, H. C., Masjuki, H. H., Silitonga, A. S., Chen, W.-H., Kusumo, F., Dharma, S., & Sebayang, A. H. (2018). Optimization of biodiesel production by microwave irradiation-assisted transesterification for waste cooking oil-*Calophyllum inophyllum* oil via response surface methodology. *Energy Conversion and Management*, 158, 400–415. https://doi.org/10.1016/j.enconman.2017.12.027

Muranaka, Y., Maki, T., Hasegawa, I., Nakagawa, H., Kubota, R., Kaji, R., & Mae, K. (2023). Biodiesel fuel production from soybean oil using a microreactor. *Journal of Chemical Engineering of Japan*, 56(1), 2213283. https://doi.org/10.1080/00219592.2023.2213283

Nath, S. (2024). Biotechnology and biofuels: Paving the way towards a sustainable and equitable energy for the future. *Discover Energy*, 4(1), 8. https://doi.org/10.1007/s43937-024-00032-w

Pagano, M. C., & Miransari, M. (2016). 1 - The importance of soybean production worldwide. In M. Miransari (Ed.), *Abiotic and Biotic Stresses in Soybean Production* (pp. 1–26). Academic Press. https://doi.org/10.1016/B978-0-12-801536-0.00001-3

Prabhu, C., Navaneetha Krishnan, B., Prakash, T., Rajasekar, V., Balasubramanian, D., Le, V. V., Linh Le, N. V., Phong Nguyen, P. Q., & Nguyen, V. N. (2023). RETRACTED: Biodiesel unsaturation and the synergic effects of hydrogen sharing rate on the characteristics of a compression ignition engine in dual-fuel mode. *Fuel*, 334, 126699. https://doi.org/10.1016/j.fuel.2022.126699

Pradhan, A., S. Shrestha, D., McAloon, A., Yee, W., Haas, M., & Duffield, J. A. (2011). Energy life-cycle assessment of soybean biodiesel revisited. *Transactions of the ASABE*, 54(3), 1031–1039. https://doi.org/10.13031/2013.37088

Pratama, J. H., Rahmawati, Z., Widyanto, A. R., Gunawan, T., Abdullah, W. N. W., Jamari, N. L. A., Hamzah, A., & Fansuri, H. (2024). Advancements in green diesel production for energy sustainability: A comprehensive bibliometric analysis. *RSC Advances*, 14(48), 36040–36062. https://doi.org/10.1039/D4RA06262K

Qu, J., & Chen, J. (2024). Pathways toward a pollution-free planet and challenges. *Frontiers of Environmental Science & Engineering*, 18(6), 67. https://doi.org/10.1007/s11783-024-1827-0

Rahman, U., Younas, Z., Ahmad, I., Yousaf, T., Latif, R., Rubab, U., Hassan, H., Shafi, U., & Mashwani, Z.-R. (2024). Enhancing health and therapeutic potential: Innovations in the medicinal and pharmaceutical properties of soy bioactive compounds. *Frontiers in Pharmacology, 15.* https://doi.org/10.3389/fphar.2024.1397872

Rajak, U., Chaurasiya, P. K., Verma, T. N., Dasore, A., Ağbulut, Ü., Meshram, K., Saleel, Ca., Saboor, S., Cuce, E., & Mian, Z. (2024). Optimizing soybean biofuel blends for sustainable urban medium-duty commercial vehicles in India: An AI-driven approach. *Environmental Science and Pollution Research*, 31(22), 32449–32463. https://doi.org/10.1007/s11356-024-33210-3

Ramlal, A., Nautiyal, A., Lal, S. K., & Chigeza, G. (2023). Editorial: A wonder legume, soybean: prospects for improvement. *Frontiers in Plant Science, 14.* https://doi.org/10.3389/fpls.2023.1294185

Rosen, M., & Farsi, A. (2022). CHAPTER THREE - Sustainability and sustainable energy. In M. Rosen & A. Farsi (Eds.), *Sustainable Energy Technologies for Seawater Desalination* (pp. 107–132). Academic Press. https://doi.org/10.1016/B978-0-323-99872-7.00007-3

Scott, W. A. (2025). Cost and carbon-intensity reducing innovation in biofuels for road transportation. *Energy Policy*, 197, 114416. https://doi.org/10.1016/j.enpol.2024.114416

Shelare, S. D., Belkhode, P. N., Nikam, K. C., Jathar, L. D., Shahapurkar, K., Soudagar, M. E. M., Veza, I., Khan, T. M. Y., Kalam, M. A., Nizami, A.-S., & Rehan, M. (2023). Biofuels for a sustainable future: Examining the role of nano-additives, economics, policy, internet of things, artificial intelligence and machine learning technology in biodiesel production. *Energy*, 282, 128874. https://doi.org/10.1016/j.energy.2023.128874

Silitonga, A. S., Riayatsyah, T. M. I., Kalam, M. A., Sarifudin, A., Fattah, I. M. R., Muraza, O., Putra, N. S. D., Sebayang, A. R., Sebayang, A. H., & Hermawan, H. (2025). Status, developments, and sustainability of biowaste feedstock: A review of current progress. *Renewable and Sustainable Energy Reviews*, 217, 115769. https://doi.org/10.1016/j.rser.2025.115769

Sundaramahalingam, M. A., Amrutha, C., Sivashanmugam, P., & Rajeshbanu, J. (2021). An encapsulated report on enzyme-assisted transesterification with an allusion to lipase. *3 Biotech, 11*(11), 481. https://doi.org/10.1007/s13205-021-03003-3

Tian, Y., Liimatainen, J., Alanne, A.-L., Lindstedt, A., Liu, P., Sinkkonen, J., Kallio, H., & Yang, B. (2017). Phenolic compounds extracted by acidic aqueous ethanol from berries and leaves of different berry plants. *Food Chemistry*, 220, 266–281. https://doi.org/10.1016/j.foodchem.2016.09.145

Wan Osman, W. N. A., Rosli, M. H., Mazli, W. N. A., & Samsuri, S. (2024). Comparative review of biodiesel production and purification. *Carbon Capture Science & Technology*, 13, 100264. https://doi.org/10.1016/j.ccst.2024.100264

Yan, Z., Li, G., & Wan, S. (2023). Oil crops: A potential source of biodiesel. *Engineering*, 29, 39–41. https://doi.org/10.1016/j.eng.2023.07.011

Yuui, K., Kudo, R., Kasuda, S., & Ueshima, S. (2025). Cyanidin 3-O-β-d-glucoside from black bean inhibits the migration of vascular smooth muscle cells in the presence of adipocytes. *Pharmacognosy Magazine*, 09731296251313616. https://doi.org/10.1177/09731296251313616

17 Oilseed Rape as an Efficient Biofuel Producer

A Critical Analysis for Sustainable Uses

Min Zhang and Yeit Haan Teow

17.1 INTRODUCTION

Against the backdrop of the continuous growth of global energy demand and the increasing concern about the environmental impact of traditional fossil fuels, the development of renewable and clean energy alternatives has become an urgent task (Velusamy et al. 2024; Bujnovský, Holíčková and Ondrejíčková 2020). As a sustainable energy option, biofuels can effectively reduce greenhouse gas emissions, decrease the dependence on imported fossil fuels, and provide support for the stability and sustainability of the energy supply (Antar et al. 2021). Oilseed rape, as one of the oilseeds crops widely cultivated around the world, demonstrates great potential in the field of biofuel production due to its unique biological characteristics and high oil yield (Wang et al. 2025b; Qi et al. 2024a). To this end, it is necessary to improve the marginal land utilization rate through technological innovations such as the cultivation of resistant varieties and the optimization of intercropping models, build a circular agricultural system that covers cake feed and straw energy, and balance energy and food security with sustainable certification standards and low-carbon technology subsidy policies.

A thorough study of the sustainable utilization strategies of oilseed rape as a biofuel producer holds great practical significance for promoting the development of the biofuel industry (Saleem 2022). First of all, by optimizing planting technology, such as promoting crop rotation system and applying precision agriculture management, the use of fertilizers and pesticides can be reduced while ensuring oilseed rape production, and the negative impact on the ecological environment can be reduced. Second, it is crucial to build a resource cycle system of the whole industrial chain: the cake meal after oil extraction can be returned to the farmland as a high-protein feed or organic fertilizer, and the straw is converted into clean energy through biomass power generation or pyrolysis technology, forming a closed-loop model of "plant-processing-waste recycling". In addition, biofuel production needs to strengthen technological innovation, such as the development of low-cost oil conversion processes and efficient catalysts, improve the production efficiency of biodiesel and jet fuel, and explore algae-rape collaborative production models to break through feedstock constraints. In the future, it is necessary to build oilseed rape biofuel development model that considers ecological protection, economic benefits, and social benefits through multidisciplinary collaborative research, to promote it to become the core path of green energy transformation.

17.2 ADVANTAGES OF RAPE AS A BIOFUEL FEEDSTOCK

17.2.1 High Fat Content

The oil content of rapeseed is usually between 35% and 50%, and for some high-oil varieties, it can even exceed 55%. This high oil content makes oilseed rape an excellent raw material for the production of biofuels such as biodiesel. Compared with some other oilseed crops, such as soybeans

DOI: 10.1201/9781003600169-17

(with an oil content of approximately 15% to 20%), oilseed rape is able to produce more oil per unit area, thereby improving the production efficiency of biofuels. For example, with the same planting area, the amount of biodiesel produced by oilseed rape may be several times that produced by soybeans. This is of great significance for large-scale biofuel production. In addition, oilseed rape has a high oil production efficiency (Arvidsson et al. 2011). Approximately 1,300 to 1,500 L of biodiesel can be produced per hectare of rapeseed, which far exceeds that of soybeans and sunflowers. Its efficiency is close to that of palm oil, but it does not require planting in tropical rainforest areas. Although palm oil has a high oil yield, it can only be planted in tropical regions, and it is often accompanied by the problem of deforestation (Indonesia loses approximately 500,000 ha of rainforest each year due to palm oil cultivation). However, oilseed rape can be cultivated on a large scale in temperate regions (such as the three prairie provinces in Canada and northern Europe), without the need to damage the original ecosystem.

In conclusion, the high oil content of oilseed rape makes it the most oil-efficient temperate oilseed crop per unit of land. By integrating the comprehensive utilization of byproducts and technological innovation, the production cost of biofuels can be significantly reduced, and the environmental benefits can be enhanced. In the future, through the cultivation of stress-tolerant varieties, the promotion of low-carbon processing technologies, and the implementation of circular agriculture models, oilseed rape is expected to play a more crucial role in the energy transition. However, it is necessary to be vigilant against the ecological risks of excessive expansion and balance the goals of "energy security" and "sustainable development". Through the cultivation of stress-tolerant varieties (such as salt-alkali tolerance and drought resistance) and the circular agriculture model (simultaneous production of oil, feed, and fertilizer), its status as a sustainable biofuel raw material can be further consolidated. In the future, it is necessary to prioritize the promotion of its cultivation on non-grain marginal lands to avoid the risk of "competing with food crops for land". Table 17.1 shows a detailed comparison of the oil content of major oilseed crops worldwide and the oil yield per unit area. Combined with scientific research and agricultural practices, it highlights the core advantages of rapeseed.

17.2.2 Wide Adaptability

Oilseed rape has strong adaptability and can grow under different climatic and soil conditions. Whether in the cold high-latitude regions or the warm and humid subtropical regions, there are oilseed rape varieties suitable for planting. The adaptability of oilseed rape is not only the basis for it to become a "universal crop", but also the key solution to solve the multiple challenges faced by global agriculture such as food security, climate change, and resource constraints. In the future, enabled by intelligent breeding, digital farming, and other technologies, it will further expand to extreme environments such as arid areas and polar margins, and become a "green cornerstone" supporting the sustainable development of human civilization (Table 17.2).

In terms of soil, oilseed rape has relatively low requirements for soil fertility and can grow well in relatively infertile soils such as slightly saline-alkali land. This wide adaptability enables oilseed rape to be planted on a large scale worldwide without being restricted by regions, providing a stable raw material supply for the biofuel industry. For example, in China, from the Northeast Plain to the Yangtze River Basin and then to the mountainous areas in the southwest, oilseed rape is planted over a large area, making full use of the land resources in different regions (Aubert et al. 2024).

In terms of climate, oilseed rape has strong cold tolerance. Different varieties of rapeseed have different ranges of temperature adaptation, but they can generally tolerate a certain degree of low temperature. For example, winter rapeseed can overwinter in areas where winters are relatively cold. Some varieties can even survive in low temperatures ranging from −5°C to −10°C and continue to grow after the temperature rises in spring. At the same time, oilseed rape can also grow in a certain high-temperature environment. During the flowering and pod-setting period, it can generally

TABLE 17.1
Comparison Table of Oil Content and Oil Production Efficiency of Major Oil Crops in the World

Crop	Castor-Oil Plant (%)	Per Unit Area Yield (L/ha)	Suitable Climate	Planting Cycle (Month)	Major Oil Producing Countries	Characteristic Analysis
Rape	40–45	1,300–1,500	Temperate and subtropical zone	4–6	Canada, the European Union, China	Adapted to temperate climate, rapeseed meal as a byproduct has a high value; Europe dominates biodiesel use
Palm (pulp)	45–50	4,000–6,000	Tropics	Year-round (perennial)	Indonesia, Malaysia	The most efficient oil production, but requires a tropical climate and faces deforestation controversy; sustainability certification (e.g., RSPO) is required
Sunflowers	25–35	800–1,000	Temperate and arid regions	3–4	Russia, Ukraine, Argentina	Strong drought resistance and high oil quality (high vitamin E), but the yield fluctuates greatly
Soybeans	18–20	400–500	Temperate and subtropical zone	3–5	United States, Brazil, Argentina	The world's most widely cultivated, but low oil content; Soybean meal feed demand drives planting
Peanut	45–50	600–800	Temperate and subtropical zone	4–5	China, India, Nigeria	The oil content is high, but the yield per unit is low. Used more for food than for biofuels
Coconut (dried coconut meat)	60–65	2,200–2,500	Tropical coastal region	Year-round (perennial)	Philippines, Indonesia, India	The oil content is high, but the yield per unit is low. Used more for food than for biofuels
Castor-oil plant	45–55	600–900	tropical and subtropical	6–8	India, China, Brazil	The oil content is usually between 40% and 60%, depending on the variety, climate, and soil conditions (Arvidsson et al. 2011)
Corn	3–5	100–200	Temperate and subtropical zone	6–9	United States, China, Brazil	Oil content is very low, mainly extracted from the germ. The byproducts are mainly starch and feed, and the economy of oil production is poor

TABLE 17.2
Adaptability Performance of Rapeseed and Corresponding Environmental Conditions

Environmental Conditions	Classification	Yield (kg/mu)	Adaptability of Rape	Typical Case
Weather conditions	Temperature range	Cold Temperate Zone: 100–150 Subtropical zone: 150–250	It can be grown from the cold temperate zone (−5°C–15°C) to the tropical zone (15°C–35°C), and the cold-tolerant varieties can survive at −10°C	The yield of spring rapeseed (Qingza No. 5) on the Qinghai-Tibet plateau of China is 120 kg/mu, and the yield of early maturing varieties in high altitude of Southeast Asia is 180 kg/mu
	Precipitation demand	Arid region: 80–120 Humid region: 150–250	It can grow with an annual precipitation of 200–1,500 mm, and drought-tolerant varieties can survive in arid areas (such as northwest China) through drip irrigation (Aubert et al. 2024)	In Aksu, Xinjiang, rapeseed is intercropped with walnuts. With drip irrigation technology, the yield reaches 150 kg/mu, which is 30% higher than that of traditional flood irrigation
	Light requirement	Long-day regions: 200–250 Short-day regions: 120–180	Long daylight hours (>12 hours per day) promote flowering and fruiting, while short-day varieties can adapt to low-light environments	Winter rapeseed in the Yangtze River Basin grows during the low-light period and matures from late spring to early summer, with an average yield of 180 kg/mu
Edaphic condition	pH valuc	pH 4,5–5.5: 120–180 pH 8.0–9.0: 80–120	It adapts to soil with a pH of 4.5–8.5, and salt-tolerant varieties can be grown in saline-alkali land with a pH of 8.0–9.0	Salt-tolerant varieties along the coast of Jiangsu yield 100 kg/mu in soil with a pH of 8.5, while reducing the soil pH value by 0.3–0.5
	Agrotype	The Barrens: 80–120 The fertile region: 200–300	Rapeseed can be grown in sandy loam, clay, red-yellow soil, chernozem, etc. Sandy soil needs to retain water and fertilizers, while clayey soil requires enhanced drainage	In the Sichuan Basin, high-ridge cultivation is adopted in clayey soil to prevent waterlogging and root rot. The yield reaches 180 kg/mu, which is 20% higher than that of flat cultivation
	Soil fertility	The Barrens: 80–120 The fertile region: 200–300	It has strong barren-tolerance ability and can grow in barren land with organic matter content less than 1%. In fertile soil, fertilizer application needs to be controlled to prevent lodging	Rapeseed is planted on the slopes of the Yunnan-Guizhou Plateau. After combining green manure and ploughing under, its yield has increased from 100 to 150 kg/mu

(Continued)

TABLE 17.2 (*Continued*)
Adaptability Performance of Rapeseed and Corresponding Environmental Conditions

Environmental Conditions	Classification	Yield (kg/mu)	Adaptability of Rape	Typical Case
Adaptitude	Utilization of marginal land	Saline and alkaline land: 80–100 Uncultivated land: 60–90	It can be planted on marginal lands such as saline-alkali land, wastelands, and terraced fields, which help to improve the ecological environment	Rapeseed is grown on saline-alkali land in Inner Mongolia, with a yield of 80 kg/mu, and at the same time, the soil pH value is reduced by 0.5–1.0
	Stress resistance	Cold-resistant varieties: 100–120 kg/mu Waterlogging-tolerant varieties: 150–200 kg/mu	It has outstanding abilities in cold resistance, drought resistance, lodging resistance, and waterlogging tolerance, and some varieties can withstand short-term waterlogging	The rapeseed variety in the Jianghan Plain of Hubei Province recovered to a yield of 140 kg/mu in 2024 after the flood (the original yield was 200 kg/mu)
Environment-friendly measures	Variety optimization	Fertilizer-reducing varieties: 180–220 kg/mu (10%–15% higher yield than traditional varieties)	Breed varieties with strong stress resistance (such as Zhongyouza series and Huayouza series) to reduce reliance on pesticides and chemical fertilizers	The selenium-enriched rapeseed bolt variety in Dali, Yunnan, still achieves a yield of 200 kg/mu after a 30% reduction in fertilizers, which is 20% higher than that of ordinary varieties
	Agronomic management	Drone seeding: 150–180 kg/mu (efficiency increased by 50%)	Models such as crop rotation (e.g., rice-rapeseed rotation), intercropping and relay cropping (e.g., walnut+rapeseed), and straw returning to the field improve resource utilization efficiency	In the hilly and mountainous areas of Tongnan, Chongqing, rapeseed is sown by drones. The straw returning rate is 100%, and the yield is 160 kg/mu, which is 15% higher than that of manual sowing

adapt to temperatures ranging from 15°C to 25°C, and some varieties can tolerate short-term high temperatures of around 30°C.

In terms of planting systems, oilseed rape can be rotated with a variety of crops, such as rice, wheat, and corn. In the Yangtze River Basin, the common rotation pattern is rice-rapeseed rotation. After the rice is harvested, oilseed rape is planted, and after the oilseed rape is harvested, rice is planted again. In this way, the land resources and seasonal climate conditions can be fully utilized, and the land utilization rate can be increased. At the same time, it can also improve the soil structure and reduce the occurrence of diseases and pests. In addition, oilseed rape can also be interplanted with other crops, such as cotton and vegetables. When oilseed rape is interplanted in cotton fields, in the early growth stage, oilseed rape can make use of the space and sunlight between cotton plants. In the later stage, after the cotton is harvested, oilseed rape can fully receive sunlight and nutrients, achieve a double harvest of the two crops, and increase the land output rate and economic benefits.

17.2.3 Short Growth Cycle

The growth cycle of oilseed rape is relatively short, generally about 150–200 days. Take winter rapeseed as an example, which is sown in autumn and harvested in the following spring (Batchelor, Booth and Walker 1995). In this way, it can achieve the efficient utilization of land without affecting the growth seasons of other major crops. Compared with some perennial energy crops, oilseed rape can generate economic benefits more quickly and bring income to farmers. At the same time, the relatively short growth cycle also means that the planting strategy can be adjusted more quickly. According to market demand and climate change, the planting area and varieties can be flexibly arranged to ensure a stable supply of raw materials for biofuels. This biology also reduces the risk of land idling, reduces soil erosion and weed breeding by shortening planting intervals, and reduces energy consumption by 35% when combined with no-till livestream technology (Yao, Bilichak and Kovalchuk 2023). Under the promotion of the policy, the Canadian province of Saskatchewan has established a three-year rotation system of canola, flax, and barley to maintain soil fertility and ensure a continuous supply of biofuel raw materials. At present, the annual sowing area of oilseed rape in the world reaches 36 million ha, of which 60% use multi-cropping, providing a stable raw material guarantee for the biofuel industry (Li et al. 2022).

17.2.4 Environmentally Friendly

During the growth process, oilseed rape absorbs carbon dioxide and releases oxygen through photosynthesis, which helps to improve air quality. In addition, the waste such as straws generated during the oilseed rape planting process can be comprehensively utilized (Gabrielle et al. 2014). For example, through biomass fermentation technology, it can be converted into energy sources such as biogas, realizing the recycling of resources. Compared with the exploitation and use of traditional fossil energy, the entire life cycle of oilseed rape as a raw material for biofuels has less negative impact on the environment, which is in line with the concept of sustainable development. For example, canola is widely used in canola cultivation in Canada. After the oilseed rape is harvested, the stalks and leaves left in the field are crushed and returned to the field. These residues are rich in nutrients such as nitrogen, phosphorus, and potassium, which can be slowly released into the soil after microbial decomposition, providing nutrient support for subsequent crops and reducing the amount of fertilizer used. Moreover, stubble can also improve the physical properties of soil, increase the content of soil organic matter, and improve the ability of soil to retain water and fertilizer.

17.3 RAPESEED CULTIVATION AND SUSTAINABILITY

17.3.1 Optimal Planting Techniques

17.3.1.1 Rational Close Planting

Reasonably, close planting is one of the important measures to increase the yield and oil content of oilseed rape. By scientifically calculating the planting density and determining the optimal plant spacing and row spacing according to different varieties, soil fertility, and climatic conditions, land resources and light conditions can be fully utilized to increase the biomass per unit area. For example, in areas with high soil fertility and abundant sunlight, appropriately increasing the planting density can promote competition among plants, making oilseed rape plants grow more robustly and improving the efficiency of oil synthesis. However, if the planting is too dense, it will lead to poor ventilation and light transmission, as well as aggravate pests and diseases, which will instead affect the yield and quality. Therefore, it is necessary to accurately grasp the planting density to achieve high yield and high efficiency in rapeseed cultivation.

China needs to focus on variety improvement, integrate precision agriculture technology and ecological subsidy policies, and promote the transformation of oilseed rape cultivation to "high density, high efficiency and sustainable". Learn from Canada's variety advantages, Australia's no-till model, and Germany's circular economy concept, combined with local practices (such as Sun Wancang's team cold-resistant varieties), and ultimately achieve a win-win situation of grain and oil security and ecological protection.

17.3.1.2 Precision Fertilization

Oilseed rape has a high demand for nutrients during its growth. Precise fertilization can not only meet the growth needs of oilseed rape but also reduce fertilizer waste and environmental pollution. Based on the results of soil nutrient testing and combined with the fertilizer requirements of oilseed rape at different growth stages, a personalized fertilization plan is formulated. In the early growth stage of rapeseed, nitrogen fertilizer is mainly applied to promote the vegetative growth of plants. During the flowering and pod-setting stages, increasing the application amount of phosphorus and potassium fertilizers is beneficial for improving the plumpness of oilseed rape and the oil content. At the same time, the rational use of organic fertilizers can improve the soil structure, enhance soil fertility, and reduce the amounts of chemical fertilizers used. For example, in some areas, the use of organic-inorganic compound fertilizers not only ensures the nutrients required for the growth of oilseed rape but also reduces the pollution to the soil and water bodies.

17.3.1.3 Integrated Pest Control

Pests and diseases are important factors affecting the yield and quality of oilseed rape. The adoption of comprehensive prevention and control measures, including the combination of agricultural control, physical control, biological, control and chemical control, can effectively control the occurrence and spread of pests and diseases. In terms of agricultural control, measures such as reasonable crop rotation and deep ploughing and sunning the soil are taken to reduce the breeding environment for pests and diseases. For physical control, methods like hanging yellow sticky boards and installing insecticidal lamps can be adopted to trap pests. In terms of biological control, natural enemy insects and microorganisms are utilized to control pests and diseases. For example, ladybugs (*Coccinella septempunctata*) are released to control aphids. When conducting chemical control, it is necessary to strictly follow the pesticide application regulations. Select highly effective, low-toxicity, and low-residue pesticides, and control the dosage and application time to reduce the pollution of pesticides to the environment and agricultural products. For instance, in some oilseed rape-planting regions, trap crops are planted to attract pests, thereby reducing the damage to rapeseed. At the same time, in combination with biological control methods, good pest and disease control results have been achieved.

17.3.2 Land Resources Utilization

17.3.2.1 Intercropping and Intercropping

Rapeseed can be intercropped and interplanted with many crops to improve the utilization efficiency of land resources (Liu et al. 2018). For example, one study in France interplanted oilseed rape with lupine, clover, or veal peas. Compared with single cropping, lupine, clover, and vetpea intercropping with rapeseed increased the nitrogen uptake from the atmosphere by 34%, 140%, and 290%, respectively (Yang et al. 2024). When intercropping with clover and lupine, the soil nitrogen pool at harvest was higher than other treatments. When intercropping with clover and lupine, the soil nitrogen pool at harvest was higher than other treatments. In addition, the intercropping of oilseed rape with other crops can also reduce the occurrence of diseases and pests and reduce the risk of the spread of diseases and pests by changing the field ecological environment. On some farms in Canada, canola is intercropped with beans. Legume plants can attract beneficial insects such as parasitic wasps, which will prey on or parasitize pests on rape, such as rape insects, thereby reducing the damage of pests to rape. Moreover, the intercropping of beans and oilseed rape improved the lighting and ventilation conditions of the field and reduced the risk of rapeseed downy mildew and other diseases.

17.3.2.2 Marginal Land Utilization

Marginal land refers to land that is not suitable for traditional crop cultivation due to poor natural conditions or remote geographical location. Oilseed rape has strong adaptability and can grow well on some marginal land, such as mild saline-alkali land and barren slope land. Using these marginal lands to grow oilseed rape can not only increase the supply of biofuel raw materials but also play a role in improving soil and preventing soil erosion. Germany, for example, is a poster child for green energy agriculture (Andert, Ziesemer and Zhang 2021). In the early 1990s, Germany began to pay attention to the development of energy cash crops. According to German scientists working on oilseed rape and other directional breeding, oilseed rape is one of the most important energy crops in Germany, because it can be used not only as chemical raw materials, but also as refined plant diesel, instead of mineral diesel used as power fuel, which greatly increased the supply of biofuel raw materials (Hoffmann, Jacobs and Whitbread 2015). When oilseed rape is planted on marginal land, its roots can penetrate the soil during the growth process, increase the soil permeability and air permeability, help to improve the soil structure, reduce soil erosion, and prevent soil erosion (Dessi et al. 2022). After harvest, the residue of oilseed rape can increase soil organic matter content, improve soil fertility and improve soil quality.

17.3.3 Water Resource Management

17.3.3.1 Water-Saving Irrigation Technique

The growth of oilseed rape requires an appropriate amount of water, but the water demand is different in different growth stages. The use of water-saving irrigation technologies, such as drip irrigation and sprinkler irrigation, can accurately supply water according to the water demand of rape and reduce the waste of water resources (Abdollahi sisi et al. 2024). Drip irrigation technology can directly transport water to the roots of oilseed rape, improve water use efficiency, reduce field humidity, and reduce the risk of disease and insect pests. Sprinkler irrigation technology is suitable for large-area rapeseed planting and can provide water for oilseed rape uniformly. Israel lacks water resources but relies on advanced technology to achieve efficient water conservation in agriculture. In oilseed rape cultivation, the country uses a precision irrigation system, utilizing developed Internet of Things, big data, and artificial intelligence technology. Through the irrigation software systems such as CropX, AgrIOT, and Saturas in the mobile phone application, farmers can timely monitor the precise water demand in a specific period and small plots of rapeseed land, thus controlling irrigation intelligently (Schmidt and Schröder 2011). All data is synchronized to the cloud for algorithmic analysis, which can provide farmers with the most optimized and accurate irrigation solutions, not only saving water resources, but also preventing water pollution and soil damage caused by excessive fertilization.

17.3.3.2 Rainwater Collection and Utilization

In the process of oilseed rape planting, rainwater collection and utilization can be actively carried out, which can make full use of natural precipitation and relieve the pressure of irrigation water. Through the construction of rainwater collection ponds, cisterns, and other facilities, the precipitation of the rainy season is collected for irrigation during the oilseed rape growth. At the same time, reasonable planning of field drainage system to avoid excessive rain is caused by the accumulation of water in the field, affecting the growth of oilseed rape. In some of the mountain oilseed rape growing areas of Provence, local farmers build a series of rainwater collection trenches along the contours of the slopes. These collection trenches are arranged in a stepped pattern, which can effectively trap and collect rainwater. The end of the collection trench is connected to an underground reservoir where rainwater is deposited and filtered before being stored. When oilseed rape growth requires irrigation, water from the reservoir is transported to the field through a system of pipes. In addition, sprinkler irrigation and drip irrigation systems are also set up in the planting area, combined with the rainwater collection system; according to the water demand of oilseed rape at different growth stages, accurate irrigation is carried out, which greatly improves the utilization efficiency of rainwater and ensures the yield and quality of oilseed rape.

17.4 PROCESS OF TRANSFORMING OILSEED RAPE INTO BIOFUEL

17.4.1 Biodiesel Production Technology

Transesterification is the most used method to produce biodiesel. With rapeseed oil as the raw material, under the action of catalyst, fatty acid methyl ester or fatty acid ethyl ester is formed by transesterification reaction with short chain alcohols such as methanol or ethanol, i.e., biodiesel, and glycerol is produced as a byproduct (Li et al. 2024). Europe is the pioneer of converting rapeseed oil into biodiesel, and it has been widely used in European countries. European countries had planned to increase the rapeseed oil content of diesel fuel from 4.8% to 20% in 2020 to encourage the use of more rapeseed oil to replace diesel (Joseph Antony Sundarsingh et al. 2024). For example, in Germany, many gas stations provide fuel with a certain proportion of biodiesel, many of which are produced by transesterification reactions using rapeseed oil as a raw material (Mathew et al. 2014). Germany's agricultural planting area of rapeseed is large, which provides rich raw materials for biodiesel production. The biodiesel transesterification process using rapeseed as a raw material should be combined with its advantages of high oil content and low sulfur, and enhance its competitiveness by pretreatment optimization, an efficient catalytic system, and byproduct resource utilization. In the future, the focus will be on the development of high acid value resistant catalysts, low temperature and high efficiency processes, and the whole industry chain technology to promote the development of rapeseed biodiesel in the direction of low carbon and high value.

To improve the production efficiency and quality of biodiesel, it is necessary to optimize the transesterification process. On the one hand, by improving the reaction equipment, such as the use of continuous production process, it strengthens the mass and heat transfer process and improves the reaction rate and conversion rate (Qi et al. 2024b). On the other hand, the pretreatment of raw materials reduces the content of impurities in raw materials and improves the purity and stability of raw materials (Zhu et al. 2023). In addition, the activity and service life of catalysts can be improved by optimizing the use of catalysts, such as supported catalysts and heterogeneous catalysts (Abdul Wahab et al. 2025). For example, the transesterification reaction using supercritical methanol technology, without catalyst, fast reaction speed, simple product separation, can effectively improve the production efficiency and quality of biodiesel (Table 17.3).

TABLE 17.3
Summary of Key Optimization Directions

Technological Type	Optimization Focus	Conversion Efficiency	Carbon Emission Reduction	Economic Improvement
Continuous ester interchange (Germany)	The reaction time is shortened, and the catalyst circulates	98%	40%–50%	Glycerol increment + energy consumption reduction by 20%
HVO (Canada)	Raw material compatibility, product compatibility	99%	90%	Hydrogen costs reduced by 40%
Enzyme Catalysis (Sweden)	Low temperature, low energy consumption, wastewater emission reduction	96%	96%	Enzyme costs reduced by 50%
Supercritical process (Australia)	No catalyst, fast reaction	99%	99%	Reduced pretreatment costs by 60%

TABLE 17.4
Biofuel Conversion Pathways

Reaction Stage	Primary Reactant	Reaction Condition	Reaction Equation	Products
Pretreatment and hydrolysis	$(C_6H_{10}O_5)_n$, H_2O	Acid or enzyme catalysis	$(C_6H_{10}O_5)_n+nH_2O=nC_6H_{12}O_6$	$C_6H_2O_6$
Fermentation	$(C_6H_{12}O_6)$, Yeast	Oxygen-free environment, suitable temperature (about 30°C)	$C_6H_{12}O_6=2C_2H_5OH+2CO_2\uparrow$	C_2H_5OH, CO_2

17.4.2 Other Biofuel Conversion Pathways

17.4.2.1 Bioethanol Production

The starches and sugars in rapeseed can be converted into bioethanol by fermentation. First, oilseed rape is pretreated, broken, cooked, etc., so that starch and sugar substances are released. Then, microorganisms such as saccharifying enzymes and yeast are added to ferment at the right temperature and pH conditions to convert sugars into ethanol. After fermentation, the ethanol was purified by distillation and other methods to obtain bioethanol products that met the requirements. Microwave pretreatment of oilseed rape straw can significantly enhance the saccharification efficiency of cellulose, with a sugar yield of 25.34%, providing technical support to produce bioethanol (Saleem 2022; Hoang et al. 2021). Bioethanol can be used as a fuel additive and mixed with gasoline to improve the octane number of gasoline and reduce exhaust emissions. The United States, through policies such as the Renewable Fuel Standard (RFS), requires a certain percentage of biofuels, including bioethanol, to be incorporated into transportation fuels to reduce dependence on imported oil and lower greenhouse gas emissions (Velusamy et al. 2024). The specific reactions are shown in Table 17.4.

17.4.2.2 Biomass Gasification and Liquefaction

Biomass such as rape stalks can be converted into energy products through gasification and liquefaction technologies. Biomass gasification is the conversion of biomass into combustible gases, such as carbon monoxide, hydrogen, and methane, under the condition of high temperature and oxygen deficiency (Ahmad Zamri et al. 2024). These gases can be used directly as fuel, as well as for power generation, heating, and so on. Biomass liquefaction is the conversion of biomass into liquid fuels, such as bio-oil. Bio-oil has a high energy density and can be used as an alternative fuel for industrial boilers, internal combustion engines, etc. (Fan et al. 2014). For example, the local canola planting area in Canada is large, with more than 6 million ha of canola planting land and a lot of canola straw resources. The researchers carried out pyrolysis treatment of oilseed rape crop straw and transformed rape straw into bio-oil through pyrolysis. The yield of bio-oil can reach 30%–50%, which provides a new way for efficient utilization of biomass energy.

17.5 ENVIRONMENTAL IMPACTS OF RAPESEED BIOFUEL

17.5.1 Greenhouse Gas Emission Reduction

Oilseed rape biofuel has a significant greenhouse gas reduction effect throughout its life cycle (Thers et al. 2019). Oilseed rape absorbs large amounts of carbon dioxide as it grows, fixing it in the body through photosynthesis. When oilseed rape biofuel is burned and used, the amount of carbon dioxide released is basically the same as the amount of carbon dioxide absorbed during the growth process of oilseed rape, and the approximate "zero emission" of carbon is realized (Meena et al.

2024). Compared with traditional fossil fuels, such as coal and oil, the combustion process will release a lot of greenhouse gases such as carbon dioxide, aggravating global climate change. The EU has long been concerned about the greenhouse gas reduction benefits of biofuels. The production of biodiesel using oilseed rape as raw material can achieve a reduction rate of 40%–60%, and the potential of fermentation of rapeseed starch to ethanol through genetic improvement technology has also been included in the research (Vinzent et al. 2017).

17.5.2 Effect on Soil Quality

Reasonable cultivation of oilseed rape combined with biofuel production can significantly improve soil quality through straw resource utilization and byproduct recycling management. Oilseed rape straw can be used as biomass power generation or bioethanol raw materials to replace traditional incineration to reduce air pollution, while retaining part of the crushed straw to return directly to the field, which can provide organic matter and avoid excessive removal resulting in soil carbon loss (Köbke et al. 2018). The cake meal (containing 4%–6% nitrogen and 1%–2% phosphorus) after oil extraction from rapeseed can replace 20%–30% fertilizer after composting and fermentation and reduce the risk of soil acidification (Neira et al. 2024). Plant ash after biomass combustion is rich in potassium and calcium (including K_2O 5%–10%), which can effectively adjust the pH value of acidic soil (Li et al. 2025). In addition, oilseed rape absorbs 10–15 tons of CO_2/ha, its biofuel can replace fossil energy to reduce carbon emissions, combined with root carbon sequestration, and organic residue storage can further stabilize the soil carbon pool, forming a "plant-energy conversion in soil remediation" sustainable closed loop (Wang et al. 2025a). The results showed that soil with long-term cultivation of oilseed rape and straw returning to field had significantly improved organic matter content and soil aggregate stability, which was conducive to the sustainable development of agriculture.

17.5.3 Impact on Water Resources

The impact of oilseed rape cultivation and biofuel production on water resources is mainly reflected in irrigation water and production water. If reasonable water-saving irrigation technology is adopted, such as drip irrigation and spray irrigation mentioned above, it can effectively reduce the waste of irrigation water and reduce the pressure on water resources (Menhas et al. 2022). In the biofuel production process, by optimizing the production process and improving the recycling rate of water resources, the demand for water resources can also be reduced. For example, in the production of biodiesel, washing water in the reaction process is recycled for raw material pretreatment and other links, which can greatly reduce the consumption of production water (Kniuipytė et al. 2023). Overall, the impact of oilseed rape biofuel industry chain on water resources can be optimized through variety selection, water-saving technology, and closed-loop water management to reduce the negative impact of oilseed rape biofuel industry on water resources.

17.6 ECONOMIC AND SOCIAL FACTORS

17.6.1 Production Cost and Economic Benefit

The production cost of oilseed rape biofuel mainly includes raw material cost, planting cost, processing cost, and transportation cost. At present, due to the price fluctuation of oilseed rape and the high cost of biofuel production process, the production cost of oilseed rape biofuel is still higher than that of traditional fossil fuels. However, with the continuous progress of planting technology, the optimization of biofuel production process, and the appearance of scale effect, the production cost of rapeseed biofuel is expected to gradually reduce. From the perspective of economic benefits, the development of the oilseed rape biofuel industry can drive the development of related industries, such as agricultural machinery manufacturing, biofuel processing equipment manufacturing, and transportation logistics, to create more jobs and promote the development of the local economy. Many biodiesel production enterprises have gained considerable profits through large-scale production and sales of biodiesel (Wang et al. 2024). Some enterprises also increase additional income by optimizing the production process and improving the recycling efficiency of byproducts such as glycerin.

17.6.2 Market Demand and Policy Support

With the growing global demand for renewable energy, the market for biofuels is promising (Seo et al. 2022). Governments around the world have introduced policies to encourage the production and use of biofuels. The European Union has set strict renewable energy targets, requiring that by 2030, renewable energy will account for 40% of energy consumption, of which the proportion of biofuels used in the transport sector will be significantly increased. The United States has also promoted the development of the biofuel industry through tax incentives, subsidies, and other policies. The Chinese government has actively supported the development of the biofuel industry, introducing a series of policies to encourage the cultivation of energy crops such as rape and the production of biofuels. The introduction of these policies provides a good market environment and policy guarantee for the development of rape biofuel industry (Cao et al. 2022).

17.6.3 Impact on Rural Development

The cultivation of rapeseed is mainly concentrated in rural areas, and the development of biofuel industry can bring many positive effects to rural areas (Bello et al. 2024). On the one hand, it increases the source of income for farmers. By planting oilseed rape and participating in the work related to biofuel production, farmers can obtain planting income and labor income. On the other hand, it has promoted the construction and improvement of rural infrastructure, and to meet the needs of rape cultivation and biofuel production, infrastructure such as transportation and water conservancy in rural areas has been improved. In addition, the development of the biofuel industry can also attract young people to return home to start businesses and employment, promote the diversified development of the rural economy, and alleviate the problem of aging and hollowing out of the rural population. Rape is grown on a large scale in many rural areas of Germany to provide raw material for the biodiesel industry. Farmers grow rapeseed and cooperate with biodiesel companies to generate income from stable sales of oilseed rape. For example, in some rural areas of Bavaria, the proportion of farmers' income from oilseed rape in agricultural income is increasing year by year. At the same time, rape planting has also driven the prosperity of local agricultural production and sales, agricultural machinery services, and other industries, as well as promoted the diversified development of the rural economy.

17.7 DEVELOPMENT TREND AND CHALLENGE

17.7.1 Technological Innovation Direction

In the future, technological innovation in the field of oilseed rape biofuel will mainly focus on the following aspects. The first is to cultivate new varieties of oilseed rape with high yield, high oil, and strong stress resistance, as well as to improve the yield and quality of oilseed rape and reduce the production cost through gene editing, molecular marker-assisted breeding, and other technical means. The second is to optimize the biofuel production process and develop more efficient, environmentally friendly, and low-cost conversion technologies, such as the research and development of new catalysts and the application of biorefining integration technology, so as to improve the production efficiency and quality of biofuels (Abdullah et al. 2024). The third is to strengthen the comprehensive utilization technology research of biomass such as rape straw, improve the utilization efficiency of biomass energy, and achieve the maximum utilization of resources. For example, the use of biotechnology means to introduce genes with stress resistance into rape and breeding varieties that can adapt to drought, salt, high temperature, low temperature, and other harsh environments. Drought tolerance genes were transferred into oilseed rape by genetic engineering technology to improve the survival ability and yield of rapeseed under drought conditions. High-value utilization technology of oil cake is developed, transforming it into biofertilizers, biological feeds, bio-based materials, and other products; fermentation technology was used to transform oil cake into biofertilizers rich in amino acids and trace elements to improve soil fertility and crop yield (Xu et al. 2025).

17.7.2 Market Competition and Cooperation

As the biofuel market continues to grow, canola biofuels will face competition from other biofuel feedstocks (such as corn, soybean, and palm oil) as well as emerging energy technologies (such as solar, wind, and hydrogen) (Chilakamarry et al. 2021). In order to gain an advantage in market competition, the oilseed rape biofuel industry needs to strengthen cooperation with upstream and downstream enterprises to form a complete industrial chain. Farmers, processing enterprises, sales enterprises, etc., should strengthen communication and cooperation, jointly reduce costs, and improve product quality and market competitiveness (Foong et al. 2022). For example, oilseed rape growers and biofuel producers can establish long-term and stable cooperative relationships. The planting enterprises shall, in accordance with the requirements of the production enterprises, provide rapeseed conforming to the quality standards. Production enterprises provide technical support and market guarantee for planting enterprises. At the same time, biofuel production enterprises can also cooperate with oil sales enterprises to bring the produced biodiesel to the market through the channels of sales enterprises to achieve mutual benefit and win-win results. At the same time, we actively carry out international cooperation, introduce foreign advanced technology and management experience, and expand the international market.

17.7.3 Improve Policies and Regulations

Policies and regulations play an important role in guiding and regulating the development of the oilseed rape biofuel industry. In the future, it is necessary to further improve relevant policies and regulations and increase support for the oilseed rape biofuel industry (Pahlmann et al. 2013). In terms of subsidy policies, the subsidy standards and methods should be reasonably adjusted according to the development of the industry, and the accuracy and effectiveness of subsidies should be improved. In terms of quality standards, it is necessary to formulate strict biofuel quality standards, strengthen market supervision, and ensure the quality and safety of biofuel products (Thamsiriroj and Murphy 2011). In addition, the environmental assessment and monitoring system of the biofuel industry should be established and improved, and the environmental supervision of the production and use of biofuels should be strengthened. The Brazilian government has formulated a development plan for the biofuel industry, which has clarified the development goals and key directions of the biofuel industry. The plan includes increasing investment in biofuel technology research and development, supporting the development of biofuel production enterprises, and strengthening the regulation of the biofuel market. The European Union has developed the Common Agricultural Policy (CAP), which includes subsidies and support measures for canola cultivation. Through direct payment, area subsidies, and other ways to encourage farmers to plant rape, the yield and quality of rape are improved.

17.8 CONCLUSION

As an efficient biofuel producer, oilseed rape has many advantages and has a great potential for sustainable use. The yield and quality of rapeseed biofuel can be improved effectively. The production cost can be reduced and the impact on the environment can be reduced by optimizing planting technology, rationally utilizing land resources and improving the oil conversion process. At the same time, the development of the oilseed rape biofuel industry also has a positive role in promoting the economy and society, which can promote rural development, increase employment opportunities, and alleviate the energy crisis. However, the industry still faces some challenges in the process of development, such as technological innovation, market competition, and policy improvement. In the future, the government, enterprises, and scientific research institutions need to work together to strengthen technology research and development and innovation, improve policies and regulations, and promote the sustainable development of the rape biofuel industry, so that it can play a more important role in the global energy transition.

ACKNOWLEDGMENT

The authors gratefully acknowledge the funding for this work by Dana Impak Perdana 2.0 (DIP 2.0) (DIP-2024-019) and Skim Geran Penyelidikan Fundamental (FRGS) (FRGS/1/2023/TK05/UKM/02/7).

COMPETING INTERESTS

The authors declare that they have no competing interest.

DECLARATION OF GENERATIVE AI AND AI-ASSISTED TECHNOLOGIES IN THE WRITING PROCESS

During the preparation of this work, the book chapter author(s) have not taken any assistance from AI, and the book chapter author(s) reviewed and edited the content as needed and take(s) full responsibility for the content of the publication.

REFERENCES

Abdollahi sisi, N., E. Herzog, A. Abbadi, R. J. Snowdon & A. A. Golicz (2024) Analysis of the winter oilseed rape recombination landscape suggests maternal–paternal bias. *Genome*, 67, 445–453.

Abdul Wahab, R., M. H. Z. Abidin, H. L. Teo, F. Z. Huyop & E. Susanti (2025) Biofuels. In *Encyclopedia of Green Chemistry* (First Edition), ed. B. Török, 441–462. Oxford: Elsevier.

Abdullah, M., Z. Ali, M. T. Yasin, K. Amanat, F. Sarwar, J. Khan & K. Ahmad (2024) Advancements in sustainable production of biofuel by microalgae: Recent insights and future directions. *Environmental Research*, 262, 119902.

Ahmad Zamri, M. F. M., S. H. A. Hassan, S. K. Tiong, J. Milano, R. Bahru, I. M. R. Fattah & T. M. I. Mahlia (2024) Progress and challenges of mesoporous catalysts in upgraded pyrolysis of biomass for biofuel production. *Journal of Analytical and Applied Pyrolysis*, 182, 106651.

Andert, S., A. Ziesemer & H. Zhang (2021) Farmers' perspectives of future management of winter oilseed rape (*Brassica napus* L.): A case study from north-eastern Germany. *European Journal of Agronomy*, 130, 126350.

Antar, M., D. Lyu, M. Nazari, A. Shah, X. Zhou & D. L. Smith (2021) Biomass for a sustainable bioeconomy: An overview of world biomass production and utilization. *Renewable and Sustainable Energy Reviews*, 139, 110691.

Arvidsson, R., S. Persson, M. Fröling & M. Svanström (2011) Life cycle assessment of hydrotreated vegetable oil from rape, oil palm and Jatropha. *Journal of Cleaner Production*, 19, 129–137.

Aubert, M., V. Clouet, F. Guilbaud, S. Berardocco, N. Marnet, A. Bouchereau & Y. Dellero (2024) Sink-source driven metabolic acclimation of winter oilseed rape leaves (*Brassica napus* L.) to drought. *Journal of Plant Physiology*, 303, 154377.

Batchelor, S. E., E. J. Booth & K. C. Walker (1995) Energy analysis of rape methyl ester (RME) production from winter oilseed rape. *Industrial Crops and Products*, 4, 193–202.

Bello, M., A. K. Modu, D. E. A. Boryo, A. A. Mahmoud & P. Ranganathan (2024) Sustainable algal biorefinery: A review on current perspective on technical maturity, supply infrastructure, business and industrial opportunities. *Journal of Environmental Management*, 370, 122208.

Bujnovský, R., M. Holíčková & P. Ondrejíčková (2020) Spring *Camelina sativa* – Perspective cultivation as biofuel feedstock in Slovakia. *Industrial Crops and Products*, 154, 112634.

Cao, B., C. Bai, M. Zhang, Y. Lu, P. Gao, J. Yang, Y. Xue & G. Li (2022) Future landscape of renewable fuel resources: Current and future conservation and utilization of main biofuel crops in China. *Science of the Total Environment*, 806, 150946.

Chilakamarry, C. R., A. M. M. Sakinah, A. W. Zularisam, A. Pandey & D.-V. N. Vo (2021) Technological perspectives for utilisation of waste glycerol for the production of biofuels: A review. *Environmental Technology & Innovation*, 24, 101902.

Dessi, F., S. Ariccio, T. Albers, S. Alves, N. Ludovico & M. Bonaiuto (2022) Sustainable technology acceptability: Mapping technological, contextual, and social-psychological determinants of EU stakeholders' biofuel acceptance. *Renewable and Sustainable Energy Reviews*, 158, 112114.

Fan, Y., Y. Cai, X. Li, H. Yin, N. Yu, R. Zhang & W. Zhao (2014) Rape straw as a source of bio-oil via vacuum pyrolysis: Optimization of bio-oil yield using orthogonal design method and characterization of bio-oil. *Journal of Analytical and Applied Pyrolysis*, 106, 63–70.

Foong, S. Y., Y. H. Chan, A. C. M. Loy, B. S. How, A. M. Tamothran, A. J. K. Yip, R. K. Liew, W. Peng, A. K. O. Alstrup, S. S. Lam & C. Sonne (2022) The nexus between biofuels and pesticides in agroforestry: Pathways toward United Nations sustainable development goals. *Environmental Research*, 214, 113751.

Gabrielle, B., N. Gagnaire, R. S. Massad, K. Dufossé & C. Bessou (2014) Environmental assessment of biofuel pathways in Ile de France based on ecosystem modeling. *Bioresource Technology*, 152, 511–518.

Hoang, A. T., S. Nižetić, H. C. Ong, M. Mofijur, S. F. Ahmed, B. Ashok, V. T. V. Bui & M. Q. Chau (2021) Insight into the recent advances of microwave pretreatment technologies for the conversion of lignocellulosic biomass into sustainable biofuel. *Chemosphere*, 281, 130878.

Hoffmann, M. P., A. Jacobs & A. M. Whitbread (2015) Crop modelling based analysis of site-specific production limitations of winter oilseed rape in northern Germany. *Field Crops Research*, 178, 49–62.

Joseph Antony Sundarsingh, T., F. Ameen, J. Ranjitha, S. Raghavan & V. Shankar (2024) Engineering microbes for sustainable biofuel production and extraction of lipids – Current research and future perspectives. *Fuel*, 355, 129532.

Kniuipytė, I., A. Dikšaitytė, M. Praspaliauskas, N. Pedišius & J. Žaltauskaitė (2023) Oilseed rape (*Brassica napus* L.) potential to remediate Cd contaminated soil under different soil water content. *Journal of Environmental Management*, 325, 116627.

Köbke, S., M. Senbayram, B. Pfeiffer, H. Nacke & K. Dittert (2018) Post-harvest N_2O and CO_2 emissions related to plant residue incorporation of oilseed rape and barley straw depend on soil NO_3-content. *Soil and Tillage Research*, 179, 105–113.

Li, X., Y. Li, X. Kang, J. Yu, S. Gao, J. Zhang, H. Wang, H. Pan, Q. Yang, Y. Zhuge & Y. Lou (2022) Effective utilization of weak alkaline soils with Cd-contamination by wheat and rape intercropping. *Ecotoxicology and Environmental Safety*, 248, 114335.

Li, X., Y. Wang, W. Tang, B. Fan, Y.-C. He & C. Ma (2024) Co-production of reducing sugars and xylo-oligosaccharides from rape straw through sodium dodecyl sulfate-mediated *p*-toluenesulfonic acid pretreatment. *Industrial Crops and Products*, 222, 119888.

Li, Y., L. Guo, M. Wu & C. Li (2025) Effects of long-term conservation tillage on soil aggregate carbon fractions and microbial characteristics under rice-oilseed rape rotation systems. *Applied Soil Ecology*, 206, 105848.

Liu, L., X. Xu, Y. Hu, Z. Liu & Z. Qiao (2018) Efficiency analysis of bioenergy potential on winter fallow fields: A case study of rape. *Science of the Total Environment*, 628–629, 103–109.

Mathew, A. K., M. Crook, K. Chaney & A. C. Humphries (2014) Continuous bioethanol production from oilseed rape straw hydrosylate using immobilised *Saccharomyces cerevisiae* cells. *Bioresource Technology*, 154, 248–253.

Meena, P. K., S. D. Awale, S. Kumar, D. Kumar & N. Kumar (2024) Thermochemical conversion of agricultural residue for the production of hydrogen, methane, and biofuels: A comprehensive overview. *Industrial Crops and Products*, 221, 119340.

Menhas, S., X. Yang, K. Hayat, A. Ali, E. F. Ali, M. Shahid, S. M. Shaheen, J. Rinklebe, S. Hayat & P. Zhou (2022) Melatonin enhanced oilseed rape growth and mitigated Cd stress risk: A novel trial for reducing Cd accumulation by bioenergy crops. *Environmental Pollution*, 308, 119642.

Neira, P., J. M. Blanco-Moreno, M. Olave, B. Caballero-López & F. X. Sans (2024) Effects of agricultural landscape heterogeneity on pollinator visitation rates in Mediterranean oilseed rape. *Agriculture, Ecosystems & Environment*, 363, 108869.

Pahlmann, I., U. Böttcher, K. Sieling & H. Kage (2013) Possible impact of the Renewable Energy Directive on N fertilization intensity and yield of winter oilseed rape in different cropping systems. *Biomass and Bioenergy*, 57, 168–179.

Qi, H., F. Li, S. Wang, F. Hou, M. Sui & F. Lu (2024a) Waste heat utilization and kinetic analysis of aerobic fermentation with cattle manure and rape straw. *Process Safety and Environmental Protection*, 189, 587–597.

Qi, H., F. Li, S. Wang, M. Sui & F. Lu (2024b) Pyrolysis and co-pyrolysis of cattle manure, rape straw, and their blend: Physicochemical characterization, kinetic triplets, reaction mechanism, and thermodynamic analysis. *Energy*, 292, 130520.

Saleem, M (2022) Possibility of utilizing agriculture biomass as a renewable and sustainable future energy source. *Heliyon*, 8, e08905.

Schmidt, G. & W. Schröder (2011) Regionalisation of climate variability used for modelling the dispersal of genetically modified oil seed rape in Northern Germany. *Ecological Indicators*, 11, 951–963.

Seo, J. Y., D. Tokmurzin, D. Lee, S. H. Lee, M. W. Seo & Y.-K. Park (2022) Production of biochar from crop residues and its application for biofuel production processes – An overview. *Bioresource Technology*, 361, 127740.

Thamsiriroj, T. & J. D. Murphy (2011) A critical review of the applicability of biodiesel and grass biomethane as biofuels to satisfy both biofuel targets and sustainability criteria. *Applied Energy*, 88, 1008–1019.

Thers, H., S. N. Djomo, L. Elsgaard & M. T. Knudsen (2019) Biochar potentially mitigates greenhouse gas emissions from cultivation of oilseed rape for biodiesel. *Science of the Total Environment*, 671, 180–188.

Velusamy, K., J. B. Isabel, S. Periyasamy, A. Thiruvenkadam, H. Ravikumar, S. K. Gupta & E. A. López-Maldonado (2024) Role of biochar as a greener catalyst in biofuel production: Production, activation, and potential utilization – A review. *Journal of the Taiwan Institute of Chemical Engineers*, 106, 105732.

Vinzent, B., R. Fuß, F.-X. Maidl & K.-J. Hülsbergen (2017) Efficacy of agronomic strategies for mitigation of after-harvest N_2O emissions of winter oilseed rape. *European Journal of Agronomy*, 89, 88–96.

Wang, F., X. Jia, Y. Zhao, X. Yang & X. Feng (2025a) Two strains of cadmium (Cd)-resistant bacteria isolated from soils and their ability to promote oilseed rape (*Brassica juncea* L.) to grow and absorb Cd in soils. *Environmental Pollution*, 366, 125549.

Wang, H., H. Li, C. K. Lee, N. S. M. Nanyan & G. S. Tay (2024) A systematic review on utilization of biodiesel-derived crude glycerol in sustainable polymers preparation. *International Journal of Biological Macromolecules*, 261, 129536.

Wang, L., Q. Liu, F. Zhao, J. Yang, J. Fu & X. Liao (2025b) An integrated three-crop rotation of oilseed rape–rice–rice enables the safe utilization and sustainable remediation of Cd-contaminated farmland. *Environmental Technology & Innovation*, 38, 104123.

Xu, R., J. Chen, N. Yan, B. Xu, Z. Lou & L. Xu (2025) High-value utilization of agricultural residues based on component characteristics: Potentiality and challenges. *Journal of Bioresources and Bioproducts*, 89, 26590.

Yang, L., C. Gu, W. Huang, H. Chang, Y. Gao, Y. Li, J. Dai, X. Li, W. Hu, W. Cao, X. Liao & L. Qin (2024) Legume and maize intercropping enhances subsequent oilseed rape productivity and stability under reduced nitrogen input. *Field Crops Research*, 319, 109644.

Yao, Y., A. Bilichak & I. Kovalchuk (2023) Dicer mutants and rdr6 mutant are impaired in an intergenerational increase in homologous recombination frequency in response to the infection with oilseed rape mosaic virus. *Biocatalysis and Agricultural Biotechnology*, 51, 102792.

Zhu, L., W. Tang, C. Ma & Y.-C. He (2023) Efficient co-production of reducing sugars and xylooligosaccharides via clean hydrothermal pretreatment of rape straw. *Bioresource Technology*, 388, 129727.

18 Sunflower as an Efficient Biofuel Producer

A Critical Analysis for Sustainable Uses

Isa khan, Bashir Ullah, Kashif Ali Khan, Muhammad Suliman, Ambreen Ali, Hanif Ullah, Yi Chen, and Zul Kamal

18.1 INTRODUCTION TO SUNFLOWER BIOFUELS

18.1.1 Overview of Biofuels and Their Importance in Sustainable Energy

Biofuels are energy-enriched chemicals that can be produced directly through biological processes or synthesized from biomass of living organisms that may include microalgae, plants, and bacteria, offering a sustainable option to reduce the consumption of fossil fuels in energy sectors. They can be categorized into first-generation biofuels made from the food part of the crops and second-generation biofuels produced from the non-food part of the crops (Barontini et al. 2015; De Marco et al. 2016). Biofuels play a decisive role in sustainable energy by reducing emissions of greenhouse gases. As of 2015, they have helped reduce carbon emission by 589.3 million tons, indicating their potential in combating climate change. The production of biofuels from various feedstocks including non-edible lignocellulosic biomass (which does not compete with food production) helps in ensuring the availability of sustainable energy while addressing food security concerns. Biofuel reduces dependence on fossil fuels, which can lead to stable energy prices and enhance national security by reducing vulnerability to fluctuations in the global oil market. The demand for biofuels encourages farmers to grow energy crops, which can lead to increased agricultural productivity and sustainability (Y. Liu et al. 2021).

18.1.2 Historical Context of Sunflower Cultivation for Oil Production

Sunflower (*Helianthus annus*) is presumed to have been domesticated by Native Americans around 3,000–5,000 years ago and was primarily used as a source of edible seeds. In the early 16th century, sunflower was introduced to Europe beginning its journey as a cultivated crop outside its native range. Later, the cultivation of sunflowers expanded significantly in Russia where modern oilseed cultivars were developed and produced on an industrial scale. In the mid-20th century, the Russian oilseed cultivars were introduced back to North America. This re-establishment prompted commercial sunflower production in the Americas, commencing a new era of sunflower breeding that was focused on oil content (Park and Burke 2020). The introduction of commercial sunflower varieties to the United States occurred in the 1960s notably through immigrants from the former Soviet Union. Following the establishment of commercial production in the United States, by 1979, the sunflower acreage grew to 5.6 million acres, driven by increased commercial interest and the development of hybrid varieties that enhanced yield and disease resistance. The Food Agricultural Conservation and Trade Act of 1990 offered enticing economic benefits for minor oilseeds including sunflower, which temporarily boosted acreage in the early 1990s. This legislative support played a vital role in crop commercialization and production dynamics (M. K. Gupta 2014).

DOI: 10.1201/9781003600169-18

18.1.3 Significance of Sunflower as a Biofuel Source

Sunflowers can be a significant source of biofuels due to their high oil contents, i.e., sunflower seeds contain about 45% oil, making them an excellent source for biodiesel production. Studies have shown that sunflowers are resistant to droughts, which make them suitable for cultivation under diverse environmental conditions. They can be embedded into food and energy systems, enabling concurrent production of food crops and biofuels. This integration promotes sustainable farming practices and better resource utilization. The economic viability of sunflower cultivation can help decrease dependence on fossil fuels while maximizing the use of agricultural resources such as water and fertilizers. The extraction of sunflower oil is considered cost-effective as compared to other oil crops, making it a significant contributor to sustainable agricultural practices and biofuel production (P. Debaeke et al. 2014). Although biofuels including those derived from sunflower tend to be more expensive than fossil fuels, the fact that they possess the potential to reduce greenhouse gas emissions may justify these costs. Life cycle assessments have compared the emissions from fossil fuels to those produced from sunflower oil. The findings have deduced that biodiesel from sunflowers can provide a more sustainable alternative to conventional fossil fuels, aiding in the reduction of greenhouse gas emissions (De Marco et al. 2016).

18.2 BOTANY AND AGRONOMY OF SUNFLOWER

18.2.1 Botanical Characteristics of Sunflower (*Helianthus annus*)

Sunflower, scientifically known as *Helianthus annus*, is a distinct member of the Asteraceae family, which is one of the largest flowering families on earth. The Asteraceae family includes about 23,000 species across 1,600 generas (Nguyen et al. 2021). The name "Helianthus" is derived from the Greek word "Helios" which means sun and "anthos" means flower. This represents the plant's characteristic of turning the flowering head toward the sun. It is an annual herb and is diploid with a chromosome number of $2n = 34$, which is typical for many flowering plants (Jocić, Miladinović, and Kaya 2015). Sunflower is typically a tall, herbaceous plant that can grow up to 3 m in height. Their stem is sturdy, upright that may be branched or unbranched often covered in fine hairs. They primarily develop a deep taproot system that can extend several feet into the ground, allowing the plant to access moisture and nutrients from deeper soil layers. The additional lateral roots spread horizontally that anchors the plant in the soil stabilizing the towering plant. The leaves of sunflowers are large, coarse, and oval-shaped with a rough texture and can grow up to 30 cm (12 inches) long. They are known for their large, bright yellow flower heads, which can be up to 30 cm (12 inches) in diameter. Each flower head is made up of densely packed numerous small florets. The outer florets are typically sterile and serve to attract pollinators, while the inner florets are fertile and produce seeds. Each flower head can produce hundreds of seeds, which are rich in oil and protein (Faria and Rani 2021).

18.2.2 Agronomic Practices for Optimal Sunflower Cultivation

Agronomy is the science and practice of soil management, crop production, and water management. These practices are oriented toward improving yield, oil quality, and overall performance of the crop. In order to attain maximal sunflower cultivation and sustainability in oilseed production, several agronomic practices must be taken into account. Comprehending the function of sunflowers within various cropping systems is of paramount importance. Scholarly research emphasizes the utilization of modeling tools to facilitate informed decision-making regarding crop rotation and crop management practices. The impact of cover crops and soil tillage on the establishment and initial growth of sunflowers is notably significant. Proper soil management practices can ameliorate seedbed conditions, thereby fostering better germination and establishment of sunflower plants. The selection of an appropriate combination of sunflower variety, environmental conditions, and management strategies is essential. This includes selecting varieties that are optimally adapted to particular regions and conditions, which can augment productivity and endurance (Debaeke et al. 2020). Determining the optimal sowing date is essential and must be aligned with the crop cycle to circumvent water deficits during critical periods that influence oil yield and quality, especially during the grain-filling stage. In circumstances with expected water deficits, a reduction in plant density or increase in row spacing may assist in delaying the onset of elevated leaf area index. Sunflower cultivation requires a well-balanced provision of macronutrients (N, K, P, Ca, Mg, S) and micronutrients (Fe, Mn, Zn, Cu, B, Cl, Mo). Nitrogen is particularly vital for leaf development and overall vegetative growth. Ensuring adequate nitrogen supply before floral initiation is essential for maximizing grain quantity and root biomass. Improving water use efficiency through practices that enhance soil moisture retention and minimize evaporation can considerably influence sunflower yield, particularly in rain-fed agricultural systems. Practices such as selecting plots with adequate soil depths are critical for root growth and water utilization (Alberio, Izquierdo, and Aguirrezábal 2014).

18.2.3 Factors Affecting Yield and Oil Content

The yield and oil content of sunflowers are influenced by various critical factors particularly the type of fertilizers used, soil properties, water availability, nutrient management, and genetic variations including several other factors. Sunflowers are sensitive to water deficit stress, which can occur at various growth stages. Studies have shown that a severe lack of irrigation (40% of potential evapotranspiration, PET) resulted in a hefty reduction in yield and oil content, with sunflower experiencing a 16.3% reduction in seed yield and 22.5% reduction in oil yield on average (Ebrahimian et al. 2019). Organic fertilizers such as compost coupled with biofertilizer (CCB) and filter mud cake substantially improved sunflower yield and oil content compared to inorganic fertilizers, which potentially lowered these parameters. Studies have found that organic treatments led to potential increases in seed yield, oil yield, and stalk yield. The use of organic fertilizer optimizes soil properties by increasing nutrient availability and improving soil structure. This leads to elevated nutrient absorption by plants, which is critical for growth and yield. Sunflower requires a nutrient-rich environment especially Nitrogen (N), Phosphorus (P), and Potassium (K) during the early growth stages for optimal yield. Improved growth parameters such as leaf area and chlorophyll density offer higher biomass production and increased grain yield (Alzamel et al. 2022). Different sunflower genotypes exhibit notable differences in oil content. For example, the oil content of female lines ranged from 19.12% to 52.46%, while male lines ranged from 13.16% to 54.05%. This variability suggests that selecting the right genotype can be crucial for optimizing oil contents (Pekcan et al. 2022). Yield is directly related to features like head diameter, weight of 100 seeds, and total plant weight. Studies have found that higher yields are correlated with larger head diameters and heavier seed weights, indicating that breeding programs should take these features into account. Sowing date has a big influence on yield and oil content. Higher yield potential was the result of early sowing, particularly on April 20. This can be explained by the plant's extended life cycle, leading to the production of more photo-assimilates and their conversion into seeds, raising the weight of the seeds and the total yield per feddan (1 acre=0.96 feddans) (M. A. Ahmed and Abdelsatar 2020). The application of different mycorrhizal species positively affects both grain yield and oil content. Mycorrhizal inoculation significantly increases oil content compared to control treatments. Among the mycorrhizal species tested, *Glomus etunicatum* was effective in improving grain yield and nutrient content, whereas *Glomus mosseae* had the highest effect on increasing oil contents. External factors such as weather conditions and temperature affect oil synthesis. For instance, oil synthesis in sunflower can cease under stress conditions including water deficits and higher temperatures (Heidari and Karami 2014). The moisture content of sunflower collets significantly affects oil yield and quality. A moisture level of 12% is optimal, promoting high tocopherol content and lower phospholipid content, reducing degumming intensity. Low moisture content (below 10%) reduces oil yield, while high moisture level (above 12%) leads to different oil composition, including higher phospholipid levels and lower tocopherol and wax concentrations. Extraction techniques play a crucial role in oil yield. Studies suggest that batch extraction using hexane is less effective with low moisture content, and a minimum moisture content of 10% is required for accurate quantification of total oil content using Soxhlet extraction (Pérez et al. 2019).

18.3 BIOCHEMICAL COMPOSITION OF SUNFLOWER OIL

18.3.1 Composition of Sunflower Oil: Fatty Acids and Other Components

Sunflower oils are primarily composed of triglycerides, which are molecules made up of glycerol and three fatty acid chains. Sunflower oil contains relatively low amount of saturated fatty acids (around 8%–15%), with a high percentage of unsaturated fatty acids. For standard type sunflower, the main unsaturated fatty acids including linoleic acids (polyunsaturated fatty acid, PUFA) are about 48.3%–74%, while oleic acids (monounsaturated fatty acids) are 14%–39.4%. High oleic sunflower oils have high proportion of oleic acids compared to standard sunflower oil. The saturated fatty acids such as palmitic acid concentration in standard type sunflower oil range from 5.0% to 7.6%, while that of stearic acid ranges from 2.7% to 6.5% (Akkaya 2018). Table 18.1 depicts a clear picture of the fatty acid composition of sunflower oils. However, the values of sunflower fatty acid contents vary significantly with different varieties and regions. Research conducted by Gagour et al. (2022) suggested that the contents of sunflower oils were slightly different from the above normal values, which indicates that fatty acid contents of sunflower oil may be different depending on the variety of sunflower, cultivation, and environmental conditions. Apart from fatty acid contents, sunflower oils are rich in vitamin E (especially alpha tocopherol), proteins, and some minerals such as magnesium, zinc, and selenium (Sheeba et al. 2015).

18.3.2 Comparison with Other Vegetable Oils Used in Biofuels

The composition and properties of vegetable oils used for biofuel production are influenced by the type of plant, cultivation conditions, and oilseed processing technologies, with saturated and unsaturated fatty acids being the primary components. Vegetable oils that are rich in saturated and unsaturated fatty acids and can be used for biofuel production include sunflower oil, soybean oil, palm oil, rapeseed oil, jatropha oil, and many others. Soybean oil (*glycine max*) is known for its significant content of PUFA. It contains about 48%–59.0% linoleic acid along with 17%–30% oleic acid. The saturated fatty acids including palmitic acid range from 8% to 13.5%, while stearic acids are about 2.0%–5.4% (Markov et al. 2021). The fatty acid composition of soybean oil can be seen in Table 18.1.

Palm oil (*Elaeis guineensis*) is an edible vegetable oil and can be used in biofuel production. It contains about 49% saturated fatty acids including 42.31% palmitic acids and 4.27% stearic acids. The unsaturated fatty acids are about 51%, including 40.90% oleic acids and 10.07% linoleic acids.

TABLE 18.1
Fatty Acid Composition of Different Vegetable Oils Used in Biofuel Production

Vegetable	Saturated Fatty Acids (%)		Unsaturated Fatty Acids (%)		
Oils	Palmitic Acid	Stearic Acid	Oleic Acid	Linoleic Acid	References
Sunflower	5.0 … 7.6	2.7 … 6.5	14.0 … 39.4	48.3 … 74.0	Akkaya (2018)
Soybean	8.0 … 13.5	2.0 … 5.4	17.0 … 30.0	48.0 … 59.0	Markov et al. (2021)
Palm	42.31	4.27	40.90	10.07	Giakoumis (2018)
Jatropha	14.39	5.83	42.05	35.90	Giakoumis (2018)
Rapeseed	2.5 … 7.0	0.8 … 3.0	51.0 … 70.0	15.0 … 30.0	Gagour et al. (2022)

Jatropha oil (*Jatropha curcas*) contains 20% saturated fatty acids with 14.39% palmitic acid and 5.83% stearic acid. About 80% of its content is unsaturated fatty acid including 35.90% linoleic acid and 42.05% oleic acid (Giakoumis 2018). Table 18.1 clearly shows the fatty acid composition of Palm oil and Jatropha oil. Another important vegetable oil used for biofuel production is rapeseed oil (*Brassica napus*). The saturated fatty acid composition of rapeseed oil includes 2.5%–7.0% palmitic acid and 0.8%–3.0% stearic acid, while the unsaturated fatty acids include 51.0%–70.0% oleic acid and 15%–30% linoleic acid (Gagour et al. 2022).

Compared to soybean and other vegetable oils, sunflower is relatively high in unsaturated fatty acids, which makes it a significant candidate for the selection of biofuel production. The degree of unsaturation in fatty acids plays a vital role in biofuel performance, affecting its cold flow property, viscosity, iodine value, and heating values, which are all critical for optimal engine performance. As suggested by Folayan et al. (2019), optimum degree of unsaturation results in better cold flow behavior and lower viscosity, both of which are essential for engine efficiency. However, it is crucial to balance these properties with potential drawbacks such as lower heating values and lubrication issues.

18.3.3 Nutritional and Industrial Application of Sunflower Oil

Sunflower oil is a nutritious option in the diet, as it is primarily composed of unsaturated fatty acids that are beneficial for cardiac health. The World Health Organization (WHO) suggests that fat intake should mainly come from unsaturated sources, making sunflower oil an optimal choice for the diet. Sunflower oil's most significant nutritional advantage is its α-tocopherol (Vitamin E) content, which ranges from 270 to 1240 mg/kg of oil. Vitamin E is essential for preventing neurological diseases and combating oxidative stress. The neutral taste and high smoke point of sunflower oil make it suitable for various cooking methods, enhancing both flavors and nutritional value (Nakonechna et al., 2024). The WHO recommends a balanced intake of Omega-6 and Omega-3 fatty acids in a ratio of 5 to 10. As sunflower oils are rich in essential fatty acids, they can help achieve this balance thus contributing to overall human health. Apart from being rich in essential fatty acids, sunflower oil also provides protein (15%–21%), making it a valuable addition to the human diet (Sheeba et al. 2015).

Sunflower oil is widely utilized in various industrial applications due to its favorable properties. Industries tending toward sustainable energy sources are more attracted to the use of biofuels, making sunflower oil an interesting source of biofuels for them. Due to their non-toxic nature and good insulating properties, they can be utilized as transformer oils in electrical applications. The moisturizing properties of sunflower oil make it a desirable ingredient in lotions, creams, and other beauty products (M. K. Gupta 2014). As the industries move toward bio-lubricants, sunflower oil is a good contender to replace conventional mineral oils, which are less eco-friendly. Because of its superior tribological characteristics, they can be used to improve lubricant performance. Sunflower oil is gaining momentum in industrial applications due to its biodegradable nature, effective performance as an additive, and the potential for cost-effective synthesis methods (Ghosh and Karmakar 2014).

18.4 BIODIESEL PRODUCTION FROM SUNFLOWER OIL

18.4.1 Transesterification Process: Mechanism and Catalyst Used

Transesterification is a chemical reaction used in biodiesel production that involves the reaction of triglycerides (fats) with short chain alcohol such as methanol or ethanol, using a catalyst to produce monoalkyl ester (biodiesel) and glycerol as a byproduct. Biodiesel is a significant substitute to diesel, with substantial environmental advantages over petroleum-based fuel. Vegetable oils being a key biofuel feedstock are hindered by their high viscosity, low volatility, and poor cold flow characteristics. Transesterification is the most feasible method for reducing the viscosities of vegetable oils and producing alkyl esters that have characteristics similar to diesel (Salaheldeen et al. 2021). The process involves three main steps. In the first step, triglyceride reacts with alcohol (ethanol) in the presence of a catalyst, such as Sodium hydroxide (NaOH), leading to the formation of diglyceride and release of one molecule of glycerol. The diglyceride reacts with another molecule of alcohol in the presence of catalyst, resulting in the formation of monoglyceride and another molecule of glycerol. Finally, the monoglycerides react with ethanol to produce glycerol and fatty acid ethyl ester (biodiesel) (Reyero et al. 2015). Studies conducted by Khan et al. (2023) suggested that the reaction is reversible, so a higher molar ratio of alcohol is needed in order to shift the equilibrium toward the biodiesel production. The study found that ideal molar ratio of alcohol to oil for transesterification of sunflower oil was 18:1, with a reaction time of 40 minutes at 50°C in the presence of catalyst such as potassium hydroxide (KOH). In transesterification, catalysts are crucial to enhance the reaction rate and biodiesel yield. They can either be homogenous or heterogenous acid or base catalyst. The choice of catalyst depends on the feedstock's characteristics, especially the free fatty acid (FFA) content. The common catalysts used by different researchers for the transesterification of vegetable oils are heterogenous catalysts, such as sodium hydroxide (NaOH) and potassium hydroxide (KOH) due to their environmental-friendly and non-corrosive characteristics.

18.4.2 Laboratory Methods and Pilot-Scale Production Techniques

Biodiesel production from sunflower oil can be achieved by different laboratory methods, each with distinct processes, efficiencies, and limitations. These methods can be categorized based on the type of catalyst used, i.e., base-catalyzed, acid-catalyzed, enzyme-catalyzed, and supercritical methods. Studies conducted by Miyuranga et al. (2023) show that base-catalyzed transesterification using bentonite-supported sodium methoxide (BSSM) catalyst is an effective method for biodiesel production. The study found that the BSSM catalyst significantly reduced the activation energy, i.e., 40.98 kJ/mol, indicating the effectiveness of this catalyst compared to other methods reported in the literature. Basic catalysts are environmental-friendly, non-corrosive, and allows for a faster and more efficient reaction with the triglyceride molecules present in the oil. On the other hand, the acid-catalyzed transesterification is associated with slower reaction rates, corrosiveness, and environmental concerns. However, Ibrahim et al. (2021) treated the S.Coal (synthetic coal derived from raw coal) with sulfuric acid at a concentration of 95% and deduced that the S.Coal catalyst achieved a remarkable biodiesel yield of 99.5% at 120°C in just 30 minutes. Acidic catalysts are effective in transesterification of vegetable oils with high FFAs. The soap formation is a common issue in traditional alkali- or acid-catalyzed process. To overcome this problem, Mandari and Kumar (2022) concluded that although the use of enzyme catalyst such as lipase formulations for biodiesel production is a costly method, it can be used to overcome the soap formation. Their miscibility with the reactants reduces the mass transfer limitations and increase reaction rate. Another laboratory method used for the transesterification of sunflower oil is the noncatalytic method under supercritical conditions. In this method, alcohol (methanol or ethanol) and triglycerides are heated under supercritical conditions, which means pressure above 5 MPa and temperature above 200°C. Under these conditions, the alcohol acts as a solvent, enhancing the reaction. This method converts the triglycerides directly into biodiesel without the formation of soap, which is common in catalytic processes (Makareviciene and Sendzikiene 2021).

Different types of reactors are used for pilot-scale biodiesel production. The pilot-scale batch reactor used by Barontini et al. (2015) for transesterification of sunflower oil achieved a significant yield of 99.3%. The batch reactor operates under specific conditions, including a 6:1 molar ratio of methanol to oil and a temperature of 70°C for duration of 2 hours. However, the batch reactors have several disadvantages including less output, poor heat control, and labor intensive. J. Gupta, Agarwal, and Dalai (2020) assessed different types of continuous reactors for the transesterification process. Among them were the continuous flow reactor, packed bed reactor, bubble column reactor, oscillatory flow reactor, microchannel reactor, plug flow reactor, laminar flow reactor separator, and coil reactor. The study reported a maximum biodiesel conversion, i.e., 99% in continuous micro coil reactor in 180 seconds residence time under moderate conditions. Also, the study concluded that the reverse flow helical coil reactor is the most efficient choice to overcome disadvantages associated with the other continuous reactors such as mass transfer resistance, longer reaction time, and low conversion of biodiesel.

18.4.3 Yield Optimization Through Various Parameters

The biodiesel yield of sunflower oil is influenced by various parameters including temperature, the type of catalyst used and its concentration, molar ratio of alcohol, and FFA content. To achieve the maximum biodiesel yield, the type and amount of catalyst must be carefully optimized. It is generally understood that different catalyst types, e.g., homogenous, heterogenous, or enzymatic catalysts, can yield varying results. Homogenous catalysts like sodium hydroxide or potassium hydroxide often provide higher yield, but may require more exhaustive purification processes. The heterogeneous bi-functional catalyst (CaO/Al_2O_3) used by Simbi et al. (2022) for the transesterification process, demonstrated a higher yield of 98.23% under suitable conditions, including a catalyst loading of 2.5 wt.%, a reaction temperature of 60°C, and a reaction time of 5 hours. The study indicated that the higher catalyst loading (such as 5.5 wt.%) resulted in decreased yield, suggesting that optimum catalyst concentration is necessary for maximum biodiesel yield. Kibar et al. (2023) compared the activity of different types of catalysts, such as homogeneous, heterogeneous, and enzymatic catalysts. The study suggested that heterogeneous catalysts are generally more favorable for the transesterification process due to their higher yield of biodiesel, cost-effectiveness, and reusability compared to homogenous and enzymatic catalysts (Figure 18.1).

Temperature significantly influences biodiesel yield of sunflower oil, increasing its efficiency and mass transfer while reducing viscosity of the reactants. A maximum yield of 99.2% was achieved when Devaraj Naik and Udayakumar (2023) raised the temperature to 65°C, using catalyst of 3 wt.%, 18:1 methanol oil ratio, and 120 minutes reaction time. Lowering the temperature adversely

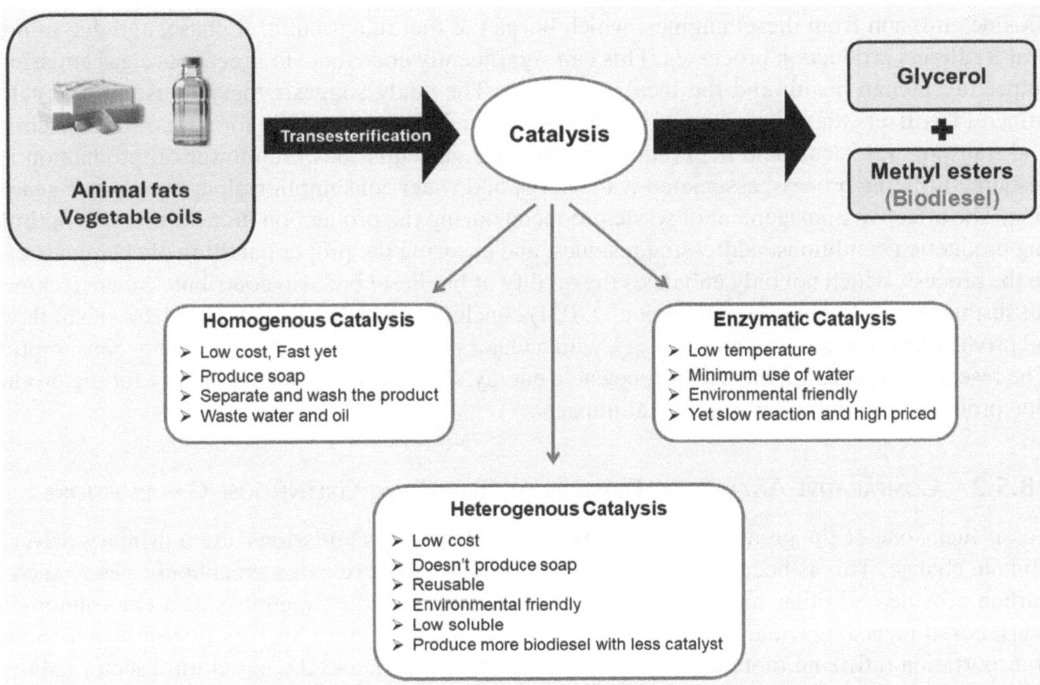

FIGURE 18.1 Comparison of catalysts used in transesterification process.

affects the yield due to increased viscosity, reduced mass transfer, and insufficient kinetic energy for effective reactant interactions. Feyzi et al. (2017) carried out the transesterification of sunflower oil at varying temperatures, and found out that at 49.85°C (323 K), the biodiesel yield was 79.1%. However, when the temperature was raised to 64.85°C (338 K), the yield reached an impressive 96.9%. This indicates a strong positive correlation between temperature and biodiesel yield. The study further revealed that the molar ratio of alcohol to oil plays a critical role in biodiesel production. The optimal molar ratio was identified as 18:1 with a significant yield of 96.9%.

18.5 ENVIRONMENTAL IMPACT ASSESSMENT

18.5.1 Life Cycle Assessment (LCA) of Biodiesel Production from Sunflower Oil

Life Cycle Assessment (LCA) of biodiesel production from sunflower oil is a cradle to grave analysis method that is used to assess the environmental impact associated with the entire process, from sunflower cultivation and oil extraction to biodiesel production, distribution, and disposal. Although biodiesel from sunflower oil presents several advantages, its production process has significant environmental impacts that need to be addressed through improved agricultural practices and consideration of alternative feedstocks (Sanz Requena et al. 2011). The agricultural stages of biodiesel production, including land use, fertilizers, and pesticides usage, are the most impactful in terms of environmental emissions. The land usage for cultivation of sunflower can result in competition with food crops, disrupting the biodiversity and ecosystem. Fertilizers usage can lead to nutrients runoff and contribute to environmental issues like eutrophication. In biofuel production process, greenhouse gases are emitted due to land use changes, cultivation of feedstock, their harvesting, and their processing into biofuels (Correa et al. 2019). Additionally, the author discussed that the carbon dioxide emissions occur due to the production and transportation of the required input (e.g., fertilizers), the transportation of biofuels to fuel stations, and also during the final burning of biofuels for energy production. De Marco et al. (2016) discussed that the higher emissions of biodiesel production are due to the high "land consumption" in agricultural phase, the carbon dioxide emission from diesel engines, which burns the fuel in agricultural phase, and due to mineral fertilizers production processes. This can significantly contribute to greenhouse gas emissions, impacting human health and the local ecosystem. The study suggests that the use of alternative mineral fertilizers that are environmentally friendly and biodiesel as fuel for agricultural machines and transport vehicles could help reduce greenhouse gas emissions. Sunflower oil production is a resource-utilizing process, associated with energy and water consumption along with waste generation. The effective management of waste produced during the production process involves optimizing production conditions, addressing residues, and ensuring the proper handling of chemicals used in the process, which not only enhances the quality of biodiesel but also contributes to environmental sustainability. Nabavi-Pelesaraei et al. (2021) concluded that the agricultural phase of sunflower oil production consumes the most energy, with a major portion attributed to electricity consumption. The research emphasizes the use of renewable energy sources, such as solar energy for the production process, to mitigate environmental impacts.

18.5.2 Comparative Analysis of Fossil Fuels Regarding Greenhouse Gas Emissions

Fossil fuels, one of the greatest contributors to greenhouse gas emissions, are a primary driver of climate change. This is because the combustion of fossil fuels releases greenhouse gases, such as carbon dioxide and other toxic gases including carbon monoxide, methane, and chlorofluorocarbons. Fossil fuels are primarily used in almost all the energy sectors, such as electricity generation, transportation (utilizing more than 70% of all petroleum fuel), and the agriculture sector (Malode et al. 2021). The overuse of fossil fuels has raised concerns about well-being and global warming. As of 2020, it has been estimated that the world energy-related carbon dioxide emissions have increased to 35.6 billion metric tons, and by 2040, this number will rise to 43.2 billion metric tons, leading to global warming (Li, Luo, and Lu 2017). To reduce the dependence on fossil fuels, biofuels are introduced, which act as a substitute for fossil fuels because of their sulfur-free, biodegradable, and nontoxic nature originating from renewable sources (Ali and Abuhabaya 2013). Although biofuels are considered environment-friendly, their production process presents social and environmental challenges, from crop cultivation to biodiesel production (Hanaki and Portugal-Pereira 2018). For instance, careful considerations of agricultural practices, land use changes, and production processes are necessary to avoid emissions and ensure sustainable energy transitions. Xu et al. (2022) suggest that the greenhouse gas emissions of biofuels produced from oilseed crops can be 63%–77% lower than petroleum diesel, indicating the importance of biofuels in combating the climate changes. This is because of biogenic carbon cycle and the renewable nature of biofuels.

18.5.3 Land Use Implications and Biodiversity Consideration

The recent advances in biofuel technologies have led to an increase in demand for biofuel feedstocks, primarily oilseed crops. Due to the increasing demands, oilseed crops such as sunflower and soybean, are being cultivated on large scale, leading to the expansion of cropland, land occupation, and competition with food crops. In terms of environmental impact, the production of sunflower biofuels has a notable influence on land use. Sanz Requena et al. (2011) studied that sunflower seed production contributes 75.07% of the total impact in biofuel production process. This is due to the larger land area required per kilogram of seed produced and the associated agricultural practices. The use of organic fertilizers in sunflower cultivation can lead to nitrous oxide emissions, affecting soil quality and leading to nutrient imbalances. The repeated application of mineral fertilizers may not sustain soil organic carbon levels, which could affect soil quality in long term (Pilecco et al. 2020). The cultivation of sunflower crop on large scale significantly impacts the local biodiversity and ecosystem. The expansion of cultivated land can result in habitat loss and fragmentation, increasing extinction risks for local species. Correa et al. (2017) found that first-generation biofuel production is associated with significant negative impact on biodiversity and ecosystem. This includes the introduction of invasive species and genetic pollution, which can replace the local species and disrupt the ecosystem. The increased water consumption can cause water depletion, affecting aquatic ecosystem. Besides, the greenhouse gas emissions due to the cultivation processes contribute to global warming altering the habitats and local environment. To mitigate the land use implications and biodiversity impacts associated with sunflower biofuel production, it is crucial to adopt efficient agronomic practices and advanced technologies, utilize marginal land to reduce competition with food crops, and adopt policies and guidelines that considers land use, water, and biodiversity impacts (F. Sher et al. 2024).

18.6 ECONOMIC VIABILITY OF SUNFLOWER BIOFUELS

18.6.1 Cost Analysis of Biodiesel Production from Sunflower Oil

The cost of biodiesel production is affected by various factors, such as the price of raw materials, processing expenses, and market circumstances. The price of raw materials for biodiesel production depends on location and market conditions. The major producers of sunflower oils are Russia and Ukraine including other countries, such as Bulgaria, Netherland, and Romania. As of 2021, the leading exporters of sunflower seed oils are Ukraine (USD 5.61 billion), Russia (USD 2.42 billion), Bulgaria (USD 626 million), and the Netherlands (USD 522 million) (Nurrochmat et al. 2024). While the major biodiesel producers include United States, Brazil, some European countries, and Indonesia (Tamás Mizik and Gábor Gyarmati 2021). According to the results of the study conducted by Mohammadshirazi et al. (2014), the total expenditure on the process of biodiesel production is $1.201 per liter and the gross production value is $2.499 per liter. The study found that the highest share of cost was attributed to the feedstock, which accounted for 54.81% of the total cost, followed by human labor cost, making up 27.75% of the total expenses, while the machinery expenses were 17%. In the case of fossil fuels Aguilera (2014) found that, for conventional oil, the production cost ranges from $5 per barrel to around $30 per barrel. While for unconventional oil, the cost range is somewhere between $20 and $90 per barrel. From the analysis, it is observed that biofuels cost is higher than that of fossil fuels, but the environmental sustainability of biofuels neutralizes this impact. F. Sher et al. (2024) stated that fluctuations in the price of fossil fuels impact the market potential of biofuels. An increase in the price of fossil fuels can trigger the buyers to tend toward biofuels. However, with the decrease in cost of fossil fuels, biofuels may struggle to retain its value in the market.

18.6.2 Market Trends and Economic Incentives for Biofuel Production

In recent years, biofuels have gained momentum as a viable and sustainable substitute for fossil fuels due to growing concerns about global climate changes, energy security, high oil prices, and declining oil reserves (Datta 2022). Back in 2000, the global biofuel production rate was 180 thousand barrels per day, but with the recent advances in technologies and awareness about the use of fossil fuels, biofuel production rate has inclined significantly to 1914 thousand barrels per day as of 2022 (El-Araby 2024). The need to mitigate climate changes and the rising energy consumption have led to the increase in demand for renewable energy sources. Biofuels have the potential to reduce carbon emissions significantly. As of 2015, biofuels have reduced carbon emissions by over 589.3 million tons, which are compelling incentives for consumers and producers to shift toward biofuels (Y. Liu et al. 2021). Currently, the production cost for biofuel is much higher, which possesses a barrier to their widespread adoption. To be cost effective, biofuel production must approach cost equivalence with traditional fossil fuels. Effective government policies and financial incentives such as subsidies for sustainable biofuel production can provide economic incentives by creating jobs, promoting diversification in land use, and preventing deforestation (Correa et al. 2019).

18.6.3 Role of Government Policies in Promoting Sunflower Biofuels

Government policies play a crucial role in promoting sunflower biofuels by providing financial incentives, supporting R&D, and ensuring sustainability goals. The United States has enacted a series of laws and regulations to promote biofuel production. This includes the energy tax act of 1978, which provided investment tax credit for renewable energy investors, and the American clean energy and security act of 2009, which consisted of four parts: clean energy, energy efficiency, reduction of global warming pollution, and transition to clean energy (Su, Zhang, and Su 2015). In 2000, the EU proposed two directives for renewable energy sources, which included the promotion of renewable electricity and the development of biofuels in transport. The biofuel directive required member states to set national indicative target for biofuel consumption in transport, starting with a 2% share by the end of 2005 and then increasing to 5.75% by the end of 2010. To complete these requirements, the member states utilized tax exemptions, biofuel obligations, and an "energy crop payment" of EUR 45 per hectare to incentivize biofuel production. In 2009, the European parliament adopted a climate policy package in which the EU committed to reduce GHG emissions by 20% by the year 2020 and increasing the share of renewable energy in total energy production from 8.5% to 20% (Gradziuk et al. 2021). The EU policies have successfully promoted biofuel production, while addressing GHG emissions and climate changes through carbon sequestration and carbon pricing. Blending methods in biofuels are implemented through government policies that mandate the mixing of specific percentage of biofuels with conventional fuels to create more sustainable energy sources. The Brazilian government enacted blending targets through various programs such as National Alcohol Commission, which were responsible for ensuring compliance with blending targets. As a result, Brazil saw a 6% growth in ethanol production in 2015, reaching 30 billion liters, and by 2020, the ethanol production increased to 7.5%, demonstrating the effectiveness of these policies in boosting production (Sajid, Da Silva, and Danial 2021). To promote sunflower biofuel production, governments can offer tax credits, grants and subsidies, and low interest loans to sunflower cultivators, which will reduce the cost of the biofuel feedstock. Besides funding the technological advancement, research and development funding is crucial for new biofuel technologies to reduce the production cost. To address the greenhouse gas emissions associated with sunflower biofuel production process, policies for carbon pricing, sustainable agricultural practices, efficient conversion technologies are of utmost importance.

18.7 TECHNOLOGICAL INNOVATION IN SUNFLOWER BIOFUEL PRODUCTION

18.7.1 Advances in Biotechnology for Enhancing Oil Yield

Advances in molecular genetics and genetic engineering have provided vital assistance for improving sunflower oil content. Enhancing oil content through breeding in oilseed crops is an attractive way to improve the oil yield, and this method calls for a complete knowledge of gene networks engaged in oil biosynthesis during seed development (Bu et al. 2023). There are over 429 genes that are linked to 12 distinct pathways, which contribute to oil synthesis (Badouin et al. 2017). Studies have shown that sunflower oil content has been diversified through empirical recombination, selection methods, and mutation breeding programs in different regions. Rauf et al. (2016) studied that variant of genes product (DGAT, SLC-1), which catalyzes the final step of triacylglycerol formation, is responsible for increasing the oil contents in various crop species. The study suggested that such genes could be used to enhance the oil contents of sunflower species. However, sunflower oil content is a polygenic trait, necessitating the selection of various approaches to increase the oil yield. The study focused on various approaches including transgenic approaches, gene editing approaches such as CRISPR-Cas method, and marker-assisted selection to improve the oil content in sunflowers. Fatty acid synthase (FAS) and diacylglycerol acyltransferase (DGAT) are key enzymes that play a crucial role in increasing the oil yield. Overexpression of genes like DGAT has been shown to significantly increase oil content in transgenic plants ranging from 29% to 35% (Bu et al. 2023). Similarly, altering genes involved in the Kennedy pathways can improve triacylglycerol production, leading to higher oil yields. Techniques such as chemical mutagenesis have been used to create high oleic sunflower cultivars. For instance, the cultivar "Pervenets" was produced by an induced mutation that significantly boosted the oleic acid content, achieving levels of 80%–90% oleic acid in the oil (Rani, Sheoran, and Sharma 2017).

18.7.2 Use of Synthetic Biology in Developing High-Yield Sunflower Varieties

Synthetic biology uses biotechnology techniques such as DNA sequencing and genome editing to modify organisms and to develop synthetic varieties. Compared to traditional breeding methods, the development of synthetic varieties has demonstrated significant gains in sunflower yield. In the development of synthetic varieties, multiple parenteral lines are intercrossed, which allows for the expression of hybrid vigor, leading to significant improvement in the yield of sunflowers compared to the original population (Attia et al. 2020). Molecular markers are vital in identifying and selecting high-yield traits in sunflower breeding programs. PCR-based simple-sequence repeat (SSR) markers are frequently used molecular markers that are used for fingerprinting, genome mapping, population studies, and marker-assisted selection of sunflowers (H. G. M.-D. Ahmed et al. 2022). The SSR-based markers facilitate more efficient selection processes and allow for the identification of beneficial traits that are linked to yield and disease resistance. To enhance breeding efficiency, genomic selection is used to leverage genomic information (Dimitrijevic and Horn 2018). It includes estimating individuals breeding values based on genomic data, which can considerably speed up the development of high-yield sunflower varieties.

18.7.3 Innovations in Processing Technologies to Improve Efficiency

There are four standard primary techniques through which raw feedstocks can be treated for use as a substitute to fossil fuels: (i) direct use of blended oils, (ii) micro-emulsion of oils, (iii) thermal cracking (pyrolysis) of oils, and (iv) transesterification (Babadi et al. 2022). All these techniques reduce oil viscosity, reduce stability against oxidation, and enhance volatility. The most favorable method used for converting sunflower oil into biodiesel is transesterification due to effective viscosity reduction. In different studies, this process is optimized through various parameters such as the alcohol to oil ratio and the type of catalyst used. Khan et al. (2023) signify that adjusting the alcohol to oil molar ratio of 18:1 for sunflower oil using a catalyst such as KOH can maximize biodiesel yield by up to 98%. The use of different nanocatalysts such as metal-oxide nano catalysts made from oxidized metals like Mg, Zn, and Ca has gained significant attention due to their large active surface area and high catalytic efficiency (Farouk et al. 2024). Microreactors offer unique advantages and are found very useful for carrying out highly exothermic reactions, two-phase reactions, and reactions that have a tendency of explosion. They can be used to have a high conversion of triglycerides (>90%) in residence time as short as 1–2 minutes (Bhoi et al. 2014). The application of machine learning-based technologies is one of the remedies aimed at optimizing the challenges associated with biodiesel production. These technologies include artificial neural networks (ANN), response surface methodologies (RSM), and adaptive network-based fuzzy interference systems (ANFIS). Awogbemi and Kallon (2023) assessed these technologies and found that the usage of these technologies raised biodiesel yield by 84%–98%. On the other hand, the liquid-phase plasma discharge (LPPD) method is one of the emerging processing technologies that use electrical discharges in the liquid phase to boost chemical reactions, achieving a conversion efficiency of 99.5% in the case of soybean oil (Wu et al. 2019). However, the use of non-thermal plasma is still at the laboratory scale and needs a better understanding of the physical and chemical mechanisms of plasma and its interaction with chemical catalysts (Asghari et al. 2023). The combination of non-thermal plasma jet and hydrodynamic reactors can enhance the efficiency and effectiveness of biodiesel production from sunflower oil. Ansari Samani et al. (2023) combined the plasma jet and hydrodynamic reactor and achieved a conversion efficiency of up to 96% under specific conditions.

18.8 CHALLENGES AND LIMITATIONS

18.8.1 Technical Challenges in Large-Scale Biodiesel Production

Biodiesel production on a large scale faces many challenges while considering cost-effectiveness and environmental sustainability. The key challenges include the availability of feedstock and its quality, managing large volumes of feedstock and product, optimizing the conversion process, ensuring consistent product quality, and dealing with potential impurities in the feedstock. Balan (2014) highlighted many challenges faced by large corporations in commercializing biofuel production technologies. These bottlenecks included challenges associated with the development of energy-efficient technologies, with several steps in the process being inefficient or costly (pretreatment, hydrolysis, microbial fermentation, and biofuel separation). The large-scale production of biodiesel requires the establishment of a large plant and an increased amount of biomass as a feedstock. A suitable feedstock should have the following properties: abundant availability, high energy content, cost-effectiveness, and no competition with food or land for cultivation (Alherbawi et al. 2021). Sunflower as a first-generation biomass lacks most of the above properties and compromises land, water, and food resources for energy production. Moreover, storage is a critical challenge for advanced biofuel feedstock, indicating a capital expense for this facility. Water or moisture content affects the cost of supply because higher moisture content increases the volume of biomass, consequently increasing the number of vehicles needed for transportation (Sarwer et al. 2022). The location of the biorefinery is an important factor because if the sources are far from the facility, this can lead to an increase in transportation costs and subsequently increased production costs. Berazneva, Woolf, and Lee (2021) reported that the transportation cost can be 10%–25% of the total feedstock cost, and it can go up to 65%. Besides public awareness is an important factor because scaling up the production of biofuels requires an increase in market demand for biofuels, which of course is possible if there is public awareness about the use of biofuels. The land use changes and greenhouse gas emissions associated with the crop stages, transportation, and during the conversion process can have a negative environmental impact.

18.8.2 Economic Barriers and Market Competition with Fossil Fuels

Sunflower biofuel production faces significant economic challenges including shortage of feedstock supplies, production cost, incomplete pricing system, transportation cost, and feedstock cost (L. Zhang and Bai 2021). Among these costs, the feedstock cost comprises 75%–80% of the total biodiesel cost (Gude et al. 2013). The transportation cost is an important contributor to the total feedstock stock cost, which can be 10%–25% and can go up to 65% (Berazneva, Woolf, and Lee 2021). The incomplete pricing system of biofuels discourages the producers to expand their production scale (H. Zhang et al. 2017). Sunflower as a biofuel feedstock requires numerous inputs in planting and harvesting, which includes resources, capital, and labor, limiting progress in commercial promotion. The fluctuations in market prices of sunflower oil and seed can impact the investment in sunflower biofuel production due to uncertainty in returns on investments. Moreover, sunflower competes with other oilseed crops such as rapeseed and soybean, which may offer higher yield and better market prices, thus affecting biofuel production supply (Paul, Mazumder, and Mukherjee 2024). In the current scenario, fossil fuels lead the global energy market due to their widespread use and well-established infrastructure. Recent studies have shown that fossil fuels still account for 81% of global energy demand (Londoño-Pulgarin et al. 2021), indicating the dominance of fossil fuels in the global energy market. However, the continuous depletion of fossil fuels and the environmental damages associated with their production processes and usage have pushed the lawmakers and societies to move toward renewable energy sources (Li, Luo, and Lu 2017). In 2021, the global biofuel market was USD 109.96 billion, and it is projected to escalate to USD 201 billion by 2030 (Awogbemi and Kallon 2024). This increase in market value of biofuel is associated with the improving global interest in renewable energy sources, which has enhanced due to the environmental hazards of fossil fuels usage and fluctuations in its global market (Figure 18.2).

18.8.3 Socio-Environmental Concerns Related to Monoculture Practices

When a single crop such as sunflower is cultivated extensively year after year in the same location, this practice is termed as monoculture. Monoculture practice is adopted for cash crops to get potential high returns but doing so may lead to production loss per hectare over time. In order to overcome the production loss, farmers increase the necessary inputs to keep the yield stable, thus polluting the environment. Additionally, such practices accelerate soil erosion, pollute water resources, raise carbon level in atmosphere, and decrease biodiversity (Demirdogen, Guldal, and Sanli 2023). Frequent cultivation of a single crop year after year increases the biodiversity of agrophages (pests, weeds, and fungal pathogens), which causes an increase in usage of pesticides, increasing the production cost and harming the environment (Andrzej Woźniak 2019). Moreover, adverse changes may occur in the soil including the loss of organic matter and organic carbon, diminishing the biological and enzymatic activity of the soil. The analysis of monoculture soil test sample has shown that such practices lower the contents of organic carbon and the number of earthworms, as the presence of earthworms increases organic carbon content in the soil and its bioactivity (A. Woźniak and Soroka 2018). Biodiversity is badly affected by this culture as various plants and animal species are eliminated that would otherwise coexist in a more diverse ecosystem. Continuously planting the same crop depletes specific nutrients from the soil, which can lead to soil degradation. In such cases, large

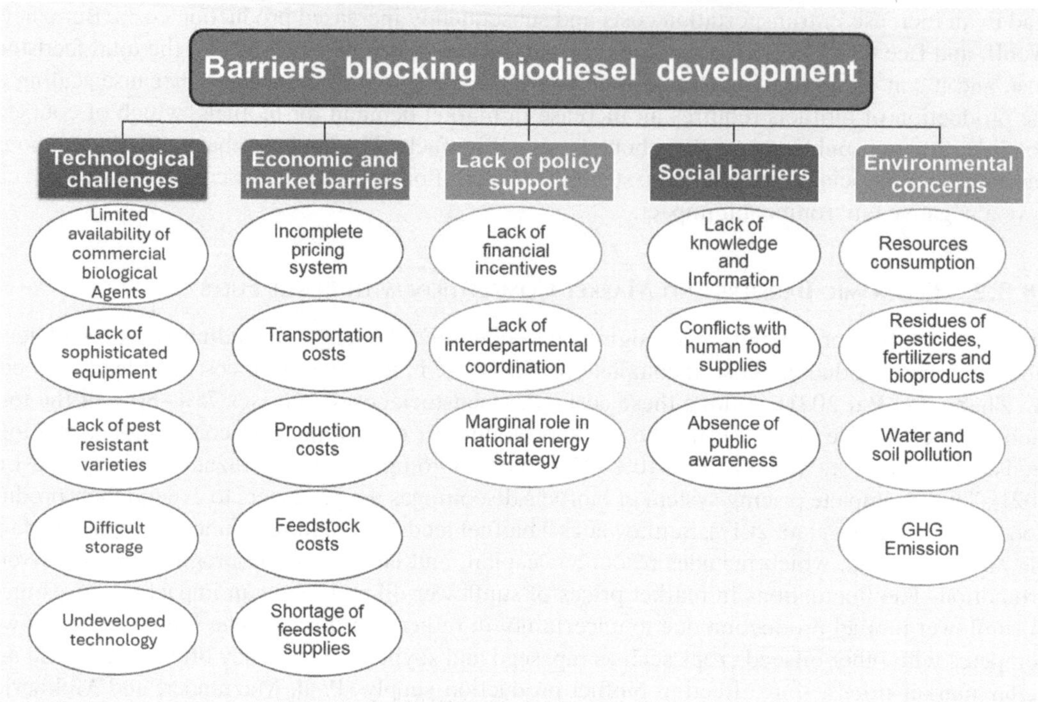

FIGURE 18.2 Barriers and challenges in biodiesel production.

bulk of fertilizers are needed to compensate for the nutrient loss, further harming the soil condition. Cultivation of the same plant in regions where water is scarce can lead to over-extraction of water, leading to water scarcity in the area. Studies have found that monoculture systems are prone to pests and diseases. This can lead to the overuse of pesticides, harming environment and human health. The hefty use of fertilizers and pesticides in monoculture practices can lead to water and nutrients runoff, affecting waterways and aquatic ecosystems (Belete and Yadete 2023).

18.9 FUTURE PROSPECTS FOR SUNFLOWER BIOFUELS

The main phases involved in sunflower biofuel production are the agricultural phase, oil extraction phase, conversion of biomass into biofuels, distribution, and waste management (De Marco et al. 2016). In the agricultural phase, research in the field of ecology and agronomy allows for the advancement of innovative methods for soil cultivation, fertilization, and water resource management, strengthening the resilience of agricultural systems (Drobitko 2024). The climate changes in recent years have negatively impacted the agricultural sectors throughout the world. A thorough approach toward the adaptability of agriculture to climate change is necessary. This calls for the development of hybrids that are resistant to climate changes with increased yield and oil content (Duca et al. 2022). To overcome the food competition, different parts of the sunflower, besides the seeds, can be used as biofuel feedstock. Ebrahimian et al. (2022) studied that about 1.77 kg of sunflower residue is produced per kg of harvested sunflower seeds. This huge amount of sunflower residue, a major low-cost source of holocellulose (cellulose and hemicellulose), can be converted into biofuels, thus overcoming the shortage of edible feedstocks. Different extraction technologies are used for oil extraction such as pressing and organic solvent extraction technologies. However, J. Liu et al. (2016) discussed that enzyme-assisted extraction processing (EAEP) is a promising alternative method for oil extraction because of its enzyme specificity and eco-friendly nature.

But this method still has weaknesses, which are the high cost of enzymes and lower oil yield than that of the solvent extraction. Research studies are conducted on this topic to overcome its drawbacks. The primary method used for converting sunflower oils into biofuel is transesterification. However, as discussed above, the conversion of sunflower oil into biodiesel is a costly method, and hence, researchers have conducted various studies to optimize the transesterification process and reduce the production cost (Aguilera 2014). The production and use of biofuel generates emissions such as carbon monoxide, carbon dioxide, nitrogen oxides, hydrocarbon, and volatile organic oxides. These emissions are lower than those produced by fossil fuels. However, the nitrous oxide emissions during the agricultural phase are of major concern. Various studies are conducted on the use of sustainable agricultural technologies along with other factors such as the use of optimized fertilization and nitrification inhibitors to reduce nitrous oxide emissions (Neupane 2022).

In sustainable agricultural systems, farmers focus on sustainable farming practices to ensure safe and high-quality food production, while also protecting the economy, environment, and ecosystem. There are four pillars of sustainable agriculture: land management, resource management, the human interface, and the ecosystem interface (Shelef et al. 2018). Land management is an effort to use, develop, and maintain land resources using three crop management principles: minimum soil disturbance, cover crops, and crop rotation (Pittelkow et al. 2015). As discussed above, sunflower cultivation has a notable influence on land use, indicating that maximum efforts are required to mitigate the land use implications in the cultivation of sunflower. Conservation tillage, i.e., minimal tillage or no tillage, is one of the techniques used to reduce the land use implications while improving the productivity of sunflowers under low moisture availability (A. Sher et al. 2021). Researchers have suggested that the no-tillage technique alone may be insufficient and can reduce the yield, but when combined with other conservation agricultural principles such as residue retention and crop rotation, its negative impacts are minimized (Pittelkow et al. 2015). Resource management refers to reducing the use of inputs such as energy, water, nutrients, and pesticides while utilizing renewable sources that are environmentally sustainable and economically stable. For sunflower cultivation, the scarce use of resources can be implemented by using the bio-solarization method to kill pests and weeds and improve soil condition, supporting circular economics (Oldfield et al. 2017). Similarly, choosing a proper sunflower cultivar can improve crop resistance to abiotic stresses, pathogens, and diseases. Other methods of sustainable resource utilization include using biofertilization, compositing, and drip irrigation. Industrial agriculture is considered an enormous portion of pollution, especially in the case of sunflower, which involves large amounts of waste and pesticides impacting human life and health (Shelef et al. 2018). This also affects the local ecosystem due to land fragmentation, which may lead to a negative impact on wildlife and the ecosystem. However, these negative effects can be mitigated by integrating the sunflower into sustainable agricultural systems.

The biofuel industry is at the developing stage and is less competitive than traditional energy in terms of high production cost, immature technology, poor supporting infrastructure, etc. However, with the support of government policies, the world biofuel ethanol production has increased from 6.4 to 23.4 billion gallons from 2003 to 2013, indicating the importance of policies in the development of the biofuel industry (Su, Zhang, and Su 2015).

After careful examination of various research papers on the biofuel industry, the following policies can be implemented to foster and develop a sustainable biofuel industry (Gradziuk et al. 2021; Qiu et al. 2012; Rodrigues and Petr 2021; Sajid, Da Silva, and Danial 2021):

Financial Support: Providing low interest loans and subsidies to address high production cost.

Research and Development: Directing the industries to invest in research and development.

Certification and Standardization: Developing clear standards for biofuels quality to build consumers' confidence.

Advanced Feedstock: Promoting the use of agricultural residues and non-crop feedstock to avoid competition with food.

Biorefinery Infrastructure: Developing biorefineries that are capable of producing multiple products handling diverse feedstocks.

Environmental Regulations: Establishing regulations on carbon emissions and waste disposal for biofuel industries.

Renewable Energy Targets: Setting specific targets for the adoption of biofuels.

Blending Mandates: Implantation of blending mandates like the Brazil 15% biodiesel blending target for 2025.

International Cooperation: Encouraging global cooperation to boost bioeconomy.

18.10 CONCLUSION AND FUTURE PROSPECTS

Sunflower seeds have a high oil content of about 45%. Compared to other vegetable oils, it is relatively high in unsaturated fatty acids, making it a significant candidate as a biofuel feedstock. Sunflowers can be embedded into both food and energy systems, promoting farming practices and better resource utilization. Their diverse nature allows for its utilization in various industrial applications such as a biofuel feedstock, cosmetics, and the food industry. Transesterification method is used to convert sunflower oils into biodiesel, which can significantly replace fossil fuels. Catalysts play an important role in the transesterification process. As discuss earlier, base-catalyzed transesterification is most commonly used for sunflower oils. However, heterogenous catalyst is said to be more favorable compared to homogenous and enzymatic catalysts. A maximum yield of 99.2% was achieved at a temperature of 65°C, indicating a suitable temperature for the transesterification of sunflower oil. Production of biofuels from organic feedstock faces several challenges, from cultivation phase to extraction, conversion, distribution, and waste management. The main focus is the emission of greenhouse gases, which are the largest contributor to global warming. However, as discussed above, the greenhouse gas emissions from biofuel production process are 63%–77% lower than the petroleum diesel. Although the biofuels/biodiesel industries are still at their early stages of advancement, the need to mitigate environmental destruction associated with the use of fossil fuels calls for the development of biofuels at their earliest. Nonetheless, biofuel production faces economic challenges and market competition with fossil fuels. The planting, harvesting, extraction, and conversion (to biodiesel) of sunflower oils require resources, capital, and labors, making the production of biofuels costlier. Compared to conventional oil which costs from $5 per barrel to around $30 per barrel, the biofuel production cost is high, i.e., between $20 and $90 per barrel. However, government policies can help lower the economic burden. This includes tax credits, subsidies, low interest loans, funding technological advancement, and investing in research and development in the field of biofuels. As indicated by various scientists and researchers, the world is running out of fossil fuels and its usage poses a threat to the whole world and human health. We emphasize on a united and global action to promote and encourage the use of biofuels. Research in the biofuel industry is of utmost importance, especially in reducing the cost of biofuel production and the environmental challenges associated with it. The introduction of various policies by the United Nations to encourage production and usage of biofuels would allow for a global and complete action toward sustainability and stability of the biofuel market.

COMPETING INTERESTS

The authors declare that they have no competing interest.

DECLARATION OF GENERATIVE AI AND AI-ASSISTED TECHNOLOGIES IN THE WRITING PROCESS

During the preparation of this work, the book chapter author(s) have not taken any assistance from AI, and the book chapter author(s) reviewed and edited the content as needed and take(s) full responsibility for the content of the publication.

REFERENCES

Aguilera, Roberto F. 2014. "Production Costs of Global Conventional and Unconventional Petroleum." *Energy Policy* 64: 134–40. doi:10.1016/j.enpol.2013.07.118.

Ahmed, Hafiz Ghulam Muhu-Din, Muhammad Rizwan, Muhammad Naeem, Muhammad Ahsan Khan, Faheem Shehzad Baloch, Sangmi Sun, and Gyuhwa Chung. 2022. "Molecular Characterization and Validation of Sunflower (*Helianthus Annuus* L.) Hybrids Through SSR Markers" ed. Muhammad Abdul Rehman Rashid. *PLoS One* 17(5): e0267383. doi:10.1371/journal.pone.0267383.

Ahmed, M. A., and Mohamed Ali Abdelsatar. 2020. "Effect of Sowing Dates on Yield and Yield-Attributes of Some Sunflower Hybrids." *Agriculture* 2(1): 131–44.

Akkaya, M. R. (2018). "Fatty Acid Compositions of Sunflowers (Helianthus annuus L.) Grown in East Mediterranean Region." *Rivista Italiana Delle Sostanze Grasse* 95(4): 239–247.

Alberio, C., N. G. Izquierdo, and L. A. N. Aguirrezábal. 2014. "Sunflower: Chemistry, Production, Processing, and Utilization." *Sunflower Crop Physiology and Agronomy*. AOCS Press. doi:10.1016/B978-1-893997-94-3.50009-X.

Alherbawi, Mohammad, Gordon McKay, Hamish R. Mackey, and Tareq Al-Ansari. 2021. "Jatropha Curcas for Jet Biofuel Production: Current Status and Future Prospects." *Renewable and Sustainable Energy Reviews* 135: 110396. doi:10.1016/j.rser.2020.110396.

Ali, J. A., and A. Abuhabaya. 2013. "Sunflower Biodiesel: Efficiency and Emissions." *WIT Transactions on Engineering Sciences* 81: 25–36. doi:10.2495/978-1-78466-034-5/017.

Alzamel, Nurah M., Eman M. M. Taha, Abeer A. A. Bakr, and Naglaa Loutfy. 2022. "Effect of Organic and Inorganic Fertilizers on Soil Properties, Growth Yield, and Physiochemical Properties of Sunflower Seeds and Oils." *Sustainability* 14(19). doi:10.3390/su141912928.

Ansari Samani, Marziyeh, Bahram Hosseinzadeh Samani, Mahdi Ghasemi-Varnamkhasti, Sajad Rostami, and Rahim Ebrahimi. 2023. "Biodiesel Production from Sunflower Oil Using a Combined Atmospheric Cold Plasma Jet-Hydrodynamic Reactor." *Biofuels* 14(9): 895–905. doi:10.1080/17597269.2023.2190569.

Asghari, Mohammadreza, Bahram Hosseinzadeh Samani, Rahim Ebrahimi, Sajad Rostami, and Ebrahim Fayyazi. 2023. "Optimization of a Novel Liquid Phase Capillary Discharge Plasma Reactor for Continuous Methyl Ester Synthesis." *Energy Conversion and Management* 277: 116667. doi:10.1016/j.enconman.2023.116667.

Attia, Mahrous, R. Bakheit, A. Abo-Elwafa, and A. El-Shimy. 2020. "Development and Evaluation of Sunflower Synthetic Varieties (*Helianthus annuus* L.)." *Journal of Environmental Studies* 22(1): 28–39. doi:10.21608/jesj.2020.205259.

Awogbemi, Omojola, and Daramy Vandi Von Kallon. 2023. "Application of Machine Learning Technologies in Biodiesel Production Process—A Review." *Frontiers in Energy Research* 11(February). doi:10.3389/fenrg.2023.1122638.

Awogbemi, Omojola, and Daramy Vandi Von Kallon. 2024. "Recent Advances in the Application of Nanomaterials for Improved Biodiesel, Biogas, Biohydrogen, and Bioethanol Production." *Fuel* 358: 130261. doi:10.1016/j.fuel.2023.130261.

Babadi, Arman Amani, Shahrooz Rahmati, Rafieh Fakhlaei, Bahram Barati, Shuang Wang, William Doherty, and Kostya (Ken) Ostrikov. 2022. "Emerging Technologies for Biodiesel Production: Processes, Challenges, and Opportunities." *Biomass and Bioenergy* 163: 106521. doi:10.1016/j.biombioe.2022.106521.

Badouin, Hélène, Jérôme Gouzy, Christopher J. Grassa, Florent Murat, S. Evan Staton, Ludovic Cottret, Christine Lelandais-Brière, et al. 2017. "The Sunflower Genome Provides Insights into Oil Metabolism, Flowering and Asterid Evolution." *Nature* 546(7656): 148–52. doi:10.1038/nature22380.

Balan, Venkatesh. 2014. "Current Challenges in Commercially Producing Biofuels from Lignocellulosic Biomass." *ISRN Biotechnology* 2014(i): 1–31. doi:10.1155/2014/463074.

Barontini, Federica, Marco Simone, Federico Triana, Andrea Mancini, Giorgio Ragaglini, and Cristiano Nicolella. 2015. "Pilot-Scale Biofuel Production from Sunflower Crops in Central Italy." *Renewable Energy* 83: 954–62. doi:10.1016/j.renene.2015.05.043.

Belete, Tegegn, and Eshetu Yadete. 2023. "Effect of Mono Cropping on Soil Health and Fertility Management for Sustainable Agriculture Practices: A Review." *Journal of Plant Sciences*. doi:10.11648/j.jps.20231106.13.

Berazneva, Julia, Dominic Woolf, and David R. Lee. 2021. "Local Lignocellulosic Biofuel and Biochar Co-Production in Sub-Saharan Africa: The Role of Feedstock Provision in Economic Viability." *Energy Economics* 93: 105031. doi:10.1016/j.eneco.2020.105031.

Bhoi, Rohidas, Nirvik Sen, K. K. Singh, Sanjay M. Mahajani, K. T. Shenoy, Hanmanth Rao, and S. K. Ghosh. 2014. "Transesterification of Sunflower Oil in Microreactors." *International Journal of Chemical Reactor Engineering* 12(1): 47–62. doi:10.1515/ijcre-2013-0105.

Bu, Mengjia, Wei Fan, Ruonan Li, Bing He, and Peng Cui. 2023. "Lipid Metabolism and Improvement in Oilseed Crops: Recent Advances in Multi-Omics Studies." *Metabolites* 13(12): 1170. doi:10.3390/metabo13121170.

Correa, Diego F., Hawthorne L. Beyer, Joseph E. Fargione, Jason D. Hill, Hugh P. Possingham, Skye R. Thomas-Hall, and Peer M. Schenk. 2019. "Towards the Implementation of Sustainable Biofuel Production Systems." *Renewable and Sustainable Energy Reviews* 107(March): 250–63. doi:10.1016/j.rser.2019.03.005.

Correa, Diego F., Hawthorne L. Beyer, Hugh P. Possingham, Skye R. Thomas-Hall, and Peer M. Schenk. 2017. "Biodiversity Impacts of Bioenergy Production: Microalgae vs. First Generation Biofuels." *Renewable and Sustainable Energy Reviews* 74(October 2016): 1131–46. doi:10.1016/j.rser.2017.02.068.

Datta, Bipasa. 2022. "An Economic Analysis of Biofuels: Policies, Trade, and Employment Opportunities." *Handbook of Biofuels*: 3–29. doi:10.1016/B978-0-12-822810-4.00001-4.

De Marco, Iolanda, Salvatore Miranda, Stefano Riemma, and Raffaele Iannone. 2016. "Biodiesel Production from Sunflower: An Environmental Study." *Chemical Engineering Transactions* 49(2010): 331–36. doi:10.3303/CET1649056.

Debaeke, Philippe, Emmanuelle Bret-Mestries, Jean Noël Aubertot, Pierre Casadebaig, Luc Champolivier, Jean François Dejoux, Pierre Maury, and Célia Seassau. 2020. "Sunflower Agronomy: 10 Years of Research in Partnership within the 'Sunflower' Technological Joint Unit (UMT) in Toulouse." *OCL – Oilseeds and Fats, Crops and Lipids* 27(173). doi:10.1051/ocl/2020006.

Debaeke, P., E. Mestries, M. Desanlis, and C. Seassau. 2014. Effects of Crop Management on the Incidence and Severity of Fungal Diseases in Sunflower. In J. I. Arribas (Ed.), *Sunflowers: Growth and Development, Environmental Influences and Pests/Diseases*. Nova Science Publishers.

Demirdogen, Alper, Huseyin Tayyar Guldal, and Hasan Sanli. 2023. "Monoculture, Crop Rotation Policy, and Fire." *Ecological Economics* 203: 107611. doi:10.1016/j.ecolecon.2022.107611.

Devaraj Naik, B., and M. Udayakumar. 2023. "Kinetics and Thermodynamic Analysis of Transesterification of Waste Cooking Sunflower Oil Using Bentonite-Supported Sodium Methoxide Catalyst." *Biomass Conversion and Biorefinery* 13(11): 9701–14. doi:10.1007/s13399-021-01836-9.

Dimitrijevic, Aleksandra, and Renate Horn. 2018. "Sunflower Hybrid Breeding: From Markers to Genomic Selection." *Frontiers in Plant Science* 8. doi:10.3389/fpls.2017.02238.

Drobitko, Antonina. 2024. "Formation of Sunflower Hybrid Productivity by Resource Saving Cultivation Technologies in Southern Ukraine Mykolaiv National Agrarian University Mykolaiv National Agrarian University PhD in Agriculture, Associate Professor PhD in Sustainable Rural Develo." 28(3): 9–18. doi:10.56407/bs.agrarian/3.2024.09.

Duca, Maria, Angela Port, Ion Burcovschi, Maria Joiţa-Păcureanu, and Mihaela Dan. 2022. "Environmental Response in Sunflower Hybrids: A Multivariate Approach." *Romanian Agricultural Research* 39(39). doi:10.59665/rar3914.

Ebrahimian, Elham, Joeri F.M. Denayer, Mortaza Aghbashlo, Meisam Tabatabaei, and Keikhosro Karimi. 2022. "Biomethane and Biodiesel Production from Sunflower Crop: A Biorefinery Perspective." *Renewable Energy* 200: 1352–61. doi:10.1016/j.renene.2022.10.069.

Ebrahimian, Elnaz, Seyyed Mohammad Seyyedi, Ahmad Bybordi, and Christos A. Damalas. 2019. "Seed Yield and Oil Quality of Sunflower, Safflower, and Sesame under Different Levels of Irrigation Water Availability." *Agricultural Water Management* 218(February 2018): 149–57. doi:10.1016/j.agwat.2019.03.031.

El-Araby, R. 2024. "Biofuel Production: Exploring Renewable Energy Solutions for a Greener Future." *Biotechnology for Biofuels and Bioproducts* 17(1). doi:10.1186/s13068-024-02571-9.

Faria, Laila Al, and Rony Rani. 2021. "Asteraceae: A Taxonomically and Medicinally Important Sunflower Family." *American International Journal of Biology and Life Sciences* 3(1): 1–17. doi:10.46545/aijbls.v3i1.307.

Farouk, Sabah Mohamed, Aghareed M. Tayeb, Shereen M. S. Abdel-Hamid, and Randa M. Osman. 2024. "Recent Advances in Transesterification for Sustainable Biodiesel Production, Challenges, and Prospects: A Comprehensive Review." *Environmental Science and Pollution Research* 31(9): 12722–47. doi:10.1007/s11356-024-32027-4.

Feyzi, Mostafa, Nahid Hosseini, Nakisa Yaghobi, and Rohollah Ezzati. 2017. "Preparation, Characterization, Kinetic and Thermodynamic Studies of MgO-La_2O_3 Nanocatalysts for Biodiesel Production from Sunflower Oil." *Chemical Physics Letters* 677: 19–29. doi:10.1016/j.cplett.2017.03.014.

Folayan, Adewale Johnson, Paul Apeye Lucky Anawe, Adeyemi Emman Aladejare, and Augustine Omoniyi Ayeni. 2019. "Experimental Investigation of the Effect of Fatty Acids Configuration, Chain Length, Branching and Degree of Unsaturation on Biodiesel Fuel Properties Obtained from Lauric Oils, High-Oleic and High-Linoleic Vegetable Oil Biomass." *Energy Reports* 5: 793–806. doi:10.1016/j.egyr.2019.06.013.

Gagour, Jamila, Moussa Nid Ahmed, Hasna Ait Bouzid, Samira Oubannin, Laila Bijla, Mohamed Ibourki, Ahmed Hajib, et al. 2022. "Proximate Composition, Physicochemical, and Lipids Profiling and Elemental Profiling of Rapeseed (*Brassica napus* L.) and Sunflower (*Helianthus annuus* L.) Grown in Morocco." *Evidence-Based Complementary and Alternative Medicine* 2022. doi:10.1155/2022/3505943.

Ghosh, Pranab, and Gobinda Karmakar. 2014. "Evaluation of Sunflower Oil as a Multifunctional Lubricating Oil Additive." *International Journal of Industrial Chemistry* 5(1): 1–10. doi:10.1007/s40090-014-0007-7.

Giakoumis, Evangelos G. 2018. "Analysis of 22 Vegetable Oils' Physico-Chemical Properties and Fatty Acid Composition on a Statistical Basis, and Correlation with the Degree of Unsaturation." *Renewable Energy* 126: 403–19. doi:10.1016/j.renene.2018.03.057.

Gradziuk, Piotr, Krzysztof Jończyk, Barbara Gradziuk, Adrianna Wojciechowska, Anna Trocewicz, and Marcin Wysokiński. 2021. "An Economic Assessment of the Impact on Agriculture of the Proposed Changes in EU Biofuel Policy Mechanisms." *Energies* 14(21): 1–21. doi:10.3390/en14216982.

Gude, Veera G., Georgene E. Grant, Prafulla D. Patil, and Shuguang Deng. 2013. "Biodiesel Production from Low Cost and Renewable Feedstock." *Central European Journal of Engineering* 3(4): 595–605. doi:10.2478/s13531-013-0102-0.

Gupta, Jharna, Madhu Agarwal, and A. K. Dalai. 2020. "An Overview on the Recent Advancements of Sustainable Heterogeneous Catalysts and Prominent Continuous Reactor for Biodiesel Production." *Journal of Industrial and Engineering Chemistry* 88: 58–77. doi:10.1016/j.jiec.2020.05.012.

Gupta, Monoj K. 2014. "Sunflower Oil: History, Applications and Trends." *Lipid Technology* 26(11–12): 260–63. doi:10.1002/lite.201400068.

Hanaki, K., and Portugal-Pereira, J. (2018). The Effect of Biofuel Production on Greenhouse Gas Emission Reductions. In *Biofuels and Sustainability: Holistic Perspectives for Policy-Making* (pp. 53–71). Tokyo: Springer Japan.

Heidari, Mostafa, and Vahid Karami. 2014. "Effects of Different Mycorrhiza Species on Grain Yield, Nutrient Uptake and Oil Content of Sunflower Under Water Stress." *Journal of the Saudi Society of Agricultural Sciences* 13(1): 9–13. doi:10.1016/j.jssas.2012.12.002.

Ibrahim, Sherouk M., Ahmed M. El-Sherbeeny, Jae Jin Shim, Ali A. AlHammadi, and Mostafa R. Abukhadra. 2021. "-SO_3H-Functionalization of Sub-Bituminous Coal as a Highly Active Acidic Catalyst during the Transesterification of Spent Sunflower Oil; Characterization, Application, and Mechanism." *Energy Reports* 7: 8699–710. doi:10.1016/j.egyr.2021.11.139.

Jocić, Siniša, Dragana Miladinović, and Yalcin Kaya. 2015. "Sunflower: Chemistry, Production, Processing, and Utilization." *Breeding and Genetics of Sunflower*. AOCS Press. doi:10.1016/B978-1-893997-94-3.50007-6.

Khan, Elena, Kadir Ozaltin, Damiano Spagnuolo, Andres Bernal-Ballen, Maxim V. Piskunov, and Antonio Di Martino. 2023. "Biodiesel from Rapeseed and Sunflower Oil: Effect of the Transesterification Conditions and Oxidation Stability." *Energies* 16(2). doi:10.3390/en16020657.

Khan, Sheeba, Choudhary Saumya, Pandey Anamika, Khan Mohd Kamran, and Thomas George. 2015. "Sunflower Oil: Efficient Oil Source for Human Consumption." *Emergent Life Sciences Research* 1(1): 1–3.

Kibar, Murat Efgan, Lina Hilal, Begüm Tuna Çapa, Buse Bahçıvanlar, and Brahim Ben Abdeljelil. 2023. "Assessment of Homogeneous and Heterogeneous Catalysts in Transesterification Reaction: A Mini Review." *ChemBioEng Reviews* 10(4): 412–22. doi:10.1002/cben.202200021.

Li, Meihui, Na Luo, and Yi Lu. 2017. "Biomass Energy Technological Paradigm (BETP): Trends in this Sector." *Sustainability* 9(4): 1–28. doi:10.3390/su9040567.

Liu, Jun-jun, Mohammed Abdalbasit A. Gasmalla, Pengfei Li, and Ruijin Yang. 2016. "Enzyme-Assisted Extraction Processing from Oilseeds: Principle, Processing and Application." *Innovative Food Science & Emerging Technologies* 35: 184–93. doi:10.1016/j.ifset.2016.05.002.

Liu, Yuzhong, Pablo Cruz-Morales, Amin Zargar, Michael S. Belcher, Bo Pang, Elias Englund, Qingyun Dan, Kevin Yin, and Jay D. Keasling. 2021. "Biofuels for a Sustainable Future." *Cell* 184(6): 1636–47. doi:10.1016/j.cell.2021.01.052.

Londoño-Pulgarin, Diana, Giovanny Cardona-Montoya, Juan C. Restrepo, and Francisco Muñoz-Leiva. 2021. "Fossil or Bioenergy? Global Fuel Market Trends." *Renewable and Sustainable Energy Reviews* 143(February). doi:10.1016/j.rser.2021.110905.

Makareviciene, Violeta, and Egle Sendzikiene. 2021. "Noncatalytic Biodiesel Synthesis under Supercritical Conditions." *Processes* 9(1): 1–15. doi:10.3390/pr9010138.

Malode, Shweta J., K. Keerthi Prabhu, Ronald J. Mascarenhas, Nagaraj P. Shetti, and Tejraj M. Aminabhavi. 2021. "Recent Advances and Viability in Biofuel Production." *Energy Conversion and Management: X* 10(December 2020): 100070. doi:10.1016/j.ecmx.2020.100070.

Mandari, Venkatesh, and Santhosh Kumar. 2022. "Biodiesel Production Using Homogeneous, Heterogeneous, and Enzyme Catalysts via Transesterification and Esterification Reactions : A Critical Review." *BioEnergy Research*: 935–61. doi:10.1007/s12155-021-10333-w.

Markov, Vladimir, Vyacheslav Kamaltdinov, Sergey Devyanin, Bowen Sa, Anatoly Zherdev, and Viktor Furman. 2021. "Investigation of the Influence of Different Vegetable Oils as a Component of Blended Biofuel on Performance and Emission Characteristics of a Diesel Engine for Agricultural Machinery and Commercial Vehicles." *Resources* 10(8): 39–41. doi:10.3390/resources10080074.

Miyuranga, K. A. Viraj, Udara S. P. R. Arachchige, T. M. M. Marso, and Gamunu Samarakoon. 2023. "Biodiesel Production through the Transesterification of Waste Cooking Oil over Typical Heterogeneous Base or Acid Catalysts." *Catalysts* 13(3). doi:10.3390/catal13030546.

Mohammadshirazi, Ahmad, Asadolah Akram, Shahin Rafiee, and Elnaz Bagheri Kalhor. 2014. "Energy and Cost Analyses of Biodiesel Production from Waste Cooking Oil." *Renewable and Sustainable Energy Reviews* 33: 44–49. doi:10.1016/j.rser.2014.01.067.

Nabavi-Pelesaraei, Ashkan, Hossein Azadi, Steven Van Passel, Zahra Saber, Fatemeh Hosseini-Fashami, Fatemeh Mostashari-Rad, and Hassan Ghasemi-Mobtaker. 2021. "Prospects of Solar Systems in Production Chain of Sunflower Oil Using Cold Press Method with Concentrating Energy and Life Cycle Assessment." *Energy* 223: 120117. doi:10.1016/j.energy.2021.120117.

Nakonechna, K., Ilko, V., Berčíková, M., Vietoris, V., Panovská, Z., and Doležal, M. 2024. "Nutritional, Utility, and Sensory Quality and Safety of Sunflower Oil on the Central European Market." *Agriculture* 14(4): 536.

Neupane, Dhurba. 2022. "Biofuels from Renewable Sources, a Potential Option for Biodiesel Production." *Bioengineering* 10(1): 29. doi:10.3390/bioengineering10010029.

Nguyen, Duyen Thi Cam, Thuong Thi Nguyen, Hanh T.N. Le, Thi Thanh Thuy Nguyen, Long Giang Bach, Trinh Duy Nguyen, Dai Viet N. Vo, and Thuan Van Tran. 2021. "The Sunflower Plant Family for Bioenergy, Environmental Remediation, Nanotechnology, Medicine, Food and Agriculture: A Review." *Environmental Chemistry Letters* 19(5): 3701–26. doi:10.1007/s10311-021-01266-z.

Nurrochmat, N. A., I. Indrawati, U. Adzkia, and M. Ekayani. 2024. "Market Competitiveness of Sunflower Seed and Oil Products." *IOP Conference Series: Earth and Environmental Science* 1379(1). doi:10.1088/1755-1315/1379/1/012024.

Oldfield, Thomas L., Yigal Achmon, Kristen M. Perano, Ruth M. Dahlquist-Willard, Jean S. VanderGheynst, James J. Stapleton, Christopher W. Simmons, and Nicholas M. Holden. 2017. "A Life Cycle Assessment of Biosolarization as a Valorization Pathway for Tomato Pomace Utilization in California." *Journal of Cleaner Production* 141: 146–56. doi:10.1016/j.jclepro.2016.09.051.

Park, Brian, and John M. Burke. 2020. "Phylogeography and the Evolutionary History of Sunflower (Helianthus Annuus L.): Wild Diversity and the Dynamics of Domestication." *Genes* 11(3). doi:10.3390/genes11030266.

Paul, Sonali, Camellia Mazumder, and Susmita Mukherjee. 2024. "Challenges Faced in Commercialization of Biofuel from Biomass Energy Resources." *Biocatalysis and Agricultural Biotechnology* 60: 103312. doi:10.1016/j.bcab.2024.103312.

Pekcan, Veli, M. Ibrahim Yilmaz, Goksel Evci, Ayse Nuran Cil, Vakas Sahin, Oguz Gunduz, Hasan Koc, and Yalcin Kaya. 2022. "Oil Content Determination on Sunflower Seeds in Drought Conditions." *Journal of Food Processing and Preservation* 46(10): 1–11. doi:10.1111/jfpp.15481.

Pérez, Ethel E., Erica R. Baümler, Guillermo H. Crapiste, and Amalia A. Carelli. 2019. "Effect of Sunflower Collets Moisture on Extraction Yield and Oil Quality." *European Journal of Lipid Science and Technology* 121(2): 1–7. doi:10.1002/ejlt.201800234.

Pilecco, Getúlio Elias, Martin H. Chantigny, Douglas Adams Weiler, Celso Aita, Marie Noëlle Thivierge, Raquel Schmatz, Bruno Chaves, and Sandro José Giacomini. 2020. "Greenhouse Gas Emissions and Global Warming Potential from Biofuel Cropping Systems Fertilized with Mineral and Organic Nitrogen Sources." *Science of the Total Environment* 729. doi:10.1016/j.scitotenv.2020.138767.

Pittelkow, Cameron M., Xinqiang Liang, Bruce A. Linquist, Lees Jan Van Groenigen, Juhwan Lee, Mark E. Lundy, Natasja Van Gestel, et al. 2015. "Productivity Limits and Potentials of the Principles of Conservation Agriculture." *Nature* 517(7534): 365–68. doi:10.1038/nature13809.

Qiu, Huanguang, Laixiang Sun, Jikun Huang, and Scott Rozelle. 2012. "Liquid Biofuels in China : Current Status, Government Policies, and Future Opportunities and Challenges." *Renewable and Sustainable Energy Reviews* 16(5): 3095–3104. doi:10.1016/j.rser.2012.02.036.

Rani, Reena, R.K. Sheoran, and Bunty Sharma. 2017. "Perspectives of Breeding for Altering Sunflower Oil Quality to Obtain Novel Oils." *International Journal of Current Microbiology and Applied Sciences* 6(8): 949–62. doi:10.20546/ijcmas.2017.608.117.

Rauf, Saeed, Nazia Jamil, Sultan Ali Tariq, Maria Khan, Maria Kausar, and Yalcin Kaya. 2016. "Progress in Modification of Sunflower Oil to Expand its Industrial Value." *Journal of the Science of Food and Agriculture* 97(7): 1997–2006. doi:10.1002/jsfa.8214.

Reyero, Inés, Gurutze Arzamendi, Silvia Zabala, and Luis M. Gandía. 2015. "Kinetics of the NaOH-Catalyzed Transesterification of Sunflower Oil with Ethanol to Produce Biodiesel." *Fuel Processing Technology* 129: 147–55. doi:10.1016/j.fuproc.2014.09.008.

Rodrigues, A. C. C., and Do Petr. 2021. "Policy, Regulation, Development and Future of Biodiesel Industry in Brazil." *Cleaner Engineering and Technology* 4: 100197. doi:10.1016/j.clet.2021.100197.

Sajid, Zaman, Maria Aparecida Batista Da Silva, and Syed Nasir Danial. 2021. "Historical Analysis of the Role of Governance Systems in the Sustainable Development of Biofuels in Brazil and the United States of America (USA)." *Sustainability (Switzerland)* 13(12). doi:10.3390/su13126881.

Salaheldeen, Mohammed, Abdalbasit Adam Mariod, Mohamed Kheireddine Aroua, S. M. Ashrafur Rahman, Manzoore Elahi M. Soudagar, and I. M. Rizwanul Fattah. 2021. "Current State and Perspectives on Transesterification of Triglycerides for Biodiesel Production." *Catalysts* 11(9): 1121. doi:10.3390/catal11091121.

Sanz Requena, J. F., A. C. Guimaraes, S. Quirós Alpera, E. Relea Gangas, S. Hernandez-Navarro, L. M. Navas Gracia, J. Martin-Gil, and H. Fresneda Cuesta. 2011. "Life Cycle Assessment (LCA) of the Biofuel Production Process from Sunflower Oil, Rapeseed Oil and Soybean Oil." *Fuel Processing Technology* 92(2): 190–99. doi:10.1016/j.fuproc.2010.03.004.

Sarwer, Asma, Murid Hussain, Ala'a H. Al-Muhtaseb, Abrar Inayat, Sikander Rafiq, M. Shahzad Khurram, Noaman Ul-Haq, et al. 2022. "Suitability of Biofuels Production on Commercial Scale from Various Feedstocks: A Critical Review." *ChemBioEng Reviews* 9(5): 423–41. doi:10.1002/cben.202100049.

Shelef, Oren, Jesús D. Fernández-Bayo, Yonatan Sher, Valeria Ancona, Heather Slinn, and Yigal Achmon. 2018. "Elucidating Local Food Production to Identify the Principles and Challenges of Sustainable Agriculture." *Sustainable Food Systems from Agriculture to Industry*. Elsevier: 47–81. doi:10.1016/B978-0-12-811935-8.00002-0.

Sher, Ahmad, Muhammad Yasir Arfat, Sami Ul-Allah, Abdul Sattar, Muhammad Ijaz, Abdul Manaf, Abdul Qayyum, et al. 2021. "Conservation Tillage Improves Productivity of Sunflower (*Helianthus annuus* L.) under Reduced Irrigation on Sandy Loam Soil" ed. Shahid Farooq. *PLoS One* 16(12): e0260673. doi:10.1371/journal.pone.0260673.

Sher, Farooq, Narcisa Smječanin, Harun Hrnjić, Amar Karadža, Rasim Omanović, Elma Šehović, and Jasmina Sulejmanović. 2024. "Emerging Technologies for Biogas Production: A Critical Review on Recent Progress, Challenges and Future Perspectives." *Process Safety and Environmental Protection* 188: 834–59. doi:10.1016/j.psep.2024.05.138.

Simbi, Ines, Uyiosa Osagie Aigbe, Oluwaseun Oyekola, and Otolorin Adelaja Osibote. 2022. "Optimization of Biodiesel Produced from Waste Sunflower Cooking Oil over Bi-Functional Catalyst." *Results in Engineering* 13(December 2021): 100374. doi:10.1016/j.rineng.2022.100374.

Su, Yujie, Peidong Zhang, and Yuqing Su. 2015. "An Overview of Biofuels Policies and Industrialization in the Major Biofuel Producing Countries." *Renewable and Sustainable Energy Reviews* 50: 991–1003. doi:10.1016/j.rser.2015.04.032.

Tamás Mizik, and Gábor Gyarmati. 2021. "Clean Technologies Economic and Sustainability of Biodiesel Production — A Systematic Literature Review." *Economic and Sustainability of Biodiesel Production—A Systematic Literature Review* 3: 19–36.

Woźniak, A., and M. Soroka. 2018. "Effect of Crop Rotation and Tillage System on the Weed Infestation and Yield of Spring Wheat and on Soil Properties." *Applied Ecology and Environmental Research* 16(3): 3087–96. doi:10.15666/aeer/1603_30873096.

Woźniak, Andrzej. 2019. "Effect of Crop Rotation and Cereal Monoculture on the Yield and Quality of Winter Wheat Grain and on Crop Infestation with Weeds and Soil Properties." *International Journal of Plant Production* 13(3): 177–82. doi:10.1007/s42106-019-00044-w.

Wu, Sarah, Shaobo Deng, Jun Zhu, Muhammad Aamir Bashir, and Forrest Izuno. 2019. "Optimization of a Novel Liquid-Phase Plasma Discharge Process for Continuous Production of Biodiesel." *Journal of Cleaner Production* 228: 405–17. doi:10.1016/j.jclepro.2019.04.311.

Xu, Hui, Longwen Ou, Yuan Li, Troy R. Hawkins, and Michael Wang. 2022. "Life Cycle Greenhouse Gas Emissions of Biodiesel and Renewable Diesel Production in the United States." *Environmental Science & Technology* 56(12): 7512–21. doi:10.1021/acs.est.2c00289.

Zhang, Huiming, Zhidong Xu, Dequn Zhou, and Jie Cao. 2017. "Waste Cooking Oil-to-Energy under Incomplete Information: Identifying Policy Options through an Evolutionary Game." *Applied Energy* 185: 547–55. doi:10.1016/j.apenergy.2016.10.133.

Zhang, Long, and Wuliyasu Bai. 2021. "Sustainability of Crop-Based Biodiesel for Transportation in China: Barrier Analysis and Life Cycle Ecological Footprint Calculations." *Technological Forecasting and Social Change* 164(December 2020): 120526. doi:10.1016/j.techfore.2020.120526.

19 Switchgrass, *Miscanthus*, and *Brachypodium* as Efficient Biofuel Producers

A Critical Analysis for Sustainable Uses

Ghulam Murtaza, Muhammad Baqir Hussain, Javed Iqbal, and Maleeha Shafqat

19.1 INTRODUCTION

19.1.1 Background

The increasing demand for renewable power has caused people to pay more attention to biofuels as a source of energy than fossil energy. Climate change and energy security are two burning issues that have attracted the attention of some crops for biomass production. Out of all the plants, switchgrass, *Miscanthus*, and *Brachypodium* were identified to be viable feed options. They identified these as perennials capable of growing alongside all categories of soil type and weather strength, making them ideal for biofuel production. They have the potential to contribute greatly to sustainable agricultural production systems because they are capable of fixing atmospheric carbon as well as improving soil quality (Bhatia et al., 2023).

A significant reduction in greenhouse gas emissions can be achieved using such crops according to science. Switchgrass, for example, plays an important role in preventing climate change impacts by effectively sequestering carbon through its roots (Bai et al., 2022). *Miscanthus* belongs to a group of species with high biomass production rates and has high efficiency of CO_2 to energy conversion that proved it to be a perspective resource for bioenergy (Lask et al., 2019). While *Brachypodium* is less talked about, it provides different benefits for genetic study and bioengineering, which may increase tempers speed for biofuel-related characters (Bhatia et al., 2019). The combination of these grasses offers a promising strategy to produce biofuels that are feasible and conform to the requirements of climate change and the environment.

19.1.2 Problem Statement

Biofuel production using switchgrass, *Miscanthus*, and *Brachypodium* is possible, but there are some barriers. Challenges and risks defined as major obstacles refer to dissimilarities in growth rates of biomass, yield, as well as the financial feasibility of biomass production and processing for energy use. For instance, even though it is understood that switchgrass is capable of growing on poor soils, yields derived from the plant are a function of the climate of the region and the methods that are used in farming the plant (Bhatia et al., 2023). For example, switchgrass can be grown in comparatively poor quality soils, but the amount of biomass that can be produced in any given year

 DOI: 10.1201/9781003600169-19

depends on climate conditions and farm management practices at the time of harvest (Cooney et al., 2017). While *Miscanthus* has high biomass production potential some *Miscanthus* varieties are pure types that may only grow in specific sone sandy loam soils, and the establishment costs are expensive for the farmers despite the yields (Ouattara et al., 2020).

Other concerns related to these crops include sustainability issues that accompany large-scale cultivation, conversion of land use, and loss of biological diversity. Surrey believes that the expansion of land for the production of feedstocks for biofuels implies the destruction of their habitat and thus displacement of indigenous species (Verdade et al., 2015). In this case, a critical analysis is needed to discuss the advantages, complexity, and disadvantages, as well as opportunities and challenges for their use as sources of biofuels. Perceptions of this nature are useful when it comes to creating measures that can improve the utility of these crops to as high a degree as is possible without damaging the natural environment.

19.1.3 Focus of the Study

Biofuel crops that are likely to be used for biofuel include switchgrass, *Miscanthus*, and *Brachypodium*. An analysis of the agronomic characteristics, costs, and impacts of these species is presented in this paper, which illustrates how they can positively impact the energy sector. Several relevant studies and cases, as well as the opinions of professionals, will be analyzed to gain a fuller understanding of their potential and disadvantages.

Consequently, we will examine specific factors that affect the growth of each species, such as the type of soil, water, and pest control, which are crucial to their growth (Bhatia et al., 2023). Economic feasibility will be examined using cost-benefit analyses of establishment costs for the crop and future returns from biomass yield and market swings (Viccaro et al., 2024). Additionally, we will assess the biophysical consequences of the large-scale production of such crops to, for example, influence habitat, soil conditions, and carbon management to foster a system understanding of these crops in bioenergy systems as identified by Winberg et al. (2023). Thus, this chapter integrates the available information to educate the relevant clients, such as the policymakers, farmers, and researchers, on the prospects, possibilities, and general considerations of incorporating these biofuel crops in existing agricultural systems.

19.1.4 Scope and Limitations

This chapter will contain numerous issues concerning these biofuel crops and their production. However, the places of focus will include performance, economic issues, and the effects on the environment. This study may find it helpful to mark that the results of the experiments can be different on the regional level due to the difference in climate, kinds of soil, and methods of agriculture. For instance, switchgrass might do very well in the Midwest but could give far different results in the Southeast because conditions such as rainfall and temperature differ significantly (He et al., 2024). Like the first crop, *Miscanthus* also has optimal requirements for the type of soil suitable for its growth, which may not be readily available in many areas (Cappelli et al., 2024).

Furthermore, although the chapter shall focus on research and statistics related to the topic, several aspects will be deemed outside its scope; equally, the chapter cannot contain information regarding all forms and types of biofuels or potential crops for their production. Much attention will be paid to the three chosen species. However, it should be noted that other plants, such as giant reed or energy cane, can also become potential candidates for biofuel production (Ortega et al., 2024). The purpose is to offer a comprehensible and succinct synthesis, contributing to more extensive debate and discovery in this essential field. Consequently, this chapter aims to highlight these biofuel crops' advantageous and disadvantageous features, believing that this knowledge will provide a better perspective for developing sustainable energy solutions.

19.2 SWITCHGRASS (*PANICUM VIRGATUM*): A COMPREHENSIVE OVERVIEW

19.2.1 Botanical Characteristics and Growth Habits

Switchgrass, botanically called *Panicum virgatum*, is a perennial grass originating from North America. It is often an herbaceous plant that ranges from 91 cm to 3 m in height due to variations in type and climate. The plant has a well-developed root system, which can grow down to the ground in search of water, and the plant can grow well in any soil – sand to clay. The tree has slender, extended leaves of a bluish-green color, which change to gold at the onset of autumn.

This species is common on prairies and open fields; the switches grow in dense stands. Apart from aiding it in competing for scarce resources with other vegetation, it also offers shelter to several wildlife species. The plant can effectively be grown in the hot and humid climates of the Southeast, as well as the comparatively dry Great Plains regions, which increases the chances of seizing the biofuel market. However, some recently published research indicates that switchgrass improves soil characteristics and increases organic matter and microbial activity (Zegada-Lizarazu et al., 2022). In addition, the plant has a long tap root, which is very useful in checking soil erosion; hence, the plant is environmentally worthwhile.

Switchgrass exhibits two primary growth forms: lowland and upland types. Switchgrass in the lowlands and uplands differs in their height and biomass, while switchgrass in the lowlands and uplands is more suited to dry conditions and marginal soils (Liu et al., 2022). As a result of such growth patterns, various cultivation practices may be selected to produce commanding yields according to the conditions at hand. Switchgrass is an important biofuel feedstock and agricultural crop due to its versatility and ecosystem-friendly characteristics.

19.2.2 Biomass Yield and Quality

Biofuel crops such as switchgrass have higher biomass yields than most, which is important to consider. In ideal conditions, the plant can yield between 5 and 10 tons per acre per annum of dry biomass. Yield, the quantity of crop produced, is influenced by soil fertility and the length of time the soil has been moist, among other practices like weeding. Available data also reveal that, while benefiting from optimum fertilization and irrigation practices, yields as high as 12 tons of produce per acre have been reported to be achievable (Tóth & Dupľák, 2023).

Quantitative aspects of the switchgrass include a desirable lignocellulosic composition from a quality perspective ideal for additional conversion processes. Due to its high cellulose content, the candidate can be converted to cellulosic ethanol and be used for direct combustion. Also, the ash content of the plant material used is lower than the values recorded in other biomass resources, thus making the plant material efficient for fuel combustion. The results have indicated that, after the consolidation, switchgrass has an energy content ranging between 18 and 20 MJ/kg, a position that makes it rank among the standard fossil energies (Wang et al., 2023).

19.2.3 Biofuel Applications

Switchgrass can primarily be used in cellulosic bioenergy, particularly ethanol production due to its highly acknowledged properties. In other words, making it possible to break down the plant's cellulose into fermentable sugars for ethanol preparation becomes realizable. Switchgrass also contributes to less fossil fuel use, thereby realizing the lowest greenhouse gas emissions. Ethanol from switchgrass was estimated to produce about 80 gallons per ton of biomass and fairly competitive to market price (Unyay et al., 2024). Besides, this biomass is also useful for direct combustion in heat and power applications. Usually having heat values varying between 18 and 20 MJ/kg, it is still useful for biomass power plants (Masum et al., 2020). There is also concern with switchgrass as a source of biogas through anaerobic digestion as improvements in modern technology occur, thus broadening the biofuel value of the product. This process produces electricity from renewable resources and delivers the digestate, which is an organic fertilizer that can replace the chemicals in the agricultural system to finish the nutrient chain (Bužinskienė et al., 2023).

The plants themselves are highly flexible and can be readily integrated into existing agricultural production systems since switchgrass. For instance, it can be added as a crop that is grown besides others or as one tree out of many trees for crop production to ensure that the gross outcome of crops within a farm is boosted and more than one source of income for people in agriculture. Thus, meeting the growing demand for research on the utilization of renewable energy sources, switchgrass is a potential organ for energy security and environment sustainability in feedstock for biofuel production.

19.2.4 Environmental Impacts

Switchgrass has several environmental advantages that make it suitable for biofuel feedstock. The timber possessed a sophisticated root system that enhances soil conservation against deterioration and cycling. Through previous research, it has been evident that through growing and incorporating switchgrass, the soil organic matter can be improved alongside the microbial community (Bai et al., 2022). Also, it showed that switchgrass could potentially store carbon, consequently enhancing climate change efforts. According to the estimations, switchgrass can capture about 1.5 tons of carbon per acre in a year – a significant resource in the carbon storage program (Collins et al., 2020).

However, some disadvantages should not be overlooked. Switchgrass can, for example, cause the complete disappearance of native species if it is not well managed, mainly where native ecosystems exist. As with many single-crop crops associated with industrial-scale, switchgrass cultivation can reduce habitat heterogeneity of the area in question and interact with native plant and animal populations (Lask et al., 2020). In addition, large-scale production using fertilizers and pesticides affects the water quality in the region because nutrient runoff and algal blooms will be experienced in water sources. These factors are flexible enough to maintain the gains of switchgrass without adverse effects. In practice, switching cultivation to switchgrass can pose various risks related to the degradation of the ecological base of agriculture. However, using indicators of sustainable agriculture, including crop rotation and integrated management of pest species, can reduce these risks and improve the efficiency of switchgrass cultivation.

19.2.5 Economic Viability

Switchgrass as a biofuel crop is analyzed in terms of its economic feasibility, including production cost, the market, and government policies. Switchgrass has high initial costs because the crops require establishment, and the planting information is higher than that of most annual crops. However, switchgrass is a perennial crop in the long run and would not need to be replanted as frequent as annual crops. It has been estimated that once established, switchgrass can produce a yield annually for over 10 years, or even more, under optimal conditions, a factor that makes switchgrass very cheap for farmers. The average biomass yield of switchgrass can vary between 5 and 10 tons/acre depending on soil quality and management, making it more economical to produce this feedstock (Jacot et al., 2021).

Demand for renewable energy sources is increasing in the market, driven by customer preferences and state initiatives in the case of carbon emissions reduction. This trend is favorable for switchgrass producers significantly as governments increase policies surrounding the use of biofuels. For example, the US Renewable Fuel Standard (RFS) requires adding renewable fuel to the transportation fuel stock, thus providing a dependable market for biofuels made from crops such as switchgrass (Wang, 2022).

However, then again, there is never a dearth of economic problems. Market risks include volatile prices of fossil fuel as well as competition from other feedstock crops. For instance, high oil prices make it attractive for producers to invest in switchgrass as biofuel demand reduces due to high production costs (Merritt & Barragán-Ocaña, 2023). Further, the distribution and production systems of biofuels are not yet fully developed, which may lead to some constraints in the chain. As a result, there are several challenges that farmers may require to consider using other strategies to counteract the unsteady market, such as processing or cultivating crop variety.

19.3 *MISCANTHUS* (*MISCANTHUS* × *GIGANTEUS*): A DETAILED ANALYSIS

19.3.1 Botanical Characteristics and Growth Habits

Giant *Miscanthus*, or *Miscanthus* × *giganteus*, is a herbaceous perennial grass that may grow up to 3 m tall. It has extended and curved stalks and large leaves, which can significantly enhance the beauty of the compound in which it is planted. That is why the habitat of this species is very diverse: it may grow in nearly all types of soil but prefers nutrient and well-drained soils. *Miscanthus* thrives in growth, and this species is noted to spread fast and form a dense cover, which can suppress the growth of many weeds (Figure 19.1).

The ornamental plant is usually reproduced through rhizomes to spread out and produce quality roots quickly. Figure 19.2 illustrates the propagation material used in experiments (Hussain et al., 2021–2024). These characteristics are helpful to its drought tolerance, and they suggest eligibility for variable climates. *Miscanthus* also appears to require fewer inputs than conventional crops, which could be a further advantage regarding inputs. There is evidence that it can thrive in areas that would otherwise be considered unproductive for conventional crops, so one of the advantages of energy crops like *Miscanthus* is that they can be produced in often suboptimal agricultural lands. *Miscanthus* can outcompete weeds in specific marginal environments that other crops cannot (Figure 19.3). Second, Giant *Miscanthus* is highly adaptable, and its phenology differs from most cultivars already tested. It starts germinating in the early spring, grows in summer, and gets fully ripe in the fall. It will also enable it to capture sunlight during the growing period, allowing it to produce high biomass. Besides, efficient photosynthesis under different light intensities increases

FIGURE 19.1 Growth stages and reproductive characteristics of *Miscanthus* × *giganteus*. (Hussain et al., 2021–2024.)

FIGURE 19.2 *Miscanthus* × *giganteus* rhizomes, which are used to illustrate the propagation material used in experiments. (Hussain et al., 2021–2024.)

FIGURE 19.3 *Miscanthus* × *giganteus* pot trial for salinity and drought tolerance, depicting resilience under challenging environmental conditions. (Hussain et al., 2021–2024.)

its growth possibility, making it reliable for bioenergy production (Urrego et al., 2021). Essentially, the botanical features and growth patterns of *Miscanthus* × *giganteus* place it effectively in the list of probable sustainable feedstock for biofuel production.

19.3.2 Biomass Yield and Quality

There are various types of biomasses, but *Miscanthus* × *giganteus* is the most preferred. The literature indicates that it can provide yields biomass of 10–20 ton of dry matter/ha/year depending on the climatic regime and/or management practices. The mix also has a high yield accompanied by a good carbon to nitrogen ratio, making it suitable to be used for biofuel. The carbon-to-nitrogen ratio of *Miscanthus* is typically around 60:1, which in turn increases microbial activities during decomposition and improves soil health according to Fosu and Davis (2024). About the characteristics of the biomass, it is interesting to note the following fact. This is especially crucial for biofuel production because *Miscanthus* is composed of a high lignocellulosic content. Being fibrous, it forms one of the biggest contributions to energy density. It should also be good for further use as a feedstock for several biofuel processes such as combustion, gasification, and fermentation. In different works, it was ascertained that the energy density ranged from 6 to 18 MJ/kg for *Miscanthus* biomass (Zhao et al., 2021).

The harvest and processing of *Miscanthus* can also increase its potential for biofuel production (Figure 19.4). The harvesting time, for example, contributes to biomass and energy content by increasing the level of frost and sugar concentration after the first harvest (Magenau et al., 2022). Figure 19.5 highlights post-harvest processing to improve biomass usability (Hussain et al., 2021–2024). *Miscanthus*, with its high-quality biomass, is well positioned to compete with the established bioenergy sector and benefit from the increasing demand for renewable sources of energy.

FIGURE 19.4 The harvesting of *Miscanthus* × *giganteus* plants. (Hussain et al., 2021–2024.)

FIGURE 19.5 Post-harvest processing to improve biomass usability. (Hussain et al., 2021–2024.)

19.3.3 Biofuel Applications

It also has versatility in the utilization of the resulting *Miscanthus × giganteus* in the production of biofuel. Such plant biomass can also be utilized directly in burning in power plants as a renewable form of energy, thus decreasing the use of fossil energy. In this case, the levels of utilization of the *Miscanthus* biomass include electrical energy and heat necessary for district heating. The latter makes it more desirable as a RE source where both electricity and heat are needed (Mamirova & Pidlisnyuk, 2024). Besides, it can also be beaten into pellets or briquettes, which are easier to transport and store comparatively. The process of compaction of *Miscanthus* biomass to form pellets increases the energy density of the pellets needed in transporting pellets over long distances. It also improves the combustion characteristic of the biomass; this brings efficiency to the biomass and shows low pollutants once it has undergone the burning process (Szufa et al., 2021).

Besides direct combustion, *Miscanthus* is suitable for conversion into bioethanol and biogas. To investigate the feasibility of utilizing *Miscanthus* for these purposes, different experiments have been conducted. Hence, its stems containing sugar can be easily fermented to produce ethanol, while biogas – a useful sort of energy – may be produced through anaerobic digestion. Certain previous research studies suggest that the fermentation stage can be enhanced to enhance ethanol yields; therefore, *Miscanthus* is the appropriate feedstock for bioethanol production (Zhao et al., 2021). These applications demonstrate that *Miscanthus* has the potential to significantly contribute to the supply of renewable energy and biofuels in the future, with the potential for its usage to expand in line with future energy requirements.

19.3.4 Environmental Impacts

The cultivation of *Miscanthus × giganteus* has advantages and disadvantages from the environmental point of view. On a positive note, it has potentially enhanced the quality of the soil by tilling the soil deeply, thus helping prevent soil erosion (Figure 19.6). In addition, it helps to aerate the soil. The spread of *Miscanthus* roots also helps to maintain soil structure and increase rhizosphere RH, which is prospective for developing a multitude of part soil inhabitants implied in nutrient cycling (Kane et al., 2023). Also, *Miscanthus* has been revealed to accumulate carbon, which helps to combat climate change. One significant characteristic of this water-growing crop is its potential to absorb atmospheric CO_2 during photosynthesis and put it back into the soil, thus making it an eco-friendly crop in the war against global warming (Fosu & Davis, 2024).

Nonetheless, its usage has roused controversy regarding its intrusiveness in some areas. Control of *Miscanthus* is challenging, and if left unattended, it can spread to unwanted areas, posing a threat to the habitats in these locations. Research has suggested that *Miscanthus* may be invasive to native plant species; this may significantly reduce the number of species in an ecosystem (Lowry et al., 2022). The best management practices should be adopted to avoid these risks and ensure that it can still be cultivated. This ranges from suppressing its spread, the method of growing sterile cultivars, to growing *Miscanthus* as part of a crop rotation system to cause minimal disruption to ecosystem balance. This paper outlines the challenges that must be met to ensure that *Miscanthus*' positive environmental attributes are fully realized while potential detrimental effects are addressed.

19.3.5 Economic Viability

The feasibility of *Miscanthus × giganteus* as a biofuel feedstock is realistic but depends on the production location. Long-term establishment costs are low, while initial costs could be high; the need to plant rhizomes and land preparation usually causes this. These costs are variable and mostly range between \$500 and \$1,500/ha, contingent on existing local conditions and practices (Winkler et al., 2020). Still, if developed, the low management demands and high biomass productivity return high investment rates. *Miscanthus* has an almost insignificant demand for fertilizers and pesticides

FIGURE 19.6 Soil regeneration potential through rhizome sprouting. (Hussain et al., 2021–2024.)

throughout its life cycle, which makes it economically less expensive in terms of operational cost (Panoutsou & Chiaramonti, 2020).

Market demand for renewable energy sources is increasing; therefore, *Miscanthus* has promising years ahead in the biofuel market. However, the escalating global climate change and energy security concerns have created a good market for biofuels worldwide. For instance, the European Union has given very high targets for using renewable energy sources in the next one and a half decades or so, including biomass (Vera et al., 2021). Furthermore, any improvement in the economic attractiveness of bioenergy may be complemented by government incentives and policies toward bioenergy. Specifically, they argue that government policies such as biofuel production subsidies and tax incentives promoting the financial viability of renewable energy projects will assist farmers in cultivating *Miscanthus* (Ford et al., 2024).

19.4 *BRACHYPODIUM DISTACHYON*: A MODEL SYSTEM AND EMERGING POTENTIAL

19.4.1 Botanical Characteristics and Growth Habits

Brachypodium distachyon, abbreviated as *Brachypodium*, is a perennial grass with great potential to be developed as a model organism for generating bioenergy crops. Most of the time, this grass grows from 30 to 100cm tall, and it has thin, elongated stems and narrow leaves that make it look lovely. Another interesting phenomenon is that *Brachypodium* is versatile when it comes to the type of climate within which it can grow, which makes it a complying candidate for cultivation in quite many different agricultural conditions – for instance, both in the temperate climate and in the Mediterranean one (Silva et al., 2022).

The vigor of the various growth habits of *Brachypodium* species is quite remarkable. It also demonstrates fast growth and can get fixated in all types of soils, including poor soils. This is nicely illustrated by the plant's limited root mass, enabling it to find moisture, whether wet or dry quickly. *Brachypodium* is better viewed as another biofuel crop candidate, not excluding its great importance as a genetic and physiological model for the study of grass species. Recently, it has been reported that it can be genetically modified to produce various enhanced characteristics such as drought tolerance and biomass production (Ludwig et al., 2023).

19.4.2 Biomass Yield and Quality

Larger grasses such as *Miscanthus* might have higher biomass yield potential than *B. distachyon* to meet this target; nevertheless, relatively high yields make it an ideal biofuel source. The biomass yield of *Brachypodium* when cultivated under optimal conditions ranges from 5 and 10 dry biomass tons/ha/year. This yield is low in comparison to other common large crops but can become essential when given the plant can thrive under poor conditions (Ludwig et al., 2023). Special regard should be given to the biomass quality, in which cellulose, hemicellulose, and lignin of the plants used are suitable for conversion into biofuels.

Previous work has brought the identification of genes and mutants on their genetic tractability for breeding and genetic regulation for improving biomass yield and quality to light. For instance, previous studies have found some genes that would improve the biosynthesis of lignin and could be adjusted to allow easy penetration of the biomass (Wang et al., 2022). Further, the viability of its growth in marginal lands, or lands not suitable for arable crops, supports its suitability for biofuel use as *Brachypodium*. Through the exploitation of such a marginal and relatively underutilized resource, a crop such as *Brachypodium* could help maintain sustainable agriculture and at the same time act as a renewable energy source (Qaseem & Wu, 2020). Further advancement of an organism for use in research entails, by and large, the provisions of accomplishing *B. distachyon* as a biofuel plant.

19.4.3 Biofuel Applications

Biofuels can be made from *B. distachyon*, a plant with several promising applications. Bioethanol is a great way to use sugarcane. The plant biomass comprises sugars that can be fermented to produce ethanol, an environmentally friendly fuel to replace gasoline. This process provides an opportunity with another power source and less emission of greenhouse gases as compared to fossil energy sources (Geem et al., 2021). Besides, *Brachypodium* can be gasified, a process in which the organic material is turned into syngas, which is localized as hydrogen and carbon monoxide. Similarly, such syngas can be upgraded to liquid hydrocarbons such as diesel or jet fuel, giving this grass the biofuel diversification potential factor (Constantinescu-Aruxandei & Oancea, 2023).

Therefore, an overview of the biological parameters of *Brachypodium*, which is a small-lived plant with a relatively short generation time, suggests intercropping or possibly a cover crop for crop rotation. It also aids farm production in the broadest sense while, at the same time, being advantageous to both species' protection and the nutrient stocks of the soil. Therefore, the implementation of *Brachypodium* into the pluralistic composite agrosystems can improve the soil structure and nutrient media that are important to agroecosystem food production (Soriano et al., 2023). Because studies are still being conducted on genetics and uses of the plant *Brachypodium*, its use in the diversification of feeding stocks will improve in the future. It has also been identified as one of the most promising biofuel crops. The progressive advancement of biotechnological techniques may make it even more effective.

19.4.4 Environmental Impacts

The environmental effects of growing *B. distachyon* are also desirable and less damaging than the effect of cultivating conventional crop plants. A significant strength of this tree is that it grows freely in substandard soils; this, therefore, can be grown without replacing fertile soils used to grow food crops. This characteristic is of particular value in contexts where arable land is becoming increasingly strained by factors like uncontrollable population growth and climate change. In other words, *Brachypodium* could be advantageous because it is grown on land resources that are less suitable for agricultural purposes.

Thus, *Brachypodium* can improve land use and possibly improve soil properties, too. These create a substantive beneficial effect on the physical status of the soil: the root system becomes more substantial and more developed, making the structure of the subsoil layer more fertile, and the supply of the matter increases; erosion decreases. All these factors enhance water retention and nutrient cycling, which are important for protecting ecosystems (Soriano et al., 2023). Also, growing the *Brachypodium* can foster habitat support for diverse organisms, meaning there would be a better and healthier ecosystem for crop production.

Therefore, proper management practices must be adopted to counteract the possible ill effects resulting from its production. For example, overapplication of nutrients can easily cause soil erosion and water pollution. Water consumption must also be controlled, as excessive watering weakens the water sources in that specific region (Grace et al., 2020). Thus, farmers can continue reaping the benefits of growing *Brachypodium* through crop rotation, cover cropping, Integrated Pest Management, etc., so that it does not become a reason for environmental pollution.

19.4.5 Economic Viability

The idea of producing biofuel from *B. distachyon* is under discussion, and research regarding its economic prognosis has many positive indications. Next, they found a lower initial establishment cost than with other large biomass crops such as *Miscanthus* or switchgrass to support the utilization of *Brachypodium*. It also means that since the financial risk is almost insignificant, it is the ideal venture for farmers who require an additional source of income other than investing so much (Jayawardena et al., 2023). Besides, the growth cycle of *Brachypodium* is comparatively short, meaning investors are washed for their investments faster, which will be helpful for those farmers working under small profits.

Given the growing demand for renewable energy feedstock globally, biofuels produced from grasses such as *Brachypodium* should also find a market. This growth is supported by growing governmental incentives to promote the generation of renewable energy and reduce CO_2 emissions. For example, there are various policies and schemes relating to the production of bioenergy crops that would raise the economic utility of *Brachypodium* (Grace et al., 2020). Furthermore, sustained financial investment in ongoing research and development into biofuel technology is likely to result in additional improvements in the processing methodologies of the *Brachypodium* biomass to biofuels to reduce the overall cost.

The problem of meeting market demand, however, still limits the advancement of such production. The importance of prestigious supply lines and processing installations will be critical to *Brachypodium*'s economic viability as a biofuel feedstock. This market has some limitations, too: market instabilities and competition from other biodiesel feedstocks are also possible. However, these problems are not critical to *Brachypodium*, where there are many opportunities to be realized, such as good germination in various environments, as well as high biomass production, which makes it an important factor in the bioenergy segment of the future. Since the area of research remains rather embryonic, it is possible to examine the position of the *B. distachyon* from the economic point of view of the biofuel market in the future.

19.5 COMPARATIVE ANALYSIS OF SWITCHGRASS, *MISCANTHUS*, AND *BRACHYPODIUM*

19.5.1 Yield Comparison

Miscanthus × *giganteus* has superior biomass yields, averaging between 10 and 20 tons of biomass/ha when growing under optimal conditions. The reason for this impressive yield is its strong growth characteristics and adaptability to a variety of soil types. As a result, it is widely recommended for commercial use in biofuel production (Jayawardena et al., 2023). Switchgrass is second, and yields usually fluctuate between 5 and 15 ton/ha depending on the type of switchgrass and use practices. Climatic and soil conditions influence how the plant can be grown, yet flexibility in farming is the key to increasing yields (Ricketts et al., 2023).

It is effective on *B. distachyon*, which averages only 5–10 ton/ha. While it may not appear to be as high as some other feedstock plants, *Brachypodium* has more excellent value due to its short generation time and genetic manipulability, which can potentially be exploited for better yield through breeding and genetic engineering (Ludwig et al., 2023). The difference in yield between these crops becomes important when discussing commercial production of biofuels in food farming capacity because higher yield equals higher energy production. However, the authors need to bear in mind the quantity of biomass generated and the sustainability of the practices used in cultivation. A detailed comparison of the key characteristics of the three biofuel crops, including their biomass yield potential, is summarized in Table 19.1. The new developments in the biofuel market will require proper information on the possible yields obtained from these crops that are most appropriate for use in renewable energy systems.

19.5.2 Biomass Quality Comparison

However, for biomass quality, all three crops, *Miscanthus*, switchgrass, and *Brachypodium*, have their advantages that suit different bioconversion processes. *Miscanthus* contains a very high proportion of lignocellulosic components that are important in the conversion processes including thermochemical process and enzymatic hydrolysis. This high lignin gives the plant rigidity as well as density, making biomass energy generation from the plant preferable (Mironova et al., 2023). Switchgrass is also characterized by a fairly high cellulose to hemicellulose ratio, which is a necessary criterion for bioethanol feedstock. Its composition allows for different conversion processes, and further research is still being carried out to improve such processes in order to improve ethanol yields (Morales et al., 2021). Due to the flexibility of growing conditions, switchgrass even becomes a more useful biofuel crop than other crops.

TABLE 19.1
Key Characteristics Comparison of Major Biofuel Crops

Characteristic	Switchgrass	*Miscanthus*	*Brachypodium*
Annual biomass yield (dry tons/ha)	5–15	10–20	5–10
Height (m)	0.9–3	2–4	0.3–1
Establishment cost ($/ha)	300–500	500–1,500	200–400
Energy content (MJ/kg)	18–20	17–19	16–18
Carbon sequestration (tons C/ha/year)	1.5–2.0	2.0–2.5	0.8–1.2
Lifespan (years)	10–20	15–25	1–3

Sources: Compiled from Grippi et al. (2020), Mahmood et al. (2022), Larnaudie et al. (2022), Lewandowski et al. (2000), Draper et al. (2001).

TABLE 19.2
Biochemical Composition of Biofuel Crops (%Dry Weight)

Component	Switchgrass	*Miscanthus*	*Brachypodium*
Cellulose	35–45	40–50	30–40
Hemicellulose	25–35	20–30	20–30
Lignin	15–20	10–15	15–25
Ash	4–6	2–4	3–5
Extractives	5–10	5–8	8–12

Sources: Compiled from Mitchell and Schmer (2014), Sorek et al. (2014), El-Desouky et al. (2022).

Research is being conducted on *Brachypodium* to produce genetically improved feed, despite its lower biomass than other species. Their quality suggests that they are well suited for biofuel uses, especially as a reference organism for bioaugmentation in other grasses. This high genetic tractability makes *Brachypodium* a helpful system for comparing these modifications and their potential to enhance biomass quality, which is an important consideration in the plant's continued development for future biofuels (Pancaldi et al., 2023). The biochemical composition of these biofuel crops, which plays a critical role in their suitability for different bioconversion technologies, is presented in Table 19.2. However, as the biofuel industry develops, knowledge and appreciation of the quality attributes of these energy crops will be critical in the efficient incorporation of the crops in renewable energy technologies.

19.5.3 Environmental Impact Comparison

From an environmental point of view, all three crops have pros and cons. Figure 19.7 Illustrates potential environmental benefits of using biofuels from these crops. However, having an outlook to use them as a feedstock for a biorefinery, I believe they all hold certain advantages but, on the other hand, certain drawbacks. For instance, *Miscanthus* is well-documented regarding its impact on soil health and carbon sequestration, which can signify energy toward climate change. It also has a spread root canopy, which improves the physical and chemical properties of the soil, while also increasing nutrient cycling (Kane et al., 2023). Switchgrass is noted for its wide adaptability to marginal soil types, making it a good choice for sustainable agriculture. They also concluded that crops can maintain wildlife and soils to enhance the growth of various wildlife and maintain the growth of deep root systems to foster the growth of various crops, reduce soil erosion, and support water reserves (Zhao et al., 2022). However, there is also information that switchgrass can be grown in crop rotation, thus significantly increasing its environmental advantages.

Brachypodium can grow in almost any condition and improve the soil while not turning people out of their gardens or fields and replacing food crops. Since it can be grown in marginal areas, its use may support the improvement and proper usage of land and reduce soil erosion (Schneider et al., 2024). However, great care has to be exercised not to harm the ecosystem of the regions where the three crops are grown. This includes specific vigilance for invasive tendencies, nutrients, and specific guarantees regarding the use of water resources. Increasing the use of biofuels will call for a significant understanding of the environmental consequences of such crops to advance sound measures in farming. The environmental impacts of these biofuel crops, including their effects on soil health, carbon sequestration, and adaptability to marginal lands, are summarized in Table 19.3.

19.5.4 Economic Viability Comparison

It is also seen that economic viability is sustainable in different ways depending on the plant species like *Miscanthus*, switchgrass, and *Brachypodium*, which depends on aspects like expenses for the establishment of plantations, continual maintenance, and demand in the market. The initial costs associated

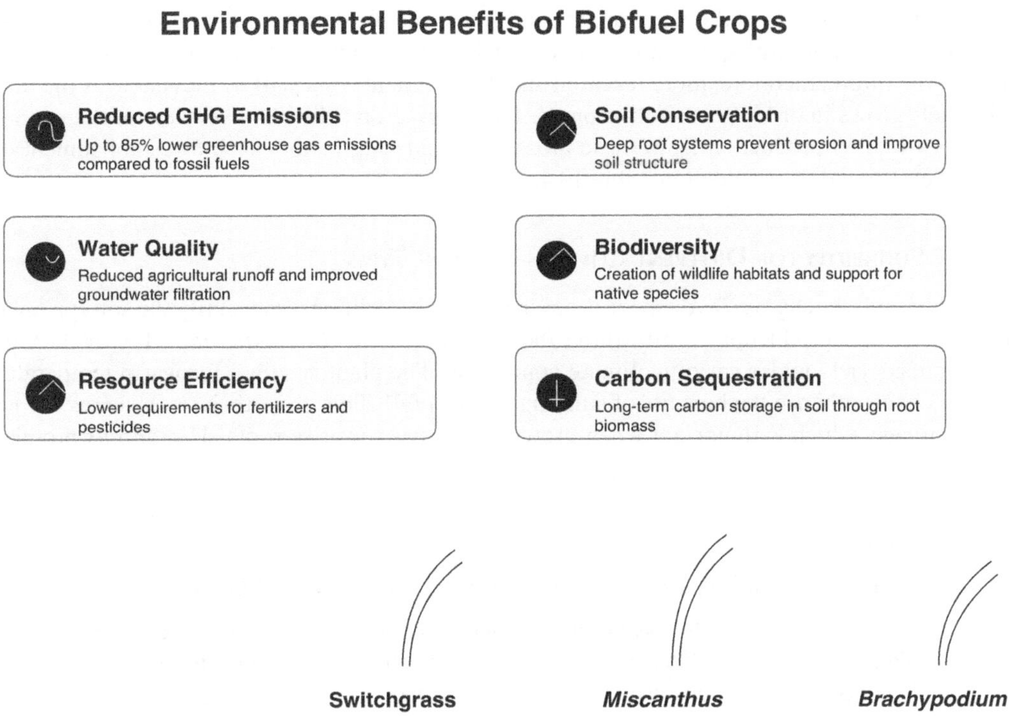

FIGURE 19.7 Potential environmental benefits of using biofuels from switchgrass, *Miscanthus*, and *Brachypodium*.

TABLE 19.3
Environmental Impact Assessment of Biofuel Crops

Environmental Factor	Switchgrass	*Miscanthus*	*Brachypodium*
Soil erosion control	High	High	Moderate
Water use efficiency	High	Moderate	High
Biodiversity support	High	Moderate	Moderate
Invasive potential	Low	High	Low
Nutrient requirements	Low	Low	Very low
GHG reduction potential	High	Very high	Moderate
Soil carbon storage	High	Very high	Moderate

Sources: Compiled from Searchinger et al. (2008), Draper et al. (2001), Sorek et al. (2014), Templeton et al. (2016).

with *Miscanthus* may be higher than those associated with establishing sugarcane crops because Miscanthus has to be propagated through rhizomes. However, its high yields can ensure its profitability, thus being a good investment for farmers willing to invest in its establishment (Mamirova & Pidlisnyuk, 2024). Besides, other advantages may make *Miscanthus* cultivation more economically viable in the long run, including the crop's ability to fix carbon in the soil and enhance the soil's health.

Brachypodium, currently in the experimental phase concerning its feasibility, may rise to the levels of cheap feedstock as more research is carried out and markets for biofuels develop. Biofuel production is a high-value application that can be easily integrated into the bioreactor and, therefore, represents relatively low initial costs and high potential for value in the future (Shrestha et al., 2022). All in all, the overall economic characteristics of these crops are constantly shifting, and the locality factors, including soil type, climate conditions, and the degree of market accessibility, shall be deemed important prerequisites to define which of the plants is the most valuable for certain zones. In the future, therefore, more research and development work will be inevitable as the market for biofuels grows in order to capitalize on the bioenergy crop fully. An overview of the economic viability of biofuel production from these grasses, considering factors such as establishment costs and market demand, is provided in Table 19.4.

19.5.5 Suitability for Different Climates and Soil Types

Miscanthus and switchgrass favor specific environmental conditions, especially the soil type, which might affect their production significantly; the same applies to *Brachypodium*. Figure 19.8 highlights biomass yield under varying climate conditions. The plant is suited for warm temperate climate regions to enhance the benefits of sunlight and rainfall. This grass thrives on slopes, allowing better drainage, which is important when growing as it has sensitive roots. *Miscanthus* production

TABLE 19.4
Economic Viability of Biofuel Production from Selected Grasses

Crop	Establishment Cost ($/acre)	Annual Maintenance Cost ($/acre)	Expected Return on Investment (%)
Switchgrass	200–400	50–100	15–25
Miscanthus	300–600	100–150	10–20
Brachypodium	150–300	40–80	Variable

Sources: Compiled from Hus Farm (2023), Mohammed et al. (2019), Sahoo et al. (2022).

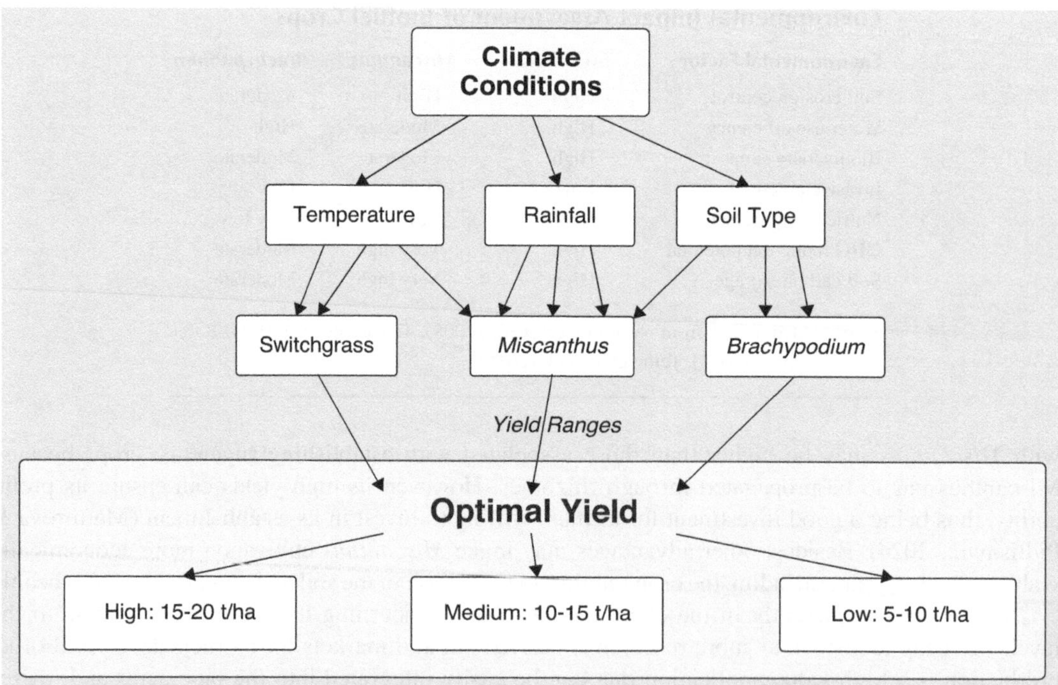

FIGURE 19.8 Biomass yield versus climate conditions.

FIGURE 19.9 Response of *Miscanthus × giganteus* to salinity and drought. Visual evidence of suitability for marginal lands (Hussain et al., 2021–2024).

can thrive where adequate rainfall is available, like in some parts of Europe and the southeastern United States (Jayawardena et al., 2023) (Figure 19.9).

While switchgrass is thus found to be somewhat particular about the climate it grows in, being comfortable only in regions of high humidity on the one hand and comparatively low humidity on the other, pap/post's complexity and adaptability mean that it poses fewer such problems. It can grow in different types of soils provided and poorer soils that may not favor other crops (Liu et al., 2021). It has an extensive root system to tap water and nutrients from other deeper areas; hence, it resists unsuitable environments. It is advantageous in this characteristic given the problem of drought in most of the regions where switchgrass can grow and produce yields.

19.6 CHALLENGES AND FUTURE DIRECTIONS

19.6.1 Technological Advancements

For switchgrass, *Miscanthus*, and *Brachypodium* as future biofuel crops, major technological factors will significantly impact their production, processing, and efficiency. Some of the techniques elaborated in harvesting and processing can go a long way in vastly improving efficiency and even over costs. For example, technologies like enzymatic hydrolysis and gasification can improve the efficiency of biofuel extraction from these crops, making the whole process more economically viable (Wang et al., 2024). These methods enhance the rate of converting plant material to usable energy and help reduce wastage, which is key to improving biofuel production systems.

Moreover, applications in precision agriculture such as drones, soil sensors, and satellite imagery will assist in determining the conditions and resources that will enhance growth. These enable the farmer to keep track of the crop health status, soil water status, and nutrient status in the field to make better decisions throughout the productivity cycle (Ali et al., 2024). For instance, those that use multispectral cameras to analyze health conditions and map out potential matches for care would lead to more excellent production rates.

Adopting these technologies will enhance the productivity of crops grown to produce biofuels. As the industry develops further, it will be essential to implement innovative technologies into farming to improve the efficiency of biofuel production and make it more sustainable. Further investment in research and its application will guarantee that this progress can reach farmers and be easily incorporated into different farming systems.

19.6.2 Genetic Improvement

There is also an area of future improvement in the cultivation of biofuel crops, which is a genetic improvement. They can select and modify those favorable characteristics that increase biomass yields, disease, and stress tolerances using expanded knowledge in genomics and other biotechnological fields. For instance, the MMPs can target breeding efforts toward growing switchgrass for specific regional adaptability. At the same time, the *Miscanthus* for improved LCC composition is important in achieving a high yield in the bioconversion process (Bhatia et al., 2023). These could result in crops with higher biomass yield and, at the same time, more minor demands for fertility inputs and water, which will, in the long run, reduce the detrimental effects of crop growing.

As a grass model, *Brachypodium* will thus allow the investigation of genetic changes that could be useful for other grasses. This is attributed to its small genome yet defined gene traits, giving it the capacity for genetic studies involving desirable attributes (Lei et al., 2023). For example, when used for research purposes, it is possible to target particular genes that enable growth and stress response and hence create robust biofuel crops.

Some of these crops include wheat, rice, cassava maize, and sorghum. Continued investment in research and development will be necessary to realize the benefits fully. Academic collaboration with government agencies and biofuel industries can help advance the rate of genetic enhancement so that energy crops are market-competitive. Many people need an efficient energy source that will ensure they get enough power without risking the natural resources and the environment in the long run, thus the significance of genetic improvement in biofuel crops.

19.6.3 Policy and Market Incentives

Switchgrass, *Miscanthus*, and *Brachypodium* feedstocks used in biofuel production depend highly on policy support and markets. Governments can also contribute significantly to this cause by providing subsidies for the production of biofuels or even tax rebates, and farmers can borrow from national and international microfinance companies. These financial enticements can make it easier for farmers producing biofuel crops to pay for growth, which is incredibly costly in the developmental phases as ecological risks remain comparatively high (Lazaro et al., 2023). Moreover, making renewable energy mandatory by establishing a stable biofuel market can also lead to their cultivation. These policies can help to establish a fixed market for biofuels to producers, allowing the farmer to predict their business plans and access credits.

Biofuels should also be labeled as environmentally friendly fuels that can be advocated as part of measures to decrease greenhouse gas emissions. This way, governments can gain public support, which in turn means further investment in R&D projects to promote the production of biofuels for climate change goals. Developing a biofuel market will require a suitable strategy and the closer cooperation of state bodies, the enterprise sector, and research institutions. For example, partnerships can enhance cooperation in getting and spreading knowledge to improve cultivation practices and technologies in processing (Wang et al., 2022).

However, international cooperation can help develop the global biofuel market, as countries can exchange experience and innovations. This is especially true as the world searches for sustainable sources of energy, hence the need for efficiency in policy framing to ensure that biofuel crops can fit in the energy market.

19.6.4 Sustainability Considerations

Therefore, Sustainability is essential as we embrace the culture of cultivating crops for biofuels. Nonetheless, benefits from switchgrass, *Miscanthus*, and *Brachypodium* production must be achieved without compromising the ecosystem or food production. Some techniques include rotating crops, associate planting, and using organic manure, which can also go a long way in conserving the soil and identifying itself to other uses as needed. For instance, biofuel crops can add value to well-established agricultural systems, improving soil health and decreasing pest pressures (Viccaro et al., 2024).

Moreover, further investigations into these crops' perennial effects will also help establish their life cycle sustainability analysis. Using life cycle assessment (LCA) makes it easier to measure the impact of biofuel production for better crop management and selection for better environmental effects (Canter et al., 2022). Thus, we can maintain ecological sustainability when using biofuel crops to meet the world's energy needs.

Additionally, there is a need for close cooperation with the people in the target areas over the production of biofuel crops. This engagement can go a long way in helping us define such conflicts between food production and local landscapes and help put biofuel production in a rightful perspective of sustainable production. With each of these components in mind, it is possible to imagine and develop systems for biofuel crop production that not only respond to energy needs but also preserve the environment and improve the quality of life of all members of society.

19.7 CONCLUSION

Here, we have discussed the possibilities of using switchgrass, *Miscanthus*, and *Brachypodium* as biofuel and biomass and its pros and cons. Of all the crops, *Miscanthus* has been found to have high biomass yield and quality, hence being favorable to biofuel production. Suitable for harvesting an estimated 10–20 tons of dry biomass ha^{-1} per annum, it received optimum growing conditions and is among the top bioenergy contenders to meet the increasing energy demands. Switchgrass has multiple purposes and is highly adaptable. Specifically, it can grow well in various climates and types of soil, and it is beneficial for biodiversity. The former is highly resistant to marginal soils, while the latter increases nutrient uptake by maintaining deep root systems that improve soil health. Agricultural practices using the former are particularly useful in sustainable farming (York et al., 2022). Even though it produces lower biomass, *Brachypodium* has excellent prospects for genetic research and improvement, improving its chances of becoming a bioenergy plant in the future. Genetic tractability also makes it possible to seek other genetic changes capable of enhancing the characteristics of other species of grasses, including biomass yield and quality (Lei et al., 2023).

Restating the conclusion pragmatically, switchgrass and *Miscanthus* are more competitive in producing biofuels if they have many advantages. *Brachypodium*, owing to the possibility of genetic improvement, may serve as one of the sustainable feedstocks for long-term biofuel production crops; however, additional investigation is needed to accomplish *Brachypodium* crops' biofuel potential. This consists of investigating how they can withstand different environmental factors as well as probing the chances of higher biomass yields through selective breeding.

Several aspects should be examined and studied to realize these crops' potential to produce sustainable biofuels fully. Future studies should be directed toward enhancing yield and biomass quality via genetic modification and enhancing processing technologies. For example, improving enzymatic hydrolysis and gasification could significantly improve biomass conversion to biofuels (Singh et al., 2024). Furthermore, establishing and implementing innovation adoption, innovation-monitoring, technological support, and policy-enabling approaches would also be important when addressing climate change issues. Policymakers should design incentives to support the development of biofuels so that farmers have adequate support to grow such crops using environmentally friendly methods (Lazaro et al., 2023).

If researchers, policymakers, and farmers work together, the growth and development of biofuels will be more efficient for energy demand and the satisfaction of sustainable environmental standards. This will, therefore, not only improve the profitability of biofuel crops but also aim to develop policies that will help reduce greenhouse gases in the human atmosphere and add to the principles of ecology.

ACKNOWLEDGMENT

The authors are thankful to Bulleh Shah Packaging Pvt. Ltd. Lahore, Pakistan for financial assistance.

COMPETING INTERESTS

The authors declare that they have no competing interest.

DECLARATION OF GENERATIVE AI AND AI-ASSISTED TECHNOLOGIES IN THE WRITING PROCESS

During the preparation of this work, the book chapter author(s) have not taken any assistance from AI, and the book chapter author(s) reviewed and edited the content as needed and take(s) full responsibility for the content of the publication.

REFERENCES

Adu Fosu, S. and Davis, S.C. 2024. Decadal change in soil carbon and nitrogen with a *Miscanthus × giganteus* crop on abandoned agricultural land in southeast Ohio. *Global Change Biology Bioenergy*, 16(7): 13171.

Ali, A., Hassan, M.U. and Kaul, H.P. 2024. Broad scope of site-specific crop management and specific role of remote sensing technologies within it—A review. *Journal of Agronomy and Crop Science*, 210(4): 12732.

Bai, J., Luo, L., Li, A., Lai, X., Zhang, X., Yu, Y., Wang, H., Wu, N. and Zhang, L. 2022. Effects of biofuel crop switchgrass (*Panicum virgatum*) cultivation on soil carbon sequestration and greenhouse gas emissions: A review. *Life*, 12(12): 2105.

Bhatia, R., Dalton, S., Roberts, L.A., Moron-Garcia, O.M., Iacono, R., Kosik, O., Gallagher, J.A. and Bosch, M. 2019. Modified expression of ZmMYB167 in *Brachypodium distachyon* and Zea mays leads to increased cell wall lignin and phenolic content. *Scientific Reports*, 9(2019): 8800.

Bhatia, R., Timms-Taravella, E., Roberts, L.A., Moron-Garcia, O.M., Hauck, B., Dalton, S., Gallagher, J.A., Wagner, M., Clifton-Brown, J. and Bosch, M. 2023. Transgenic ZmMYB167 *Miscanthus sinensis* with increased lignin to boost bioenergy generation for the bioeconomy. *Biotechnology for Biofuels and Bioproducts*, 16(1): 29.

Bužinskienė, R., Miceikienė, A., Venslauskas, K. and Navickas, K. 2023. Assessment of energy–economy and environmental performance of perennial crops in terms of biogas production. *Agronomy*, 13(5): 1291.

Canter, C.E., Zolton, K., Cacho, J.F., Negri, M.C., Zumpf, C.R. and Quinn, J.J. 2022. Impact of landscape design on the greenhouse gas emissions of shrub willow bioenergy buffers in a US Midwest corn-production landscape. *Biofuels, Bioproducts and Biorefining*, 16(3): 629–639.

Cappelli, G.A., Ginaldi, F., Fanchini, D., Ceotto, E. and Donatelli, M. 2024. How much energy can giant reed and *Miscanthus* produce in marginal lands across Italy? A modelling solution under current and future scenarios. *Global Change Biology Bioenergy*, 17(1): 13186.

Collins, H.P., Kimura, E., Polley, W., Fay, P.A. and Fransen, S. 2020. Intercropping switchgrass with hybrid poplar increased carbon sequestration on a sand soil. *Biomass and Bioenergy*, 138: 105558.

Constantinescu-Aruxandei, D. and Oancea, F. 2023. Closing the nutrient loop—The new approaches to recovering biomass minerals during the biorefinery processes. *International Journal of Environmental Research and Public Health*, 20(3): 2096.

Cooney, D., Kim, H., Quinn, L., Lee, M.S., Guo, J., Chen, S.L., Xu, B.C. and Lee, D.K. 2017. Switchgrass as a bioenergy crop in the Loess Plateau, China: Potential lignocellulosic feedstock production and environmental conservation. *Journal of Integrative Agriculture*, 16(6): 1211–1226.

Draper, J., Mur, L.A., Jenkins, G., Ghosh-Biswas, G.C., Bablak, P., Hasterok, R. and Routledge, A. P. 2001. *Brachypodium distachyon*. A new model system for functional genomics in grasses. *Plant Physiology*, 127(4): 1539–1555.

El-Desouky, M.G., Khalil, M.A., El-Bindary, A.A. and El-Bindary, M.A. 2022. Biological, biochemical and thermochemical techniques for biofuel production: An updated review. *Biointerface Research in Applied Chemistry*, 12(3): 3034–3054.

Ford, J.S., Bale, C.S. and Taylor, P.G. 2024. The factors determining uptake of energy crop cultivation and woodland creation in England: Insights from farmers and landowners. *Biomass and Bioenergy*, 180: 107021.

Geem, K.R., Song, Y., Hwang, I., Bae, H.J. and Lee, D.W. 2021. Production of *Gloeophyllum trabeum* endoglucanase Cel12A in *Nicotiana benthamiana* for cellulose degradation. *Frontiers in Plant Science*, 12: 696199.

Grace, O.M., Lovett, J.C., Gore, C.J., Moat, J., Ondo, I., Pironon, S., Langat, M.K., Pérez-Escobar, O.A., Ross, A., Suzan Abbo, M. and Shrestha, K.K. 2020. Plant power: Opportunities and challenges for meeting sustainable energy needs from the plant and fungal kingdoms. *Plants, People, Planet*, 2(5): 446–462.

Grippi, D., Clemente, R. and Bernal, M.P. 2020. Chemical and bioenergetic characterization of biofuels from plant biomass: Perspectives for Southern Europe. *Applied Sciences*, 10(10): 3571. https://doi.org/10.3390/app10103571

He, Y., Jaiswal, D., Long, S.P., Liang, X.Z. and Matthews, M.L. 2024. Biomass yield potential on US marginal land and its contribution to reach net-zero emission. *Global Change Biology Bioenergy*, 16(2): 13128.

Hus Farm. 2023. Grassrootsgrowth: The economic viability of grass-based biofuels. *Hus Farm*. https://husfarm.com/article/grassroots-growth-the-economic-viability-of-grass-based-biofuels.

Hussain, M. B., Iqbal, S., Khanam, P. and Hanif, S. 2021–2024. *Establishment and propagation of Miscanthus as a pulp and energy crop [project images]*. MNS University of Agriculture Multan, Pakistan.

Jacot, J., Williams, A.S. and Kiniry, J.R. 2021. Biofuel benefit or bummer? A review comparing environmental effects, economics, and feasibility of North American native perennial grass and traditional annual row crops when used for biofuel. *Agronomy*, 11(7): 1440.

Jayawardena, D.M., Robertson, G.P., Sanford, G.R. and Thelen, K.D. 2023. Comparative productivity of six bioenergy cropping systems on marginal lands in the Great Lakes Region, United States. *Agronomy Journal*, 115(5): 2451–2468.

Kane, J.L., Schartiger, R.G., Daniels, N.K., Freedman, Z.B., McDonald, L.M., Skousen, J.G. and Morrissey, E.M. 2023. Bioenergy crop *Miscanthus × giganteus* acts as an ecosystem engineer to increase bacterial diversity and soil organic matter on marginal land. *Soil Biology and Biochemistry*, 186: 109178.

Larnaudie, V., Ferrari, M.D. and Lareo, C. 2022. Switchgrass as an alternative biomass for ethanol production in a biorefinery: Perspectives on technology, economics and environmental sustainability. *Renewable and Sustainable Energy Reviews*, 158: 112115.

Lask, J., Magenau, E., Ferrarini, A., Kiesel, A., Wagner, M. and Lewandowski, I. 2020. Perennial rhizomatous grasses: Can they really increase species richness and abundance in arable land?—A meta-analysis. *Global Change Biology Bioenergy*, 12(11): 968–978.

Lask, J., Wagner, M., Trindade, L.M. and Lewandowski, I. 2019. Life cycle assessment of ethanol production from *miscanthus*: A comparison of production pathways at two European sites. *Global Change Biology Bioenergy*, 11(1): 269–288.

Lazaro, L.L.B., Grangeia, C.S., Santos, L. and Giatti, L.L. 2023. What is green finance, after all?–Exploring definitions and their implications under the Brazilian biofuel policy (RenovaBio). *Journal of Climate Finance*, 2: 100009.

Lei, L., Gordon, S.P., Liu, L., Sade, N., Lovell, J.T., Rubio Wilhelmi, M.D.M., Singan, V., Sreedasyam, A., Hestrin, R., Phillips, J. and Hernandez, B.T. 2023. The reference genome and abiotic stress responses of the model perennial grass *Brachypodium sylvaticum*. *G3: Genes, Genomes, Genetics*, 14(1): 245.

Lewandowski, I., Clifton-Brown, J. C., Scurlock, J. M. O. and Huisman, W. 2000. *Miscanthus*: European experience with a novel energy crop. *Biomass and Bioenergy*, 19(4): 209–227.

Liu, Y., Chen, S., von Cossel, M., Xu, B., Gao, H. and Jiang, R. 2021. Evaluating the suitability of marginal land for a perennial energy crop on the Loess Plateau of China. *Global Change Biology Bioenergy*, 13: 1388–1406.

Liu, Y., Hastings, A., Chen, S. and Faaij, A. 2022. The development of a new crop growth model Switch for yield mapping of switchgrass. *Global Change Biology Bioenergy*, 14(12): 1281–1302.

Lowry, C.J., Matlaga, D.P., West, N.M., Williams II, M.M. and Davis, A.S. 2022. Estimating local eradication costs for invasive *Miscanthus* populations throughout the eastern and midwestern United States. *Invasive Plant Science and Management*, 15(3): 115–121.

Ludwig, E., Sumner, J., Berry, J., Polydore, S., Ficor, T., Agnew, E., Haines, K., Greenham, K., Fahlgren, N., Mockler, T.C. and Gehan, M.A. 2023. Natural variation in *Brachypodium distachyon* responses to combined abiotic stresses. *The Plant Journal*, 117(6): 1676–1701.

Magenau, E., Clifton-Brown, J., Awty-Carroll, D., Ashman, C., Ferrarini, A., Kontek, M., Martani, E., Roderick, K., Amaducci, S., Davey, C. and Jurišić, V. 2022. Site impacts nutrient translocation efficiency in intraspecies and interspecies *miscanthus* hybrids on marginal lands. *Global Change Biology Bioenergy*, 14(9): 1035–1054.

Mahmood, A.R., Abdullah, M.R. and Khalaf, H. S. 2022. Biofuel production and its impact on global food security: A review article. *Iraqi Journal of Industrial Research*, 9(1): 75–83.

Mamirova, A. and Pidlisnyuk, V. 2024. The economic and environmental aspects of *Miscanthus × giganteus* phytomanagement applied to non-agricultural land. *Agronomy*, 14(4): 791.

Masum, M.F.H., Dwivedi, P. and Anderson, W.F. 2020. Estimating unit production cost, carbon intensity, and carbon abatement cost of electricity generation from bioenergy feedstocks in Georgia, United States. *Renewable and Sustainable Energy Reviews*, 117: 109514.

Merritt, H. and Barragán-Ocaña, A. 2023. The impact of market factors on the development of eco-friendly energy technologies: The case of bioethanol. *Clean Technologies and Environmental Policy*, 25(2): 313–321.

Mironova, G.F., Budaeva, V.V., Skiba, E.A., Gismatulina, Y.A., Kashcheyeva, E.I. and Sakovich, G.V. 2023. Recent advances in *miscanthus* macromolecule conversion: A brief overview. *International Journal of Molecular Sciences*, 24(16): 13001.

Mitchell, R. and Schmer, M. 2014. An overview of switchgrass. In: *Compendium of Bioenergy Plants* (pp. 15–38). CRC Press.

Mohammed, I.Y., Abakr, Y.A. and Mokaya, R. 2019. Biofuel and valuable products recovery from Napier grass pre-processing: Process design and economic analysis. *Journal of Environmental Chemical Engineering*, 7(2): 102962.

Morales, M., Arvesen, A. and Cherubini, F. 2021. Integrated process simulation for bioethanol production: Effects of varying lignocellulosic feedstocks on technical performance. *Bioresource Technology*, 328: 124833.

Ortega, Z., Bolaji, I., Suárez, L. and Cunningham, E. 2024. A review of the use of giant reed (*Arundo donax* L.) in the biorefineries context. *Reviews in Chemical Engineering*, 40(3): 305–328.

Ouattara, M.S., Laurent, A., Barbu, C., Berthou, M., Borujerdi, E., Butier, A., Malvoisin, P., Romelot, D. and Loyce, C. 2020. Effects of several establishment modes of *Miscanthus × giganteus* and *Miscanthus sinensis* on yields and yield trends. *Global Change Biology Bioenergy*, 12(7): 524–538.

Pancaldi, F., van Loo, E.N., Senio, S., Al Hassan, M., van der Cruijsen, K., Paulo, M.J., Dolstra, O., Schranz, M.E. and Trindade, L.M. 2023. Syntenic cell wall QTLs as versatile breeding tools: Intraspecific allelic variability and predictability of biomass quality loci in target plant species. *Plants*, 12(4): 779.

Panoutsou, C. and Chiaramonti, D. 2020. Socio-economic opportunities from *Miscanthus* cultivation in marginal land for bioenergy. *Energies*, 13(11): 2741.

Qaseem, M.F. and Wu, A.M. 2020. Marginal lands for bioenergy in China; an outlook in status, potential and management. *Global Change Biology Bioenergy*, 13(1): 21–44.

Ricketts, M.P., Heckman, R.W., Fay, P.A., Matamala, R., Jastrow, J.D., Fritschi, F.B., Bonnette, J. and Juenger, T.E. 2023. Local adaptation of switchgrass drives trait relations to yield and differential responses to climate and soil environments. *Global Change Biology Bioenergy*, 15(5): 680–696.

Sahoo, G., Wani, A.M., Swamy, S.L., Sharma, A. 2022. Environmental impact and economic benefits of biofuel production. In: Chowdhary, P., Khanna, N., Pandit, S. and Kumar, R. (eds) *Bio-Clean Energy Technologies: Volume* 1. *Clean Energy Production Technologies*. Springer, Singapore. https://doi.org/10.1007/978-981-16-8090-8_16

Schneider, M.V., Oberholzer, S. and Speranza, C.I. 2024. Revegetation is key for soil organic carbon sequestration on abandoned and degraded land in northern Spain. *Geoderma Regional*, 38: 00835.

Searchinger, T., Heimlich, R., Houghton, R. A., Dong, F., Elobeid, A., Fabiosa, J., Tokgoz, S., Hayes, D. and Yu, T. H. 2008. Use of US croplands for biofuels increases greenhouse gases through emissions from land-use change. *Science*, 319(5867): 1238–1240.

Shrestha, H.K., Fichman, Y., Engle, N.L., Tschaplinski, T.J., Mittler, R., Dixon, R.A., Hettich, R.L., Barros, J. and Abraham, P.E. 2022. Multi-omic characterization of bifunctional peroxidase 4-coumarate 3-hydroxylase knockdown in *Brachypodium distachyon* provides insights into lignin modification-associated pleiotropic effects. *Frontiers in Plant Science*, 13: 908649.

Silva, T.N., Thomas, J.B., Dahlberg, J., Rhee, S.Y. and Mortimer, J.C. 2022. Progress and challenges in sorghum biotechnology, a multipurpose feedstock for the bioeconomy. *Journal of Experimental Botany*, 73(3): 646–664.

Singh, S., Morya, R., Jaiswal, D.K., Keerthana, S., Kim, S.H., Manimekalai, R., de Araujo Pereira, A.P. and Verma, J.P. 2024. Innovations and advances in the enzymatic deconstruction of biomass and their sustainability analysis: A review. *Renewable and Sustainable Energy Reviews*, 189: 113958.

Sorek, N., Yeats, T.H., Szemenyei, H., Youngs, H. and Somerville, C.R. 2014. The implications of lignocellulosic biomass chemical composition for the production of advanced biofuels. *Bioscience*, 64(3): 192–201. https://doi.org/10.1093/biosci/bit037

Soriano, M.A., Cabezas, J.M. and Gómez, J.A. 2023. Field evaluation of selected autochthonous herbaceous species for cover crops in Mediterranean woody crops. *European Journal of Agronomy*, 143: 126723.

Szufa, S., Piersa, P., Adrian, Ł., Czerwińska, J., Lewandowski, A., Lewandowska, W., Sielski, J., Dzikuć, M., Wróbel, M., Jewiarz, M. and Knapczyk, A. 2021. Sustainable drying and torrefaction processes of *miscanthus* for use as a pelletized solid biofuel and biocarbon-carrier for fertilizers. *Molecules*, 26(4): 1014.

Templeton, D. W., Wolfrum, E. J., Yen, J. H. and Sharpless, K. E. 2016. Compositional analysis of biomass reference materials: Results from an interlaboratory study. *Bioenergy Research*, 9: 303–314.

Tóth, Š. and Dupľák, Š. 2023. Effect of a soil-applied humic ameliorative amendment on the yield potential of switchgrass *Panicum virgatum* L. cultivated under central European continental climate conditions. *Agronomy*, 13(4): 1095.

Unyay, H., Perendeci, N.A., Piersa, P., Szufa, S. and Skwarczynska-Wojsa, A. 2024. Harnessing switchgrass for sustainable energy: Bioethanol production processes and pretreatment technologies. *Energies*, 17(19): 4812.

Urrego, J.P.F., Huang, B., Næss, J.S., Hu, X. and Cherubini, F. 2021. Meta-analysis of leaf area index, canopy height and root depth of three bioenergy crops and their effects on land surface modeling. *Agricultural and Forest Meteorology*, 306: 108444.

Vera, I., Hoefnagels, R., Junginger, M. and van der Hilst, F. 2021. Supply potential of lignocellulosic energy crops grown on marginal land and greenhouse gas footprint of advanced biofuels—A spatially explicit assessment under the sustainability criteria of the Renewable Energy Directive Recast. *Global Change Biology Bioenergy*, 13(9): 1425–1447.

Verdade, L.M., Piña, C.I. and Rosalino, L.M. 2015. Biofuels and biodiversity: Challenges and opportunities. *Environmental Development*, 15: 64–78.

Viccaro, M., Romano, S., Rosalia, I. and Cozzi, M. 2024. Searching for the profitability of energy crops: An agroecological–economic land use suitability (AE-landUSE) model. *Environments*, 11(5): 91.

Wang, D., Huangfu, Y., Dong, Z. and Dong, Y. 2022. Research hotspots and evolution trends of carbon neutrality—Visual analysis of bibliometrics based on cite space. *Sustainability*, 14(3): 1078.

Wang, M., Wang, Y., Liu, J., Yu, H., Liu, P., Yang, Y., Sun, D., Kang, H., Wang, Y., Tang, J. and Fu, C. 2024. Integration of advanced biotechnology for green carbon. Green Carbon, 2(2): 164–175.

Wang, N., Torralbo, F., Naumann, H.D. and Fritschi, F.B. 2023. Tissue composition and higher heating values differ among six bioenergy grasses. *Industrial Crops and Products*, 205: 117437.

Wang, W. 2022. Agricultural and forestry biomass for meeting the renewable fuel standard: Implications for land use and GHG emissions. *Energies*, 15(23): 8796.

Wang, Y., Gui, C., Wu, J., Gao, X., Huang, T., Cui, F., Liu, H. and Sethupathy, S. 2022. Spatio-temporal modification of lignin biosynthesis in plants: A promising strategy for lignocellulose improvement and lignin valorization. *Frontiers in Bioengineering and Biotechnology*, 10: 917459.

Winberg, J., Smith, H. G. and Ekroos, J. 2023. Bioenergy crops, biodiversity and ecosystem services in temperate agricultural landscapes—A review of synergies and trade-offs. *Global Change Biology Bioenergy*, 15(10): 1204–1220.

Winkler, B., Mangold, A., von Cossel, M., Clifton-Brown, J., Pogrzeba, M., Lewandowski, I., Iqbal, Y. and Kiesel, A. 2020. Implementing *miscanthus* into farming systems: A review of agronomic practices, capital and labour demand. *Renewable and Sustainable Energy Reviews*, 132: 110053.

York, L.M., Cumming, J.R., Trusiak, A., Bonito, G., von Haden, A.C., Kalluri, U.C., Tiemann, L.K., Andeer, P.F., Blanc-Betes, E., Diab, J.H. and Favela, A. 2022. Bioenergy underground: Challenges and opportunities for phenotyping roots and the microbiome for sustainable bioenergy crop production. *The Plant Phenome Journal*, 5(1): 20028.

Zegada-Lizarazu, W., Zanetti, F., Di Virgilio, N. and Monti, A. 2022. Is switchgrass good for carbon savings? Long-term results in marginal land. *Global Change Biology Bioenergy*, 14(7): 814–823.

Zhao, C., Hou, X., Guo, Q., Yue, Y., Wu, J., Cao, Y., Wang, Q., Li, C., Wang, Z. and Fan, X. 2022. Switchgrass establishment can ameliorate soil properties of the abandoned cropland in Northern China. *Agriculture*, 12(8): 1138.

Zhao, J., Wang, D., Pidlisnyuk, V. and Erickson, L.E. 2021. *Miscanthus* biomass for alternative energy production. In: Erickson, L.E. and Pidlisnyuk, V. (eds) *Phytotechnology with Biomass Production: Sustainable Management of Contaminated Sites* (Chap. 10, pp. 177–200). CRC Press.

20 Mahua (*Madhuca longifolia*) as an Efficient Biofuel Producer

A Critical Analysis for Sustainable Uses

Syed Uzma Jalil and Mohammad Israil Ansari

20.1 INTRODUCTION

Increasing global energy demand coupled with the emergent need for climate change mitigation and a reduction in dependency on fossil fuels is driving the move against renewable and sustainable energy sources (Jayabal 2024). Among the many alternatives available, biofuels have become highly popular due to their potential to effect cleaner combustion, lower greenhouse gas (GHG) emissions, and security in energy provision, especially in developing economies (El-Araby 2024). However, the large-scale use of biofuels from food crops raises issues of food security, land use change, and ecological sustainability, leading to the concept of second-generation biofuels obtained from non-edible oil sources. The most important candidate for such sustainable biofuel production is Mahua (*Madhuca longifolia*), a native tree species of the Indian subcontinent (Dhairiyasamy and Gabiriel 2025). Mahua has multiple uses ranging from edible flowers and oil-rich seeds to cultural and medicinal uses. Therefore, this tree source has great potential as a non-edible biodiesel feedstock that has been largely untapped (Devi et al. 2024). The seeds produce an oil of high viscosity and give biodiesel comparable to conventional diesel after it has undergone the transesterification process. Drought tolerance, marginal land adaptation, and successful cultivation across many agro-climatic zones make Mahua a very promising candidate for a sustainable agroforestry and bioenergy system (Sinha et al. 2024). Moreover, the incorporation of Mahua into rural economies of tribal communities will be an upside toward inclusive development, livelihood generation, and decentralized energy access (Tamilselvi et al. 2011). Despite all these advantages, a number of technical, economic, and policy-related challenges exist that hinder the large-scale adoption and commercialization of Mahua-based biofuels. This chapter gives a detailed and critical exposition on *M. longifolia*, the biofuel resource. The botanical, biochemical, and bioenergy attributes of Mahua are described; the efficiency and quality of biodiesel derived from Mahua would be reviewed for its environmental, socioeconomic, and policy implications.

20.2 BOTANICAL AND ETHNOBOTANICAL PROFILE OF MAHUA

Mahua is a member of the Sapotaceae family, usually deciduous, and this striking plant is usually 10–15 m high. This species is easily recognized by the widely round, high and thick shady crown of the tree. The bark is a brown leathery surface with cracks and fissures which on cutting reveals inner bark of reddish colour exuding white milky latex. The leaves are broad elliptical in shape and measure about 15–25 cm in length and 8–15 cm in width. They have a thick leathery texture, having hairy surface on the under-side and having distinctly and prominently veined showing about 12 pairs of prominent lateral veins with oblique tertiary vein patterns. The leaves are entire but slightly wavy at margins and the petiole is reddish, which is usually 2–4 cm long. Fruiting the

DOI: 10.1201/9781003600169-20

tree, there are clusters of fleshy, white flowers at the tip of the branches. Each flower is about 2 cm long, has a pointed shape, and bears a very strong sweet odour, reminiscent of perspiration, thus being rich in nectar and very important in the ecology to the food of pollinators. The fruit is fleshy, greenish in colour, ovoid drupe measuring about 2 to 4 cm in diameter. Generally, the fruit contains between 1 and 4 seeds. These are elongated, brown, smooth and glossy, measuring approx. 2 cm long (Sinha et al. 2017; Tushar et al. 2022).

Mahua is widely distributed across the Indian subcontinent, particularly in the central and southern regions (Devanand et al. 2024). Native to India, Sri Lanka, and Myanmar, Mahua is most extensively found in India, especially in the states of Madhya Pradesh, Chhattisgarh, Jharkhand, Odisha, Maharashtra, Bihar, Uttar Pradesh, Andhra Pradesh, Telangana, and Tamil Nadu (Pradhan et al. 2024). These areas predominantly fall under tropical dry deciduous forest zones, where the species is found either in the wild, semi-domesticated in village groves, or integrated into agroforestry systems. Its widespread distribution is largely due to its remarkable adaptability to a range of climatic and soil conditions. Mahua thrives in regions receiving 500–1200 mm of annual rainfall and can tolerate extreme temperatures ranging from 5°C to 47°C, which enables it to survive in both dry and semi-arid climates (Aransiola et al. 2019; Nayak et al. 2023). The species is well adapted to various soil types, including red loamy, sandy, lateritic, and shallow rocky soils, often with poor nutrient content. It also exhibits tolerance to a wide soil pH range, from slightly acidic (pH 5.5) to moderately alkaline (pH 8.5) (Nayak et al. 2023). This versatility allows Mahua to grow on marginal and degraded lands, making it an ecologically valuable species for reforestation and land rehabilitation efforts.

In terms of habitat, Mahua may be found in mixed deciduous forests, in forest fringes, degraded woodlands, village commons, and traditional agroforestry systems, and sometimes with crops like millets and legumes, according to Chacko and Narvekar in 2023. The deep root system makes it drought resistant, as Mahua can tap into underground water reserves. The system also significantly contributes to soil stability and erosion control (Sinha et al. 2024). Mahua is a high-staple species in many dry land ecosystems, ecologically speaking, because of its importance in biodiversity support (Kacker 2022). The flowers, fruits, and foliage of Mahua act as food for many birds, insects, and mammals, while its litter adds organic matter to improve soil fertility (Kacker 2022). The tree contributes to water conservation by canopies and improving infiltration, which is important for sustaining rural livelihoods and the ecosystem health of semi-arid regions (Chacko and Narvekar 2023).

Mahua is indeed culturally very important, and these roots are therapeutic in nature as well since mostly tribal communities associate it with religious rituals and spiritual practices (Tamilselvi et al. 2011). The flowers have a wide range of traditional uses in medicine and have been credited for having effects such as pain relief, diuretic activity, cooling effects, and tonic benefits. They have also been traditionally used as an aphrodisiac, carminative, astringent, and demulcent (Cohen 1992). Folk medicine used Mahua flowers in the treatment of helminthic infections, in bronchitis, and in acute and chronic tonsillitis. Furthermore, they also enhance lactation in nursing mothers and improve male reproductive health by increasing seminal fluid (Akshatha et al. 2013). The fruit of the Mahua tree is astringent and is used to cure chronic throat symptoms like pharyngitis and tonsillitis (Kamal 2014). Another valuable part of the plant is its seeds that have been traditionally used for the treatment of skin disorders, rheumatism, headaches, constipation, and even piles (Sinha et al. 2017). Mahua is also thought to be helpful in local traditional medicine for snakebites and diabetes (Singh and Kaushal 2018). Interestingly enough, the remaining seed cake is developed into a natural poison to fish. The fruit kernels yield a greenish-yellow oil that is primarily used for cooking in tribal households, sometimes mixed with ghee as an adulterant (Dinesh 2001). The bark of this tree is boiled to prepare a decoction reputed to aid in blood sugar regulation. In contrast, the leaves are used for treating the skin ailments known as eczema, and the bark extract will treat spongy or bleeding gums (Jyothi and Seshagiri 2012). Rich in reducing sugars, Mahua flowers also feature as natural sweeteners in several tribal communities' traditional recipes (Jha and Mazumder 2018).

20.3 MAHUA OIL: PHYSICOCHEMICAL PROPERTIES AND SUITABILITY FOR BIOFUEL

Mahua oil is increasingly being researched as a potential non-edible biodiesel source due to its oil content and availability (Dhairiyasamy and Gabiriel 2025). It can be extracted from seeds and availably processed through different methods each having their distinct advantages and disadvantages (Lawson 1995; Ramadan et al. 2006). Mechanical extraction with an expeller is a prominent and affordable approach that avoids chemicals. However, the oil recovery rate of 30%–35% is low, along with higher residual oil content. With the use of solvents like hexane, oil recovery can be increased to 45%–50% (Bharath and Nikkam 2022). However, the increased recovery comes with the need to remove solvent residues that pose a potential environmental threat. Cold pressing is deemed "cleaner" as it does not alter the chemical structure of the oil, making it suitable for high-quality oil, despite the lower yield.

Mahua oil physicochemical features deeply influence its suitability for biodiesel production. One of its critical characteristics is its high kinematic viscosity of around 40–55 mm^2/s at 40°C, which requires transesterification to reduce viscosity to improve flow; this is significantly higher than conventional diesel fuel (Dugala et al. 2021). Furthermore, Mahua oil has a calorific value between 32 and 36 MJ/kg, which denotes reasonable energy potential similar to fossil diesel (Acharya et al. 2017). The cetane number of raw Mahua oil is modest (35–45) while positively improving post-transesterification biodiesel fuel, often exceeding 50, enhancing ignition quality and combustion efficiency (Ahmad et al. 2023). Mahua oil has also high free fatty acid ("FFA") content of 5%–20% in non-edible or degraded seeds (Ghadge and Raheman 2005). This leads to necessity of two-step esterification: acid esterification and thereafter base-catalysed transesterification to prevent soap formation and effectively convert to biodiesel. Other relevant properties are a density of 0.91–0.93 g/cm^3 and a flash point of 220°C–240°C, giving Mahua oil an edge on safety as high ignition temperature allows safer handling and storage (Dugala et al. 2021).

Among the non-edible plant oils, Mahua oil is the best source for biodiesel when compared to Jatropha, Pongamia, and neem. Mahua seeds contain around 35%–50% oil, which qualifies them for being a rich source for biodiesel (Ahmad et al. 2023). Mahua oil contains more FFA than Jatropha oil or Pongamia oil in its composition. This makes the processing of this oil somewhat tricky; however, this can be solved through some pre-fuelling treatment (Dugala et al. 2021). As far as viscosity and flash points are concerned, Mahua oil closely resembles other common oils. When biodiesel is produced through transesterification, it meets the required fuel quality standards, including the cetane number, which is a good parameter for the smooth functionality of an engine. Mahua oil is abundantly produced because of the wide availability of Mahua trees grown across India, most of which go waste, thus making it possible to obtain this source sustainably and economically for biodiesel production. However, mainstream application would require better handling of high amounts of FFAs and better consistency quality for the biodiesel produced with that approach.

20.4 BIODIESEL PRODUCTION FROM MAHUA OIL

Mahua oil serves as a promising feedstock for biodiesel production due to its high oil content, renewable availability, and rich triglyceride composition (Devi et al. 2024). However, like most non-edible oils, it requires chemical modification to reduce its inherent viscosity and improve its suitability for engine applications. The most widely adopted method for converting Mahua oil into biodiesel is transesterification, which involves reacting the oil with an alcohol (commonly methanol or ethanol) in the presence of a catalyst to produce fatty acid methyl esters (FAME) and glycerol as a byproduct (Ghadge and Raheman 2006).

20.4.1 Transesterification Process: Catalysts, Optimization, and Process Parameters

The transesterification of Mahua oil is influenced by several key factors including the type of catalyst, alcohol-to-oil ratio, reaction temperature, time, and the presence of FFA (Sabariswaran et al. 2014). Both alkaline and acidic catalysts are used, depending on the initial quality of the oil. When the FFA content is low (<2%), base catalysts such as sodium hydroxide (NaOH) or potassium hydroxide (KOH) are typically employed (Díaz-Muñoz et al. 2022). For high FFA oils that are common in Mahua due to storage-induced degradation, acid-catalysed esterification (using sulfuric acid) is preferred as a pretreatment step to reduce FFA, followed by base-catalysed transesterification (Sabariswaran et al. 2014).

The ideal transesterification conditions for Mahua oil have been optimized in various studies, often suggesting a molar ratio of methanol to oil of 6:1, a catalyst concentration of 0.5%–1%, a reaction temperature around 55°C–60°C, and a reaction time of 60–90 minutes (Bai et al. 2021). The use of heterogeneous catalysts (CaO, zeolites, and supported solid acids) and enzyme-based biocatalysts has also gained interest for sustainable and reusable processing, though they are often costlier and slower (Rabiei 2025).

20.4.2 Yield and Quality of Mahua Biodiesel

The biodiesel yield from Mahua oil can be as high as 90%–95% (Ahmad et al. 2023). Mahua biodiesel meets international fuel quality standards set by ASTM (United States) and EN (Europe) (Acharya et al. 2017) though some parameters may need to change based on the feedstock FFA and moisture content (Agarwal and Das 2001).

Mahua biodiesel has a cetane number above 50 and good oxidation stability, along with good cold flow properties. However, the pour point may change based on the geographical source of the oil (Acharya et al. 2017). Also, viscosity, density, flash point, and calorific value are at acceptable levels for biodiesel. Some post-treatment processes may involve washing and drying to remove excess glycerol, alcohol, and catalyst (Acharya et al. 2017).

20.4.3 Glycerol and Byproduct Utilization

About 10% of the total product volume is made up of glycerol, the primary byproduct of the transesterification process (Bagnato et al. 2017; Khan et al. 2025). Glycerol has a lot of industrial potential despite being frequently regarded as a waste product (Figure 20.1). According to Kaur et al. (2020), crude glycerol can be refined and utilized in food additives, cosmetics, pharmaceuticals, and even microbial fermentation processes to create valuable chemicals like citric acid, ethanol, and propanediol.

There are also chances for reuse with other byproducts like seed cake and methanol residues. Following oil extraction, the leftover seed cake can be utilized as organic manure or, following detoxification, as a possible feedstock for the production of biogas (Sudalai et al. 2023). The overall sustainability and economics of the biodiesel production process are improved by integrating these byproduct utilization strategies.

20.4.4 Engine Performance, Combustion, and Emission Characteristics Using Mahua Biodiesel Blends

Several engine trials have demonstrated that Mahua biodiesel blends (B10, B20, B50) can be used effectually in original diesel engines (Ansari et al. 2022; Chandrasekharan et al. 2024). Engine performance metrics such as brake thermal efficiency and specific fuel consumption for B20 blends are often comparable to those of conventional diesel. In some cases, a minor decrease in efficiency is

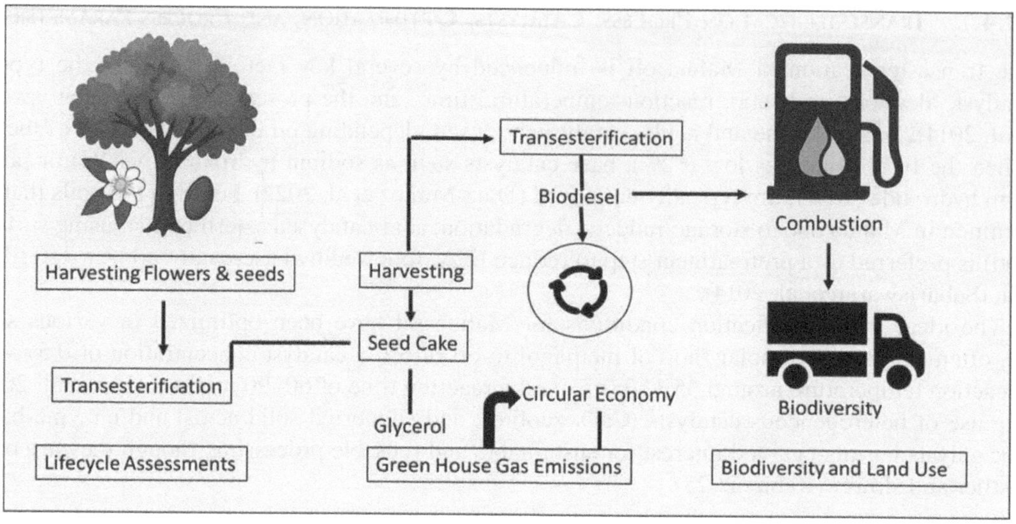

FIGURE 20.1 Schematic representation of biodiesel.

observed, primarily due to the higher viscosity and oxygen content of the biodiesel (Kshatriya et al. 2022; Dhairiyasamy and Gabiriel 2025).

Combustion analysis reveals that biodiesel generally leads to longer ignition delays but improved combustion stability due to better air–fuel mixing (Dhairiyasamy and Gabiriel 2025). Importantly, emission characteristics improve significantly with Mahua biodiesel blends that show reductions in carbon monoxide (CO), unburned hydrocarbons (HC), and particulate matter (PM), along with a slight increase in nitrogen oxides (NOx), which is typical for oxygen-rich biofuels (Dhairiyasamy and Gabiriel 2025).

Overall, the use of Mahua biodiesel not only provides a renewable alternative to fossil fuels but also supports cleaner combustion and reduced environmental impact when used in compression ignition (CI) engines.

20.5 ENVIRONMENTAL AND SUSTAINABILITY ASPECTS

Biodiesel derived from non-edible feedstocks like Mahua is a promising alternative as the world goes towards cleaner and more sustainable energy. It will help in reducing the dependence on fossil fuel and supports ecological and socioeconomic balance in the areas native for the tree. Understanding the environmental footprint and sustainability implications of Mahua-based biodiesel is essential for assessing its long-term viability. Life cycle assessment (LCA), carbon emissions, land use impact, and integration into the circular bioeconomy are key areas.

20.5.1 Life Cycle Assessment (LCA) of Mahua-Based Biodiesel

LCA denotes a life cycle approach to the assessment of product/process environmental impacts, from cradle to grave. Several researchers have shown that Mahua oil biodiesel scores lower on environmental impact indicators than petroleum diesel, particularly concerning fossil energy use and emission of contaminants (Hameed and Muralidharan 2023; Dhairiyasamy and Gabiriel 2025). Since Mahua trees grow naturally in the arid and semi-arid regions – without any intensive agricultural inputs – the energy input during cultivation is quite low (Johar and Kumar 2020). In addition, since seed collection is often done from wild or uncultivated land, there is hardly any demand for synthetic fertilizer and irrigation, and mechanized harvesting of seeds may not be required, which significantly reduces the upstream emissions. Mahua biodiesel holds a good EROI when produced by optimally, clean technologies, thus enhancing the sustainability of the energy profile (Ahmad et al. 2023).

20.5.2 Carbon Footprint and Greenhouse Gas (GHG) Emission Reduction Potential

Mahua biodiesel offers very interesting environmental benefits, especially in terms of GHG emissions. Compared to fossil diesel, Mahua biodiesel produces very much lower carbon emissions. The carbon from biodiesel burns to form CO_2. But this carbon was earlier absorbed from the atmosphere by the Mahua trees. This phenomenon is referred to as carbon neutrality (Kapilan and Prabhu 2022; Devi et al. 2024).

It has been estimated that life cycle GHG emissions can be reduced from 60% to 80% depending on processing efficiency and regional practices by switching from conventional diesel to Mahua oil biodiesel (Osman et al. 2024). In addition, collection of Mahua is done without land clearing or forest dislocation, thereby not disturbing carbon stocks in soil and vegetation and improving the overall carbon balance of the fuel. The reductions in emission thus achieved are a direct way of contributing to the global efforts of climate change mitigation under international agreements.

20.5.3 Impact on Biodiversity and Land Use

Mahua trees grow mostly wild or semi-managed; therefore, they are more sustainable in land use and biodiversity conservation than oilseed commercial cash crops, which often lead to monoculture farming and deforestation (Patel et al. 2024). Many tribal areas practice traditional agroforestry systems involving these trees, generally supporting diverse flora and fauna.

There is no food competition with Mahua, and it grows in degraded or marginal lands, making it a low-risk feedstock that does not have to deal with the "food versus fuel" dilemma. However, a lot of precaution has also to be exercised so that the demand for Mahua seeds does not end up over-collection or disturbance of the local ecosystems because many wildlife species depend on Mahua flowers and fruits. Regulation is in place, and there are community-based seed collection practices and ecological monitoring, which are influenced by this practice.

20.5.4 Waste Valorization and Circular Bioeconomy Potential

One important aspect of using Mahua oil for biodiesel is that it contributes to a circular bioeconomy. The production process generates useful byproducts: of greatest importance for this purpose is crude glycerol, which could be purified and utilized in various industries, including pharmaceuticals, cosmetics, and bioplastics (Tan et al. 2013; Sudalai et al. 2023). In addition, although after oil extraction the seed cake is toxic in its raw form, it may profitably be used as an organic fertilizer, soil conditioner, or even a substrate in biogas production systems by anaerobic digestion after adequate treatment (Gupta et al. 2013).

The effective utilization of these byproducts reduces environmental waste, improves process economics, and promotes resource efficiency. Moreover, the integration of waste streams into new value chains not only enhances sustainability but also creates opportunities in rural entrepreneurship and green jobs, especially within tribal and forest-dwelling communities where Mahua is native (Sudalai et al. 2023).

20.6 CONCLUSION

Mahua proves to be an oilseed that has a bright future in complementing biodiesel generation-past forays into edible oil. With an oil concentration in its seeds and tolerant to arid and semi-arid regions, India has the fortune of being thickly populated with Mahua trees, making this excellent alternative feedstock source against non-edible oil sources. Thus, relatively high levels of FFA can be changed to bring about the needed feedstock types and qualities using pretreatment and processing technologies. Its byproducts make biodiesel on its own. Glycerol and seed cake would constitute a waste-reducing measure, a GHG emission-reducing method, and serving a circular economy. For example, promotion of a Mahua-based biofuel leads to interesting socioeconomic impacts in rural and tribal communities, creating wealth and enhancing the use of unexploited potential natural resources. Standardizing fuel quality, advancing processing technologies, and undertaking complete life cycle analysis are planned as future milestones towards large-scale commercialization. Ultimately, this Mahua-based biodiesel should prove to be a good substitute for fossil fuel, and indeed, part of the global direction towards cleaner and renewable energies.

COMPETING INTERESTS

The authors declare that they have no competing interest.

DECLARATION OF GENERATIVE AI AND AI-ASSISTED TECHNOLOGIES IN THE WRITING PROCESS

During the preparation of this work, the book chapter author(s) have not taken any assistance from AI, and the book chapter author(s) reviewed and edited the content as needed and take(s) full responsibility for the content of the publication.

REFERENCES

Acharya, N., Nanda, P., Panda, S. & Acharya, S. (2017). Analysis of properties and estimation of optimum blending ratio of blended Mahua biodiesel. *Engineering Science and Technology, an International Journal*, 20(2), 511–517. https://doi.org/10.1016/j.jestch.2016.12.005

Agarwal, A.K. & Das, L.M. (2001). Biodiesel development and characterization for use as a fuel in compression ignition engines. *Transactions of ASME*, 123, 440–447.

Ahmad, A., Yadav, A.K. & Singh, A. (2023). Biodiesel production from Mahua oil: Characterisation, optimisation, and modelling with a hybrid statistical approach. *International Journal of Ambient Energy*, 44(1), 2600–2609. https://doi.org/10.1080/01430750.2023.2262468

Akshatha, K.N., Murthy, S.M. & Lakshmidevi, N. (2013). Ethnomedical uses of *Madhuca longifolia* – A review. *International Journal of Life Science & Pharma Research*, 3(1), 44–53.

Ansari, K., Goga, G. & Mohan, R. (2022). Performance and emission characteristics of Mahua blended biodiesel. *Materials Today: Proceedings*, 71(Part 2), 293–299. https://doi.org/10.1016/j.matpr.2022.09.154

Aransiola, E.F., Ehinmitola, E.O., Adebimpe, A.I., Shittu, T.D. & Solomon, B.O. (2019). Advances in eco-fuels for a sustainable environment, prospects of biodiesel feedstock as an effective eco-fuel source and their challenges. In *Woodhead Publishing Series in Energy* (pp. 53–87). Edited by Kalam Azad.Duxford, United Kingdom; Cambridge, MA : Woodhead Publishing, an imprint of Elsevier.

Bagnato, G., Iulianelli, A., Sanna, A. & Basile, A. (2017). Glycerol production and transformation: A critical review with particular emphasis on glycerol reforming reaction for producing hydrogen in conventional and membrane reactors. *Membranes (Basel)*, 7(2), 17. https://doi.org/10.3390/membranes7020017

Bai, M.T., Swarna, U., Mahalakshmi, R., Raju, Ch.A.I. & Sridevi, V. (2021). Production of methyl ester from Mahua oil: Characterization and optimization by using RSM. *Materials Today: Proceedings*, 44(Part 1), 1609–1616. https://doi.org/10.1016/j.matpr.2020.11.815

Bharath, K.L. & Nikkam, S. (2022). Mahua oil as an alternative biodegradable collector for the flotation of low-rank high-ash oxidized coals based on kinetic studies. *International Journal of Coal Preparation and Utilization*, 43(3), 448–467. https://doi.org/10.1080/19392699.2022.2055003

Chacko, I.S. & Narvekar, N.V. (2023). Unlocking the power of Mahua: A comprehensive guide. *International Research Journal of Plant Science*, 14(2), 1–5.

Chandrasekharan, T., Muthuswamy, P., Jayapalan, S., Balsing, S.M., Sekar, P. & Kuppuswami, S. (2024). Performance and emission evaluation of Mahua oil biodiesel blends in a common rail direct injection diesel engine. *AIP Conference Proceedings*, 3042(1), 020035. https://doi.org/10.1063/5.0194212

Cohen, M.L. (1992). Epidemiology of drug resistance: Implications for a post antimicrobial era. *Science*, 257, 1050–1055.

Devanand, P.S., Radha, P., Hemaprabha, K., Kumar, P., Raja, N., Sivakumar, B., Kiruba, M., Utharasu, S. & Revathi, R. (2024). *Madhuca longifolia*: A multifaceted tree rich in nutritional factors. *International Journal of Research in Agronomy*, 7(8), 209–213. https://doi.org/10.33545/2618060X.2024.v7.i8c.1223

Devi, N., Phani, M.B.S.S.D., Vara Prasad, R., Karthik, K. & Tejeswar, M. (2024). Production and characterization of Mahua biodiesel for sustainable energy applications. *International Journal of Creative Research Thoughts*, 12(3), 766–770.

Dhairiyasamy, R. & Gabiriel, D. (2025). Performance and emission characteristics of Mahua oil biodiesel in low heat rejection diesel engines with retarded injection timing. *Chemical Thermodynamics and Thermal Analysis*, 17, 100158. https://doi.org/10.1016/j.ctta.2024.100158

Díaz-Muñoz, L.L., Reynel-Ávila, H.E., Mendoza-Castillo, D.I., Bonilla-Petriciolet, A. & Jáuregui-Rincón, J. (2022). Preparation and characterization of alkaline and acidic heterogeneous carbon-based catalysts and their application in vegetable oil transesterification to obtain biodiesel. *International Journal of Chemical Engineering*, 2022, 7056220. https://doi.org/10.1155/2022/7056220

Dinesh, C. (2001). Analgesic effect of aqueous and alcoholic extracts of *Madhuca longifolia* (Koeing). *Indian Journal of Pharmacology*, 33(2), 108.

Dugala, N.S., Goindi, G.S. & Sharma, A. (2021). Evaluation of physicochemical characteristics of Mahua (*Madhuca indica*) and Jatropha (*Jatropha curcas*) dual biodiesel blends with diesel. *Journal of King Saud University - Engineering Sciences*, 33(6), 424–436. https://doi.org/10.1016/j.jksues.2020.05.006

El-Araby, R. (2024). Biofuel production: Exploring renewable energy solutions for a greener future. *Biotechnology for Biofuels*, 17, 129. https://doi.org/10.1186/s13068-024-02571-9

Ghadge, S.V. & Raheman, H. (2005). Biodiesel production from mahua (*Madhuca indica*) oil having high free fatty acids. *Biomass Bioenergy*, *28*(6), 601–605. https://doi.org/10.1016/j.biombioe.2004.11.009

Ghadge, S.V. & Raheman, H. (2006). Process optimization for biodiesel production from Mahua (*Madhuca indica*) oil using response surface methodology. *Bioresource Technology*, 97(3), 379–384. https://doi.org/10.1016/j.biortech.2005.03.014

Gupta, A., Kumar, A., Sharma, S., & Vijay, V.K. (2013). Comparative evaluation of raw and detoxified Mahua seed cake for biogas production. *Applied Energy*, 102, 1514–1521. https://doi.org/10.1016/j.apenergy.2012.09.017

Hameed, A.Z. & Muralidharan, K. (2023). Performance, emission, and catalytic activity analysis of AL_2O_3 and CeO_2 nano-additives on diesel engines using Mahua biofuel for a sustainable environment. *ACS Omega*, 8(6), 5692–5701. https://doi.org/10.1021/acsomega.2c07193

Jayabal, R. (2024). Towards a carbon-free society: Innovations in green energy for a sustainable future. *Results in Engineering*, 24, 103121. https://doi.org/10.1016/j.rineng.2024.103121

Jha, D. & Mazumder, P.M. (2018). Biological, chemical and pharmacological aspects of *Madhuca longifolia*. *Asian Pacific Journal of Tropical Medicine*, 11(1), 9.

Johar, V., & Kumar, B. (2020). Mahua: A versatile Indian tree species. *Journal of Pharmacognosy and Phytochemistry*, 9(6), 1926–1931.

Jyothi, K.S. & Seshagiri, M. (2012). In-vitro activity of saponins of *Bauhinia purpurea*, *Madhuca longifolia*, *Celastrus paniculatus* and *Semecarpus anacardium* on selected oral pathogens. *Journal of Dentistry (Tehran, Iran)*, 9(4), 216.

Kacker, S. (2022). Mahua – A keystone tree. *Sanctuary Asia*, 42(6).

Kamal, A. (2014). Qualitative phytochemical analysis of *Madhuca longifolia. Indian Journal of Plant Sciences*, 3(4), 38–41.

Kapilan, N. & Prabhu, S. (2022). Studies on Mahua oil biodiesel as the sustainable fuel in compression ignition engine. *Acta Technologica Agriculturae*, 25(2), 85–91. https://doi.org/10.2478/ata-2022-0014

Kaur, J., Sarma, A.K., Jha, M.K. & Gera, P. (2020). Valorisation of crude glycerol to value-added products: Perspectives of process technology, economics and environmental issues. *Biotechnology Reports*, 27, e00487. https://doi.org/10.1016/j.btre.2020.e00487

Khan, S.A., Arabidian, V.D., da Silveira Junior, N., de Farias, B.S., Kessler, F., Paes, R.L., de Almeida Pinto, L.A. & Tito Cadaval Junior, R.S. (2025). Valorization of glycerol produced as a by-product in the biodiesel industry: An insight into technical and economic studies. In Naushad, M., Mohamed Kutty, S.R., Hossain, M.S., Birniwa, A.H. & Jagaba, A.H. (Eds.), *Biorefinery of Industrial Effluents for a Sustainable Circular Economy* (pp. 307–316). Elsevier. https://doi.org/10.1016/B978-0-443-21801-9.00021-5

Kshatriya, A.S., Tiwari, P., M, S., Yunus Khan, T.M., Abdul Khadar, S.D., Mansour, M. & Feroskhan, M. (2022). Investigations into the combined effect of Mahua biodiesel blends and biogas in a dual fuel engine. *Energies*, 15, 2057. https://doi.org/10.3390/en15062057

Lawson, H. (1995). Sources of oils and fats. In Lawson, H. (Ed.), *Food Oils and Fats, Technology, Utilization and Nutrition* (pp. 39–48). Chapman & Hall.

Nayak, M.R., Dash, S.R., Khuntia, A. & Paramaguru, S. (2023). Traditional uses and nutritional properties of Mahua (*Madhuca longifolia*): A multipurpose tropical tree grown by tribals of Odisha. *The Pharma Innovation Journal*, 12(2), 2593–2598.

Osman, W.N.A., Rosli, M.H., Mazli, W.N.A. & Samsuri, S. (2024). Comparative review of biodiesel production and purification. *Carbon Capture Science & Technology*, 13, 100264. https://doi.org/10.1016/j.ccst.2024.100264

Patel, B., Sonkar, A., Patel, S.K., Maurya, A.K. & Gurudev, N. (2024). Mahua: An eco-friendly solution for economic and environmental sustainability. *Biotica Research Today*, 6(6), 306–308.

Pradhan, M., Malakar, A. & Sinha, A. (2024). Appraisal of the potential habitat distribution of *Madhuca longifolia* manifested remarkable resilience under various socio-climatic scenarios pan-India. *Modeling Earth Systems and Environment*, 10, 2435–2446. https://doi.org/10.1007/s40808-023-01913-0

Rabiei, K. (2025). Recent application of clay-based heterogeneous catalyst in organic reactions. *Journal of Inorganic and Organometallic Polymers and Materials*. https://doi.org/10.1007/s10904-024-03519-0

Ramadan, M.F., Sharanabasappa, G., Parmjyothi, S., Seshagri, M. & Moersel, J.-T. (2006). Profile and levels of fatty acids and bioactive constituents in Mahua butter from fruit seeds of buttercup tree (*Madhuca longifolia*). *European Food Research and Technology*, 222, 710–718. https://doi.org/10.1007/s00217-005-0155-2

Sabariswaran, K., Selvakumar, S. & Kathirselvi, A. (2014). Biodiesel production from Mahua oil by using two-step trans-esterification process. *Chemical Science Review and Letters*, 3(9), 52–57.

Singh, K. (2018). *Madhuca indica* J.F. Gmel: A multipurpose tree of traditional agroforestry systems of Bundelkhand. *International Journal of Forestry & Crop Improvement*, 9(2), 44–47.

Sinha, A., Malakar, A., Pradhan, M., Minz, A., Banerjee, S. & Sarkar, A. (2024). Tree improvement programme for Mahua (*Madhuca longifolia*) in Eastern India. *Biotica Research Today*, 6(10), 452–455.

Sinha, J., Singh, V., Singh, J. & Rai, A.K. (2017). Phytochemistry, ethnomedical uses and future prospects of Mahua (*Madhuca longifolia*) as a food: A review. *Journal of Nutrition & Food Sciences*, 7(573), 2.

Sudalai, S., Rupesh, K.J., Devanesan, M.G. & Arumugam, A. (2023). A critical review of *Madhuca indica* as an efficient biodiesel producer: Towards sustainability. *Renewable and Sustainable Energy Reviews*, 188, 113811. https://doi.org/10.1016/j.rser.2023.113811

Tamilselvi, N., Krishnamoorthy, P., Dhamotharan, R. & Sivakumar, S. (2011). Studies on phytochemical constituents of *Indigofera aspalathoides* Vahl ex DC. *Journal of Pharmaceutical Research*, 4, 3052–3053.

Tan, H.W., Abdul Aziz, A.R. & Aroua, M.K. (2013). Glycerol production and its applications as a raw material: A review. *Renewable and Sustainable Energy Reviews*, 27, 118–127. https://doi.org/10.1016/j.rser.2013.06.035

Tushar, S.D., Kumbhar, U.J. & Shah, N. (2022). *Madhuca longifolia*: Ethnobotanical, phytochemical studies, pharmacological aspects with future prospects. *Interdisciplinary Journal of Applied and Basic Subjects*, 2(7), 1–9. https://identifier.visnav.in/1.0002/ijacbs-22d-22006/

Index

Note: **Bold** page numbers refer to tables and *italic* page numbers refer to figures.

C

H

I

J

N

O

P

R

S